ADVANCES IN STRUCTURAL OPTIMIZATION

SOLID MECHANICS AND ITS APPLICATIONS
Volume 25

Series Editor: **G.M.L. GLADWELL**
Solid Mechanics Division, Faculty of Engineering
University of Waterloo
Waterloo, Ontario, Canada N2L 3G1

Aims and Scope of the Series

The fundamental questions arising in mechanics are: *Why?*, *How?*, and *How much?*
The aim of this series is to provide lucid accounts written by authoritative researchers giving vision and insight in answering these questions on the subject of
mechanics as it relates to solids.

The scope of the series covers the entire spectrum of solid mechanics. Thus it
includes the foundation of mechanics; variational formulations; computational
mechanics; statics, kinematics and dynamics of rigid and elastic bodies; vibrations
of solids and structures; dynamical systems and chaos; the theories of elasticity,
plasticity and viscoelasticity; composite materials; rods, beams, shells and
membranes; structural control and stability; soils, rocks and geomechanics;
fracture; tribology; experimental mechanics; biomechanics and machine design.

The median level of presentation is the first year graduate student. Some texts are
monographs defining the current state of the field; others are accessible to final
year undergraduates; but essentially the emphasis is on readability and clarity.

For a list of related mechanics titles, see final pages.

Advances in
Structural Optimization

Edited by

JOSÉ HERSKOVITS

Mechanical Engineering Program,
COPPE/Federal University of Rio de Janeiro,
Rio de Janeiro, Brazil

SPRINGER-SCIENCE+BUSINESS MEDIA, B.V.

Library of Congress Cataloging-in-Publication Data

Advances in structural optimization / edited by José Herskovits.
      p.   cm. -- (Solid mechanics and its applications ; v. 25)
    ISBN 978-0-7923-2510-9   ISBN 978-94-011-0453-1 (eBook)
    DOI 10.1007/978-94-011-0453-1
    1. Structural optimization.   I. Herskovits, José.   II. Series.
  TA658.8.A39   1995
  624.1'7--dc20   95-13513

ISBN 978-0-7923-2510-9

TABLE OF CONTENTS

PREFACE

Modern design techniques seek the best design to perform the desired tasks. Structural Optimization deals with the optimal design of structural elements and systems employed in several engineering fields. It was first applied in the aerospace industry, where reducing the structural weight is of utmost importance. Nowadays, use of Structural Optimization is rapidly growing in automotive, aeronautical, mechanical, civil, nuclear, naval and off-shore engineering. This is due to the increase of technological competition and the development of strong and efficient techniques for several practical applications. These techniques integrate CAD tools for geometric modelling, general analysis methods, like the finite element method, and methods of design sensitivity analysis with mathematical programming or optimality criteria methods. The increasing speed and capacity of digital computers makes large scale structural optimization possible and profitable in large number of applications.

This book contains a collection of papers by leading experts describing in a didactic way the foundations and recent advances in Structural Optimization. The techniques for a wide set of applications are presented, ranging from the historically first studied problems of size and shape optimization to topology and material optimization. It considers structural models using discrete element, such as trusses or beams, or based on the finite element method. The structural materials can be the classic ones or new materials, like composite or sandwich. Lastly, some emerging methodologies are also addressed, such as Automatic Differentiation, Intelligent Structures Optimization, Integration of Structural Optimization in Concurrent Engineering Environments and Multidisciplinary Optimization.

This volume should interest design engineers and researchers. It should be useful as an introductory text to Structural Optimization and also as a contemporary view of the subject. Even if it is not a text book, it can be a useful reference book for advanced undergraduate or graduate courses on Structural Optimization and Optimum Design.

The first four papers give a modern view of the models and methods for classical structural optimization. An introductory contribution of N. Olhoff and E. Lund presents basic techniques for finite element based Structural Optimization. J. S. Arora develops a unified approach for design sensitivity analysis and J. Herskovits explains in a simple way most of the modern mathematical programming methods. Efficient methods for large scale optimization are then discussed by G. Vanderplaats.

Next, G. Rozvany and N. Kikuchi present two different approaches for topology optimization, P. Pedersen shows the scope of material optimization from an engineering point of view, and A. Sepulveda makes a presentation of intelligent structures and their optimal design.

A variational approach for shape optimization with contact is studied by J. Haslinger. Z. Mróz and J. Piekarski present a paper on sensitivity analysis of nonlinear structures. Rousselet et al. study the optimal design of linear and nonlinear elastic shells.

M. Masmoudi et al. give an overview of automatic differentiation techniques and apply it to shape optimization. An example of a concurrent engineering environment that includes Structural Optimization tools is described by K. K. Choi et al.. The last paper, by J. Sobieszczanski-Sobieski, introduces the reader to Multidisciplinary Design Optimization, an emerging discipline that studies the design of complex systems governed by mutually interacting physical phenomena and/or composed of interacting subsystems.

José Herskovits

FINITE ELEMENT BASED ENGINEERING DESIGN SENSITIVITY ANALYSIS AND OPTIMIZATION

NIELS OLHOFF and ERIK LUND
Institute of Mechanical Engineering
Aalborg University, DK-9220 Aalborg, Denmark

Abstract: The aim of this paper is to present basic concepts and selected finite element based methods and tools for sensitivity analysis and rational engineering design and optimization of mechanical structures and components. The main emphasis is devoted to sensitivity and optimization problems that involve shape and sizing design variables.

The methods are selected from conditions of versatility, computational efficiency, and suitability for integration into an engineering design optimization system which realizes the design process as an iterative solution procedure of a multicriterion optimization problem based on the concept of integration of finite element analysis, sensitivity analysis, and optimization by mathematical programming. Typical design objectives that may enter into the multicriterion problem formulation include, e.g., structural cost, displacements and stresses from static and thermal loads, and structural vibration frequencies and buckling loads.

For these types of objectives, expressions for design sensitivities are derived, and the paper presents a new approach to semi-analytical design sensitivity analysis which is easy to implement as an integral part of the finite element analysis. The new approach is computationally inexpensive, and completely eliminates severe sensitivity errors that may occur in problems of semi-analytical shape design sensitivity analysis and optimization of certain types of structures. Such errors have been reported in several papers in recent years.

Use of the methods presented in this paper is illustrated by examples.

1. INTRODUCTION

Engineering activity has always involved endeavours towards optimization, and this particularly holds true for the field of engineering design. Earlier, engineering design was conceived as a kind of "art" that demanded great ingenuity and experience of the designer, and the development of the field was characterized by *gradual evolution* in terms of continual improvement of *existing* types of engineering designs. The design process generally was a sequential "trial and error" process where the designer's skills and experience were most important prerequisites for successful decisions for the "trial" phase.

However, nowaday's strong technological competition which requires reduction of design time and costs of products with high quality and functionality, and current emphasis on saving of energy, saving and re-use of material resources, consideration of environmental problems, etc., often involve creation of new products for which prior engineering experience is totally lacking. Development of such products must naturally lend itself on application of scientific methods.

1

J. Herskovits (ed.), Advances in Structural Optimization, 1–45.
© 1995 *Kluwer Academic Publishers.*

Hence, during recent decades, engineering design has changed from "art" and "evolution" to scientifically based methods of rational design and optimization. This development has been strongly boosted by the advent of reliable general analysis methods like the finite element method, methods of design sensitivity analysis, and methods of mathematical programming, along with the exponentially increasing speed and capacity of digital computers. Thus, methods of rational design and optimization are now finding widespread use in aeronautical, aerospace, mechanical, nuclear, civil, and off-shore engineering.

The development has been supported by vigourous research activities in the fields of design sensitivity analysis and optimum design, see, e.g., the review papers by Haug (1981), Schmit (1982), Olhoff & Taylor (1983), Haftka & Grandhi (1986), Ding (1986), and the proceedings from various conferences and symposia published by Haug & Cea (1981), Morris (1982), Eschenauer & Olhoff (1983), Atrek *et al* (1984), Bennett & Botkin (1986), Mota Soares (1987), Rozvany & Karihaloo (1988), Eschenauer & Thierauf (1989), Eschenauer *et al* (1991), Bendsøe & Mota Soares (1993), Haug (1993), Pedersen (1993a), Rozvany (1993), Herskovits (1993) and Gilmore *et al* (1993).

Rational engineering design and optimization based on the concept of integration of finite element analysis, design sensitivity analysis, and optimization by mathematical programming, was early undertaken by Kristensen & Madsen (1976), Pedersen (1981, 1983), Pedersen & Laursen (1983), Esping (1983), Braibant & Fleury (1984), Eschenauer (1986) and Santos & Choi (1989), and this work has paved the road for development of large, practice-oriented optimization systems, see, e.g., Braibant & Fleury (1984), Sobieski & Rogers (1984), Bennett & Botkin (1985), Esping (1986), Stanton (1986), Eschenauer (1986), Botkin *et al* (1986), Hörnlein (1987), Arora (1989), Rasmussen (1989, 1990), Choi & Chang (1991), and Rasmussen, Lund, Olhoff & Birker (1992, 1993).

It is the objective of the current paper to present basic concepts for problems of engineering design treated as multicriterion optimization problems, and to give a technical survey of selected enabling methods for finite element based analysis and design sensitivity analysis. The methods selected for presentation in this paper reflect those that constitute part of the backbone of the Optimum DESign SYstem ODESSY, see Rasmussen, Olhoff & Lund (1992), Olhoff, Lund & Rasmussen (1993), Lund (1993) and Rasmussen, Lund & Olhoff (1993a,b), which is being developed at the Institute of Mechanical Engineering of Aalborg University. The present paper will be mainly devoted to issues of design sensitivity analysis and optimization pertaining to shape and sizing design variables, i.e., for reasons of brevity, the paper will not cover other important concepts and features of ODESSY like the CAD-integration and the interactive design capabilities of the system as developed by Rasmussen (see the papers just cited) on the basis of his earlier experiences with the system CAOS (Computer Aided Optimization System), see Rasmussen (1990, 1991) and Rasmussen, Lund, Birker & Olhoff (1993). As to CAOS' and ODESSY's facilities for topology optimization, and the integration of topology and shape optimization in the CAD-environment, the reader is also referred to the literature, Rasmussen (1991), Olhoff, Bendsøe & Rasmussen (1991), Rasmussen, Lund, Olhoff & Birker (1993), and Olhoff, Krog & Thomsen (1993).

The development of the general solid modeling based engineering design optimization system ODESSY is carried out in a project within the "Programme of Research on Computer Aided Design" under the auspices of the Danish Technical Research Council. The project is carried out in fruitful cooperation with Profs. M.P. Bendsøe and P. Pedersen of the Technical University of Denmark, and provides a framework as well for their well-known research on development of new methods for optimization of structural topology and material, see, e.g., Bendsøe & Kikuchi (1988), Bendsøe (1989) and a comprehensive monograph by Bendsøe (1994), and optimization of and with advanced materials, see, e.g., Pedersen (1989, 1990, 1991, 1993b). The participants in our joint project gratefully acknowledge the inspiring cooperation with many colleagues and friends from abroad.

Chapter 2 of this paper gives an account of basic concepts for finite element based analysis, design sensitivity analysis and engineering design optimization, and Chapter 3 presents the mathematical formulation for multicriterion optimization.

Chapter 4 briefly discusses the very simple, but computationally inefficient method of sensitivity analysis which is usually termed the Overall Finite Difference (OFD) approach.

In Chapter 5, the so-called semi-analytical method of sensitivity analysis is selected for a detailed presentation from among other methods of design sensitivity analysis. This method, which is simple to implement and computationally efficient, is exemplified for various design objectives such as displacements, stresses and compliance of elastic structures under static and thermal loads, and for simple as well as multiple eigenvalues in the form of structural vibration frequencies and buckling loads.

Chapter 6 is devoted to the problem that, starting with Barthelemy & Haftka (1988), it has been shown by way of examples in several recent papers that the semi-analytical method of design sensitivity analysis may yield severely inaccurate results for certain problems involving shape design variables. It is by now well understood that these severe inaccuracies can be traced back to small errors in the approximate finite difference calculation of element matrix derivatives which is inherent in the semi-analytical method of sensitivity analysis. However, studies of Cheng & Olhoff (1991, 1993) led to the conclusion that the severe error problem is restricted to problems where the displacement field is characterized by rigid body rotations which are large relative to actual deformations, i.e., typically bending problems for long-span beam-like structures and for plate and shell structures. Therefore, in order to cater for all types of structures, a new method for "exact" semi-analytical design sensitivity analysis, which completely eliminates inaccuracies associated with the traditional approach, has been developed by Olhoff, Rasmussen & Lund (1993). This method, which is implemented in the optimization system ODESSY, is described and illustrated by an example in Chapter 6.

Chapter 7 presents an illustrative example of how design sensitivity results displayed on the computer screen, and so-called *what-if* studies, can be used by the designer to improve his engineering designs.

Chapter 8 contains two examples of engineering design optimization which are carried out by means of ODESSY and illustrate that a powerful optimization system may be obtained by integration of the methods discussed here.

Finally, Chapter 9 presents the conclusions of the paper.

2. BASIC CONCEPTS

In this chapter the distinctions between usual analysis, redesign or sensitivity analysis, and optimization of structures will be made clear, and the basic concepts pertaining to optimum design will be outlined.

In a usual *analysis problem*, the structural design is given, together with relevant properties of the material(s) to be used and the support conditions of the structure. Also, one set or more of loading is specified, that is, completely specified in deterministic problems, or given in terms of probabilities in probabilistic problems. For each set of loading, the relevant set of equilibrium (or state) equations, compatibility conditions, and boundary conditions, are then used for determining the structural *response*, e.g., the state of stress, strain and deflection, natural vibration frequencies, and load factors for elastic instability or plastic collapse.

Redesign (or *sensitivity analysis*) refers to the type of problem where the design, material, or support parameters are changed (or varied), and where the corresponding changes (or variations) of the structural response are determined via a repeated (or special) analysis.

The label *engineering design optimization* identifies the type of design problem where the set of structural parameters is subdivided into so-called *preassigned parameters* and *design variables*, and the problem consists in determining optimum values of the design variables such that they *maximize* or *minimize* a specific function termed the *objective* (or *criterion*, or *cost*) *function*, while satisfying a set of *geometrical* and/or *behavioural requirements* which are specified prior to design, and are called *constraints*.

It should be noted that a *conventional design procedure* normally consists of a series of repeated changes of the structural parameters followed by analyses, and this process is continued until a structure is found that fulfills the behavioural requirements and is reasonable in cost. However, in general, the changes are only decided by kind of guesswork based on information obtained from the previous analyses, and a structure obtained by the conventional procedure will not necessarily be any better than other possible alternatives. Only if in each design step the changes of the structural parameters are determined rationally as the best possible ones by sensitivity analysis and use of an optimizer, the procedure would identify one of optimum design.

2.1 Typical Analysis Problems

In the sequel, different types of finite element based linearly elastic, solid mechanical analysis problems will be briefly presented with a view to give a survey of the problems and to introduce the finite element notation to be used in the following.

2.1.1 Structures subjected to static loading

The most common type of analysis is the determination of displacements, strains and stresses of linearly elastic structures under static loads. For this type of problem, the set of global equilibrium equations of a finite element discretized structure is given by

$$KD = F \tag{1}$$

where K is the global stiffness matrix, D the nodal displacement vector and F is the consistent nodal force vector from external loads. These global matrices and vectors are, as always in the finite element method, assembled from element matrices and vectors, i.e.,

$$K = \sum_{n_e} k, \qquad D = \sum_{n_e} d, \qquad F = \sum_{n_e} f \tag{2}$$

Here, k is the element stiffness matrix, d the element nodal displacement vector, f the consistent element nodal force vector, and n_e is the number of finite elements in which the structure is discretized.

2.1.2 Free vibrations

The finite element formulation of a problem of free vibrations of a structure has the form of a real, symmetric eigenvalue problem

$$K\phi_j = \omega_j^2 M\phi_j, \qquad j = 1,...,n \tag{3}$$

where K is the global stiffness matrix, M the global mass matrix, ω_j the eigenfrequency, ϕ_j the corresponding eigenvector, and n is the dimension of the problem.

2.1.3 Thermal analysis

The finite element equilibrium equation for a steady state heat conduction problem has the form

$$K^{th} T = Q \tag{4}$$

where K^{th} is the global thermal "stiffness matrix" involving contributions from element heat conduction matrices and coefficients of the temperature vector T arising from convection boundary conditions, and Q is the thermal load vector comprising forcing terms due to heat addition processes, e.g., heat flux.

2.1.4 Thermo-elastic analysis

This type of analysis consists of a steady state thermal analysis followed by a static analysis. First the temperature field T is found from the solution of Eq. 4, resulting in thermal strains ϵ^{th}. Then Eq. 1 is solved by taking the thermal effects into account.

2.1.5 Free vibrations with initial stress stiffening

When calculating eigenfrequencies of a structure, it may be necessary to take into account initial stress stiffening effects due to mechanical loading. First a static analysis is performed, resulting in element stresses σ. These stresses are used to generate element initial stress stiffness matrices (also termed geometric stiffness matrices) k_σ which represent the stress stiffening effects due to the loads applied to the structure. The conventional global stiffness matrix K is then augmented with the global initial stress stiffening matrix K_σ resulting in a modified form of Eq. 3, i.e.,

$$(K + K_\sigma)\phi_j = \omega_j^2 M\phi_j , \qquad j = 1,...,n \tag{5}$$

2.1.6 Free vibrations with thermal loading and initial stress stiffening

Eigenfrequency analysis with initial stress stiffening effects taken into account can be extended taking thermal effects into account, i.e., the initial stress stiffening effects may originate from a thermo-elastic analysis taking both thermal and mechanical loading into consideration.

2.1.7 Linearized buckling analysis

The finite element formulation of a linearized buckling problem has the form of an eigenvalue problem,

$$(K + \lambda_j K_\sigma)\phi_j = 0 , \qquad j = 1,...,n \tag{6}$$

where K is the global stiffness matrix, K_σ the initial stress stiffness matrix established from an initial static stress analysis, n the dimension of the problem, λ_j the buckling eigenvalue, and ϕ_j is the corresponding eigenvector of displacements.

2.1.8 Buckling analysis with thermal loading

The linearized buckling analysis may be extended to encompass thermal loading, i.e., buckling load factors may be both determined with respect to mechanical and thermal loads.

2.2 Design Variables

The design variables, i.e., the structural parameters which are at the choice of the designer, will be denoted by

$$a_i, \quad i = 1,..,I, \tag{7}$$

and be assembled in the vector a. The design variables can be categorized as follows:

- *Geometrical design variables*:

  - *Sizing design variables*: describe cross-sectional properties of structural components like dimensions, cross-sectional areas or moments of inertia of bars, beams, columns and arches; or thicknesses of membranes, plates and shells.

  - *Configurational design variables*: describe the coordinates of the joints of discrete structures like trusses and frames; or the form of the center-line or mid-surface of continuous structures like curved beams, arches and shells.

  - *Shape design variables:* govern the shape of external boundaries and surfaces, or of interior interfaces of a structure. Examples are the cross-sectional shape of a torsion rod, column or beam; the boundary shape of a disk, plate, or shell; or the shape of interfaces within a structural component made of different materials.

  - *Topological design variables*: describe the type of structure, number of interior holes, etc., for a continuous structure. For a discrete structure like a truss or frame, these variables describe the number, spatial sequence, and mutual connectivity of members and joints.

- *Material design variables*: represent constitutive parameters of isotropic materials, or, e.g., stacking sequence of lamina, and concentration and orientation of fibers in composite materials.

- *Support design variables*: describe the support (or boundary) conditions, i.e., the number, positions and types of support for the structure.

- *Loading design variables*: describe the positioning and distribution of external loading which in some cases may be also at the choice of the designer.

- *Manufacturing design variables*: parameters pertaining to the manufacturing process(es), surface treatment, etc., which influence the properties and cost of the structure.

2.3 The Concepts of Design Model and Analysis Model

Optimization problems are highly non-linear in general. It is therefore necessary to employ iterative numerical solution schemes and determine the optimum design through a sequence of redesign and reanalysis. This implies that the structural geometry must be repeatedly converted into a finite element model with the proper loads and boundary conditions, and that the variable structural design must be described (parametrized) in terms of a finite number of geometrical variables.

Thus, particularly for shape optimization problems, it is necessary to make a clear distinction between the *analysis model* as represented by the finite element model, and the parametrized geometric model of the variable structure which is termed the *design model*.

In the early days of structural shape design sensitivity analysis and optimization, attempts were made to use the finite element model directly as design model, i.e., to use node coordinates as design variables. However, this approach at least has four serious drawbacks:

- the number of design variables may become very large
- it is difficult to ensure compatibility and slope continuity between boundary nodes
- it is difficult to avoid distorsion of the finite element mesh during the shape updating process
- the structural shape design sensitivities may not be accurate unless high order finite element types are used.

It is these experiences that have led to the aforementioned important distinction between the *design model* and the *analysis model* which is introduced nowadays in most computer aided environments for interactive structural shape design and optimization, see, e.g., Esping (1983, 1984), Braibant & Fleury (1984), Rasmussen (1990), and Olhoff, Bendsøe & Rasmussen (1991). The design model is endowed with additional significance because it can be closely connected with a CAD model as described by Rasmussen (1990), Rasmussen, Lund, Birker & Olhoff (1993), and Olhoff, Lund & Rasmussen (1993).

The design model may consist of so-called *design elements* as presented by Braibant & Fleury (1984). The boundaries (or surfaces in case of a three-dimensional model) of the design elements can be curves of almost any character, i.e., piecewise straight lines, arcs, b-splines of any given degree of continuity, Bezier curves, Coons patches, etc. It is therefore very simple to generate relatively complicated geometries with a small number of design elements. The shapes of the boundaries are controlled by a number of control points, also often termed *master nodes*.

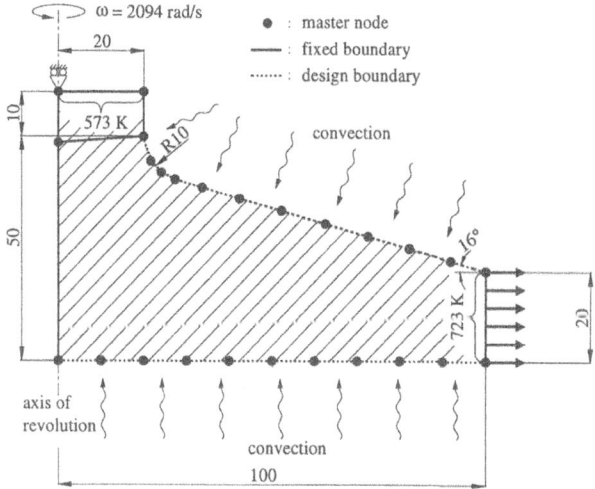

Figure 1. Design model of turbine disk.

For exemplification, let us consider the axisymmetric model in Fig. 1 of a turbine disk which rotates at 2094 rad/s (= 20000 rev/min). The blades have been replaced by a uniformly distributed load at the rim of the disk. This load represents the centrifugal forces from the blades.

Furthermore, as is also indicated in Fig. 1, the turbine disk is subjected to different temperatures. These temperatures derive from the hot exhaust gas which drives the disk. Part of the boundary is subjected to a convection boundary condition which is also due to the exhaust gas. At these boundaries the temperature of the environment is specified to be 450 °C and the convection coefficient is 0.0012 W/(mm^2·K).

The design model consists of two design elements. There are two design boundaries, i.e., boundaries whose shapes are allowed to change. Each of these shapes are defined by the positions of a number of master nodes, and this creates an evident connection between the design variables (the movements of the master nodes) and the shape of the geometry.

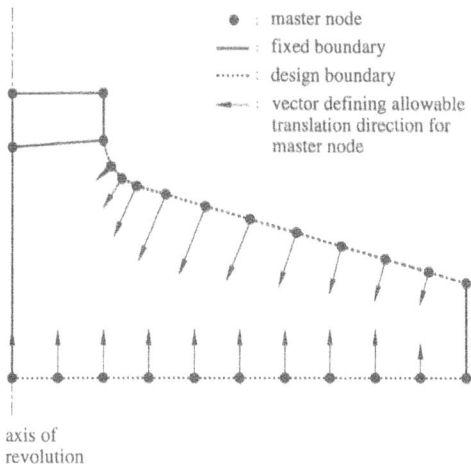

Figure 2. Variable design model.

In this example, the direction of the movement of each master node is constrained to follow some predefined move directions specified by the designer as shown in Fig. 2. Such translational transformations are very common in definitions of shape design models. Other possible modifiers may be point scaling, line scaling, rotation, etc. Thus, in this example, the design variables a_i, $i = 1,...,I$, where I is the number of design variables, are simply taken to be the sizes of the movements of the master nodes along the associated move directions.

In addition, the distribution of finite element nodes on the boundaries and the desired finite element type for each design element must be defined. All necessary specifications including loading conditions are assigned to the design model and a preprocessor with automatic mesh generation can automatically convert it into a finite element analysis model as shown in Fig. 3. The preprocessor has meshed the design elements with a mixture of 6 and 9 node isoparametric 2-D axisymmetric finite elements. All load specifications are automatically converted into consistent nodal loads.

The design model now has been converted into an analysis model, and by changing the values of the design variables, the geometry can be changed parametrically into other shapes.

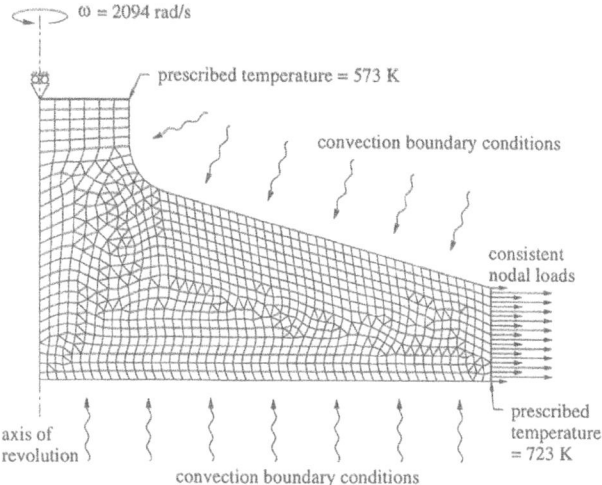

Figure 3. Finite element analysis model.

2.4 Constraints

Any set of values of the design variables defines a design of the structure and may be represented as a point in the so-called *design space*. Many designs from among the totality of possible designs will generally not be acceptable in terms of various design and performance requirements. To exclude such designs as candidates for an optimum solution, the design and performance requirements are expressed mathematically in the form of *constraints* prior to optimization. These constraints directly or indirectly impose limits on the ranges of variation of the design variables. Relative to the design space, the constraints define hyper-surfaces that encircle the set of acceptable, or so-called *feasible* or *admissible designs*. The constraints may be of two types: geometrical constraints and behavioural constraints.

Geometrical (or *side*) constraints are restrictions imposed explicitly on the design variables due to considerations such as manufacturing limitations, physical practicability, aestetics, etc. Constraints of this kind are typically *inequality constraints* that specify lower or upper bounds on the design variables, but may also be *equality constraints* like, e.g., *linkage constraints* that prescribe given proportions between a group of design variables.

Behavioural constraints are generally nonlinear and implicit in terms of the design variables, and they may be of two types. The first type consists of *equality constraints* such as state and compatibility equations governing the structural response associated with the loading condition(s) under consideration. The second type of behavioural constraints comprises *inequality constraints* that specify restrictions on those quantities that characterize the response of the structure. These constraints may impose bounds on *local* (or *point-wise*) quantities like stresses and deflections, or on *global* (or *integral*) quantities such as compliance, natural vibration frequencies, etc.

2.5 Objective Function

The *objective function*, which is also often called the *cost* or *criterion function*, must be expressed mathematically in terms of the design variables such that its value can be determined for any point in the design space. It is this function whose value is to be minimized or maximized by the optimum set of values of the design variables within the admissible design space, and it may represent the structural weight or cost, or it may be taken to represent some local or global measure of the structural performance like stress, displacement, stress intensity factor, integral stiffness, plastic collapse load, fatigue life, buckling load, natural vibration frequency, aeroelastic divergence or flutter speed, etc.

In *single-criterion* problems, the objective function represents a single performance index that is of primary concern, but the objective function may, for example, also represent a weighted sum of a number of different properties, or the minimum or maximum of a weighted set of different criteria, as may be the case in so-called *multicriteria optimization* problems (see Chapter 3).

2.6 The Notion of Problem Variables

In practical problems of optimum engineering design, it is necessary to take into account several performance and failure criteria in the problem formulation, and this means that a multicriterion approach must be adopted. In order to cater for a broad range of problem formulations, it is also required that it should be easy to interchange typical objective and constraint functions from one problem to another. In developing such a versatile mathematical formulation for multicriterion optimization, it has been found convenient to introduce the common notion "*problem variables*" for typical objective and constraint functions. From among these functions we distinguish between *global problem variables* and *local problem variables*.

2.6.1 Global problem variables

The global problem variables that enter in the multicriterion objective function will be denoted by

$$f_j, \quad j = 1,..,J, \tag{8}$$

and those entering in prescribed constraints will be denoted by

$$g_p, \quad p = 1,..,P. \tag{9}$$

The global problem variables may be categorized as *cost variables* and *global behavioural problem variables*, of which the following examples may be given:

- *Cost problem variables:*
  Structural volume or weight
  Material cost
  Manufacturing cost
  Life cycle cost
  etc.

- *Global behavioural problem variables:*
  Compliance
  Buckling load
  Plastic collapse load
  Eigenfrequency
  Dynamic response
  Fatigue life
  etc.

Note that the global problem variables are independent of the spatial variables of the structure. A cost problem variable f_j or g_p only depends on the vector a of design variables, i.e.,

$$f_j = f_j(a), \quad g_p = g_p(a) \qquad \text{(cost PVs)} \qquad (10)$$

while a global behavioural problem variable f_j or g_p also depends on the relevant displacement (or state) vector D_j or D_p associated with the behaviour variables,

$$f_j = f_j(a,D_j), \quad g_p = g_p(a,D_p) \qquad \text{(global behavioural PVs)} \qquad (11)$$

Here, D_j or D_p must be first obtained by solution of the relevant finite element equilibrium equations.

The global behavioural problem variables listed above are all of a solid mechanical nature, but extensions into, e.g., the field of fluid mechanics (e.g., drag in fluid flow) are straight-forward.

2.6.2 Local problem variables

Local problem variables will be equipped with an asterisk (*). The local problem variables entering in the multicriterion objective function will be denoted by

$$f_k^*, \quad k=1,..,K, \qquad (12)$$

and those entering in prescribed constraints by

$$g_s^*, \quad s=1,..,S. \qquad (13)$$

Examples of this category of variables are the

- *Local behavioural problem variables:*
  Stresses
  Strains
  Displacements
  etc.

These variables depend on the spatial variables of the structure in addition to their dependence on the design variables and the relevant displacement field, i.e., we have

$$f_k^* = f_k^*(a,D_k,x), \quad g_s^* = g_s^*(a,D_s,x), \quad x \in \Omega, \qquad \text{(local behavioural PVs)} \qquad (14)$$

where x designates the coordinates of any point within the structural domain Ω. Note that it is also required to solve the finite element equilibrium equations associated with the given (sets of) external loading in order to determine this type of problem variables.

3. FORMULATION FOR MULTICRITERION OPTIMIZATION

The approach adopted here for solution of multicriterion optimization problems is based on the concept of integration of software modules of finite element analysis, sensitivity analysis, and mathematical program-ming. In order to meet practical needs of versatility, the basic mathematical formulation must possess sufficient flexibility such that we can both handle problems of minimizing cost subject to several constraints, and problems of multicriterion optimization for prescribed resource (and additional constraints). To achieve this goal, the multicriterion problem is cast in scalar form by stating it as minimization of the maximum of a weighted set of the criteria. Such an interpretation of the multicriterion optimization problem can be formulated as a problem of minimizing a variable upper bound on the weighted criteria, see Bendsøe, Olhoff

& Taylor (1983), Taylor & Bendsøe (1984), and Olhoff (1989), and this *bound formulation* implies the considerable advantage that the multicriterion problem becomes differentiable.

Consider a multicriterion optimization problem in the initial form:

Objective:

$$\min_{a_i} \left[\max_{jk} \left\{ \hat{w}_j \cdot f_j \; ; \; \hat{w}_k^* \cdot \max_{x \in \Omega} f_k^* \right\} \right] \tag{15a}$$

Subject to:

Constraints for problem variables:

$$g_p \leq \hat{G}_p \; , \quad p = 1,..,P \qquad \text{(global)} \tag{15b}$$

$$g_s^* \leq \hat{G}_s^* \; , \quad s = 1,..,S, \quad \forall \, x \in \Omega \quad \text{(local)} \tag{15c}$$

Linking constraints for design variables:

$$\sum_{i=1}^{I} \hat{c}_{iq} a_i \leq \hat{c}_{oq} \; , \quad q = 1,..,Q \; , \qquad \sum_{i=1}^{I} \hat{b}_{ir} a_i = \hat{b}_{or} \; , \quad r = 1,..,R \tag{15d}$$

Side constraints for design variables:

$$\underline{a}_i \leq a_i \leq \bar{a}_i \; , \quad i = 1,..,I \tag{15e}$$

Equations for problem variables:

$$f_j = f_j(a, D_j) \; , \quad g_p = g_p(a, D_p) \tag{15f}$$

$$f_k^* = f_k^*(a, D_k, x) \; , \quad g_s^* = g_s^*(a, D_s, x) \; , \quad x \in \Omega \tag{15g}$$

Finite element state equations:

$$(KD = F)_t \; , \quad t = 1,..,T, \quad etc. \tag{15h}$$

In Eqs. (15a-h), all symbols equipped with carets and upper or lower bars are assumed to be prescribed. Thus, in (15a), $\hat{w}_j$ and $\hat{w}_k^*$ are given weighting factors for the separate design criterion functions f_j, $j = 1,..,J$, and f_k, $k = 1,..,K$, respectively; in (15b) and (15c) $\hat{G}_p$ and $\hat{G}_s^*$ denote given upper constraint values; $\hat{c}_{iq}$ and $\hat{b}_{ir}$ in (15d) are given linking coefficients; and in (15e) $\underline{a}_i$ and $\bar{a}_i$ denote prescribed lower and upper side constraint values for the design variables a_i, $i = 1,..,I$.

The problem defined by Eqs. (15a-h) pertains to determine the set of values of the design variables a_i, $i = 1,..,I$, that minimizes the maximum of the set of weighted global and local problem variables f_j, $j = 1,..,J$, and f_k^*, $k = 1,..,K$, respectively, in the multicriterion objective function (15a), subject to the constraints (15b-h).

It is assumed that each of the problem variables f_j and f_k^* that enter the scalar objective function (15a) have been suitably preconditioned for minimization, and that the finite element state equations in (15h), which may also include eigenvalue problems, etc., contain all the sets of equations that are necessary for computation of the different displacement (or state) vectors required for the variety of behavioural problem variables that needs to be calculated in (15f) and (15g) when the optimization is carried out under consideration of several loading cases and different types of response. Notice that although not stated directly, (15f) also comprise equations for cost variables, and that these generally only depend on the vector of design variables a.

Now, the objective (15a) of the multicriterion optimization problem is equivalent to

$$\min_{a_i} \left[\max_{j,(x)_k} \left\{ \hat{w}_j \cdot f_j \; ; \; \hat{w}_k^* \cdot f_k^* \right\} \right]. \tag{16}$$

To circumvent the inherent difficulty that this min-max objective function is generally *not differentiable* with respect to the design variables a_i, $i = 1,..,I$, we develop a *bound formulation* of the problem, see Bendsøe, Olhoff & Taylor (1983), Taylor & Bendsøe (1984), and Olhoff (1989). By this technique we introduce an additional variable β, which is termed the *bound parameter*, and write (16) in the form of the *differentiable problem*

$$\min_{a_i, \beta} \beta \tag{17a}$$

subject to:

$$\hat{w}_j \cdot f_j \leq \beta \;, \quad j = 1,..,J \qquad \text{(global constraints)} \tag{17b}$$

$$\hat{w}_k^* \cdot f_k^* \leq \beta \;, \quad k = 1,..,K, \quad \forall x \in \Omega \quad \text{(local constraints)} \tag{17c}$$

Thus, we simply transform the statement (16) into the constraints (17b) and (17c), and minimize the common upper bound β on these constraints according to (17a).

Notice that in this formulation, the upper bound parameter β is an additional variable which replaces the original non-differentiable functional, and is to be minimized over a constraint set in an enlarged space. The original points of non-differentiability correspond to "corners" in the constraint set of the enlarged space, and arise from intersections of differentiable constraints.

We now in Eqs. (15a-h) replace Eq. (15a) by Eqs. (17a-c) and arrive at the following *bound formulation of the multicriterion optimization problem*:

$$\min_{a_i, \beta} \beta \tag{18a}$$

Subject to:

Variable bound constraints:

$$\hat{w}_j \cdot f_j \leq \beta \;, \; j = 1,..,J \qquad \text{(global)} \tag{18b}$$

$$\hat{w}_k^* \cdot f_k^* \leq \beta \;, \; k = 1,..,K, \quad \forall x \in \Omega \qquad \text{(local)} \tag{18c}$$

Original constraints:

$$g_p \leq \hat{G}_p \;, \quad p = 1,..,P \qquad \text{(global)} \tag{18d}$$

$$g_s^* \leq \hat{G}_s^* \;, \; s = 1,..,S, \quad \forall x \in \Omega \qquad \text{(local)} \tag{18e}$$

Linking constraints for design variables:

$$\sum_{i=1}^{I} \hat{c}_{iq} a_i \leq \hat{c}_{oq} \;, \quad q = 1,..,Q \;, \qquad \sum_{i=1}^{I} \hat{b}_{ir} a_i = \hat{b}_{or} \;, \quad r = 1,..,R \tag{18f}$$

Side constraints for design variables:

$$\underline{a}_i \leq a_i \leq \overline{a}_i \;, \quad i = 1,..,I \tag{18g}$$

Equations for problem variables:

$$f_j = f_j\,(a, D_j), \quad g_p = g_p(a, D_p) \tag{18h}$$

$$f_k^* = f_k^*\,(a, D_k, x), \quad g_s^* = g_s^*(a, D_s, x), \quad x \in \Omega \tag{18i}$$

Finite element state equations:

$$(KD = F)_t, \quad t = 1,.., T, \quad etc. \tag{18j}$$

Notice that the bound approach is a very simple technique for rendering min-max problems differentiable, and that the bound formulation is very versatile in terms of handling different types of problem variables. Thus, it is very simple to switch from, e.g., a prescribed-cost to a cost-minimization formulation of a given type of problem. It is also noteworthy that the sensitivity information concerning problem variables is virtually independent of whether a particular problem variable is an objective or a constraint function. A generalized approach to the formulation of a multicriterion optimization problem has recently been published by Lund & Rasmussen (1994).

The problem variables f_j, f_k^*, g_p and g_s^* are determined using necessary finite element software behind (18h-j). Although f_j and g_p are global, and f_k^* and g_s^* are local problem variables, there is no need to make this distinction in practice. Thus, the local problem variables are just evaluated at a number of nodal points of the finite element discretized structure according to some suitable *active set strategy*, and thereby result in a number of inequality constraints that have the same mathematical form as the constraints (18b) and (18d) for the global problem variables.

While Eqs. (18h-j) represent the necessary tools for finite element analysis, Eqs. (18a-g) constitute a standard problem of mathematical programming which is solved sequentially using, e.g., a SIMPLEX algorithm, the CONLIN optimizer, see Fleury & Braibant (1986) and Fleury (1989), or the Method of Moving Asymptotes, see Svanberg (1987), all of which require first order design sensitivities of the objective and constraint functions to be calculated. For this purpose, a sensitivity analysis is performed for each design variable at each step of redesign. For shape optimization, the systems CAOS and ODESSY use the semi-analytical method which will be discussed in Chapter 5.

4. OVERALL FINITE DIFFERENCE APPROACH TO SENSITIVITY ANALYSIS

The simplest method of design sensitivity analysis is the finite difference approach. If displacements, stresses, compliance, mass, or any other problem variable calculated by one of the analysis modules is denoted by $f_j(a)$, which is a function of the design variables a_i, the overall forward finite difference approximation $\Delta f_j / \Delta a_i$ to the design sensitivity $\partial f_j / \partial a_i$ is given by

$$\frac{\partial f_j(a)}{\partial a_i} \approx \frac{\Delta f_j(a_1,...,a_I)}{\Delta a_i} = \frac{f_j(a_1,...,a_i + \Delta a_i,...,a_I) - f_j(a_1,...,a_i,...,a_I)}{\Delta a_i} \tag{19}$$

If the derivatives of the function f_j are sought for n design variables, the overall finite difference (OFD) method requires n additional analyses. Therefore this method is computationally very costly and mainly used as a reference method. Its limits with regard to accuracy is set only by the accuracy of the finite element solution procedure, the discretization, and the usual accuracy capabilities of the applied finite element.

Whenever a finite difference scheme is used to approximate derivatives, there are two sources of error: *truncation* and *condition errors*. The truncation error is a result of the neglected terms in the Taylor series expansion of the perturbed function $f_j(a_1,...,a_i + \Delta a_i,...,a_I)$. This source of error can be reduced by using a small perturbation Δa_i. The condition error is the difference between the numerical evaluation of the function and its exact value. Contributions to the condition error are, e.g., computational round-off errors or errors due to an iterative solution process which is terminated early. The round-off errors are normally small for most computers unless the perturbation Δa_i is very small. These contradictory demands to the magnitude of the perturbation may give rise to the so-called "step-size dilemma", see, e.g., Haftka & Adelman (1989). If we select the perturbation to be small, so as to reduce the truncation error, the result may be an excessive condition error. In some cases there may not even exist a perturbation which results in sufficiently small errors.

5. METHOD OF SEMI-ANALYTICAL DESIGN SENSITIVITY ANALYSIS

An efficient and reliable method of design sensitivity analysis is indispensable in a computer aided engineering environment for interactive design and optimization, and the most commonly used technique is based on implicit differentiation of the finite element discretized equilibrium or state equations for the structure with respect to the design variables.

The most general implementation of this technique, which is preferable from the viewpoint of computational cost and ease of implementation, implies application of numerical differentiation of the finite element stiffness matrices, and is termed the *semi-analytical method of design sensitivity analysis*, see Zienkiewicz & Campbell (1973), Esping (1983) and Cheng & Liu (1987). An alternate approach is based on application of the material derivative concept of continuum mechanics to relate variations in the structural shape to measures of structural performance, cf. Haug, Choi & Komkov (1985), Cea (1981), Zolesio (1981), Dems & Mroz (1983, 1984, 1993), Choi (1985), and Haber (1987). The relationship between the two methods has been studied in Yang & Botkin (1986) and Choi & Twu (1988).

In this paper, the main attention will be devoted to the semi-analytical method of design sensitivity analysis.

5.1 Sensitivity Analysis of Displacements due to Static Loads

In finite element based methods of design sensitivity analysis, the sensitivities of various problem variables entering in multiple criteria and constraints, are generally determined from the design sensitivities of pertinent displacement fields. Thus, when the displacement sensitivities are known, stress and compliance sensitivities, for example, are easily computed.

Referring to Section 2.1.1, the global equilibrium equation of a finite element discretized linearly elastic structure subjected to static loading is given by

$$KD = F \tag{20}$$

where K is the global stiffness matrix, D is the nodal displacement vector, and F is the consistent nodal force vector. The solution of Eq. 20 is conveniently carried out by Gaussian elimination reformulated in a two phase process which does not require simultaneous modification of K and F. The procedure is known as "factorization" and makes it possible to solve Eq. 20 for additional load cases, i.e. several right hand sides, without much additional computational effort. The time consuming part of solving Eq. 20 is the factorization of the global stiffness matrix, in which K is basically decomposed into the product LU, where L is a lower triangular matrix (with non-zero elements only in and below the diagonal) and U is an upper triangular matrix. Slightly different versions of this basic decomposition are, for example, the LDU, Crout, and the

Cholesky decomposition schemes, see Dhatt & Touzot (1984) or other standard finite element books for details. When the stiffness matrix K has been decomposed into the product LU, Eq. 20 can be rewritten as

$$LUD = F \tag{21}$$

Using this decomposed form of the stiffness matrix, it is only necessary to solve a triangular set of equations which is very simple. First a vector V is found by forward substitution in the triangular system of equations:

$$LV = F \tag{22}$$

and then the displacement vector D can be found by back substitution

$$UD = V \tag{23}$$

Having now determined the displacement vector D, we proceed to consider the design sensitivity analysis of displacements.

The *direct approach* to obtain design sensitivities of the displacement field is based on *implicit differentiation of the global equilibrium equation*. If Eq. 20 is differentiated with respect to a design variable a_i and the terms are rearranged, the following expression is obtained for computation of the displacement sensitivities $\partial D / \partial a_i$,

$$K(a) \frac{\partial D}{\partial a_i} = - \frac{\partial K(a)}{\partial a_i} D + \frac{\partial F}{\partial a_i} \tag{24}$$

Eq. 24 is of the same form as Eq. 20, so the factorized stiffness matrix K in the form LU can be re-used, and only the new right hand side, which is termed the *pseudo load vector*, needs to be calculated before the sensitivities $\partial D / \partial a_i$ can be found by forward and back substitution. The derivatives $\partial F / \partial a_i$ of the force vector in Eq. 24 are generally easily calculated (note that they vanish for design independent loads), and then the determination of $\partial D / \partial a_i$ in Eq. 24 only requires calculation of the design sensitivities $\partial K / \partial a_i$ of the global stiffness matrix. These derivatives are normally calculated at the element level, i.e.

$$\frac{\partial K}{\partial a_i} = \sum_{n_e} \frac{\partial k}{\partial a_i} , \qquad i = 1,...,I \tag{25}$$

where k is the element stiffness matrix and n_e is the number of finite elements.

If the design sensitivities $\partial k / \partial a_i$ are determined analytically before their numerical evaluation, the approach is called *analytical design sensitivity analysis*, and if they are determined by numerical differentiation, the method is called *semi-analytical design sensitivity analysis*, cf. Zienkiewicz & Campbell (1973), Esping (1983), Cheng & Liu (1987), and Haftka & Adelman (1989). In the semi-analytical method of design sensitivity analysis the derivatives of the element matrices are usually approximated by first order forward finite differences

$$\frac{\partial k(a)}{\partial a_i} \approx \frac{\Delta k(a_1,...,a_I)}{\Delta a_i} = \frac{k(a_1,...,a_i + \Delta a_i,...,a_I) - k(a_1,...,a_i,...,a_I)}{\Delta a_i} \tag{26}$$

The method of *analytical* design sensitivity analysis is very cumbersome to implement in a general purpose shape design system which contains many different kinds of shape design variables and finite element types. Thus, a large amount of analytical work and programming will be required in order to develop analytic expressions for derivatives of various stiffness matrices with respect to all possible design variables.

It is much more attractive to use the *method of semi-analytical design sensitivity analysis* in this context, as it is easy to implement for many different kinds of design variables and finite element types, because simple and computationally inexpensive first order finite differences are used. Therefore, the method of semi-analytical sensitivity analysis is very popular and, in most cases, this method is very efficient and reliable.

The efficiency of the design sensitivity analysis can be increased by using a so-called "active element" or "boundary layer" strategy. When the domain shape is perturbed, only finite elements located at the boundary are perturbed (and named active) as illustrated in Fig. 4 where the perturbation has been strongly exaggerated. The boundary on Fig. 4 is modeled by a quadratic b-spline and the master node shown is translated in the direction specified by the vector. The active (perturbed) finite elements (in the "boundary layer") are hatched in Fig. 4 where the unperturbed mesh also is shown. The active element strategy implies that in the assembly of element matrix derivatives, only active elements give contributions, i.e.

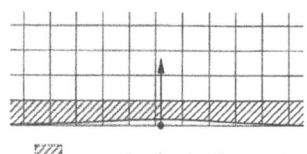

///// : perturbed element

Figure 4. Perturbed boundary.

$$\frac{\partial K}{\partial a_i} = \sum_{n_e^a} \frac{\partial k^a}{\partial a_i}, \qquad i = 1,...,I \tag{27}$$

where k^a is the element stiffness matrix for an active element, and n_e^a is the number of active finite elements.

5.2 Design Sensitivity of Stresses

Once the displacement design sensitivities have been obtained, it is straight-forward to calculate design sensitivities of stresses. The finite element expression for the element stresses σ may be written

$$\sigma(a) = E \epsilon(a) = E B(a) d(a) \tag{28}$$

Here, the constitutive matrix E is assumed to be independent of the design, B is the strain-displacement matrix, and d is the element nodal displacement vector.

One way to obtain the stress design sensitivities is to use a first order forward finite difference approximation, i.e.

$$\frac{\partial \sigma}{\partial a_i} \approx \frac{\sigma(a + \Delta a_i) - \sigma(a)}{\Delta a_i} \tag{29}$$

and calculate the stress sensitivities from the equation

$$\sigma(a + \Delta a_i) = E \epsilon(a + \Delta a_i) = E B(a + \Delta a_i) d(a + \Delta a_i) \tag{30}$$

Here, the perturbed strain-displacement matrix $B(a + \Delta a_i)$ is easily calculated, and the element displacement vector for the perturbed design can be approximated by the first order Taylor expansion

$$d(a + \Delta a_i) \approx d(a) + \frac{\partial d(a)}{\partial a_i} \Delta a_i \tag{31}$$

5.3 Design Sensitivity of Compliance

The compliance C of a structure subjected to given loads, is defined by

$$C = D^T F \tag{32}$$

By differentiating Eq. 32 the following expression for the compliance design sensitivity is obtained:

$$\frac{\partial C}{\partial a_i} = \frac{\partial D^T}{\partial a_i} F + D^T \frac{\partial F}{\partial a_i} \tag{33}$$

Here, all terms on the right hand side are known from the calculation of displacement sensitivities, so the design sensitivity of the compliance is easily evaluated.

5.4 Sensitivity Analysis of Thermo-elastic Problems

In many engineering problems, thermo-elastic effects have to be taken into account due to a temperature distribution in the structure. The finite element equilibrium equation for a steady state heat conduction problem has the form

$$K^{th} T = Q \tag{34}$$

cf. Eq. 4, where K^{th} is the global thermal "stiffness matrix", T the temperature vector, and Q is the "thermal load vector".

Eq. 34 can be solved for the temperatures T by standard solution procedures as described in Section 5.1, and once the temperature distribution is known, its influence can be taken into account in the static finite element analysis. The temperature T at a given point gives rise to thermal strains ϵ^{th} given by

$$\epsilon^{th} = \left\{ \epsilon_x^{th} \ \epsilon_y^{th} \ \epsilon_z^{th} \ \gamma_{xy}^{th} \ \gamma_{yz}^{th} \ \gamma_{xz}^{th} \right\}^T = \alpha \left\{ \hat{T} \ \hat{T} \ \hat{T} \ 0 \ 0 \ 0 \right\}^T, \quad \hat{T} = T - T_0 \tag{35}$$

where α is the thermal expansion coefficient, and T_0 is the strain-free reference temperature.

As is well known, due to possible kinematic boundary conditions and internal restraints, generally the thermal strains cannot be added directly to the mechanical strains in order to obtain the resulting stress field. Instead the thermal strains ϵ^{th} are used to calculate a consistent global nodal force vector F^{th} due to the thermal strains

$$F^{th} = \sum_{n_e} \int_{\Omega} B^T E \epsilon^{th} d\Omega \tag{36}$$

where E is the constitutive matrix, B the strain-displacement matrix, Ω the domain of the finite element and n_e is the number of finite elements. This nodal force vector associated with the thermal strains is added to the mechanical nodal force vector in Eq. 20 when solving the static equilibrium equations. This results in a displacement vector D used to calculate the element strains ϵ, see Eq. 28, and the total element stresses σ due to both the mechanical and thermal strains, are then given by

$$\sigma = E(Bd - \epsilon^{th}) \tag{37}$$

where d is the element displacement vector.

In the design sensitivity analysis, the sensitivities of the temperatures T can be found in the same way as described for displacement sensitivities in Section 5.1. Eq. 34 is differentiated with respect to a design variable a_i, $i = 1,...,I$, and rearranging terms, the following expression for the temperature sensitivities $\partial T / \partial a_i$ is obtained:

$$K^{th}(a) \frac{\partial T}{\partial a_i} = -\frac{\partial K^{th}(a)}{\partial a_i} T + \frac{\partial Q}{\partial a_i} \tag{38}$$

The factorized global thermal "stiffness matrix" can be reused as in the case of sensitivity analysis of displacements, and the new right hand side which may be termed the "*thermal pseudo load vector*" can be obtained using the semi-analytical approach as described previously for static design sensitivity analyses, see Eqs. 25 and 26. When the new right hand side has been determined, the temperature sensitivities $\partial T / \partial a_i$

can be calculated by forward and back substitution.

Having obtained the temperature sensitivities $\partial T / \partial a_i$, the perturbed thermal strains $\epsilon^{th}(a + \Delta a_i)$ and the perturbed nodal force vector $F^{th}(a + \Delta a_i)$ are easily calculated, and finite difference approximations to sensitivities of thermal strains and the corresponding nodal force vector can be evaluated. These sensitivities can be included in the static design sensitivity analysis, whereby Eq. 24 can be extended as

$$K(a) \frac{\partial D}{\partial a_i} = - \frac{\partial K(a)}{\partial a_i} D + \frac{\partial F}{\partial a_i} + \frac{\partial F^{th}}{\partial a_i} \tag{39}$$

In a similar way, the thermal strain sensitivities are included in the stress sensitivities in Eq. 29 by using Eqs. 30 and 37, and the thermo-elastic effects are thereby included in the design sensitivity analysis.

5.5 Design Sensitivity Analysis of Structural Eigenvalues

In this section, design sensitivity analysis of structural eigenvalues will be discussed. Firstly, we present some basic concepts for the eigenvalue problems to be considered, and then devote Sub-section 5.5.1 to sensitivity analysis of *simple* (or *distinct*, or *unimodal*) structural eigenvalues. As is by now well-known, design sensitivity analysis of eigenvalues is less straight-forward for cases of *multiple* (or *repeated*, or *multimodal*) eigenvalues, i.e., cases where two or more eigenvalues attain the same value. Then these eigenvalues are no longer differentiable functions of design in the normal sense, and the design sensitivity analysis of such eigenvalues is considered in Sub-section 5.5.2. The reader is, e.g., referred to the recent work by Seyranian, Lund & Olhoff (1994) for further details. Optimization of multiple eigenvalues was initiated in Olhoff & Rasmussen (1977) and comprehensive surveys of such problems can be found in, e.g., Zyczkowski (1989) and Gajewski & Zyczkowski (1988).

The eigenvalue problems under consideration may be problems of free vibrations (possibly including stress stiffening effects) or structural buckling problems. Within a finite element formulation, the problems are governed by the real, symmetric, eigenvalue problem

$$K\phi_j = \lambda_j M\phi_j, \qquad j = 1,...,n \tag{40}$$

where K and M are symmetric positive definite matrices, λ_j is the eigenvalue, and ϕ_j is the corresponding eigenvector. As discussed in Section 2.1, depending on the type of problem at hand, the global K and M matrices may consist of contributions from either element stiffness, mass or stress stiffness matrices. The dimension of the problem is denoted by n, so Eq. 40 has n solutions consisting of eigenvalues λ_j and corresponding eigenvectors ϕ_j. The eigenvalues are all real and represent squared angular vibration frequencies or structural buckling loads depending on the type of problem. The eigenvalues can be ordered by magnitude as

$$0 < \lambda_1 \leq \lambda_2 \leq ... \leq \lambda_j \leq ... \leq \lambda_n \tag{41}$$

In the following it is assumed that the eigenvectors have been M-orthonormalized, i.e.,

$$\phi_j^T M \phi_k = \delta_{jk}, \qquad j,k = 1,...,n \tag{42}$$

where δ_{jk} denotes Kronecker's delta.

If Eq. 40 is premultiplied by ϕ_j^T the following expression is obtained

$$\phi_j^T K \phi_k = \lambda_j \delta_{jk}, \qquad j,k = 1,...,n \tag{43}$$

which implies that the eigenvectors are also K-orthogonal.

5.5.1 Simple eigenvalues

Assume again that the design of the structure is governed by a set of design variables a_i, $i = 1,...,I$, and consider now the problem of determining the sensitivity of a simple eigenvalue with respect to these design variables. Since the eigenvalue is simple, the corresponding eigenvector is unique, and we shall assume that the components of the K and M matrices are smooth functions of design variables a_i.

The direct approach to obtain the eigenvalue sensitivities is to differentiate Eq. 40 with respect to a design variable a_i. Assuming that λ_j is simple, we have

$$\frac{\partial K}{\partial a_i}\phi_j + (K - \lambda_j M)\frac{\partial \phi_j}{\partial a_i} = \frac{\partial \lambda_j}{\partial a_i}M\phi_j + \lambda_j\frac{\partial M}{\partial a_i}\phi_j, \qquad i = 1,...,I \tag{44}$$

By premultiplying Eq. 44 by ϕ_j^T and making use of Eq. 40, the following expression is obtained for the sensitivity of a simple eigenvalue λ_j with respect to any of the design variables a_i, cf., e.g., Courant & Hilbert (1953) and Wittrick (1962).

$$\frac{\partial \lambda_j}{\partial a_i} = \phi_j^T\left(\frac{\partial K}{\partial a_i} - \lambda_j\frac{\partial M}{\partial a_i}\right)\phi_j, \qquad i = 1,...,I \tag{45}$$

where the term $\phi_j^T M\phi_j = 1$ due to the M-orthonormalization, Eq. 42, has been omitted.

The only unknown quantities in Eqs. 45 are the derivatives of the K and M matrices. As in the case of semi-analytical design sensitivity analysis of a linearly elastic problem with static loads, these derivatives are calculated at the element level, see Eqs. 25 and 26, and the eigenvalue design sensitivities are then easily determined.

If all the design variables a_i are changed simultaneously, then, due to the differentiability of the simple eigenvalue with respect to the design variables, the increment of the eigenvalue λ_j is determined by the scalar product

$$\Delta \lambda_j = \nabla^T \lambda_j \, \Delta a \tag{46}$$

where $\nabla \lambda_j$ denotes the gradient vector of λ_j and Δa is the vector of changes of the design variables a_i

$$\nabla \lambda_j = \left(\frac{\partial \lambda_j}{\partial a_1},...,\frac{\partial \lambda_j}{\partial a_I}\right), \qquad \Delta a = (\Delta a_1,...,\Delta a_I) \tag{47}$$

5.5.2 Multiple eigenvalues

We now consider the situation where the solution of the generalized eigenvalue problem in Eq. 40 has an N-fold multiple eigenvalue

$$\tilde{\lambda} = \lambda_j, \qquad j = 1,...,N \tag{48}$$

Here, for convenience, the repeated eigenvalues have been numbered from 1 to N. Now, the computation of the design sensitivities of the eigenvalue is no longer straight-forward. This is due to the fact that the eigenvectors ϕ_j, $j = 1,...,N$, of repeated eigenvalues are not unique. Thus, any linear combination of the eigenvectors will satisfy the original eigenvalue problem, Eq. 40.

However, a set of M-orthonormal eigenvectors which span the subspace that corresponds to a multiple eigenvalue can always be chosen. In other words, since it is assumed that $\tilde{\lambda}$ has multiplicity N, then we can choose N eigenvectors $\phi_1,...,\phi_N$, which span the N-dimensional subspace corresponding to the eigenvalues of magnitude $\tilde{\lambda}$ and satisfy the orthogonality conditions in Eqs. 42 and 43.

In the following sensitivity analysis we shall use such eigenvectors $\bar{\phi}_j$ that remain continuous with design changes, see Courant & Hilbert (1953). These eigenvectors will be defined as linear combinations of the aforementioned eigenvectors $\phi_1, \dots, \phi_N$,

$$\bar{\phi}_j = \sum_{k=1}^{N} \beta_{jk} \phi_k, \quad j = 1, \dots, N \tag{49}$$

where β_{jk} are unknown coefficients to be determined.

Works by Courant & Hilbert (1953), Wittrick (1962), and Lancaster (1964) have provided a basis for calculating the sensitivities of multiple eigenvalues. It is shown that the design sensitivities of multiple eigenvalues can be found by formulation and solution of a sub-eigenvalue problem.

Let us consider a small change $\epsilon \Delta a_i$ of a single, arbitrarily chosen design variable a_i where ϵ is a small positive parameter. Due to this change the K and M matrix will be incremented, i.e., the new matrices become

$$K + \epsilon \frac{\partial K}{\partial a_i} \Delta a_i \quad \text{and} \quad M + \epsilon \frac{\partial M}{\partial a_i} \Delta a_i, \quad i = 1, \dots, I \tag{50}$$

Then multiple eigenvalues and corresponding eigenvectors for the perturbed design can be written as

$$\lambda_j (a_i + \epsilon \Delta a_i) = \bar{\lambda} + \epsilon \mu_j (a_i, \Delta a_i) + o(\epsilon)$$

$$\phi_j (a_i + \epsilon \Delta a_i) = \bar{\phi}_j + \epsilon v_j (a_i, \Delta a_i) + o(\epsilon), \quad j = 1, \dots, N \tag{51}$$

where μ_j and v_j are unknown eigenvalue and eigenvector sensitivities, respectively, and $o(\epsilon)$ represents higher order terms.

Substituting Eqs. 50 and 51 into the main eigenvalue problem in Eq. 40, we obtain in the first approximation

$$\left(\frac{\partial K}{\partial a_i} - \bar{\lambda} \frac{\partial M}{\partial a_i} \right) \bar{\phi}_j + (K - \bar{\lambda} M) v_j = \mu_j M \bar{\phi}_j \tag{52}$$

Premultiplying this equation by ϕ_s^T gives

$$\phi_s^T \left(\frac{\partial K}{\partial a_i} - \bar{\lambda} \frac{\partial M}{\partial a_i} \right) \bar{\phi}_j = \mu_j \phi_s^T M \bar{\phi}_j, \quad s = 1, \dots, N \tag{53}$$

Here the term $\phi_s^T (K - \bar{\lambda} M) v_j = v_j^T (K - \bar{\lambda} M) \phi_s$ drops out because ϕ_s is the eigenvector corresponding to $\bar{\lambda}$.

Recalling that $\bar{\phi}_j$ is the linear combination in Eq. 49 of the original eigenvectors ϕ_k, from Eq. 53 the following system of linear algebraic equations of unknown coefficients β_{jk} is obtained

$$\sum_{k=1}^{N} \beta_{jk} \left[\phi_s^T \left(\frac{\partial K}{\partial a_i} - \bar{\lambda} \frac{\partial M}{\partial a_i} \right) \phi_k - \mu_j \delta_{sk} \right] = 0, \quad s = 1, \dots, N \tag{54}$$

where the M-orthonormalization, Eq. 42, has been used.

A nontrivial solution to these equations only exists if the determinant of the system is equal to zero

$$\det \left[\phi_s^T \left(\frac{\partial K}{\partial a_i} - \bar{\lambda} \frac{\partial M}{\partial a_i} \right) \phi_k - \mu \delta_{sk} \right] = 0, \quad s,k = 1, \dots, N, \quad i = 1, \dots, I \tag{55}$$

This is the main equation for determining the coefficients μ_j, $j = 1, \dots, N$, which represent the sensitivities of the multiple eigenvalue $\bar{\lambda}$ with respect to a change Δa_i of a single design parameter a_i, see Eq. 51. As in the case of simple eigenvalues, the derivatives of the K and M matrix, respectively, must be calculated first, and then the eigenvalue problem of Eq. 55 is easily formulated and solved.

If the off-diagonal terms in the quadratic matrix of dimension N in Eq. 55 are equal to zero, then the eigenvalues of this matrix, i.e., the directional derivatives of the multiple eigenvalue $\tilde{\lambda}$, are equal to the traditional Fréchet derivatives obtained by using Eq. 45.

Let us now consider the general case when all the design variables a_i, $i = 1,...,I$, are changed simultaneously. It should be noted that multiple eigenvalues are not differentiable in the common sense, i.e., not Fréchet-differentiable, see e.g. Haug, Choi & Komkov (1985). This means that the expression for the eigenvalue increments in Eqs. 46 are no longer valid. Thus, to find the sensitivities of multiple eigenvalues it is necessary to use *directional derivatives* in the design space.

For this purpose, for the vector of design variables $a = (a_1,...,a_I)$, a variation in the form $a + \epsilon e$ is considered, where e is an arbitrary vector of variation $e = (e_1,...,e_I)$ with the unit norm $\|e\| = \sqrt{e_1^2 + ... + e_I^2} = 1$ and ϵ is a small positive parameter. The vector e represents a direction in the design space along which the design variables a_i are changed, and ϵ represents the magnitude of the perturbation in this direction.

As a result of this perturbation of the vector a the matrices K and M are incremented and become

$$K + \epsilon \sum_{i=1}^{I} \frac{\partial K}{\partial a_i} e_i, \quad M + \epsilon \sum_{i=1}^{I} \frac{\partial M}{\partial a_i} e_i \tag{56}$$

Using expansions for λ_j and ϕ_j in the form

$$\lambda_j = \tilde{\lambda} + \epsilon \mu_j + o(\epsilon)$$
$$\phi_j = \tilde{\phi}_j + \epsilon v_j + o(\epsilon), \quad j = 1,...,N \tag{57}$$

and performing the same manipulations as earlier, instead of Eq. 55 the following equation for determining the sensitivities μ_j of the eigenvalues λ_j is obtained:

$$\det \left[\sum_{i=1}^{I} \phi_s^T \left(\frac{\partial K}{\partial a_i} - \tilde{\lambda} \frac{\partial M}{\partial a_i} \right) \phi_k e_i - \mu \delta_{sk} \right] = 0, \quad s,k = 1,...,N \tag{58}$$

If the *generalized gradient vectors* f_{sk} of dimension I are introduced

$$f_{sk} = \left(\phi_s^T \left[\frac{\partial K}{\partial a_1} - \tilde{\lambda} \frac{\partial M}{\partial a_1} \right] \phi_k, ..., \phi_s^T \left[\frac{\partial K}{\partial a_I} - \tilde{\lambda} \frac{\partial M}{\partial a_I} \right] \phi_k \right) \tag{59}$$

then Eq. 58 takes the form

$$\det \left[f_{sk}^T e - \mu \delta_{sk} \right] = 0, \quad s,k = 1,...,N \tag{60}$$

Note that $f_{sk} = f_{ks}$ due to the symmetry of the matrices K and M. Also note the notation used here for the generalized gradient vector f_{sk}; the subscripts only refer to the modes from which the vector is calculated. Thus, $f_{sk}^T e$ is a scalar product.

Thus, knowing the eigenvectors ϕ_k, $k = 1,...,N$, corresponding to the multiple eigenvalue $\tilde{\lambda}$, the generalized gradient vectors f_{sk} can be constructed, and the sensitivities $\mu = \mu_j$, $j = 1,...,N$ for any vector of variation e can be determined, i.e., for any direction in the space of the design variables. The quantities μ_j constitute the directional derivatives of the multiple eigenvalue $\tilde{\lambda}$, cf. Eq. 57. Eq. 60 was first obtained by Bratus & Seyranian (1983) and Seyranian (1987), see also Haug & Rousselet (1980), and Masur (1984, 1985), and Haug, Choi & Komkov (1985).

In many cases it is expedient to eliminate the unit vector e from Eq. 60 and establish a formula for determining the increments $\Delta\lambda_j$, $j = 1,...,N$ of the N-fold eigenvalue $\bar{\lambda}$ subject to a given vector $\Delta a = (\Delta a_1,...,\Delta a_I)$ of actual increments of the design variables a_i, $i = 1,...,I$. To this end, we multiply each of the components in Eq. 60 by ϵ, note from the foregoing that $\epsilon e = \Delta a$ and $\epsilon\mu_j = \Delta\lambda_j$, $j = 1,...,N$, and obtain

$$\det\left[\, f_{sk}^T \Delta a \,-\, \delta_{sk}\Delta\lambda \,\right] = 0, \qquad s,k = 1,...,N \tag{61}$$

If we solve this N-th order algebraic equation for $\Delta\lambda$, we obtain the increments $\Delta\lambda = \Delta\lambda_j$, $j = 1,...,N$, of the N-fold eigenvalue corresponding to the vector Δa of increments of the design variables.

6. ERROR ELIMINATION IN SEMI-ANALYTICAL SENSITIVITY ANALYSIS

It has been shown recently by, e.g., Barthelemy & Haftka (1988), Cheng, Gu & Zhou (1989), Pedersen, Cheng & Rasmussen (1989), and Cheng & Olhoff (1991, 1993), that the semi-analytical method of design sensitivity analysis is prone to large errors for certain types of problems involving shape design variables. The error problem has since then been studied theoretically and numerically by, e.g., Olhoff & Rasmussen (1991a,b), Fenves & Lust (1991), Cheng & Olhoff (1991, 1993), and Mlejnek (1992) and a new method for error elimination that is efficient and applicable for a wide range of problems has been developed. A detailed account of this new method is given by Olhoff, Rasmussen & Lund (1992) for static problems, and the method is extended to eigenvalue problems in Lund & Olhoff (1993a,b).

From the studies in Pedersen, Cheng & Rasmussen (1989), Olhoff & Rasmussen (1991a) and Cheng & Olhoff (1991, 1993), it is now well understood that the occurrence of severe sensitivity errors can be traced back to small truncation errors in the approximate finite difference calculation of element matrix derivatives, see Eq. 26.

It is important to note that it was found in Cheng & Olhoff (1991, 1993), Olhoff, Rasmussen & Lund (1992) and Lund & Olhoff (1993a,b) that the traditional semi-analytical method of sensitivity analysis performs very well for *usual* problems. The type of problems where the traditional method is prone to large shape design sensitivity errors, is problems where *the displacement field is characterized by rigid body rotations which are large relative to actual deformations*, i.e. problems that involve linearly elastic bending of long-span, beam-like structures, and of plate and shell structures.

6.1 Method of "Exact" Numerical Differentiation of Element Matrices

The accuracy of the first order finite difference approximations in Eq. 26 is strongly dependent on the chosen size of perturbation Δa_i as the element matrices generally depend non-linearly on shape design variables, and in some cases, so small perturbations are needed that computational round-off errors become the problem. In order to avoid dependence on the chosen perturbation Δa_i, it is now the goal to construct a method for "exact" numerical differentiation of element matrices based on computationally inexpensive first order finite differences.

This goal may seem unattainable, but a closer study of the functions that form the element matrices reveals that the same mathematical forms are common for large groups of finite elements. For instance, the element matrices for all isoparametric elements with translational degrees of freedom, and isoparametric Mindlin plate and shell elements depend on the same class of functions. Similarly, the element matrices of a large class of finite elements comprising Bernoulli-Euler beam, and Kirchhoff plate and shell elements have a similar mathematical structure. The members of these classes of matrices in general depend non-linearly on the design variables, but are defined within a special mathematical form. The mathematical form implies that their *approximate numerical derivatives*, computed by a usual first order finite difference scheme, *can be upgraded to "exact" derivatives* by simple multiplication by appropriate *correction factors*. The values of these correction factors can be very easily pre-computed and be used throughout the procedure of design sensitivity analysis. It follows as a remarkable side-effect of the "exactness" that the results become totally independent of the magnitude of the perturbation.

In the following, we shall assume that the element matrices of a particular finite element only depend on a given sub-set from among the total set of shape design variables a_i, $i=1,...,I$, which in this context is considered to be the *global coordinates of the finite element nodal points*. This sub-set of the design variables is renumbered and denoted by a_j, $j = 1,...,J$, where $J < I$.

A study of various types of finite elements has revealed that the element matrices can be described in terms of functions g, which are incomplete polynomia and defined by the following form:

$$g(a_1,...,a_J) = p_j(a_1,...,a_{j-1},a_{j+1},...,a_J) + q_j(a_1,...,a_{j-1},a_{j+1},...,a_J) \cdot (a_j)^{r_j},$$

$$r_j \in \mathbb{N} \cup \{0\}, \quad \text{for all } j = 1,...,J$$

(62)

Thus, a function g is such that for any j, the term p_j and the coefficient $q_j \neq 0$ are independent of a_j. The design variable a_j appears in one and only one power r_j which belongs to the set $\mathbb{N} \cup \{0\}$ of non-negative integers. Although not stated explicitly, p_j and q_j, and therefore the element function g, will generally depend on the local coordinates within the finite element. Typically, g may represent the determinant or components of the Jacobian matrix or components of other matrices in the definition of an element matrix.

As shown in Olhoff, Rasmussen & Lund (1993), it is possible to establish the following relationship of proportionality between the analytical (and exact) derivative $\partial g/\partial a_j$ and its first order finite difference approximation $\Delta g/\Delta a_j$:

$$\frac{\partial g}{\partial a_j} = c_{r_j} \cdot \frac{\Delta g}{\Delta a_j}, \quad r_j \in \mathbb{N} \cup \{0\}, \quad j = 1,...,J$$

(63)

The proportionality factors c_{r_j} may be termed *correction factors* and they only depend on the relative perturbation η_j defined as

$$\eta_j = \frac{\Delta a_j}{a_j} > 0, \quad j = 1,...,J$$

(64)

Thus, the correction factors c_{r_j} can be expressed as

$$c_o = 0$$
$$c_1 = 1$$
$$c_2 = \left(1 + \frac{1}{2}\eta_j\right)^{-1}$$
$$c_3 = \left(1 + \eta_j + \frac{1}{3}\eta_j^2\right)^{-1}$$

(65)

....

$$c_k = \left(\frac{1}{k}\sum_{p=0}^{k-1}\binom{k}{p}\eta_j^{(k-1)-p}\right)^{-1}$$

where the latter expression for c_k is given in terms of binomial coefficients.

The correction factors in Eq. 65 are independent of the actual values of the design variables and can therefore be precomputed for a selected finite value of $\eta_j > 0$. Hence, the absolute perturbation Δa_j can be eliminated in Eq. 63 which may be rewritten as

$$\frac{\partial g}{\partial a_j} = c_{r_j} \cdot \frac{\Delta g}{\Delta a_j} = \frac{c_{r_j}}{\eta_j a_j} \left[g((1+\eta_j)a_j) - g(a_j) \right], \quad r_j \in \mathbb{N} \cup \{0\}, \quad j = 1,...,J \tag{66}$$

Eqs. 65 and 66 are the main expressions to be used when calculating derivatives of finite elements functions.

6.2 "Exact" Numerical Derivatives of Isoparametric 3D Solid Finite Elements

In this section the method of "exact" numerical differentiation is used for determining "exact" numerical derivatives of element stiffness, mass, initial stress stiffness and thermal "stiffness" matrices. The method is exemplified for isoparametric 3D solid finite elements. For more details concerning the derivation of "exact" numerical derivatives of other element matrices, the reader is referred to Olhoff, Rasmussen & Lund (1993) and Lund & Olhoff (1993a,b). Here and in the following, the term "exact derivatives" implies that the derivatives have no truncation error due to neglection of higher order terms in their Taylor series expansion, and are exact except for computational round-off error.

6.2.1 Derivative of element stiffness matrix

The element stiffness matrix k for a 3D solid isoparametric finite element is given by

$$k = \int_\Omega B^T E B |J| d\Omega \tag{67}$$

Here, Ω is the domain of the finite element described in curvilinear, non-dimensional $\xi - \eta - \zeta$ coordinates for the element, see Fig. 5, and $|J|$ is the determinant of the Jacobian matrix J, which at each point defines the transformation of differentials $d\xi$, $d\eta$, and $d\zeta$ into dx, dy, and dz. Like J, the strain-displacement matrix B depends on the coordinates of the nodal points, whereas the constitutive matrix E depends only on the constitutive parameters of the assumed linearly elastic material.

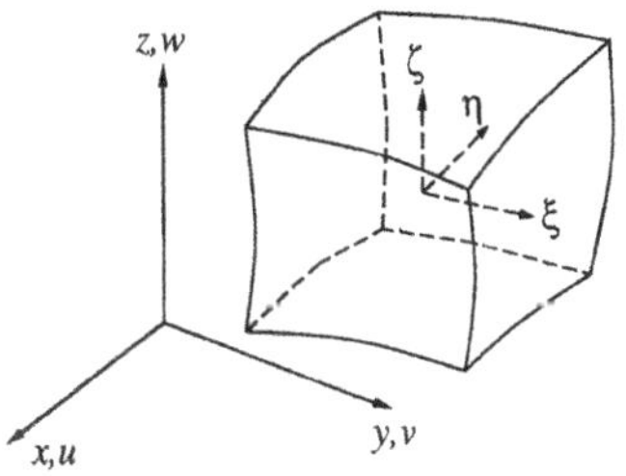

Before the derivative of the stiffness matrix is found, let us recall that within the isoparametric formulation of a finite element with an arbitrarily given number n of nodal points, the shape functions N_i only depend on the local, nondimensional coordinates ξ, η, and ζ, i.e.,

$$N_i = N_i(\xi,\eta,\zeta), \quad i = 1,...,n \tag{68}$$

Fig. 5. Domain, coordinates and nodal degrees of freedom of isoparametric 3D solid finite elements.

The same set of shape functions N_i is used for interpolation of the global coordinates x, y, z from nodal values x_i, y_i, z_i and of displacement functions u, v, w from nodal values u_i, v_i, w_i.

In terms of the vector d_i of nodal degrees of freedom

$$d_i = \{u_i \, v_i \, w_i\}^T, \quad i = 1,...,n \tag{69}$$

the element nodal vector d containing nodal displacements is

$$d = \{d_1^T \, d_2^T \, ... \, d_i^T \, ... \, d_n^T\}^T \tag{70}$$

and the strain vector ϵ is

$$\epsilon(x,y,z) = \left\{ \epsilon_x \ \epsilon_y \ \epsilon_z \ \gamma_{xy} \ \gamma_{yz} \ \gamma_{xz} \right\}^T \tag{71}$$

with their mutual relationsship defined by

$$\epsilon = \mathbf{B}\mathbf{d} \tag{72}$$

The strain-displacement matrix $\mathbf{B}$ is determined by operating on the shape functions N_i, and it is found that

$$\mathbf{B} = \begin{bmatrix} \mathbf{b}_1 \ \mathbf{b}_2 \ \cdots \ \mathbf{b}_i \ \cdots \ \mathbf{b}_n \end{bmatrix} \tag{73}$$

where the submatrix $\mathbf{b}_i$, which is associated with the nodal point i of the finite element, has the form

$$\mathbf{b}_i = \begin{bmatrix} N_{i,x} & 0 & 0 \\ 0 & N_{i,y} & 0 \\ 0 & 0 & N_{i,z} \\ N_{i,y} & N_{i,x} & 0 \\ 0 & N_{i,z} & N_{i,y} \\ N_{i,z} & 0 & N_{i,x} \end{bmatrix} , \quad i = 1,...,n \tag{74}$$

Here, the derivatives of the shape functions N_i with respect to x, y, and z are given by

$$\begin{Bmatrix} N_{i,x} \\ N_{i,y} \\ N_{i,z} \end{Bmatrix} = \begin{bmatrix} \xi_{,x} & \eta_{,x} & \zeta_{,x} \\ \xi_{,y} & \eta_{,y} & \zeta_{,y} \\ \xi_{,z} & \eta_{,z} & \zeta_{,z} \end{bmatrix} \begin{Bmatrix} N_{i,\xi} \\ N_{i,\eta} \\ N_{i,\zeta} \end{Bmatrix} = \mathbf{\Gamma} \begin{Bmatrix} N_{i,\xi} \\ N_{i,\eta} \\ N_{i,\zeta} \end{Bmatrix} , \quad i = 1,...,n \tag{75}$$

where the matrix $\mathbf{\Gamma}$ is the inverse of the Jacobian

$$\mathbf{J} = \begin{bmatrix} x_{,\xi} & y_{,\xi} & z_{,\xi} \\ x_{,\eta} & y_{,\eta} & z_{,\eta} \\ x_{,\zeta} & y_{,\zeta} & z_{,\zeta} \end{bmatrix} = \sum_{i=1}^{n} \begin{bmatrix} N_{i,\xi}x_i & N_{i,\xi}y_i & N_{i,\xi}z_i \\ N_{i,\eta}x_i & N_{i,\eta}y_i & N_{i,\eta}z_i \\ N_{i,\zeta}x_i & N_{i,\zeta}y_i & N_{i,\zeta}z_i \end{bmatrix} \tag{76}$$

Now the terms of the stiffness matrix in Eq. 67 are described, and the derivative of the stiffness matrix can be found by differentiating Eq. 67 with respect to any of the design variables a_j, $j=1,...,J$

$$\frac{\partial \mathbf{k}}{\partial a_j} = \int_\Omega \left[\frac{\partial \mathbf{B}^T}{\partial a_j} \mathbf{E}\mathbf{B} + \mathbf{B}^T \mathbf{E} \frac{\partial \mathbf{B}}{\partial a_j} \right] |\mathbf{J}| d\Omega + \int_\Omega \mathbf{B}^T \mathbf{E}\mathbf{B} \frac{\partial |\mathbf{J}|}{\partial a_j} d\Omega , \quad j = 1,...,J \tag{77}$$

Introducing the notation $[\ ]_S$ for the *operation*

$$[\mathbf{C}]_S = \frac{1}{2}(\mathbf{C}^T + \mathbf{C}) \tag{78}$$

of symmetrization of a quadratic matrix C, Eq. 77 can be rewritten as

$$\frac{\partial \mathbf{k}}{\partial a_j} = 2 \left[\int_\Omega \mathbf{B}^T \mathbf{E}\hat{\mathbf{B}}^{(j)} |\mathbf{J}| d\Omega \right]_S , \quad j = 1,...,J \tag{79}$$

where the matrix $\hat{\mathbf{B}}^{(j)}$ is defined as

$$\hat{\mathbf{B}}^{(j)} = \frac{\partial \mathbf{B}}{\partial a_j} + \frac{\mathbf{B}}{2|\mathbf{J}|} \frac{\partial |\mathbf{J}|}{\partial a_j} \tag{80}$$

Thus, the derivatives of the determinant of the Jacobian $\mathbf{J}$ and of the components in the strain-displacement matrix $\mathbf{B}$ must be determined.

Applying Eqs. 62, 65, and 66 to the above terms in Eqs. 73-76 yields the "exact" numerical derivatives based on first order finite differences. It is easily shown that the scalar $|J|$ will be either independent or a linear function of any of the shape design variables a_j, and the "exact" numerical derivative of $|J|$ is therefore simply given by

$$\frac{\partial |J|}{\partial a_j} = \frac{1}{\eta_j a_j} [\, |J((1+\eta_j) a_j)| - |J(a_j)| \,], \quad j = 1,...,J \tag{81}$$

The computation of derivatives of components of the strain-displacement matrix B requires differentiation of b_i, and hence of the derivatives of $N_{i,x}$, $N_{i,y}$, and $N_{i,z}$, with respect to a_j. This involves differentiation of the matrix Γ, and since the components of this matrix are given by $\Gamma_{qp} = |J|^{-1} cof(J_{pq})$, these components cannot be differentiated exactly on the basis of a simple polynomial approximation. This difficulty can be circumvented by differentiating the identity $\Gamma J = I$, where I is the identity matrix, which gives

$$\frac{\partial \Gamma}{\partial a_j} = -\Gamma \frac{\partial J}{\partial a_j} \Gamma , \quad j = 1,...,J \tag{82}$$

The derivatives of $N_{i,x}$, $N_{i,y}$, and $N_{i,z}$ can then be found using Eqs. 75 and 82, i.e.

$$\begin{Bmatrix} N_{i,x} \\ N_{i,y} \\ N_{i,z} \end{Bmatrix} = \Gamma \begin{Bmatrix} N_{i,\xi} \\ N_{i,\eta} \\ N_{i,\zeta} \end{Bmatrix}$$

$$\begin{aligned}
\frac{\partial}{\partial a_j} \begin{Bmatrix} N_{i,x} \\ N_{i,y} \\ N_{i,z} \end{Bmatrix} &= \frac{\partial \Gamma}{\partial a_j} \begin{Bmatrix} N_{i,\xi} \\ N_{i,\eta} \\ N_{i,\zeta} \end{Bmatrix} = -\Gamma \frac{\partial J}{\partial a_j} \Gamma \begin{Bmatrix} N_{i,\xi} \\ N_{i,\eta} \\ N_{i,\zeta} \end{Bmatrix} = -J^{-1} \frac{\partial J}{\partial a_j} \begin{Bmatrix} N_{i,x} \\ N_{i,y} \\ N_{i,z} \end{Bmatrix} \\
&= -J^{-1} \frac{1}{\eta_j a_j} [J((1+\eta_j) a_j) - J(a_j),] \begin{Bmatrix} N_{i,x} \\ N_{i,y} \\ N_{i,z} \end{Bmatrix} \quad i = 1,...,n, \quad j = 1,...,J
\end{aligned} \tag{83}$$

Now all terms needed for the "exact" numerical derivative of the stiffness matrix, cf. Eq. 77 or 79, are found, and we shall see that the same procedure can be used to calculate other element matrix derivatives.

6.2.2 Derivative of element mass matrix

The element mass matrix m for a 3D solid isoparametric finite element is given by

$$m = \int_\Omega \rho \, N^T N |J| d\Omega \tag{84}$$

where ρ is the mass density, N the vector of shape functions N_i, $|J|$ the determinant of the Jacobian matrix J, and Ω is the domain of the finite element.

The derivative of the mass matrix can be found by differentiating Eq. 84 with respect to any of the design variables a_j, $j=1,...,J$, i.e.,

$$\frac{\partial m}{\partial a_j} = \int_\Omega \rho \, N^T N \frac{\partial |J|}{\partial a_j} d\Omega , \quad j = 1,...,J \tag{85}$$

The derivative of the determinant of the Jacobian is given by Eq. 81 so all terms needed for calculating the "exact" numerical derivative of the mass matrix are known.

6.2.3 Derivative of element initial stress stiffness matrix

In the derivation of derivatives of the element initial stress stiffness matrix (also called the geometric stiffness matrix) it is convenient to reorder nodal degrees of freedom by introducing the element displacement vector d^*, where translational nodal degrees of freedom are reordered so that first all x-direction d.o.f. are given, then y, and then z as follows

$$d^* = \{u_1 \, u_2 \, \cdots \, u_i \, \cdots \, u_n \, v_1 \, v_2 \, \cdots \, v_i \, \cdots \, v_n \, w_1 \, w_2 \, \cdots \, w_i \, \cdots \, w_n \}^T \tag{86}$$

Relating d.o.f. to the reordered element vector d^*, the element initial stress stiffness matrix k_σ for a 3D solid finite element is given by

$$k_\sigma = \int_\Omega G^T S G |J| d\Omega \tag{87}$$

where G is a matrix obtained by appropriate differentiation of shape functions N_i, S is a matrix of initial stresses, $|J|$ the determinant of the Jacobian matrix J, and Ω is the domain of the finite element.

The matrix G is given by

$$G = \begin{bmatrix} g & 0 & 0 \\ 0 & g & 0 \\ 0 & 0 & g \end{bmatrix} \quad \text{with} \quad g = \begin{bmatrix} N_{i,x} \\ N_{i,y} \\ N_{i,z} \end{bmatrix}, \quad i = 1,...,n \tag{88}$$

and the initial stress matrix S is given by

$$S = \begin{bmatrix} s & 0 & 0 \\ 0 & s & 0 \\ 0 & 0 & s \end{bmatrix} \quad \text{with} \quad s = \begin{bmatrix} \sigma_x & \tau_{xy} & \tau_{xz} \\ \tau_{xy} & \sigma_y & \tau_{yz} \\ \tau_{xz} & \tau_{yz} & \sigma_z \end{bmatrix} \tag{89}$$

Here σ_x, σ_y, etc., are stresses found by an initial static stress analysis.

If Eq. 87 is differentiated with respect to any of the design variables a_j, $j = 1,...,J$, the following expression for the "exact" numerical derivative of the initial stress stiffness matrix is obtained

$$\frac{\partial k_\sigma}{\partial a_j} = \int_\Omega \left[\frac{\partial G^T}{\partial a_j} S G + G^T \frac{\partial S}{\partial a_j} G + G^T S \frac{\partial G}{\partial a_j} \right] |J| d\Omega + \int_\Omega G^T S G \frac{\partial |J|}{\partial a_j} d\Omega, \quad j = 1,...,J \tag{90}$$

Using the operation of symmetrization of a quadratic matrix, see Eq. 78, Eq. 90 can be rewritten as

$$\frac{\partial k_\sigma}{\partial a_j} = 2 \left[\int_\Omega G^T S \hat{G}^{(j)} |J| d\Omega \right]_S + \int_\Omega G^T \frac{\partial S}{\partial a_j} G |J| d\Omega, \quad j = 1,...,J \tag{91}$$

where the matrix $\hat{G}^{(j)}$ is defined as

$$\hat{G}^{(j)} = \frac{\partial G}{\partial a_j} + \frac{G}{2|J|} \frac{\partial |J|}{\partial a_j} \tag{92}$$

Thus, it is necessary to find the derivatives of the components of the stress matrix S, of the determinant of the Jacobian J, and of the components of the matrix G.

The two latter terms are already found; the "exact" numerical derivative of the determinant of the Jacobian J is given by Eq. 81, and the "exact" derivatives of the components of the matrix G are given by Eq. 83. The derivatives of the stress components of the stress matrix S have to be found by an initial static design sensitivity analysis for displacements, see Section 5.1, followed by using Eq. 29 or another expression for the stress sensitivities. Thus, all terms necessary for determining the "exact" numerical derivative of the element initial stress stiffness matrix are now found.

6.2.4 Derivative of thermal element "stiffness matrix"

The thermal element "stiffness matrix" consists of contributions from the heat conduction matrix k^{th} given by

$$k^{th} = \int_\Omega B^{th^T} \kappa B^{th} |J| d\Omega \tag{93}$$

Here, Ω is the domain of the finite element in its local coordinate system, see Fig. 5, B^{th} a matrix obtained by appropriate differentiation of shape functions N_i, κ the thermal conductivity matrix, and $|J|$ is the determinant of the Jacobian matrix J. If the material is isotropic, κ can be simply replaced by the scalar λ, the conductivity coefficient.

The matrix B^{th} is given by

$$B^{th} = \begin{bmatrix} b_1^{th} & b_2^{th} & \dots & b_i^{th} & \dots & b_n^{th} \end{bmatrix} \tag{94}$$

where the submatrix b_i^{th}, which is associated with the nodal point i of the finite element, has the form

$$b_i^{th} = \begin{bmatrix} N_{i,x} \\ N_{i,y} \\ N_{i,z} \end{bmatrix}, \quad i = 1,...,n \tag{95}$$

In case of boundary conditions in terms of convection heat transfer, the thermal "stiffness matrix" receives additional contributions given by the element matrix h

$$h = \int_\omega N^T h N |J| d\omega \tag{96}$$

Here, ω is the surface of the finite element described in curvilinear, non-dimensional $\xi-\eta$, $\eta-\zeta$, or $\xi-\zeta$ coordinates for the element, for which the convection boundary condition is applied. N contains shape functions N_i, h is the convection coefficient specified, and $|J|$ is the determinant of the Jacobian matrix J for the surface ω.

If Eqs. 93 and 96 are differentiated with respect to any shape design variable a_j, $j = 1,...,J$, the following expressions for the derivatives of the thermal "stiffness matrices" are obtained

$$\frac{\partial k^{th}}{\partial a_j} = \int_\Omega \left[\frac{\partial B^{th^T}}{\partial a_j} \kappa B^{th} + B^{th^T} \kappa \frac{\partial B^{th}}{\partial a_j} \right] |J| d\Omega + \int_\Omega B^{th^T} \kappa B^{th} \frac{\partial |J|}{\partial a_j} d\Omega, \quad j = 1,...,J \tag{97}$$

and

$$\frac{\partial h}{\partial a_j} = \int_\omega N^T h N \frac{\partial |J|}{\partial a_j} d\omega, \quad j = 1,...,J \tag{98}$$

Eq. 97 can be rewritten in a more compact form by using the operation of symmetrization of a quadratic matrix, cf. Eq. 78:

$$\frac{\partial k^{th}}{\partial a_j} = 2 \left[\int_\Omega B^{th^T} E \hat{B}^{th\,(j)} \, |J| \, d\Omega \right]_S , \quad j = 1,...,J \tag{99}$$

where the matrix $\hat{B}^{th\,(j)}$ is defined as

$$\hat{B}^{th\,(j)} = \frac{\partial B^{th}}{\partial a_j} + \frac{B^{th}}{2\,|J|} \frac{\partial |J|}{\partial a_j} \tag{100}$$

The "exact" numerical derivative of the determinant $|J|$ is given by Eq. 81, and "exact" derivatives of the components $N_{i,x}$, $N_{i,y}$, and $N_{i,z}$ in the matrix B^{th}, see Eqs. 94 and 95, are given by Eq. 83. All terms required for evaluating the "exact" numerical derivatives in Eqs. 97 and 98 are therefore found.

6.3 Example: Static Design Sensitivity Analysis of Long Cantilever Beam

Let us consider an example in which, within the usual linear theory of elasticity, the displacement field entails dominance of rigid-body rotation relative to actual deformation of the finite elements in a subdomain of the structure. The example is the cantilever beam problem used by Barthelemy & Haftka (1988) to study the inaccuracy problem associated with the traditional semi-analytical method of design sensitivity analysis (S-A method). The problem pertains to a slender cantilever beam of given length L and aspect ratio 50, as shown in Fig. 6.

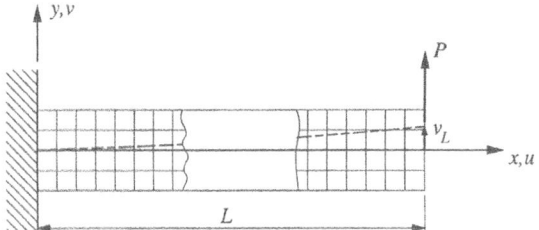

Figure 6. Finite element model of long cantilever beam.

The beam is subjected to a given tip load P at the free end and is modeled by a regular pattern of 8-node isoparametric serendipity finite elements. The length of the element sides are equal and denoted by ℓ. The design sensitivity $\Delta v_L / \Delta L$ of the tip displacement v_L with respect to the length L of the beam is studied, and the "design boundary layer" approach of using one-element-deep sensitivity information is adopted.

It has been demonstrated by numerical experiments in Barthelemy & Haftka (1988) that the displacement design derivative $\Delta v_L / \Delta L$ determined by the traditional S-A method for this beam problem is subject to severe inaccuracy problems. This fact is illustrated by the results presented in Table 1 for different values of the relative perturbation $\Delta\ell / \ell$ of the lengths ℓ of the finite elements in the "design boundary layer". The results are based on a discretization of the beam into 200 x 4 8-node isoparametric finite elements as indicated in Fig. 6, the beam has length $L = 100$ and unit thickness, the load P has unit value, Young's modulus is set to $2.1 \cdot 10^5$ MPa, and Poisson's ratio = 0.3. Since the beam is long, sensitivity results in Table 1 may be compared with analytical results for $\partial v_L / \partial L$ stated in the caption of Table 1 for a corresponding Bernoulli-Euler beam.

TABLE 1

Computed Displacement Design Sensitivities $\dfrac{\Delta v_L}{\Delta L}$ for Cantilever Beam Modeled by 200 x 4 isoparametric 8-node elements

(Computed Displacement: $v_L = 2.3807$. Bernoulli-Euler Comparison Beam: $v_L = 2.3809$ and $\dfrac{\partial v_L}{\partial L} = 0.1429$).

Bdy. Layer Element Length Perturbation $\dfrac{\Delta \ell}{\ell}$	Beam Length Perturbation $\dfrac{\Delta L}{L}$	Traditional S-A Method $\dfrac{\Delta v_L}{\Delta L}$	New S-A Method $\dfrac{\Delta v_L}{\Delta L}$
10^{-1}	2.5×10^{-2}	-3235.	0.1428
10^{-2}	2.5×10^{-3}	-324.8	0.1428
10^{-3}	2.5×10^{-4}	-32.25	0.1428
10^{-4}	2.5×10^{-5}	-3.096	0.1428
10^{-5}	2.5×10^{-6}	-0.1771	0.1428
10^{-6}	2.5×10^{-7}	0.1344	0.1428

It is seen from the results in Table 1 that for values of the relative perturbation $\Delta \ell / \ell$ as small as 10^{-5}, even the sign is wrong for the sensitivities obtained by the traditional S-A method. It can be also seen in the Table that use of the new S-A method based on "exact" numerical differentiation gives sensitivities that are independent of the perturbation $\Delta \ell / \ell$ and agree with the sensitivity value for the corresponding Bernoulli-Euler comparison beam within 0.02%. This confirms the "exactness" of the new S-A method, and, furthermore, demonstrates that the "design boundary layer" approach of using one-element-deep sensitivity information gives accurate results.

6.4 Comments on the New Method of Semi-analytical Sensitivity Analysis

It has turned out to be very easy to implement the method of "exact" numerical differentiation in a unified manner for different types of finite elements, and as an integral part of the finite element analysis. The method can even be implemented in connection with existing finite element codes where different subroutines for computation of element stiffness matrices are only available as black-box routines.

The new method of semi-analytical design sensitivity analysis may be considered as a hybrid between the semi-analytical and the analytical method. Hence, it possesses all important advantages of the semi-analytical method with regard to relative ease of implementation, applicability for a wide range of different finite elements, and adaptivity to existing finite element software, while, at the same time, it possesses the important property of being equally as accurate as the analytical method. The new method is furthermore computationally efficient, especially in the case of many design variables. If only a few shape design variables are used, the new method is normally slower than the traditional method, but with increasing number of design variables, the new method gradually becomes more efficient than the traditional one as the "exact" numerical derivatives of element matrices only need to be calculated once.

The new method of semi-analytical design sensitivity analysis has been tested for many different problems, and it has performed excellently in all cases. The reader is referred to Olhoff, Rasmussen & Lund (1992) and Lund & Olhoff (1993a,b) for a more detailed description of these results.

7. EXAMPLE OF DESIGN SENSITIVITY DISPLAY AND WHAT-IF STUDIES

With expressions for design sensitivities of displacements, stresses, compliance, and eigenvalues at hand, this chapter will briefly illustrate how design sensitivity analysis can be used to improve engineering designs. The turbine disk introduced in Section 2.3 will be used as an example.

As was mentioned in the initial part of Chapter 2, previously, when using the traditional design process, the designer was required to use intuition and trial and error procedures to find ways of improving the design. Nowadays, by using a structural analysis program which has capabilities for design sensitivity analysis, the efficiency of the design process can be highly improved. Through the use of, e.g., color stress contour plots together with color stress design sensitivity contour plots, the engineer easily identifies critical regions in which design improvements can be made.

Let us consider the turbine disk example in Section 2.3, where loads, boundary conditions, and the initial geometry are defined. The disk is made of steel with the mass density 7.75 Kg/mm^3, Young's modulus 180000 MPa, Poisson's ratio 0.3, thermal expansion coefficient $1.2 \cdot 10^{-5}$ K^{-1}, and thermal conductivity coefficient 0.027 W/(K·mm). As a starting point we would like to:

Decrease the maximum von Mises reference stress

This problem is quite complex, because the stresses depend on the design, the temperatures, and the forces, which again depend on the design. The temperature field and the von Mises stress field in the initial disk are displayed in Figs. 7 and 8, respectively. In Fig. 10 the von Mises stresses are shown in the region near the lower boundary where the maximum value is found.

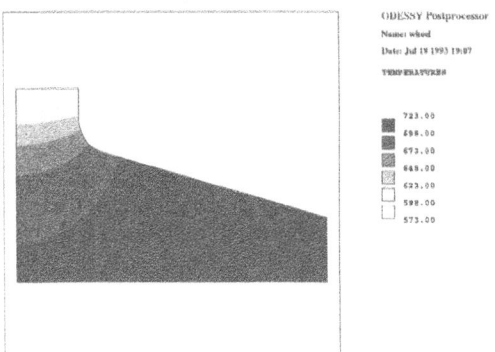

Figure 7. Temperature field in turbine disk.

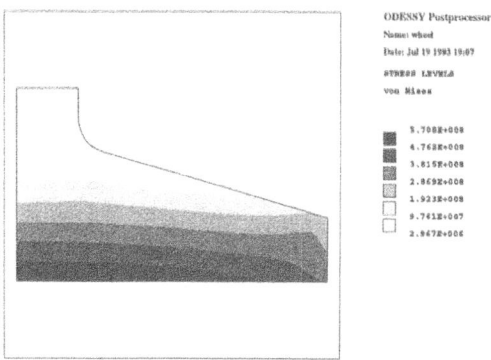

Figure 8. von Mises stress field in turbine disk.

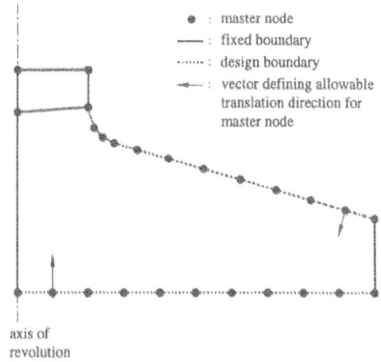

Figure 9. The two selected design variables for which sensitivities will be shown.

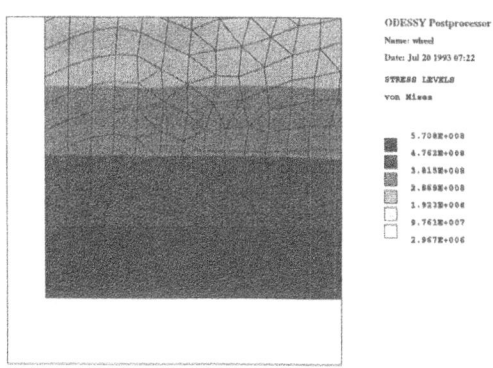

Figure 10. Zoom of region with largest von Mises stress.

As was already indicated in Fig. 2, the disk has been assigned 20 shape design variables which control the shape of the structural domain, and the geometry may be perturbed for each of these variables in order to display stress design sensitivities with color contour plots.

For the two particular design variables indicated in Fig. 9, corresponding stress design sensitivity fields are shown in Figs. 11 and 12. Each stress design sensitivity field is associated with translation of the master node in the direction indicated by the arrow.

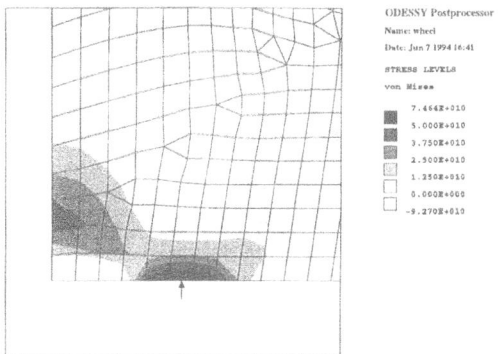

Figure 11. Stress design sensitivity plot no. 1.

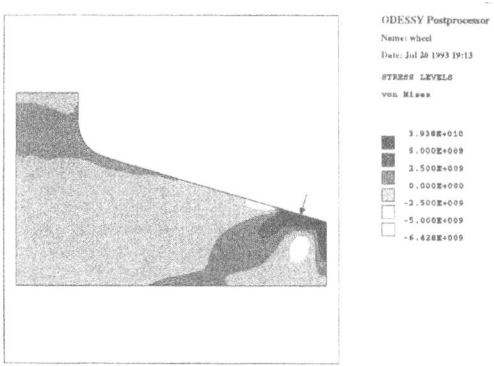

Figure 12. Stress design sensitivity plot no. 2.

Fig. 11 shows that most of the large von Mises stresses in this region will be increased if the master node is moved in the direction of the arrow. In other words, in order to decrease the large stresses in this region, the master node must be moved in the opposite direction of the arrow in Fig. 11.

Fig. 12 shows that the indicated inward movement of the master node results in negative stress sensitivities in the highly stressed region at the lower boundary, i.e., a design change of this kind has a desirable decreasing effect on the maximum von Mises stress. No doubt, this is because such a design change will imply a decrease of the centrifugal forces in the rotating disk.

When all 20 design perturbations have been performed and the corresponding design sensitivity fields determined, the designer can do a *what-if study*, where he decides on suitable changes of the values of the design variables according to the stress sensitivity information.

The first result of a what-if study on the disk example is shown in Fig. 13, where changes of all design variables have been guessed on the basis of the stress sensitivity information obtained, and without considering, e.g., a constraint on the volume of the disk. This new design implies a reduction of the maximum von Mises stress from 571 to 453 MPa.

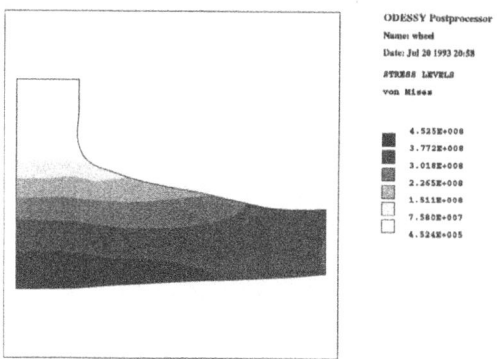

Figure 13. von Mises stresses in updated geometry obtained by what-if study.

Based on this improved design, a new design sensitivity study can be performed, and the design of the disk presumeably can be further improved, if necessary. This leads us to another use of design sensitivity analysis, namely engineering design optimization.

8. EXAMPLES OF ENGINEERING DESIGN OPTIMIZATION

This chapter illustrates by way of two examples how engineering design can be realized as a rational, iterative solution procedure for multicriterion optimum design through systematic sequences of redesign and reanalysis by application of a suitable optimization system. Thus, even if the designer has access to tools for design sensitivity analysis, if he is confronted with a problem that involves more than just a few design variables and criteria to be taken into account, it becomes impossible to survey and quantify the design sensitivity information so as to make decisions regarding values of the design variables that would satisfy design constraints and be any better than other alternative values.

Optimization systems developed for solution of broad classes of engineering design problems, are generally based on the concept of integration of software modules for finite element analysis, design sensitivity analysis, and optimization by mathematical programming.

A typical flow diagram for such a design optimization system is seen in Fig. 14. The computational procedure is quite similar to that of a what-if study, with the notable exception that the current values of the design variables a_i, the problem variables f_j, and the design sensitivities ∇f_j are used to set up a standard mathematical programming problem which is solved sequentially by an optimizer. The use of the optimizer ensures that, in general, the iterative procedure yields successively improved designs, and hence converges towards the optimum solution.

In the following examples, our shape design and optimization system ODESSY has been used, and a SIMPLEX algorithm has been chosen as the optimizer.

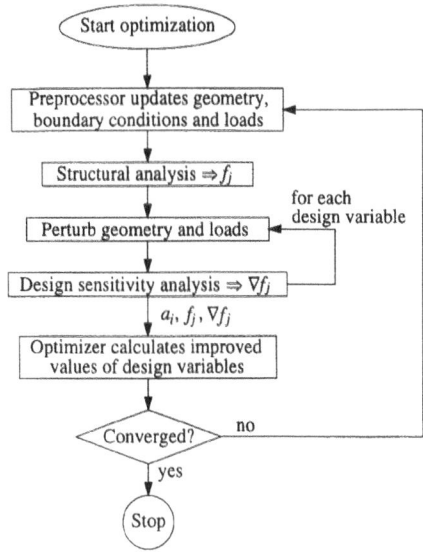

Figure 14. Flow diagram for optimization procedure.

8.1 Optimization of Turbine Disk

In this section, we undertake optimization of the turbine disk considered in Section 2.3 and Chapter 7. The objective of the optimization, however, will be different from the design objective considered in Chapter 7, since now we would like to minimize the mass moment of inertia of the disk. At the same time, we would like to specify a temperature dependent non-linear constraint on the von Mises stresses as shown in Fig. 15. This constraint is realized by normalizing the von Mises stress $\sigma_{vm,i}$ with the function shown in Fig. 15. It should be noted that this non-linear stress constraint changes during the optimization process as the temperature field changes with design. Furthermore, we specify the fabricational constraint that the radius of curvature of the boundary of the disk may nowhere be less than 3 mm.

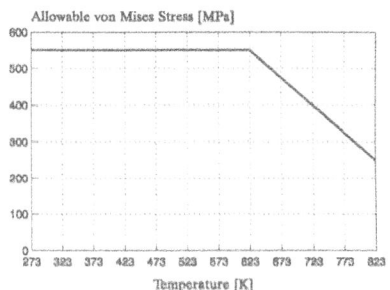

Figure 15. Allowable von Mises stress.

Thus, the definition of the optimization problem is

Minimize   the mass moment of inertia

Subject to   normalized stress
$$\begin{cases} \dfrac{\sigma_{vm,i}\ [MPa]}{550} & \text{if } T_i \le 623\ K \\[2mm] \dfrac{\sigma_{vm,i}\ [MPa]}{1484.5 - 1.5T_i} & \text{if } T_i > 623\ K \end{cases} \le 1$$

minimum boundary radius of curvature $\geq$ 3 mm

The temperature dependent stress constraint is displayed for the initial structure in Fig. 16, which shows that this constraint is violated by 34% as the maximum value is 1.34.

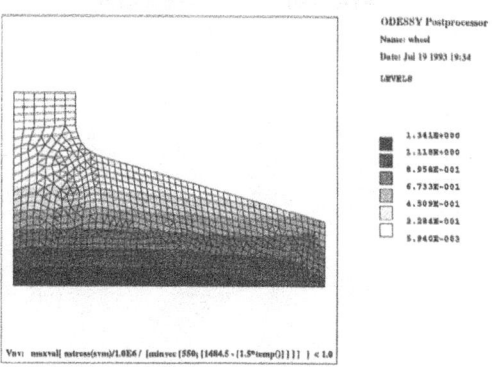

Figure 16. Temperature dependent stress constraint for initial model.

A shape optimization of the disk is now performed, whereby the mass moment of inertia is reduced from $3.12 \cdot 10^4$ to $2.33 \cdot 10^4$ Kg·mm². Moreover, the stress constraint is now fulfilled as shown in Fig. 17.

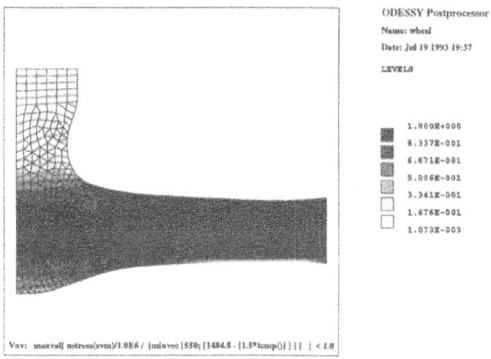

Figure 17. Temperature dependent stress constraint for final geometry.

The final minimum value of the radius of curvature of the boundary is found to be 13.5 mm, i.e., this constraint is not active.

8.2 Optimization of an Automobile Hood

This example involves shape optimization of a shell structure in the form of the hood of a Mazda 323 automobile with the objective of maximizing the fundamental frequency of free vibrations. In practical shape design optimization it is usually advantageous to start with a relatively simple model in terms of geometry and finite element representation. The model is then refined iteratively as the process converges towards a good solution and the designer acquires more knowledge about the nature of the problem. The final result often will be the last in a long sequence of models. However, due to space limitations and with the purpose of illustration, we shall start out this example with a fairly accurate geometric modeling of the original real-life structure and not attempt to generate improved models along the way.

38

We shall assume a constraint on the total amount of material such that the weight of the optimized structure does not exceed the original one. The hood is made from steel plates with thickness 1.0 mm, and we assume a Young's modulus of 210 GPa, Poisson's ratio of 0.3 and a density of 7800 kg/m^3. The overall dimensions of the hood are approximately 1.25 x 0.85 m and must remain unchanged.

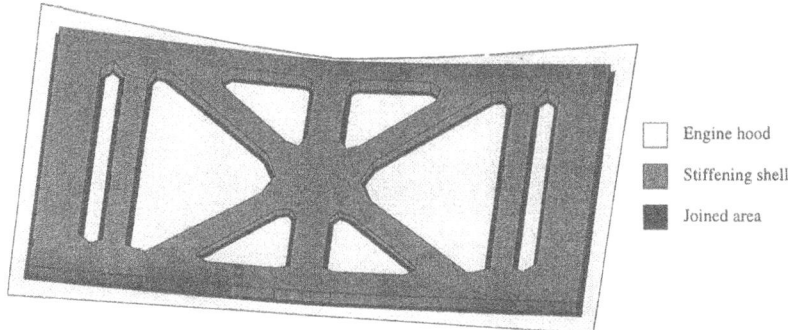

Figure 18.   *The underside of the original geometry.* **The black area is the joined region (joined by welding), the dark gray is the stiffening shell, and the light gray is the upper side of the hood.**

8.2.1 Model

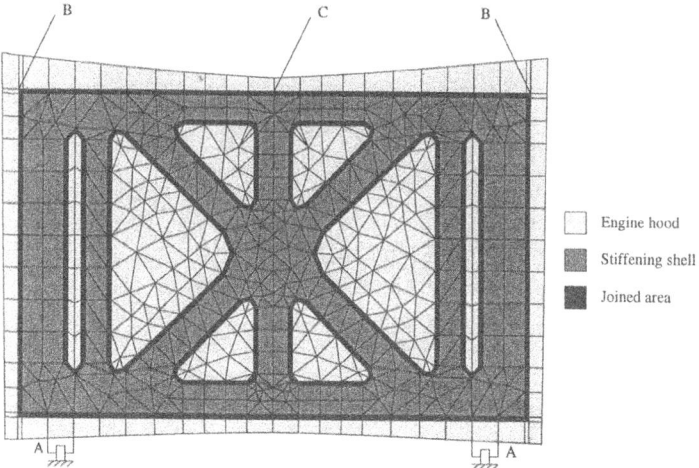

Figure 19. The original geometry with boundary conditions.

The original geometry, which is visualized in Figs. 18 and 19, consists of two doubly curved shells joined by welding. One shell is the external part of the hood, and the other shell, which is welded to the inner side of the hood, is a multi-connected set of stiffeners. A total of 496 boundaries and 238 curved surface patches form the model. The shapes of the surfaces are controlled by their boundaries which in turn depend on a number of master nodes as described in Section 2.3. Modifiers and design variables are defined such that positions, heights and widths of the stiffeners can be changed continuously along each stiffener. The outer surface of the engine hood is not changed during the design optimization as we do not want to affect the aerodynamical properties. Symmetry with respect to the vertical mid-plane in Fig. 19 is imposed on possible design changes, and the model then has a total of 32 independent design variables.

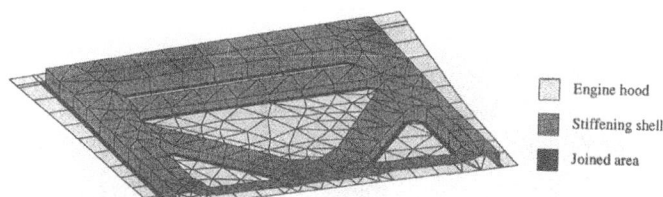

Engine hood

Stiffening shell

Joined area

Figure 20. Section through the two shells.

The finite element model (see Figs. 19 and 20) is built using a mix of mapping and free meshing of curved surfaces. Fig. 20 shows a section through the two shells illustrating their interrelation.

As seen in Fig. 19, the structure is supported by two hinges attached to the edge of the hood, and three support points on the opposite side. The hinges (A) fix all degrees of freedom except for one rotation. The supports (B) fix the out-of-plane translational degree of freedom while the support (C) fixes the out-of-plane and vertical translational degrees of freedom.

Isoparametric shell elements with 6 and 8 nodes are used for the analysis, and the model comprises a total of more than 20,000 degrees of freedom. This modeling gives a rather crude description of the geometry, but a test of the convergence shows that it is adequate for the representation of the eigenmodes at hand.

8.2.2 Analysis

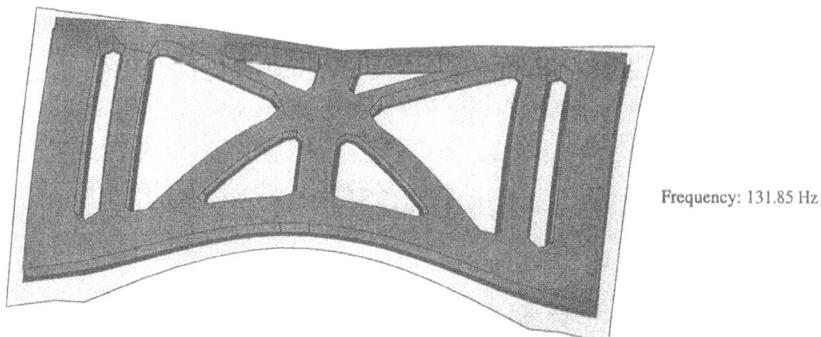

Frequency: 131.85 Hz

Figure 21. First eigenmode for initial geometry.

The analysis of the original geometry reveals the following values of the lowest three eigenfrequencies:

(1) 132 Hz
(2) 177 Hz
(3) 198 Hz

We see that none of these eigenfrequencies are multiple. The lowest eigenfrequency corresponds to the eigenmode shown in Fig. 21.

8.2.3 Result

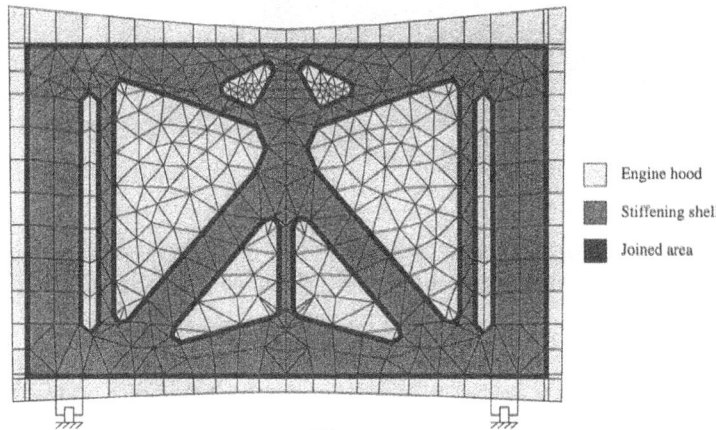

Figure 22. Shape optimized geometry.

The final geometry is shown in Figs. 22 and 23, and is obtained in 12 iterations using the SLP algorithm of ODESSY. The three lowest eigenfrequencies of the optimized design are found to be

(1)   163 Hz
(2)   182 Hz
(3)   216 Hz

The minimum eigenfrequency of 163 Hz implies an increase of 24% relative to that of the original structure, and the final volume corresponds exactly to the original one.

It is obvious that rather large geometric changes have taken place, and many of the design variables reach specified maximum (or minimum) allowable values which indicates that the topology of the initial geometry is not optimum for the case of maximizing the smallest eigenfrequency. Thus, the final geometry may only be considered optimum within this topology and the limitations we have set for variations of the design variables in the mathematical formulation of the problem.

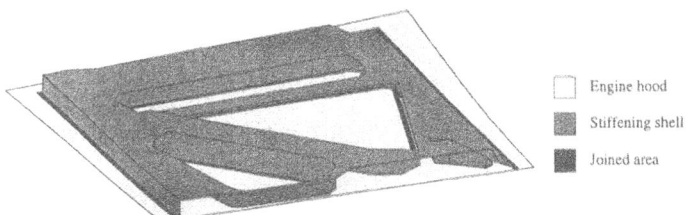

Figure 23. Section through the two shells in optimized geometry.

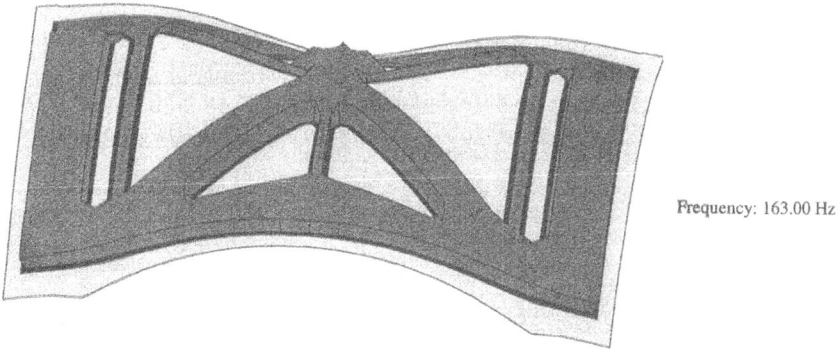

Frequency: 163.00 Hz

Figure 24. First eigenmode for optimized geometry.

9. CONCLUSIONS

Finite element based analysis, semi-analytical design sensitivity analysis and optimization of engineering structures and components are discussed in this paper. The basic concepts of the field are reviewed, and expressions for computation of a number of different design objectives and their design sensitivities are derived.

The handling of multiple criteria and constraints for a given design problem is carried out very efficiently via a versatile mathematical programming formulation for multicriterion optimization based on the so-called bound formulation.

The design sensitivities can be easily computed by means of the semi-analytical approach of sensitivity analysis. The paper shows how this approach can be augmented to yield "exact" numerical design sensitivities by a new technique developed by Olhoff, Rasmussen & Lund (1993) for static problems and extended to eigenvalue problems by Lund & Olhoff (1993a,b). By this technique, which is computationally inexpensive and very easy to implement as an integral part of the finite element analysis, one completely avoids the often severely erroneous shape design sensitivities of long-span beam, plate, and shell structures that may result from application of the traditional method of semi-analytical sensitivity analysis to such structures.

The main developments described in this paper are implemented and continually being used in the (CAD) solid modeling integrated optimization system ODESSY which is being developed at Aalborg University. Other important issues and features of this system, such as the concepts for its CAD integration, and its capabilities for generation of optimum structural topologies, are beyond the scope of the current article, and we refer the interested reader to papers cited earlier.

Acknowledgment - The developments presented in this paper received support from the Danish Technical Research Council's Programme of Research on Computer Aided Design.

REFERENCES

Arora, J.S. (1989): Interactive Design Optimization of Structural Elements. In: Discretization Methods and Structural Optimization - Procedures and Applications, (Eds. H.A. Eschenauer & G. Thierauf), pp. 10-16, Springer-Verlag, Berlin.

Atrek, E. R.; Gallagher, R.H.; Ragsdell, K.M.; Zienkiewicz, O.C. (1984) (Eds.): New Directions in Optimum Structural design, John Wiley & Sons, New York.

Barthelemy, B.; Haftka, R.T. (1988): Accuracy Analysis of the Semi-Analytical Method for Shape Sensitivity Analysis. AIAA Paper 88-2284, Proc. AIAA/ASME/ASCE/ASC 29th Structures, Structural Dynamics and Materials Conference, Part 1, pp. 562-581. Also Mechanics of Structures and Machines., Vol. 18, pp. 407-432 (1990).

Bendsøe, M.P.; Mota Soares, C.A. (1993) (Eds.): Topology Design of Structures, Kluwer, Dordrecht, The Netherlands.

Bendsøe, M.P. (1994): Methods for the Optimization of Structural Topology, Shape and Material. Mathematical Institute, Technical University of Denmark, Lyngby, Denmark, 299 pp.

Bendsøe, M.P. (1989): Optimal Shape as a Material Distribution Problem. Structural Optimization, Vol. 1, No. 4, pp. 193-202.

Bendsøe, M.P.; Kikuchi, N. (1988): Generating Optimal Topologies in Structural Design Using a Homogenization Method. Computer Methods in Applied Mechanics and Engineering, Vol. 117, pp. 197-224.

Bendsøe, M.P.; Olhoff, N.; Taylor, J.E. (1983): A Variational Formulation for Multicriteria Structural Optimization. J. Struct. Mech., Vol. 11, pp. 523-544.

Bennett, J.A.; Botkin, M.E. (1985): Structural Shape Optimization with Geometric Description and Adaptive Mesh Refinement. AIAA Journal, Vol. 23, No. 3, pp. 458-464.

Bennett, J.A.; Botkin, M.E. (1986) (Eds.): The Optimum Shape. Automated Structural Design, Plenum Press, New York.

Botkin, M.E.; Yang, R.J.; Bennett, J.A. (1986): Shape Optimization of Three-dimensional Stamped and Solid Automotive Components. In: The Optimum Shape. Automated Structural Design (Eds. J.A. Bennett & M.E. Botkin), pp. 235-262, Plenum Press, New York.

Braibant, V.; Fleury, C. (1984): Shape Optimal Design Using B-splines. Computer Methods in Applied Mechanics and Engineering, Vol. 44, pp. 247-267.

Bratus, A.S.; Seyranian, A.P. (1983): Bimodal Solutions in Eigenvalue Optimization Problems. Prikl. Matem. Mekhan., Vol. 47, No. 4, pp. 546-554; Also Applied Mathematics in Mechanics, Vol. 47, pp. 451-457.

Cea, J. (1981): Problems of Shape Optimal Design. In: Optimization of Distributed Parameter Structures (Eds. E.J. Haug & J. Cea), Vol. 2, pp. 1005-1048, Sijthoff & Nordhoff, The Netherlands.

Cheng, G.; Gu, Y.; Zhou, Y. (1989): Accuracy of Semi-Analytical Sensitivity Analysis. Finite Elements in Analysis and Design, Vol. 6, pp. 113-128.

Cheng, G.; Liu, Y. (1987): A New Computational Scheme for Sensitivity Analysis. Engineering Optimization, Vol. 12, pp. 219-235.

Cheng, G.; Olhoff, N. (1991): New Method of Error Analysis and Detection in Semi-analytical Sensitivity Analysis. Report No. 36, Institute of Mechanical Engineering, Aalborg University, Denmark, 34 pp.

Cheng, G.; Olhoff, N. (1993): Rigid Body Motion Test Against Error in Semi-Analytical Sensitivity Analysis. Computers & Structures, Vol. 46, No. 3, pp. 515-527.

Choi, K.K.; Chang, K.-H. (1991): Shape Design Sensitivity Analysis and What-If Workstation For Elastic Solids. Technical Report R-105, Center for Simulation and Design Optimization and Department of Mechanical Engineering, The University of Iowa, 33 pp.

Choi, K.K. (1985): Shape Design Sensitivity Analysis of Displacements and Stress Constraints. Journal of Structural Mechanics, Vol. 13, pp. 27-41.

Choi, K.K.; Twu, S.-L. (1988): On Equivalence of Continuum and Discrete Methods of Shape Sensitivity Analysis. AIAA Journal, Vol. 27, No. 10, pp. 1418-1424.

Courant, R.; Hilbert, D. (1953): Methods of Mathematical Physics, Vol. 1, Interscience Publishers, New York.

Dems, K.; Mróz, Z. (1983): Variational Approach by Means of Adjoint Systems to Structural Optimization and Sensitivity Analysis - I Variation of Material Parameters within Fixed Domain. International Journal of Solids and Structures, Vol. 19, pp. 677-692.

Dems, K.; Mróz, Z. (1984): Variational Approach by Means of Adjoint Systems to Structural Optimization and Sensitivity Analysis - II Structure Shape Variation. International Journal of Solids and Structures, Vol. 20, pp. 527-552.

Dems, K.; Mróz, Z. (1993): On Shape Sensitivity Approaches in the Numerical Analysis of Structures. Structural Optimization, Vol. 6, pp. 86-93.

Dhatt, G.; Touzot, G. (1984): The Finite Element Method Displayed, Wiley & Sons, New York.

Ding, Y. (1986): Shape Optimization of Structures: a Literature Survey. Computers and Structures, Vol. 24, No. 6, pp. 985-1004.

Eschenauer, H.A.; Olhoff, N. (1983) (Eds.): Optimization Methods in Structural Design, B.I.-Wissenschaftsverlag, Mannheim.

Eschenauer, H.A. (1986): Numerical and Experimental Investigations of Structural Optimization of Engineering Designs. Siegen: Bonn+Fries, Druckerei und Verlag.

Eschenauer, H.A.; Mattheck, C.; Olhoff, N. (1991) (Eds.): Engineering Optimization in Design Processes, Springer-Verlag, Berlin.

Eschenauer, H.A.; Thierauf, G. (1989) (Eds.): Discretization Methods and Structural Optimization - Procedures and Applications, Springer-Verlag, Berlin.

Esping, B.J.D. (1983): Minimum Weight Design of Membrane Structures, Ph.D. Thesis, Dept. Aeronautical Structures and Materials, The Royal Institute of Technology, Stockholm, Report 83-1.

Esping, B.J.D. (1984): Minimum Weight Design of Membrane Structures Using Eight Node Isoparametric Elements and Numerical Derivatives. Computers and Structures, Vol. 19, No. 4, pp. 591-604.

Esping, B.J.D. (1986): The OASIS Structural Optimization System. Computers & Structures, Vol. 23, No. 3, pp. 365-377.

Fenves, P.A.; Lust, R.V. (1991): Error Analysis of Semi-Analytical Displacement Derivatives for Shape and Sizing Variables. AIAA Journal, Vol. 29, pp. 271-279.

Fleury, C. (1989): CONLIN: An Efficient Dual Optimizer Based On Convex Approximation Concepts. Structural Optimization, Vol. 1, pp. 81-89.

Fleury, C.; Braibant, V. (1986): Structural Optimization: A New Dual Method Using Mixed Variables. International Journal for Numerical Methods in Engineering, Vol. 23, 409-428.

Gajewski, A.; Zyczkowski, M. (1988): Optimal Structural Design Under Stability Constraints, Kluwer Academic Publishers, Dordrecht-Boston-London.

Gilmore, B.J.; Hoeltzel, D.A.; Azarm, S.; Eschenauer, H.A. (1993) (Eds.): Advances in Design Automation. Proc. 1993 ASME Design Technical Conferences - 19th Design Automation Conference, Albuquerque, New Mexico, The American Society of Mechanical Engineers, New York.

Haber, R.B. (1987): A New Variational Approach to Structural Shape Sensitivity Analysis. In: Computer Aided Optimal Design: Structural and Mechanical Systems (Ed. C.A. Mota Soares), pp. 573-587, Springer-Verlag, Berlin.

Haftka, R.T.; Adelman, H.M. (1989): Recent Developments in Structural Sensitivity Analysis. Structural Optimization, Vol. 1, pp. 137 - 151.

Haftka, R.T.; Grandhi, R.U. (1986): Structural Shape Optimization - A Survey. Computer Methods in Applied Mechanics and Engineering, Vol. 57, pp. 91-106.

Haug, E.J. (1993) (Ed.): Concurrent Engineering: Tools and Technologies for Mechanical System Design, 998 pp., Springer-Verlag, Berlin.

Haug, E.J.; Cea, J. (1981) (Eds.): Optimization of Distributed Parameter Structures, Vol. 1: pp. 1-842, Vol. 2: pp. 843-1609, Sijthoff & Nordhoff, The Netherlands.

Haug, E.J. (1981): A Review of Distributed Parameter Structural Optimization Literature. In: Optimization of Distributed Parameter Structures (Eds. E.J. Haug & J. Cea), Vol. 1, pp. 3-74, Sijthoff & Nordhoff, The Netherlands.

Haug, E.J.; Choi, K.K.; Komkov, V. (1986): Design Sensitivity Analysis of Structural Systems, Academic Press, New York.

Haug, E.J.; Rousselet, B. (1980b): Design Sensitivity Analysis in Structural Mechanics. II: Eigenvalue Variations. Journal of Structural Mechanics, Vol. 8, No. 2, pp. 161-186.

Herskovits, J. (1993) (Ed.): Proc. Structural Optimization '93 - The World Congress on Optimal Design of Structural Systems, Vol. 1 & 2. Federal University of Rio de Janeiro, Brazil.

Hörnlein, H.R.E.M. (1987): Take-off in Optimum Structural Design. In: Computer Aided Optimal Design. Structural and Mechanical Systems (Ed. C.A. Mota Soares), NATO ASI Series F: Computer and System Sciences, Vol. 27, pp. 901-919, Springer-Verlag, Berlin.

Kristensen, E.S.; Madsen, N.F. (1976): On the Optimum Shape of Fillets in Plates Subjected to Multiple In-plane Loading Cases. International Journal for Numerical Methods in Engineering, Vol. 10, pp. 1007-1019.

Lancaster, P. (1964): On Eigenvalues of Matrices Dependent on a Parameter. Numerische Mathematik, Vol. 6, pp. 377-387.

Lund, E. (1993): Design Sensitivity Analysis of Finite Element Discretized Structures. Report No. 55, Institute of

44

Mechanical Engineering, Aalborg University, Denmark, 31 pp. Earlier version published in: Lecture Notes for COMETT course on Computer Aided Optimum Design of Structures, August 1993, Aalborg.

Lund, E.; Olhoff, N. (1993a): Reliable and Efficient Finite Element Based Design Sensitivity Analysis of Eigenvalues. In: Proc. Structural Optimization 93 - The World Congress on Optimal Design of Structural Systems, (Ed. J. Herskovits), Vol. 2, pp. 197-204, Rio de Janeiro, Brazil.

Lund, E.; Olhoff, N. (1993b): Shape Design Sensitivity Analysis of Eigenvalues Using "Exact" Numerical Differentiation of Finite Element Matrices. Report No. 54, Institute of Mechanical Engineering, Aalborg University, Denmark, 19 pp. Structural Optimization (submitted).

Lund, E.; Rasmussen, J. (1994): A General and Flexible Method of Problem Definition in Structural Optimization Systems. Report No. 58, Institute of Mechanical Engineering, Aalborg University, Denmark, 18 pp. Structural Optimization (submitted).

Masur, E.F. (1984): Optimal Structural Design under Multiple Eigenvalue Constraints. International Journal of Solids and Structures, Vol. 20, pp. 211-231.

Masur, E.F. (1985): Some Additional Comments on Optimal Structural Design under Multiple Eigenvalue Constraints. International Journal of Solids and Structures, Vol. 21, pp. 117-120.

Mlejnek, H.P. (1992): Accuracy of Semi-analytical Sensitivities and Its Improvement by the "Natural Method". Structural Optimization, Vol. 4, pp. 128-131.

Morris, A.J. (1982) (Ed.): Foundations of Structural Optimization: A Unified Approach, Chichester, England.

Mota Soares, C.A. (1987) (Ed.): Computer Aided Optimal Design: Structural and Mechanical Systems, Springer-Verlag, Berlin.

Olhoff, N. (1989): Multicriterion Structural Optimization Via Bound Formulation and Mathematical Programming. Structural Optimization, Vol. 1, pp. 11-17.

Olhoff, N.; Bendsøe, M.P.; Rasmussen, J. (1991): On CAD-integrated Structural Topology and Design Optimization. Computer Methods in Applied Mechanics and Engineering, Vol. 89, pp. 259-279.

Olhoff, N.; Lund, E.; Rasmussen, J. (1993): Concurrent Engineering Design Optimization In a CAD Environment. In: Concurrent Engineering: Tools and Technologies for Mechanical System Design (Ed. E.J. Haug), pp. 523-586, Springer-Verlag, Berlin (1993).

Olhoff, N.; Rasmussen, J. (1991a): Study of Inaccuracy in Semi-Analytical Sensitivity Analysis - A Model Problem. Structural Optimization, Vol. 3, pp. 203-213.

Olhoff, N.; Rasmussen, J. (1991b): Method of Error Elimination for a Class of Semi-Analytical Sensitivity Analysis Problems. In: Engineering Optimization in Design Processes (Eds. H.A. Eschenauer, C. Mattheck & N. Olhoff), pp. 193-200, Springer-Verlag, Berlin.

Olhoff, N.; Rasmussen, J.; Lund, E. (1993): A Method of "Exact" Numerical Differentiation for Error Elimination in Finite Element Based Semi-Analytical Shape Sensitivity Analysis. Mechanics of Structures and Machines, Vol. 21, pp. 1-66.

Olhoff, N.; Krog, L.; Thomsen, J. (1993): Bi-material Topology Optimization. In: Proc. Structural Optimization '93 - The World Congress on Optimal Design of Structural Systems (Ed. J. Herskovits), Federal University of Rio de Janeiro, Brazil, Vol.1, pp. 327-334.

Olhoff, N.; Rasmussen, S.H. (1977): On Single and Bimodal Optimum Buckling Loads of Clamped Columns. International Journal of Solids and Structures, Vol. 13, pp. 605-614.

Olhoff, N.; Taylor, J.E. (1983): On Structural Optimization. Journal of Applied Mechanics, Vol. 50, pp. 1139-1151.

Pedersen, P. (1993a) (Ed.): Optimal Design with Advanced Materials, Elsevier, The Netherlands.

Pedersen, P. (1993b): Concurrent Engineering Design with and of Advanced Materials. In: Concurrent Engineering: Tools and Technologies for Mechanical System Design (Ed. E.J. Haug), pp. 627-670, Springer-Verlag, Berlin.

Pedersen, P. (1991): On Thickness and Orientational Design with Orthotropic Materials. Structural Optimization, Vol. 3, pp. 69-78.

Pedersen, P. (1990): Bounds on Elastic Energy in Solids of Orthotropic Materials. Structural Optimization, Vol. 2, pp. 55-63.

Pedersen, P. (1989): On Optimal Orientation of Orthotropic Materials. Structural Optimization, Vol. 1, pp. 101-106.

Pedersen, P.; Cheng, G.; Rasmussen, J. (1989): On Accuracy Problems for Semi-Analytical Sensitivity Analysis. Mechanics of Structures and Machines, Vol. 17, No. 3, pp. 373-384.

Pedersen, P. (1983): A Unified Approach to Optimal Design. In: Optimization Methods in Structural Design (Eds. H.A. Eschenauer & N. Olhoff), pp. 182-187, B.I.-Wissenschaftsverlag, Mannheim.

Pedersen, P.; Laursen, C.L. (1983): Design for Minimum Stress Concentration by Finite Elements and Linear Programming. Journal of Structural Mechanics, Vol. 10, No. 4, pp. 375-391.

Pedersen, P. (1981): Design with Several Eigenvalue Constraints by Finite Elements and Linear Programming. Journal of Structural Mechanics, Vol. 10, No. 3, pp. 243-271.

Rasmussen, J. (1989): Development of the Interactive Structural Shape Optimization System CAOS (in Danish), Ph.D. thesis, Institute of Mechanical Engineering, Aalborg University, Denmark, Special Report No. 1a.

Rasmussen, J. (1990): The Structural Optimization System CAOS. Structural Optimization, Vol. 2, pp. 109-115.

Rasmussen, J. (1991): Shape Optimization and CAD. International Journal of Systems Automation: Research and Applications (SARA), Vol. 89, pp. 259-279.

Rasmussen, J.; Lund, E.; Birker, T. (1992): Collection of Examples, CAOS Optimization System, 3rd edition. Special Report No. 1c, Institute of Mechanical Engineering, Aalborg University, Denmark, 119 pp.

Rasmussen, J.; Lund, E.; Birker, T.; Olhoff, N. (1993): The CAOS System. International Series of Numerical Mathematics, Vol. 110, pp. 75-96, Birkhäuser Verlag, Basel.

Rasmussen, J.; Lund, E.; Olhoff, N. (1993a): Parametric Modeling and Analysis for Optimum Design. In: Proc. Structural Optimization 93 - The World Congress on Optimal Design of Structural Systems, (Ed. J. Herskovits), Vol. 2, pp. 407-414, Rio de Janeiro, Brazil.

Rasmussen, J.; Lund, E.; Olhoff, N. (1993b): Integration of Parametric Modeling and Structural Analysis for Optimum Design. In: Proc. Advances in Design Automation, The American Society of Mechanical Engineers, (Eds. Gilmore et. al.), Albuquerque, New Mexico, USA, 1993.

Rasmussen, J.; Lund, E.; Olhoff, N.; Birker, T. (1993): Structural Topology and Shape Optimization with the CAOS System. In Lecture Notes for COMETT course on Computer Aided Optimal Design, February 1993, Eindhoven.

Rasmussen, J.; Olhoff, N.; Lund, E. (1992): Foundations for Optimum Design System Development. In Lecture Notes for: Advanced TEMPUS course on Numerical Methods in Computer Optimal Design, 11-15 maj 1992, Zakopane, Poland.

Rozvany, G.I.N.; Karihaloo, B.L. (1988) (Eds.): Structural Optimization, Kluwer, Dordrecht, The netherlands.

Rozvany, G.I.N. (1993) (Ed.): Optimization of Large Structural Systems, Vol. 1: pp. 1-622, Vol. 2: pp. 623-1198, Kluwer, Dordrecht, The Netherlands.

Santos, J.L.T.; Choi, K.K. (1989): Integrated Computational Considerations for Large Scale Structural Design Sensitivity Analysis and Optimization. In: Discretization Methods and Structural Optimization - Procedures and Applications, (Eds. H.A. Eschenauer & G. Thierauf), pp. 299-307, Springer-Verlag, Berlin.

Schmit, L.A. (1982): Structural Synthesis - Its Genesis and Development. AIAA Journal, Vol. 20, pp. 992-1000.

Seyranian, A.P. (1987): Multiple Eigenvalues in Optimization Problems. Prikl. Mat. Mekh., Vol. 51, pp. 349-352; Also Applied Mathematics in Mechanics, Vol. 51, pp. 272-275.

Seyranian, A.P.; Lund, E.; Olhoff, N. (1994): Multiple Eigenvalues in Structural Optimization Problems. Special Report No. 21, Institute of Mechanical Engineering, Aalborg University, Denmark, 51 pp. Structural Optimization (to appear).

Sobieszanski-Sobieski, J.; Rogers, J.L. (1984): A Programming System for Research and Applications in Structural Optimization. In: New directions in optimum structural design (Eds. E. Atrek, R.H. Gallagher, K.M. Ragsdell & O.C. Zienkiewicz), pp. 563-585, John Wiley & Sons, New York.

Stanton, E.L. (1986): Geometric Modeling for Structural and Material Shape Optimization. In: The Optimum Shape. Automated Structural Design (Eds. J.A. Bennett & M.E. Botkin), pp. 365-383, Plenum Press, New York.

Svanberg, K. (1987): The Method of Moving Asymptotes - A New Method For Structural Optimization. International Journal for Numerical Methods in Engineering, Vol. 24, pp. 359-373.

Taylor, J.E.; Bendsøe, M.P. (1984): An Interpretation of Min-Max Structural Design Problems Including a Method for Relaxing Constraints. International Journal of Solids and Structures, Vol. 20, pp. 301-314.

Yang, R.J.; Botkin, M.E. (1986): The Relationship Between the Variational Approach and the Implicit Differentiation Approach to Shape Design Sensitivities. In: The Optimum Shape. Automated Structural Design (Eds. J.A. Bennett & M.E. Botkin), pp. 61-77, Plenum Press, New York.

Zienkiewicz, O.C.; Campbell, J.S. (1973): Shape Optimization and Sequential Linear Programming. In: Optimum Structural Design, Theory and Applications (Eds. R.H. Gallagher & O.C.Zienkiewicz), Wiley and Sons, London, pp. 109-126.

Zolesio, J.-P. (1981): The Material Derivative (or Speed) Method for Shape Optimization. In: Optimization of Distributed Parameter Structures (Eds. E.J. Haug & J. Cea), Vol. 2, pp. 1089-1151, Sijthoff & Nordhoff, The Netherlands.

Zyczkowski, M. (1989) (ed.): Structural Optimization Under Stability and Vibration Constraints. Springer, Wien - New York.

STRUCTURAL DESIGN SENSITIVITY ANALYSIS: CONTINUUM AND DISCRETE APPROACHES

Jasbir S. Arora
Optimal Design Laboratory
College of Engineering
The University of Iowa
Iowa City, Iowa 52242 U.S.A.

ABSTRACT. A unified approach for structural design sensitivity analysis involving both shape and sizing variables is presented. Starting with a continuum formulation and a general response functional needing sensitivity analysis, the direct variation and adjoint approaches are derived. Discretization of the continuum expressions for the two approaches is presented and the numerical implementation aspects are discussed. The discretized forms of the continuum sensitivity expressions are compared with the ones obtained by starting with the discretized model *ab initio*. This comparison shows that the two approaches give similar discretized expressions for numerical calculations. Therefore, exactly same procedures can be used for computer implementation of both the approaches. The continuum approach, however, gives certain insights that would not be possible with only the discrete approach. The presented analyses and insights lead to a unified view point for numerical implementation of design sensitivity analysis which is quite straightforward with existing or new finite element analysis codes. The explicit design variations (partial derivatives with respect to the design variables) of the internal and external nodal forces are the major calculations needed to implement the design sensitivity analysis. An implementation scheme is suggested that is quite general and simple needing minimal programming.

List of Symbols

b, b_i	design variable
$\mathbf{B}$	strain-displacement matrix
$D_{ijk\ell}$, $\mathbf{D}$	material modulus tensor
dS, $^r dS$	differential surface; differential surface in the reference domain
dV, $^r dV$	differential volume; differential volume in the reference domain
e_{ij}, $\mathbf{e}$	infinitesimal strain tensor
f_i, $\mathbf{f}$	body force per unit volume
$\mathbf{F}$	vector of node point internal forces
g	integrand of the displacement specified boundary integral in the response functional
G	integrand of the volume integral in the response functional

47

J. Herskovits (ed.), Advances in Structural Optimization, 47–70.

h	integrand of the traction specified boundary integral in the response functional
$H^1(V)$	Sobolev function space of order 1 defined on the domain V; i.e., a collection of functions that are differentiable at least once, and the functions and their derivatives are square integrable [belong to $L^2(V)$]
K	stiffness matrix
n	unit normal vector
N_{ij}, **N**	matrix of shape functions
P	a vector field
Q	unbalanced node point force vector
R	vector of node point external forces
S_u, S_T	displacement and traction specified boundaries
T_i^0, $\mathbf{T}^0$	specified surface traction vector
u_i, **u**	displacement field
u_i^0	specified displacement on the boundary S_u
U	node point displacement vector
V, ^rV	material and reference volumes
x_i, **x**	coordinates of the material point in the original configuration
X	vector of node point coordinate
Z	Jacobian matrix for the transformation $V \rightarrow {}^rV$
δ	arbitrary variation operator
$\bar{\delta}(\ )$	total design variation of ( ); i.e., $\bar{\delta}(\ ) = \dfrac{D(\ )}{Db_k}\delta b_k$
$\bar{\bar{\delta}}(\ )$	explicit design variation (partial derivative with respect to the design variable) of ( ); i.e., $\bar{\bar{\delta}}(\ ) = \dfrac{\partial(\ )}{\partial b_k}\delta b_k$ for which state fields are frozen
$\tilde{\delta}(\ )$	design variation of the fields that implicitly depend on the design variables, such as displacements, strains, and stresses; also design variation of the functionals with respect to the implicit state fields; for this variation, the explicit dependence on the design variables is frozen
δ_{ij}	Kronecker's delta function
σ_{ij}, σ	Cauchy stress tensor
ξ_i	components of intrinsic coordinates; coordinates of a point in the changed configuration for the material derivative approach
ζ, ζ_s	Jacobian determinant and area metric of the transformation $V \rightarrow {}^rV$
$^r(\ )$	quantity measured over the fixed reference domain
$(\ )^T$	transpose of ( )
$\dfrac{D(\ )}{Db}$	total design derivative of ( )
$(\ )_{,i}$	$\partial(\ )/\partial x_i$, i.e., partial derivative with respect to the coordinate x_i
$(\ )_{,b}$	$\partial(\ )/\partial b$, i.e., partial derivative with respect to b
$(\ )_{;i}$	$\partial(\ )/\partial \xi_i$, i.e., partial derivative with respect to the intrinsic coordinate ξ_i

$F_{,b}$ $k \times n$ matrix $\dfrac{\partial F^T}{\partial b}$ containing derivatives of the n dimensional vector F with respect to the k dimensional vector b; ith column of the matrix contains gradient of F_i with respect to b

1. Introduction

Development of expressions and procedures for calculating design variations (gradients) of response dependent functionals is called *design sensitivity analysis*. The response functionals are dependent on the design variables in an implicit manner because a closed-form solution of the governing equations for the response quantities is not possible for practical problems. This complicates the procedures for evaluation of design variations. Two procedures, the direct variation method and the adjoint method, have been developed and investigated in the literature. Each method has certain features that are suitable for different classes of problems. These have been thoroughly investigated and understood in the literature.

The present paper describes the direct variation and adjoint methods of design sensitivity analysis for linearly elastic structures (the terms direct variation and direct differentiation are used interchangeably; direct variation is used usually for the continuum formulation and the direct differentiation for the discrete formulation). The methods are developed for both the continuum and discrete models of the structural system. The sensitivity expressions for the continuum models are discretized and compared with those for the discrete model. In the literature, there is some confusion about the sensitivity expressions with the two models and the two approaches are considered to be different. Purpose of this paper is to study the two sensitivity expressions without reference to a particular finite element and analyze the differences and similarities between the two approaches.

Although only linearly elastic structural models subjected to static loads are considered, most of the discussions and derivations can be extended to nonlinear problems and dynamic systems (Lee et al, 1993, 1995; Ohsaki and Arora 1994; Hsieh and Arora 1984; Cardoso and Arora 1992; and many other articles cited in these references). Also, the reference domain concept is used to unify the shape and sizing design sensitivity analysis. The material derivative concept can also be used and analyzed (Dems and Mroz 1984; Yang and Choi 1985; Hou et al 1986; Haug et al 1986; Dems and Haftka 1989; Arora 1993; and many other articles cited in these references). It has been recently shown, however, that the reference domain procedure and the material derivative approach are equivalent for some formulations and under certain conditions; i.e., the final expression of one approach can be derived from the other (Arora, Lee and Cardoso 1992). Therefore there is no need to discuss the material derivative approach separately.

The equivalence between the continuum and discrete approaches has also been studied by Choi and Twu (1989) for planar truss and beam elements. It is shown for these elements and several load conditions that if linear shape functions are used for the design velocity field, then the two approaches give same sensitivity coefficients. However, if quadratic or cubic shape functions are used, the two approaches give different design sensitivity coefficients for some cases.

In the next section, the problem of design sensitivity analysis is defined and formulated. The reference domain concept is described and the problem is transformed to that domain. In Section 3, the direct variation method of design sensitivity analysis is developed using the continuum formulation. Section 4 describes the adjoint method of design sensitivity analysis. In Section 5, discretizations of the contiuum expressions for structural analysis and design sensitivity analysis are presented. Section 6 describes the design sensitivity analysis for discretized structural models. Both the direct differentiation and adjoint methods are described. Analysis of the continuum and discrete methods is presented in Section 7. It is shown that the continuum design sensitivity expression when discretized is similar to the one obtained by starting with a discrete model *ab initio*. Thus the continuum and discrete design sensitivity expressions can be implemented on the computer in an exactly the same way. In Section 8, other classes of structural problems that have been addressed in the literature are described. Finally, discussion and conclusions are presented.

2. Problem Definition and Formulation

2.1 PROBLEM DEFINITION

The problem is to determine design variations of the response dependent functionals. Considerable work has been done on this problem during the last about 30 years. Both discrete and continuum models have been treated, and nonlinear and dynamic problems have been addressed. A detailed review of the literature on the subject is beyond the scope of this paper; papers by Adelman and Haftka (1986) and Haftka and Adelman (1989), and books by Bennett and Botkin (1986), Mota Soares (1987), Rozvany (1993), and Kamat (1993) can be consulted for this purpose.

For a *continuum formulation* of the problem, either the differential form (local) of the equilibrium equation or the virtual work equation (global form or integral form) can be used as the starting point. Both formulations are equivalent under appropriate continuity conditions on the field variables. Here we use the virtual work equation (also called the weak form) governing the equilibrium state is given as

$$\int \sigma_{ij}(\mathbf{u}) \, \delta e_{ij}(\mathbf{u}) \, dV - \int f_i \, \delta u_i \, dV - \int T_i^0 \, \delta u_i \, dS_T = 0 \tag{1}$$

where a *repeated index implies summation* over its range, $\sigma_{ij}(\mathbf{u})$ is the Cauchy stress tensor, $e_{ij}(\mathbf{u})$ is the infinitesimal strain tensor, f_i is the body force field per unit volume, T_i^0 is the specified surface traction, u_i is the displacement field having appropriate smoothness (i.e., u_i are in $H^1(V)$), V is the volume occupied by the structure, S_T is the traction specified surface, and δu_i is a kinematically admissible virtual displacement field having appropriate smoothness (i.e., δu_i are in $H^1(V)$), and $\delta e_{ij}(\mathbf{u})$ is the compatible virtual strain field (i.e., $\delta e_{ij}(\mathbf{u}) \equiv e_{ij}(\delta \mathbf{u})$). Note that the surface traction is assumed to be independent of the displacement field; i.e., only the conservative force field is treated. For treatment of nonconservative force fields, papers by Poldneff et al (1993) and Arora and Lee (1993) may be consulted.

The *strain tensor and its arbitrary variation*, and the *linear stress-strain relationship* are given as

$$e_{ij}(u) = \frac{1}{2}(u_{i,j} + u_{j,i}) \tag{2}$$

$$\delta e_{ij}(u) \equiv e_{ij}(\delta u) = \frac{1}{2}(\delta u_{i,j} + \delta u_{j,i}) \tag{3}$$

$$\sigma_{ij} = D_{ijk\ell} e_{k\ell} \tag{4}$$

where $D_{ijk\ell}$ is the material modulus tensor.

In general, a response functional needing design sensitivity analysis is defined as

$$\psi = \int G(\sigma_{ij}, e_{ij}, u_i, b) \, dV + \int g(u_i^0, T_i, b) \, dS_u + \int h(u_i, T_i^0, b) \, dS_T \tag{5}$$

where u_i^0 is the specified displacement on part of the boundary S_u, and b is a design variable that is independent of x_i. The functional ψ may represent cost function for the problem or constraints on stresses, strains, displacements and reaction forces. These constraints may be imposed at a particular point in the structure or over a subdomain.

2.2 TRANSFORMATION TO THE REFERENCE DOMAIN

In the reference domain approach, all the material configurations of the deformable body for different designs are separately mapped on to the same conveniently selected fixed domain. All the field variables and integrals are transformed to this fixed reference domain before design variations are taken. A major advantage of the approach is that when the design variations are taken, variations of the fixed domain and its boundary vanish, simplifying the design sensitivity expressions for the domain and surface integrals. In numerical calculations, this approach can be viewed as an extension of the isoparametric finite element concept to the design process. There, all elements of the same type are mapped on to a parent element of fixed dimensions. Here, all design configurations are mapped on to a fixed reference domain (parent design element). The approach was developed during 1980s, first for the discrete models (Ryu et al 1985; Haririan et al 1987; Wu and Arora 1987) and then for the continuum models (Haber 1987; Arora and Cardoso 1986; Cardoso and Arora 1988; Phelan and Haber 1989; Arora and Cardoso 1989; Tsay and Arora 1990). A more detailed review of various approaches can be found in Adelman and Haftka (1986) and Haftka and Adelman (1989).

Figure 1 shows the transformation from the material configurations for designs b_1 and b_2 to a fixed domain. The mapping between the material configuration and reference domain can be written in general as $x_i = x_i(\xi, b)$. It can be seen that as the design changes, this mapping changes with it, but the reference domain never changes. Let ξ_i be the coordinates in the fixed reference domain with volume ^rV and surface ^rS that do not change with design variations. In the finite element literature, ξ_i are called the *intrinsic, isoparametric or element's native coordinates*. The differential volume and surface elements in the two coordinate systems are related as

$$dV = \zeta \, (^r dV), \quad dS = \zeta_s \, (^r dS) \tag{6}$$

where

$$\zeta = |Z|, \qquad \zeta_s = \zeta \, \|Z^{-T} n\|, \qquad Z_{ij} = \partial x_i / \partial \xi_j \tag{7}$$

and **n** is the unit outward normal to the fixed reference boundary. The left superscript r indicates quantities in the reference domain.

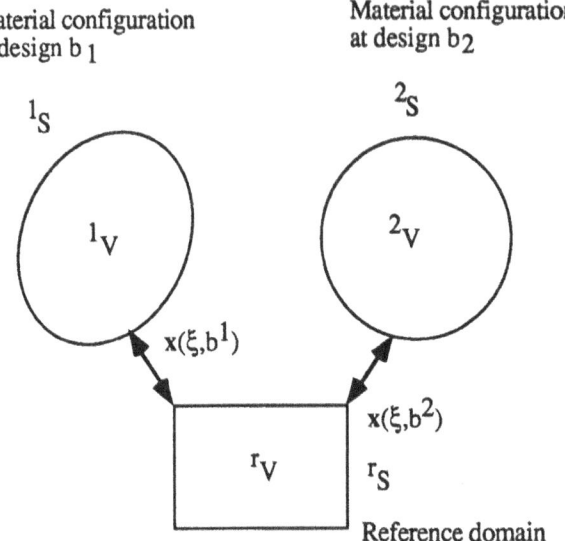

Figure 1. Transformation to Reference Domain

Substituting the transformations to the reference domain, the virtual work equations (1), and the strain tensor and its arbitrary variation given in Eqs. (2) and (3), become

$$\int (\sigma_{ij} \, \delta e_{ij} - f_i \, \delta u_i) \, \zeta \, {}^r dV - \int T_i^0 \, \delta u_i \, \zeta_s \, {}^r dS_T = 0 \tag{8}$$

$$e_{ij}(\mathbf{u}) = \frac{1}{2} \left[u_{i;k} \xi_{k,j} + u_{j;k} \xi_{k,i} \right] \tag{9}$$

$$e_{ij}(\delta\mathbf{u}) = \frac{1}{2} \left[\delta u_{i;k} \xi_{k,j} + \delta u_{j;k} \xi_{k,i} \right] \tag{10}$$

where $\xi_{k,i} = \partial \xi_k / \partial x_i$ and $u_{j;k} = \partial u_j / \partial \xi_k$. Note that the strain tensor and its arbitrary variation in Eqs. (9) and (10) now depend explicitly on the design variables through the transformation matrix $\xi_{k,j}$. They also depend implicitly on the design variables because $u_{j;k}$ depends in that way. Transforming the functional in Eq. (5) to the reference domain, we get

$$\psi = \int G(\sigma_{ij}, e_{ij}, u_i, b) \, \zeta \, {}^r dV + \int g(u_i^0, T_i, b) \, \zeta_s \, {}^r dS_u + \int h(u_i, T_i^0, b) \, \zeta_s \, {}^r dS_T \tag{11}$$

This functional depends explicitly as well as implicitly on the design variables.

3. Design Sensitivity Analysis: Direct Variation Method

In this approach, design variation of the governing equilibrium equation, i.e., the virtual work equation is taken, and after substituting for the design variation of the stress-strain law and strain-displacement relationship, design variation of the displacement field can be computed. Design variations of the stresses, strains and the response functional ψ can be then computed using the displacement variations.

In the sequel, we shall use the variational notations $\bar{\delta}$, $\bar{\bar{\delta}}$ and $\tilde{\delta}$ to represent the total, explicit and implicit design variations, respectively. Thus $\bar{\delta}\psi$ will represent

$$\bar{\delta}\psi = \left(\frac{D\psi}{Db}\right)^T \delta b \tag{12}$$

where $\frac{D\psi}{Db}$ is the desired design gradient. In the sequel, it is assumed that all the field variables are at least once continuously differentiable with respect to the design variables.

The total design variation of the response functional in Eq. (11) gives

$$\bar{\delta}\psi = \bar{\bar{\delta}}\psi + \tilde{\delta}\psi \tag{13}$$

where

$$\bar{\bar{\delta}}\psi = \int \bar{\bar{\delta}}(G\zeta)\,^r dV + \int \bar{\bar{\delta}}(g\,\zeta_s)\,^r dS_u + \int \bar{\bar{\delta}}(h\,\zeta_s)\,^r dS_T \tag{14}$$

$$\tilde{\delta}\psi = \int (\tilde{\delta}G)\zeta\,^r dV + \int (\tilde{\delta}g)\zeta_s\,^r dS_u + \int (\tilde{\delta}h)\zeta_s\,^r dS_T \tag{15}$$

$$\bar{\bar{\delta}}G = G_{,\sigma_{ij}}\,\bar{\bar{\delta}}\sigma_{ij} + G_{,e_{ij}}\,\bar{\bar{\delta}}e_{ij} + G_{,b_i}\,\delta b_i \quad \text{in } V \tag{16}$$

$$\bar{\bar{\delta}}g = g_{,u_i}\,\bar{\bar{\delta}}u_i^0 + g_{,T_i}\,\bar{\bar{\delta}}T_i + g_{,b_i}\,\delta b_i \quad \text{on } S_u \tag{17}$$

$$\bar{\bar{\delta}}h = h_{,T_i}\,\bar{\bar{\delta}}T_i^0 + h_{,b_i}\,\delta b_i \quad \text{on } S_T \tag{18}$$

$$\tilde{\delta}G = G_{,\sigma_{ij}}\,\tilde{\delta}\sigma_{ij} + G_{,e_{ij}}\,\tilde{\delta}e_{ij} + G_{,u_i}\,\tilde{\delta}u_i \quad \text{in } V \tag{19}$$

$$\tilde{\delta}g = g_{,T_i}\,\tilde{\delta}T_i \quad \text{on } S_u \tag{20}$$

$$\tilde{\delta}h = h_{,u_i}\,\tilde{\delta}u_i \quad \text{on } S_T \tag{21}$$

In the foregoing equations, the fact that ζ and ζ_s depend only explicitly on design has been used. In some formulations, these quantities may also depend on the displacement field. In that case, their implicit variations must also be included in the derivations. Note that δu_i^0 and δT_i^0 are explicit design variations that can be evaluated at the given design. δT_i on S_u (design variation of the reaction traction) can be evaluated once δu_i is known. The total design variations of the stress and strain tensors are given as

$$\bar{\delta}\sigma_{ij} = \tilde{\delta}\sigma_{ij} + \bar{\bar{\delta}}\sigma_{ij} \tag{22}$$

$$\tilde{\delta}\sigma_{ij} = D_{ijk\ell}\,\tilde{\delta}e_{k\ell} \tag{23}$$

$$\bar{\bar{\delta}}\sigma_{ij} = D_{ijk\ell}\,\bar{\bar{\delta}}e_{k\ell} \tag{24}$$

$$\bar{\delta}e_{ij} = \tilde{\delta}e_{ij} + \bar{\bar{\delta}}e_{ij} \tag{25}$$

$$\tilde{\delta}e_{ij} = \tfrac{1}{2}\left[\bar{\delta}u_{i;k}\,\xi_{k,j} + \bar{\delta}u_{j;k}\,\xi_{k,i}\right] \tag{26}$$

$$\bar{\bar{\delta}}e_{ij} = \tfrac{1}{2}\left[u_{i;k}\,\bar{\bar{\delta}}\xi_{k,j} + u_{j;k}\,\bar{\bar{\delta}}\xi_{k,i}\right] \tag{27}$$

where $\tilde{\delta}u_i = \bar{\delta}u_i$ is used in Eq. (26) since u_i depends only implicitly on the design variables.

In order to calculate $\bar{\delta}u_i$, total design variation of the virtual work equation (8) is taken as

$$-\int \bar{\delta}(\sigma_{ij}\delta e_{ij}\,\zeta)\,^r dV + \int \bar{\delta}(f_i\delta u_i\,\zeta)\,^r dV + \int \bar{\delta}(T_i^0\delta u_i\,\zeta_s)\,^r dS_T = 0 \tag{28}$$

Let the virtual displacement δu_i depend arbitrarily on design. Expanding Eq. (28), we get

$$\int\left[(\bar{\delta}\sigma_{ij})\delta e_{ij}\,\zeta + \sigma_{ij}\bar{\delta}(\delta e_{ij})\zeta + \sigma_{ij}\delta e_{ij}\,\bar{\delta}\zeta\right]\,^r dV - \int\left[\bar{\bar{\delta}}(f_i\zeta)\delta u_i + (f_i\zeta)\bar{\delta}(\delta u_i)\right]\,^r dV$$

$$-\int\left[\bar{\bar{\delta}}(T_i^0\,\zeta_s)\delta u_i + (T_i^0\,\zeta_s)\bar{\delta}(\delta u_i)\right]\,^r dS_T = 0 \tag{29}$$

Total design variation of the arbitrary strain field δe_{ij} is given as

$$\bar{\delta}(\delta e_{ij}) = \bar{\bar{\delta}}(\delta e_{ij}) + \tilde{\delta}(\delta e_{ij}) \tag{30}$$

and use of Eq. (10) gives

$$\bar{\bar{\delta}}(\delta e_{ij}) = \tfrac{1}{2}\left[\delta u_{i;k}\bar{\bar{\delta}}\xi_{k,j} + \delta u_{j;k}\bar{\bar{\delta}}\xi_{k,i}\right] \tag{31}$$

$$\tilde{\delta}(\delta e_{ij}) = \tfrac{1}{2}\left[\bar{\delta}(\delta u_{i;k})\xi_{k,j} + \bar{\delta}(\delta u_{j;k})\xi_{k,i}\right] \tag{32}$$

Substituting Eq. (30) into Eq. (29), and rearranging and collecting terms, we have

$$\left\{\int \sigma_{ij}\tilde{\delta}(\delta e_{ij})\,\zeta\,^r dV - \int (f_i\zeta)\bar{\delta}(\delta u_i)\,^r dV - \int (T_i^0\,\zeta_s)\bar{\delta}(\delta u_i)\,^r dS_T\right\}$$

$$+ \int\left[(\bar{\delta}\sigma_{ij})\delta e_{ij}\,\zeta + \sigma_{ij}\bar{\bar{\delta}}(\delta e_{ij})\,\zeta + \sigma_{ij}\delta e_{ij}\,\bar{\delta}\zeta\right]\,^r dV$$

$$- \int \bar{\bar{\delta}}(f_i\zeta)\delta u_i\,^r dV - \int \bar{\bar{\delta}}(T_i^0\,\zeta_s)\delta u_i\,^r dS_T = 0 \tag{33}$$

Now let $\bar{\delta}(\delta u_i)$ be specified as a kinematically admissible field having appropriate smoothness (i.e., in $H^1(V)$). Note from Eq. (32) that $\delta(\delta e_{ij})$ is compatible with $\bar{\delta}(\delta u_i)$, i.e., $\delta(\delta e_{ij}) = e_{ij}(\bar{\delta}(\delta u))$. The set $(\bar{\delta}(\delta u_i), e_{ij}(\bar{\delta}(\delta u)))$ can be replaced by any other compatible set, such as $(\delta u_i, e_{ij}(\delta u))$. Thus, the expression within the braces in Eq. (33) represents the equilibrium equation for the structure, so it vanishes. This expression also vanishes if we assume the admissible virtual displacement field δu_i to be independent of design, i. e., $\bar{\delta}(\delta u_i) = 0$; in this case $\delta(\delta e_{ij})$ also vanishes. Thus, the total design variation of the virtual work equation in Eq. (33), after using Eqs. (22) for $\bar{\delta}\sigma_{ij}$, and transferring all the known terms to the right hand side, becomes

$$\int (\tilde{\delta}\sigma_{ij})\delta e_{ij}\, \zeta\,^r dV = \int \bar{\bar{\delta}}(f_i\zeta)\delta u_i\,^r dV + \int \bar{\bar{\delta}}(T_i^0\,\zeta_s)\delta u_i\,^r dS_T$$
$$- \int [(\bar{\bar{\delta}}\sigma_{ij})\delta e_{ij}\,\zeta + \sigma_{ij}\bar{\bar{\delta}}(\delta e_{ij})\,\zeta + \sigma_{ij}\delta e_{ij}\,\bar{\bar{\delta}}\zeta]\,^r dV \qquad (34)$$

This equation can be used to solve for the total design variation $\bar{\delta}u_i$ of the displacement field u_i. Then Eqs. (13) to (15) can be used to calculate design variation of the response functional ψ.

4. Design Sensitivity Analysis: Adjoint Method

In order to calculate the design sensitivity of the reaction traction force T_i on S_u with the adjoint method, the virtual work equation (8) is extended to include the integral on the displacement specified boundary as (the integral vanishes because u_i^0 is a specified field on S_u):

$$\int (\sigma_{ij}\,\delta e_{ij} - f_i\,\delta u_i)\,\zeta\,^r dV - \int T_i^0\,\delta u_i\,\zeta_s\,^r dS_T - \int T_i\,\delta u_i^0\,\zeta_s\,^r dS_u = 0 \qquad (35)$$

A simple example showing calculation of design sensitivities of the reaction force is given in Tsay, Cardoso and Arora (1990).

Several procedures can be used to derive the adjoint sensitivity expression. A simple and straightforward way is to use the design sensitivity analysis principle presented by Arora and Cardoso (1992). In order to use the principle, the arbitrary state fields in the virtual work expression (35) are replaced by the so-called adjoint fields that will be determined later. Thus, the weak form of the governing equation (35) becomes:

$$W^a \equiv \int (-\sigma_{ij}e_{ij}^a + f_i u_i^a)\,\zeta\,^r dV + \int T_i^0 u_i^a\,\zeta_s\,^r dS_T + \int T_i u_i^{a0}\,\zeta_s\,dS_u = 0 \qquad (36)$$

where u_i^a is the adjoint displacement field, the adjoint strain e_{ij}^a is given using $e_{ij}(\delta u)$ in Eq. (10) as

$$e_{ij}^a = \frac{1}{2}\left[u_{i;k}^a\xi_{k,j} + u_{i;k}^a\xi_{k,i}\right] \qquad (37)$$

and u_i^{a0} is a specified displacement for analysis purposes (to be defined later). Now an augmented functional (also called the Lagrangian function in the optimization literature) is defined using Eqs. (11) and (36) as

$$L = \psi + W^a \tag{38}$$

According to the design sensitivity principle, the total design variation of the constraint functional ψ in Eq. (11) is given as the explicit design variation of the functional L in Eq. (38) as

$$\bar{\delta}\psi = \bar{\bar{\delta}}L$$

$$= \bar{\bar{\delta}}\psi + \bar{\bar{\delta}}W^a \tag{39}$$

The explicit design variation $\bar{\bar{\delta}}\psi$ is given in Eq. (14), and $\bar{\bar{\delta}}W^a$ is given from Eq. (36) as

$$\bar{\bar{\delta}}W^a = \bar{\bar{\delta}}\int -\sigma_{ij}e_{ij}^a \zeta \,^r dV + \bar{\bar{\delta}}\int f_i u_i^a \zeta \,^r dV + \bar{\bar{\delta}}\int T_i^0 u_i^a \zeta_s \,^r dS_T + \bar{\bar{\delta}}\int T_i u_i^{a0} \zeta_s \,^r dS_u \tag{40}$$

Thus the final sensitivity expression for the adjoint method is obtained by substituting Eqs. (14) and (40) into Eq. (39) as

$$\bar{\delta}\psi = \int \bar{\bar{\delta}}[(G - \sigma_{ij}e_{ij}^a + f_i u_i^a) \zeta] \,^r dV + \int \bar{\bar{\delta}}[(h + T_i^0 u_i^a) \zeta_s] \,^r dS_T$$

$$+ \int \bar{\bar{\delta}}[(g + T_i u_i^{a0}) \zeta_s] \,^r dS_u \tag{41}$$

To calculate the design sensitivity using Eq. (41), the adjoint fields are needed. According to the design sensitivity analysis principle, the governing equilibrium equation for the adjoint structure can be obtained by requiring the implicit design variation of the augmented functional in Eq. (38) to vanish; i.e.,

$$\tilde{\delta}L = 0, \quad \text{or} \quad \tilde{\delta}\psi + \tilde{\delta}W^a = 0 \tag{42}$$

To impose the condition in Eq. (42), we write $\tilde{\delta}\psi$ and $\tilde{\delta}W^a$ using Eqs. (11) and (36) as

$$\tilde{\delta}\psi = \int (G_{,\sigma_{ij}}\tilde{\delta}\sigma_{ij} + G_{,e_{ij}}\tilde{\delta}e_{ij} + G_{,u_i}\tilde{\delta}u_i) \zeta \,^r dV + \int g_{,T_i}\tilde{\delta}T_i \zeta_s \,^r dS_u$$

$$+ \int h_{,u_i}\tilde{\delta}u_i \zeta_s \,^r dS_T \tag{43}$$

$$\tilde{\delta}W^a = \left\{ \int (-\sigma_{ij}\tilde{\delta}e_{ij}^a + f_i\tilde{\delta}u_i^a) \zeta \,^r dV + \int T_i^0 \tilde{\delta}u_i^a \zeta_s \,^r dS \right\} - \int \tilde{\delta}\sigma_{ij}e_{ij}^a \zeta \,^r dV$$

$$+ \int \tilde{\delta}T_i u_i^{a0} \zeta_s \,^r dS_u \tag{44}$$

where $\tilde{\delta}T_i = 0$ on S_T and $u_i^a = u_i^{a0}$ on S_u have been used. In Eq. (44), the expression within the braces represents equilibrium equation for the primary structure, so it vanishes and $\tilde{\delta}W^a$ reduces to

$$\tilde{\delta}W^a = -\int \tilde{\delta}\sigma_{ij}e_{ij}^a \zeta \,^r dV + \int \tilde{\delta}T_i u_i^{a0} \zeta_s \,^r dS_u \tag{45}$$

Referring to Eq. (43), let us define the following quantities for the adjoint structure:

$$\text{Initial Strain:} \quad e_{ij}^{aI} = G_{,\sigma_{ij}} \quad \text{in V} \tag{46}$$

Initial Stress: $\quad \sigma_{ij}^{al} = G_{,e_{ij}} \qquad$ in V $\qquad\qquad$ (47)

Body Force: $\quad f_i^a = G_{,u_i} \qquad$ in V $\qquad\qquad$ (48)

Boundary Displacement: $\quad u_i^{a0} = - g_{,T_i} \qquad$ on S_u $\qquad$ (49)

Specified Traction: $\qquad T_i^{a0} = h_{,u_i} \qquad$ on S_T $\qquad$ (50)

Substituting Eqs. (43), and (45) to (50) into Eq. (42) and combining terms, we have

$$\int (e_{ij}^{al} \, \tilde{\delta}\sigma_{ij} + \sigma_{ij}^{al} \, \tilde{\delta}e_{ij} + f_i^a \, \tilde{\delta}u_i) \, \zeta^{\,r} dV - \int u_i^{a0} \, \tilde{\delta}T_i \, \zeta_s^{\,r} dS_u$$

$$+ \int T_i^{a0} \, \tilde{\delta}u_i \, \zeta_s^{\,r} dS_T - \int \tilde{\delta}\sigma_{ij} e_{ij}^{a} \, \zeta^{\,r} dV + \int \tilde{\delta}T_i u_i^{a0} \, \zeta_s^{\,r} dS_u = 0$$

Collecting terms and using $\tilde{\delta}\sigma_{ij} = D_{ijk\ell}\tilde{\delta}e_{k\ell}$, we have

$$\int [-(D_{ijk\ell}(e_{ij}^a - e_{ij}^{al}) - \sigma_{k\ell}^{al})\tilde{\delta}e_{k\ell} + f_i^a \, \tilde{\delta}u_i] \, \zeta^{\,r} dV + \int T_i^{a0} \, \tilde{\delta}u_i \, \zeta_s^{\,r} dS_T = 0 \qquad (51)$$

If we define the stress-strain law for the adjoint structure as

$$\sigma_{ij}^a = D_{ijk\ell}(e_{k\ell}^a - e_{k\ell}^{al}) - \sigma_{ij}^{al} \qquad\qquad (52)$$

where $D_{ijk\ell}$ is assumed to be symmetric, then Eq. (51) becomes

$$\int (-\sigma_{ij}^a \, \tilde{\delta}e_{ij} + f_i^a \, \tilde{\delta}u_i) \, \zeta^{\,r} dV + \int T_i^{a0} \, \tilde{\delta}u_i \, \zeta_s^{\,r} dS_T = 0 \qquad\qquad (53)$$

Equation (53) represents the governing equilibrium equation for the adjoint structure because δu_i can be considered as arbitrary with δe_{ij} to be compatible with it.

The adjoint method is now summarized as follows: use Eqs. (35), (4) and (9) to determine the primary field variables, use Eqs. (37), (52) and (53) to determine the adjoint fields, calculate explicit design variations of the primary stresses and strains given in Eqs. (24) and (27) respectively, calculate explicit design variations of the adjoint stresses and strains given in Eqs. (52) and (37) respectively, and then use Eq. (39) or (41) to calculate the final design sensitivity coefficients. Note that if design sensitivity of the reaction traction is not desired, then $u_i^{a0} = 0$ from Eq. (49). In this case the integral on the displacement specified boundary vanishes in Eq. (40), and the corresponding term drops out of the final sensitivity equation (41).

5. Discretization of Continuum Expressions

5.1 DISCRETIZATION OF THE ANALYSIS PROBLEM

The design sensitivity expressions of both the approaches can be discretized for numerical calculations using the standard finite element shape functions. Using the isoparametric formulation of finite element analysis, the coordinates and the displacement field are expressed as

$$\mathbf{x} = \mathbf{N}(\xi)\mathbf{X}, \quad \text{or} \quad x_i = N_{ij}X_j \tag{54}$$

$$\mathbf{u} = \mathbf{N}(\xi)\mathbf{U}, \quad \text{or} \quad u_i = N_{ij}U_j \tag{55}$$

where $\mathbf{N}(\xi)$ is a matrix of shape functions given in element's native coordinates ξ_j, and $\mathbf{X}$ and $\mathbf{U}$ are respectively the node point coordinate vector and the node point displacement vector. These quantities have appropriate dimensions depending on the element type. Using the shape functions, the strain vector and its arbitrary variation, and the stress vector are given as (note that now the vector notation is used):

$$\mathbf{e} = \mathbf{BU}, \quad \delta\mathbf{e} = \mathbf{B}\delta\mathbf{U}, \quad \text{and} \quad \sigma = \mathbf{De} = \mathbf{DBU} \tag{56}$$

where $\mathbf{B}$ is the strain-displacement matrix that is calculated by appropriately differentiating $\mathbf{N}$. Substituting appropriate equations into the virtual work Eq. (8) and using the fact that $\delta\mathbf{U}$ is arbitrary, we obtain a finite dimensional form of the equilibrium equation as

$$\mathbf{Q} \equiv \mathbf{R} - \mathbf{F} = 0 \tag{57}$$

where the equivalent node point external and internal forces $\mathbf{R}$ and $\mathbf{F}$ are given as

$$\mathbf{R} = \int \mathbf{N}^T \mathbf{f} \, \zeta \, ^r dV + \int \mathbf{N}^T \mathbf{T}^0 \, \zeta_s \, ^r dS_T \tag{58}$$

$$\mathbf{F} = \mathbf{KU}, \quad \text{with} \quad \mathbf{K} = \int \mathbf{B}^T \mathbf{DB} \, \zeta \, ^r dV \tag{59}$$

The matrix $\mathbf{K}$ in Eq. (59) is called the stiffness matrix for the structure which is assembled using the element stiffness matrices.

In order to discretize the general response functional, discretizations of Eqs. (55) and (56) are substituted in to Eq. (11). After carrying out the integrations, the discretized form of the response dependent functional needing sensitivity analysis depends on the design variables and the nodal displacements. In general, this functional can be written as

$$\psi = \psi(\mathbf{b}, \mathbf{U}) \tag{60}$$

5.2 DIRECT VARIATION METHOD

To solve Eq. (34) for $\bar{\delta}u_i$, we first introduce the foregoing finite element discretizations (although different discretizations may also be used for design sensitivity calculations) into design variation of the stress vector, and the strain vector and its arbitrary variation. This gives

$$\bar{\bar{\delta}}\sigma = \bar{\bar{\delta}}(\mathbf{DB})\mathbf{U}, \quad \tilde{\delta}\sigma = \mathbf{DB}\,\bar{\delta}\mathbf{U}, \quad \bar{\bar{\delta}}\mathbf{e} = \bar{\bar{\delta}}\mathbf{B}\,\mathbf{U} \quad \tilde{\delta}\mathbf{e} = \mathbf{B}\,\bar{\delta}\mathbf{U} \text{ and } \bar{\bar{\delta}}(\delta\mathbf{e}) = \bar{\bar{\delta}}\mathbf{B}\,\delta\mathbf{U} \tag{61}$$

Note that the matrix $\mathbf{B}$ depends only explicitly on the design variables. Substituting Eqs. (61) into Eq. (34), we get

$$\int \delta U^T B^T (DB) \, \bar{\delta} U \, \zeta \, ^r dV = \int \delta U^T N^T \bar{\bar{\delta}} (f \, \zeta) \, ^r dV$$

$$+ \int \delta U^T N^T \bar{\bar{\delta}} (T^0 \, \zeta_s) \, dS_T - \int \delta U^T B^T \bar{\bar{\delta}} (DB) U \, \zeta \, ^r dV$$

$$- \int \delta U^T (\bar{\bar{\delta}} B^T) DBU \, \zeta \, ^r dV - \int \delta U^T B^T DBU \, \bar{\bar{\delta}} \zeta \, ^r dV \qquad (62)$$

Using the facts that δU is arbitrary and the shape function matrix $N(\xi)$ does not depend on the design variables, and rearranging and combining certain terms, Eq. (62) becomes:

$$\left[\int B^T DB \, \zeta \, ^r dV \right] \bar{\delta} U = \int \bar{\bar{\delta}} (N^T f \, \zeta) \, ^r dV + \int \bar{\bar{\delta}} (N^T T^0 \, \zeta_s) \, ^r dS_T$$

$$- \int \bar{\bar{\delta}} (B^T DB) U \, \zeta \, ^r dV - \int B^T DBU \, \bar{\bar{\delta}} \zeta \, ^r dV$$

$$= \bar{\bar{\delta}} \int N^T f \zeta \, ^r dV + \bar{\bar{\delta}} \left[\int N^T T^0 \, \zeta_s \, ^r dS_T - \int B^T DBU \, \zeta \, ^r dV \right] \qquad (63)$$

Using Eqs. (58) and (59) in Eq. (63), we have

$$\bar{K} \delta U = \bar{\bar{\delta}} R - \bar{\bar{\delta}} F, \quad \text{or} \quad K \frac{DU}{Db} = R_{,b} - F_{,b} \qquad (64)$$

Equation (64) can be now solved for $\bar{\delta} U$ or $\frac{DU}{Db}$. Note that this equation has same coefficient matrix as for the analysis problem in Eq. (57), if the shape functions are the same for the response analysis and design sensitivity analysis calculations. Therefore, solution procedure for Eq. (64) does not require any additional decomposition of the coefficient matrix.

In order to calculate the total design variation of the response functional using Eq. (13), we need its explicit and implicit design variations. Substituting discretizations of Eqs. (54) to (56) and (61) into Eqs. (14) and (16) to (18), we get discretized form of the explicit design variation of the response functional as

$$\bar{\bar{\delta}} \psi = \int G \, \bar{\bar{\delta}} \zeta \, ^r dV + \int g \, \bar{\bar{\delta}} \zeta_s \, ^r dS_u + \int h \, \bar{\bar{\delta}} \zeta_s \, ^r dS_T$$

$$+ \int [G_{,\sigma} \, \bar{\bar{\delta}} (DB) U + G_{,e} \, \bar{\bar{\delta}} BU + G_{,b_i} \delta b_i] \, \zeta \, ^r dV$$

$$+ \int [g_{,u_i} \bar{\bar{\delta}} u_i^0 + g_{,T_i} \bar{\bar{\delta}} T_i + g_{,b_i} \delta b_i] \, \zeta_s \, ^r dS_u + \int [h_{,T_i} \bar{\bar{\delta}} T_i^0 + h_{,b_i} \delta b_i] \, \zeta_s \, ^r dS_T \qquad (65)$$

Using Eqs. (15) and (19) to (21), we get the implicit design variation for the response functional as

$$\tilde{\delta} \psi = \int (G_{,\sigma} \, DB \, \bar{\delta} U + G_{,e} \, B \, \bar{\delta} U + G_{,u} \, N \, \bar{\delta} U) \, \zeta \, ^r dV$$

$$+ \int (g_{,T_i} \tilde{\delta} T_i) \, \zeta_s \, ^r dS_u + \int (h_{,u} \, N \, \bar{\delta} U) \, \zeta_s \, ^r dS_T \qquad (66)$$

Note that explicit and implicit design variations of reaction force T_i on S_u can be obtained by using the Cauchy formula and the corresponding variations of the stresses.

To calculate the design sensitivity coefficients of any response functional by the direct differentiation method, we need to evaluate the right hand side of Eq. (64) which involves partial derivatives of the external and internal forces with respect to the design variables. Then using the decomposed structural stiffness matrix saved from the analysis phase, Eq. (64) can be solved for the derivatives of the displacements. Using these derivatives, derivatives of the stresses and strains can be evaluated. Substituting the foregoing derivatives into Eqs. (65) and (66), design derivative of the response functional can be evaluated.

5.3 ADJOINT METHOD

Using discretizations of Eqs. (54) and (55), the adjoint strain and stress vectors can be written as (again it is important to note that different discretizations may be used for the adjoint problem):

$$\mathbf{e}^a = \mathbf{B} \mathbf{U}^a, \quad \text{and} \quad \sigma^a = \mathbf{D}\mathbf{B}\mathbf{U}^a - \mathbf{D}\mathbf{e}^{aI} - \sigma^{aI} \tag{67}$$

Substituting these in the adjoint Eq. (53), we obtain the equilibrium equation for the adjoint structure as

$$\mathbf{Q}^a \equiv \mathbf{R}^a - \mathbf{F}^a = 0 \tag{68}$$

where the equivalent node point external and internal forces $\mathbf{R}^a$ and $\mathbf{F}^a$ are given as

$$\mathbf{R}^a = \int \mathbf{N}^T \mathbf{f}^a \, \zeta \, ^r dV + \int \mathbf{N}^T \mathbf{T}^{a0} \, \zeta_s \, ^r dS_T + \int \mathbf{B}^T [\mathbf{D}\mathbf{e}^{aI} + \sigma^{aI}] \, \zeta \, ^r dV \tag{69}$$

$$\mathbf{F}^a = \mathbf{K}\mathbf{U}^a, \quad \text{where} \quad \mathbf{K} = \int \mathbf{B}^T \mathbf{D}\mathbf{B} \, \zeta \, ^r dV \tag{70}$$

Note that the coefficient matrix for the adjoint equation (68) is the same as for the analysis problem in Eq. (57). Note also that the initial stresses and strains contribute to the loads on the adjoint structure.

Discretization of the sensitivity Eq. (39) gives (after substitution from Eq. (40))

$$\bar{\delta}\psi = \bar{\bar{\delta}}\psi + \bar{\bar{\delta}}\int -\sigma_{ij}e^a_{ij} \, \zeta \, ^r dV + \bar{\bar{\delta}}\int f_i u^a_i \, \zeta \, ^r dV + \bar{\bar{\delta}}\int T^0_i u^a_i \, \zeta_s \, ^r dS_T + \bar{\bar{\delta}}\int T_i u^{a0}_i \, \zeta_s \, ^r dS_u \tag{71}$$

Substituting the finite element discretizations into Eq. (71), we get

$$\bar{\delta}\psi = \bar{\bar{\delta}}\psi + \mathbf{U}^{aT}\bar{\bar{\delta}}\int -\mathbf{B}^T \mathbf{D}\mathbf{B}\mathbf{U} \, \zeta \, ^r dV + \mathbf{U}^{aT} \bar{\bar{\delta}}\int \mathbf{N}^T \mathbf{f} \, \zeta \, ^r dV + \mathbf{U}^{aT} \bar{\bar{\delta}}\int \mathbf{N}^T \mathbf{T} \, \zeta_s \, ^r dS \tag{72}$$

Using Eqs. (57) to (59) in the foregoing equation, we get

$$\bar{\delta}\psi = \bar{\bar{\delta}}\psi + \mathbf{U}^{aT}\bar{\bar{\delta}}\mathbf{Q}, \quad \text{or} \quad \frac{D\psi}{Db} = \psi_{,b} + \mathbf{U}^{aT} \mathbf{Q}_{,b} \tag{73}$$

If **b** is a vector representing several design variables, then Eq. (73) can be used to obtain the design sensitivity vector as

$$\frac{D\psi}{Db} = \psi_{,b} + Q_{,b} \, U^a \tag{74}$$

Note that the sensitivity equation (73) involves partial derivatives of the external and internal forces with respect to the design variables.

6. Design Sensitivity Analysis: Discrete Models

The design sensitivity analysis with discrete models is based on the discretized equilibrium equation which is written using Eqs. (57) to (59) as

$$K(b) \, U = R(b) \tag{75}$$

The discretized form of the response functional is given as

$$\psi = \psi(b,U) \tag{76}$$

6.1 DIRECT DIFFERENTIATION METHOD

The total design derivative of the response functional in Eq. (76) is given as

$$\frac{D\psi}{Db} = \psi_{,b} + \frac{DU^T}{Db} \, \psi_{,U} \tag{77}$$

To calculate $\frac{DU}{Db}$, total derivative of the equilibrium equation (75) is taken as

$$K\frac{DU}{Db} = R_{,b} - (KU)_{,b} \tag{78}$$

Note that **U** is kept fixed in calculating $(KU)_{,b}$. Thus to calculate the design derivative of ψ, one calculates partial derivatives $\psi_{,U}$, $\psi_{,b}$, $R_{,b}$ and $(KU)_{,b}$ and then calculates $\frac{DU}{Db}$ from Eq. (78). Then use of Eq. (77) completes the calculations for the design derivative of the given response functional.

6.2 ADJOINT METHOD

The adjoint method for the discretized model can be derived by using the design sensitivity analysis principle (Arora and Cardoso 1992). To use the principle, we define the augmented functional using the yet to be determined adjoint variable U^a and the equilibrium Eq. (75) as

$$L = \psi(b,U) + U^{aT} (R - KU) \tag{79}$$

The sensitivity expression is given as the partial derivatives of L with respect to **b**, as

$$\frac{D\psi}{Db} = L_{,b}$$

$$= \psi_{,b} + [R_{,b} - (KU)_{,b}] U^a \qquad (80)$$

The adjoint vector is determined by requiring the partial derivative of the augmented functional in Eq. (79) with respect to **U** to vanish; i.e., $L_{,U} = 0$ which gives the following equation:

$$\psi_{,U} - K^T U^a = 0; \qquad \text{or} \qquad KU^a = \psi_{,U} \qquad (81)$$

Thus to calculate the design derivatives with the adjoint method, one calculates $\psi_{,U}$ first and then U^a from Eq. (81). The design derivative is then calculated from Eq. (80) by using all the partial derivatives.

7. Analysis of the Continuum and Discrete Approaches

It has been stated in the literature (Haug et al 1986; Choi and Seong 1986; Belegundu and Rajan 1988; Belegundu et al 1987) that the continuum approach of shape design sensitivity analysis does not need differentiation of the element matrices whereas the discrete approach needs this calculation. Also the two approaches have been shown to be different under certain conditions (Choi and Twu 1989). To study these aspects, let us analyze and compare design sensitivity expressions for the discrete model and the corresponding expressions obtained by discretizing the continuum expressions. First, we briefly compare the direct differentiation and adjoint methods, and then the continuum and discrete methods. Finally, the computer implementation aspects are discussed.

7.1 COMPARISON OF DIRECT DIFFERENTIATION AND ADJOINT METHODS

Comparing the sensitivity expressions for the adjoint and direct differentiation methods, it is observed that both the approaches need essentially the same quantities to complete calculations for the sensitivity coefficients. For example, comparing Eqs. (13) to (18) with Eq. (41) for the continuum formulation, we observe that both the approaches need exactly the same explicit design variations. The direct differentiation approach also needs design variations of the displacement field which are calculated from Eq. (34) and the adjoint method does not need them; it however needs the adjoint fields and their explicit design variations.

Comparing the discrete models, similar observations can be made about the direct differentiation and adjoint methods. Comparing Eqs. (77) and (78) with Eqs. (80) and (81), we observe that both the approaches need exactly the same partial derivatives. The direct differentiation approach needs design derivatives of the displacement vector whereas the adjoint method does not need them. The adjoint method needs to calculate the adjoint displacement vector by solving Eq. (81). It is well known that if the number of response functionals that need sensitivity analysis is less than the number of design variables then the adjoint method is more efficient otherwise the direct differentiation method is more efficient.

7.2 COMPARISON OF CONTINUUM AND DISCRETE METHODS

For the computer implementation of the continuum design sensitivity expressions, discretizations must be introduced. In Section 5, discretization of the continuum equations was presented using the isoparametric finite element formulation. In this case, the discretized continuum expressions are identical to those obtained with the discretized model; e.g., compare Eqs. (64) and (78) for the direct differentiation method, and Eqs. (74) and (80) for the adjoint method. Thus it is concluded that the two approaches lead to essentially the same design sensitivity expressions if same discretizations are used in both the approaches. It is important, however, to note that the continuum design sensitivity expression may use discretizations that are different from the ones used in the response analysis phase. This offers flexibility for computer implementations that is not obvious with the discrete approach.

Since the discretized sensitivity expressions for the two approaches are identical, it should be possible to implement them in an exactly the same way. The partial derivatives such as $R_{,b}$, $\psi_{,b}$, and $\psi_{,U}$ can be calculated easily in the two approaches because explicit dependence of R and ψ on b and U is known. The term $(KU)_{,b}$ which is the partial derivative of the internal force with respect to b is a major calculation needing further analysis and discussion. This term can also be expressed as $K_{,b}U$ which may be implemented with or without differentiating the element matrices. In the following paragraphs, we discuss these approaches.

The first approach that we consider is the one where the element matrices are explicitly differentiated with respect to the design variables. For some finite elements, such as the one dimensional elements, explicit expressions for K in terms of b are available. In such cases, one can differentiate K with respect to b to calculate the term $K_{,b}U$. In most cases, however, the explicit form for K in terms of b is not available and numerical integration may have to be used to evaluate the element matrices. In such cases we can proceed as follows. Using Eq. (59), the term $K_{,b}U$ can be written as

$$K_{,b}U = \int B_{,b}^T DBU \zeta^r dV + \int B^T DB_{,b}U \zeta^r dV + \int B^T DBU \zeta_{,b}^r dV \qquad (82)$$

Substitute $\zeta_{,b} = \zeta v_{k,k}$ (Arora, Lee and Cardoso 1992), where v_k is the design velocity field and $v_{k,k}$ is its divergence. Rearranging Eq. (82) after the substitution, we get

$$K_{,b}U = \left[\int (B_{,b}^T DB + B^T DB_{,b} + B^T DB \, v_{k,k}) \zeta^r dV \right] U \qquad (83)$$

The expression within the square brackets gives $K_{,b}$ which can be evaluated using numerical integration such as Gaussian quadrature rules. To implement the equation, the matrices B and $B_{,b}$, ζ and $v_{k,k}$ need to be known. The matrix B is available from the analysis phase, or it can be evaluated by appropriate differentiation of the shape functions N. The matrix $B_{,b}$ can be evaluated by differentiating B. Calculations for the velocity field v_k and $\zeta_{,b}$ or $v_{k,k}$ are explained later. Thus, Eq. (83) shows that for the approach where the element matrices are explicitly differentiated, the expression within the square brackets is evaluated first and then post multiplied by the displacement vector U to complete the calculation.

To develop the second way of calculating $K_{,b}U$, Eq. (83) can be written slightly differently as follows:

$$K_{,b}U = \int B_{,b}^T \sigma \zeta^r dV + \int B^T \bar{\sigma} \zeta^r dV + \int B^T \sigma \zeta v_{k,k}{}^r dV \qquad (84)$$

where we have used $\sigma = DBU$ and $\bar{\sigma} = DB_{,b}U$. Combining terms, this equation can be written as

$$K_{,b}U = \int P \zeta^r dV \qquad (85)$$

where

$$P = \left[B_{,b}^T \sigma + B^T(\bar{\sigma} + \sigma v_{k,k}) \right] \qquad (86)$$

Equation (85) shows that another way to calculate $K_{,b}U$ is to first calculate the vector P given in Eq. (86) and then carry out the integration. Since the Gaussian quadrature rules are usually used for numerical integration, the vector P needs to be evaluated at only the Gauss points. The data needed to evaluate P are: σ which is available from the analysis phase, D, U, B, $B_{,b}$ and $v_{k,k}$. This procedure then does not explicitly differentiate the element matrices.

The basic difference between the two foregoing numerical procedures to calculate $K_{,b}U$ given in Eqs. (83) and (85) now becomes transparent: It is the stages at which the displacement vector is used and the integrations are performed. In the first procedure, $K_{,b}$ is evaluated first by carrying out the integrations and then postmultiplied by U to evaluate the partial derivative of the internal forces at the node points. In the latter procedure, the order of using U and integrations is reversed, i.e., U is used first to evaluate the stresses σ and $\bar{\sigma}$ at the Gauss points, and then the integrations are performed to calculate the partial derivative $K_{,b}U$ of the node point internal forces KU. Both the procedures can be implemented inside or outside a finite element analysis program. A count of operations for some typical finite elements shows the second procedure to be requiring fewer numerical operations and fewer integrations. Therefore, that procedure is theoretically more efficient. In actual implementation with a finite element code, programming environment and the programming effort may dictate as to which procedure is more efficient and effective in an overall sense.

Choi and Twu (1989) have also analyzed the continuum and discrete design sensitivity analysis methods. The direct differentiation approach of design sensitivity analysis is used for the discrete model. The adjoint variable method based on the material derivative concept is used for the continuum formulation. Truss and beam finite elements are analyzed where length of the element is treated as a shape design variable. The response for both the approaches is calculated using the same finite element equilibrium equation. Three loading conditions are considered: point load, uniform load and linearly varying load. For the discrete approach, the design sensitivity coefficients are obtained by analytically differentiating the element stiffness matrices. For the continuum formulation, they are obtained by analytically integrating the sensitivity expression after the finite element solutions for the response and adjoint variables are substituted. Three different shapes for the design velocity field, linear, quadratic and cubic, are used as separate cases and the discrete and continuum sensitivity coefficients are compared. It is shown that if the design velocity field is taken as a linearly varying function, then the discrete and continuum approaches give identical sensitivities; otherwise, they are different in some cases. Based on these observations, it is concluded that the discrete and continuum design sensitivity analyses are not equivalent in general.

The present analysis shows that the continuum and discrete sensitivity expressions can be interpreted in such a way that they are equivalent and can be implemented in a similar way. Therefore it is not necessary to call them as two different methods. The most efficient approach to calculate the design sensitivity coefficients is to use the second approach described in the foregoing paragraphs, where the finite element matrices are not explicitly differentiated.

7.3 COMPUTER IMPLEMENTATION ASPECTS

To implement the foregoing procedure with a finite element program, the element external force generation and internal force recovery routines can be used to calculate the right hand side of the sensitivity equation (64) or $Q_{,b}$ in Eq. (73) (for implementation outside the program, these routines will have to be developed). For this calculation, the displacement that is calculated during the analysis phase is kept fixed as only the partial derivatives with respect to the design variables need to be calculated. Since analytical derivatives of B, ζ and ζ_s with respect to the design variables may be difficult (or impossible in some cases) to calculate for all types of finite elements, the central difference approach is suggested to calculate $R_{,b}$ and $F_{,b}$. This approach is quite simple and straightforward to implement, requiring minimal programming. In this procedure, to calculate $F_{,b}$ the node point forces need to be evaluated at a slightly changed design, keeping the node point displacements U unchanged. These forces should be calculated using the equation

$$KU \equiv F = \int B^T \sigma \zeta^r dV \qquad (87)$$

instead of Eq. (59) due to the reasons explained earlier. The central finite difference procedure is independent of the finite element type which can be implemented with minimal programming. Similar procedure is also suggested for calculating the explicit design derivatives of the node point external forces.

Note that to calculate $\zeta_{,b}$ and $\zeta_{s,b}$, the central difference approach suggested in the foregoing can also be used. For this calculation, the relative movement of the finite element mesh due to the change in a boundary node is needed; i.e., the design velocity field is needed to perform the foregoing calculation. The reason is that as the boundary of the structure is changed, the mapping to the reference domain $x_i = x_i(\xi,b)$ changes and new ζ needs to be calculated from Eq. (7). For the discretized model, relative movement of the nodes in the interior of the domain is needed for use in Eq. (54). The finite element mesh generator can be used to calculate this movement of the interior nodes. Or, unit force/unit displacement approach for an auxiliary structure can be used (Belegundu and Rajan 1988).

The foregoing approach has been successfully implemented for planar elements with an existing finite element analysis program. Also, analytical differentiation as well as central difference approach for calculation of $\zeta_{,b}$, $\zeta_{s,b}$, $B_{,b}$, $R_{,b}$ and $F_{,b}$ have been implemented. In some implementations, only the elements connected to the design node that was perturbed, were used in the calculation of sensitivities. In that case the design velocity field was not needed, therefore it was not calculated. All these implementations have given quite accurate sensitivity results (Ryu et al 1985; Haririan et al 1987; Wu and Arora 1987; Phelan and Haber 1989; Arora and Cardoso 1989; Arora 1991; Arora, Lee and Kumar 1992). It is important to note, however, that in the shape optimal design process, the finite element mesh can distort considerably, especially near the design boundary. Therefore, re-meshing of the structure has to be an integral part of the shape optimal design process.

8. Other Classes of Structural Design Problems

Only linearly elastic structures under static loading conditions are considered in the present paper. There are many other classes of structural design problems for which design sensitivity analysis has been developed and demonstrated in the literature. Most of the conclusions based on the analysis of the present paper can be applied to these classes of problems. These include nonlinear problems with geometric and material nonlinearities (Ryu et al 1985; Haririan et al 1987; Wu and Arora 1987; Choi and Santos 1987; Cardoso and Arora 1988; Mukherjee and Chandra 1989; Tsay and Arora 1990; Vidal et al 1991; Zhang et al 1992; Lee et al 1993, 1995; Poldneff et al 1993; Kamat 1993; Ohsaki and Arora 1994; Arora et al 1986, 1989, 1993; and many other articles cited in these references), transient dynamic problems (Hsieh and Arora 1984; Tortorelli et al 1990,1991; Cardoso and Arora 1992; and many other articles cited in these references) and control of structures (Tseng and Arora 1989; Holtz and Arora 1994).

9. Discussion and Conclusions

Continuum and discrete approaches for shape and nonshape design sensitivity analysis of linearly elastic structures are presented and analyzed. Both direct differentiation and adjoint methods are presented. General discretizations of the final design sensitivity expressions for the continuum approach are presented and discussed without reference to a particular type of finite element. Computer implementation aspects are discussed.

Based on the study the following general conclusions are stated:

1. Adjoint method is more efficient and effective if the number of response functionals is less than the number of design variables; otherwise the direct differentiation is more efficient. Both the methods have many calculations that are common.

2. An analysis of design sensitivity expressions obtained with the *continuum and discrete forms* of the model for the same problem shows that both the expressions can be interpreted and implemented in exactly the same way if same finite element discretization procedures are used for both the models. That is, the discrete sensitivity equations can be converted to the continuum form for numerical implementation purposes. This interpretation unifies the continuum and discrete methods of design sensitivity analysis. The continuum theory, however, gives certain *insights* that would not be possible with the discrete theory, i.e.,

   (i)   The discretizations for the analysis model and the design sensitivity analysis model can be different from each other, assuming that both discretizations satisfy appropriate accuracy requirements. This offers flexibilities when the sensitivity analysis is implemented outside an established finite element analysis program.

(ii) The design sensitivity expression with both the approaches can be implemented with or without the explicit differentiation of the element matrices. The procedure that does not evaluate derivatives of the element matrices explicitly needs fewer numerical operations and is recommended for numerical implementation.

3. The control volume approach with the continuum formulation and the discrete form of the design sensitivity analysis for shape and nonshape optimization problems have shown that for numerical implementation of the sensitivity expressions, one of the major calculations is to evaluate the *explicit design variations of the internal and external forces*. Analytical procedures can be used for this calculation. However, due to their complexity for general applications, a semi-analytical approach is suggested. In this approach, it is proposed to use a central finite difference procedure at the element level to evaluate the explicit design variations of the internal and external node point forces.

4. The proposed semi-analytical procedure of design sensitivity analysis can be implemented *inside a finite element program or completely outside* of it. Each approach has certain advantages and disadvantages. Implementation inside the program requires minimal programming effort, but complete knowledge of the source code (Haririan et al 1987; Wu and Arora 1987; Arora and Cardoso 1989; Arora et al 1992; Belegundu et al 1987; Poldneff et al 1993). Therefore it is recommended for use by the analysis code developers. Implementation outside the code requires substantial programming effort and may be tedious for nonlinear structural problems (Tsay and Arora 1990; Arora et al 1992), but it offers flexibility in terms of using the same sensitivity analysis package with several different analysis codes. In addition, with this approach, a restart capability is needed for the analysis code so that Eq. (64) or (68) can be solved without complete reanalysis of the structure.

10. References

Adelman, H.M. and Haftka, R.T. (1986), "Sensitivity Analysis of Discrete Structural Systems," *AIAA Journal* 24, 823-832.

Arora, J.S. (1991), "Structural Shape Optimization: An Implementable Algorithm," O. Ural and T-L. Wong (eds.), *Electronic Computation*, Proceedings of the Tenth Conference, American Society of Civil Engineers, New York, pp. 213-222.

Arora, J.S. (1993), "An Exposition of Material Derivative Approach for Structural Shape Sensitivity Analysis," *Computer Methods in Applied Mechanics and Engineering* 105, 41-62.

Arora, J.S. and Cardoso, J. B. (1986), "Design Sensitivity Analysis with Nonlinear Response," Proceedings of the *NASA Symposium on Sensitivity Analysis in Engineering*, NASA Langley Research Center, September 25-26, NASA CP-2457.

Arora, J.S. and Cardoso, J.B. (1989), "A Design Sensitivity Analysis Principle and Its Implementation into ADINA," *Computers and Structures* 32, 691-705.

Arora, J.S. and Lee, T.H. (1993), "Shape Sensitivity Analysis of Nonlinear Nonconservative Problems", in G. Rozvany (ed.),*Optimization of Large Structural Systems*, Proceedings of the NATO Advanced Study Institute, 23 September - 4 October, 1991, Berchtesgaden, Germany, Kluwer Academic Publishers, NATO ASI Series, Series E: Applied Sciences - Vol. 231, Boston, MA, pp. 345-361.

Arora, J.S., Lee, T.H. and Cardoso, J.B. (1992), "Structural Shape Sensitivity Analysis: Relationship between Material Derivative and Control Volume Approaches" *AIAA Journal* 30, 1638-1648.

Belegundu, A.D. and Rajan, S.D. (1988), "A Shape Optimization Approach Based on Natural Design Variables and Shape Functions," *Computer Methods in Applied Mechanics and Engineering* 66, 87-106.

Belegundu, A.D., Rajan, S.D., Choi, B.K. and Budiman, J. (1987), "Shape Optimal Design Using Natural Shape Functions," *Technical Report PSU-ME-86/87-0029*, Dept. of Mechanical Engineering, College of Engineering, The Pennsylvania State University, University Park, Pa 16802.

Bennett, J.A. and Botkin, M.E. (eds.) (1986), *The Optimum Shape: Automated Structural Design*, Proceedings of International Symposium, G.M. Research Labs, Warren, MI, September 30-October 1, 1985, published by Plenum Press, New York.

Cardoso, J.B. and Arora, J.S. (1988), "Variational Method for Design Sensitivity Analysis in Nonlinear Structural Mechanics," *AIAA Journal* 26, 595-603.

Cardoso, J.B. and Arora, J.S. (1992), "Design Sensitivity analysis of Nonlinear Dynamic Response of Structural and Mechanical Systems," *Structural Optimization*. 4, 37-46.

Choi, K.K. and Seong, H.G. (1986), "A Domain Method for Shape Design Sensitivity Analysis of Built-up Structures," *Computer Methods in Applied Mechanics and Engineering*, 57, 1-15.

Choi, K.K. and Santos, J.L.T. (1987), "Design Sensitivity Analysis of Nonlinear Structural Systems. Part I: General Theory," *International Journal for Numerical Methods in Engineering*, 24, 2039-2055.

Choi, K.K. and Twu, S-L. (1989), "Equivalence of Continuum and Discrete Methods of Shape Design Sensitivity Analysis," *AIAA Journal*, 27, 1418-1424.

Dems, K. and Haftka, R.T. (1989), "Two Approaches to Sensitivity Analysis for Shape Variation of Structures," *Mechanics of Structures and Machines*, 16, 501-522.

Dems, K. and Mróz, Z. (1984), "Variational Approach by Means of Adjoint Systems to Structural Optimization and Sensitivity Analysis - II. Structure Shape Variation," *International Journal of Solids and Structures*, 20, 527-552.

Haber, R.B. (1987), "A New Variational Approach to Structural Shape Design Sensitivity Analysis," in *Computer-Aided Optimal Design*, C.A. Mota Soares (ed.), Series F: Computer and System Sciences, Vol. 27, Springer-Verlag, New York, pp. 573-587.

Haftka, R.T. and Adelman, H.M. (1989), "Recent Developments in Structural Sensitivity Analysis," *Structural Optimization*, 1, 137-152.

Haririan, M., Cardoso, J.B. and Arora, J.S. (1987), "Use of ADINA for Design

Optimization of Nonlinear Structures," *Computers and Structures*, 26, 123-134.

Haug, E.J., Choi, K.K. and Komkov, V. (1986), *Design Sensitivity Analysis of Structural Systems*, Academic Press, Orlando, Fl.

Holtz, D.L. and Arora, J.S. (1994), "A Study of Large Scale Nonlinear Optimal Control," Proceedings of the 5th AIAA/NASA/USAF/ISSMO Symposium on Multidisciplinary Analysis and Optimization, Panama City, Florida, pp. 1459-1458.

Hou, J.W., Chen, J.L. and Sheen, J.S. (1986), "Computational Method for Optimization of Structural Shapes," *AIAA Journal*, 24, 1005-1012.

Hsieh, C.C. and Arora, J.S. (1984), "Design Sensitivity Analysis and Optimization of Dynamic Response," *Computer Methods in Applied Mechanics and Engineering*, 43, 195-219.

Kamat, M. (ed.) (1993), *Structural Optimization: Status & Review*, AIAA Series in Astronautics and Aeronautics, Vol. 150, American Institute of Aeronautics and Astronautics, Washington, D.C.

Lee, T.H. and Arora, J.S. (1995), "A Computational Method for Design Sensitivity Analysis of Elastoplastic Structures," *Computer Methods in Applied Mechanics and Engineering*, to appear .

Lee, T.H., Arora, J.S. and Kumar,V. (1993), "Shape Design Sensitivity Analysis of Viscoplastic Structures," *Computer Methods in Applied Mechanics and Engineering*, 108, 237-259.

Mota Soares, C.A. (ed.) (1987), *Computer-Aided Optimal Design*, Proceedings of a NATO Advanced Study Institute, Series F: Computer and System Sciences, Vol. 27, Springer-Verlag, New York.

Mukherjee, S. and Chandra, A. (1989), "A Boundary Element Formulation for Design Sensitivity Analysis in Materially Nonlinear Problems," *Acta Mechanica*, 78, 243-253.

Ohsaki, M. and Arora, J.S. (1994), "Design Sensitivity Analysis of Elasto-Plastic Structures," *International Journal for Numerical Methods in Engineering*, 37, 737-762.

Phelan, D.G. and Haber, R.B. (1989), "Sensitivity Analysis of Linear Elastic Systems Using Domain Parameterization and Mixed Mutual Energy Principle," *Computer Methods in Applied Mechanics and Engineering*, 77, 31-59.

Poldneff, M. J. and Arora, J. S. (1993), "Design Sensitivity Analysis of Coupled Thermoviscoelastic Systems," *Intern. Journal of Solids and Structures*, 30, 607-635.

Poldneff, M.J., Rai, I.S. and Arora, J.S. (1993), "Design Variations of Nonlinear Elastic Structures Subjected to Follower Forces," *Computer Methods in Applied Mechanics and Engineering*, 110, 211-219.

Rozvany, G.I.N., (ed.) (1993), *Optimization of Large Structural Systems*, Proceedings of the NATO Advanced Study Institute, 23 September - 4 October, 1991, Berchtesgaden, Germany, Kluwer Academic Publishers, NATO ASI Series, Series E: Applied Sciences - Vol. 231, Boston, MA.

Ryu, Y.S., Haririan, M., Wu, C.C. and Arora, J.S. (1985), "Structural Design Sensitivity Analysis of Nonlinear Response," *Computers and Structures*, 21, 245-255.

Tortorelli, D.A., Lu, S.C.Y. and Haber, R. (1990), "Design Sensitivity Analysis for Elastodynamic Systems," *Mechanics of Structures and Machines*, 18, 77-106.

Tortorelli, D.A., Haber, R. and Lu, S.C.Y. (1991), "Adjoint Sensitivity Analysis for Nonlinear Dynamic Thermoelastic Systems," *AIAA Journal*, 29, 253-263.

Tsay, J.J. and Arora, J.S. (1990), "Nonlinear Structural Design Sensitivity Analysis with Path Dependent Response. Part 1: General Theory," *Computer Methods in Applied Mechanics and Engineering*, 81, 183-208.

Tsay, J.J., Cardoso, J.B. and Arora, J.S. (1990), "Nonlinear Structural Design Sensitivity Analysis with Path Dependent Response. Part 2: Analytical Examples," *Computer Methods in Applied Mechanics and Engineering*, 81, 209-228.

Tseng, C-H. and Arora, J.S. (1989), "Optimum Design of Systems for Dynamics and Controls Using Sequential Quadratic Programming," *AIAA Journal*, 27, 1793-1800.

Vidal, C.A., Lee, H.S. and Haber, R.B. (1991), "A Consistent Tangent Operator for Design Sensitivity Analysis of History-Dependent Response," *Computer Systems in Engineering*, 2, 509-523.

Wu, C.C. and Arora, J.S. (1987), "Design Sensitivity Analysis of Nonlinear Response Using Incremental Procedure," *AIAA Journal*, 25, 1118-1125.

Yang, R.J. and Choi, K.K. (1985), "Accuracy of Finite Element Based Shape Design Sensitivity Analysis," *J. of Structural Mechanics*, 13, 223-229.

Zhang, Q., Mukherjee, S. and Chandra, A. (1992), "Shape Design Sensitivity Analysis for Geometrically and Materially Nonlinear Problems by the Boundary Element Method," International Journal of Solids and Structures, 29, 2503-2525.

Acknowledgment

This paper is based on a part of the research sponsored by the U.S. National Science Foundation under the project, "Design Sensitivity Analysis and Optimization of Nonlinear Structural Systems," Grant No. CMS 9301580.

A VIEW ON NONLINEAR OPTIMIZATION

JOSÉ HERSKOVITS
Mechanical Engineering Program
COPPE / Federal University of Rio de Janeiro[1]
Caixa Postal 68503, 21945-970 Rio de Janeiro, BRAZIL

1. Introduction

Once the *concepts* of a material object are established, the act of *designing* consists on choosing the values for the quantities that prescribe the object, or *dimensioning*. These quantities are called *Design Variables*. A particular value assumed by the design variables defines a *configuration*. The design must meet *Constraints* given by physical or others limitations. We have a feasible configuration if all the constraints are satisfied.

A better design is obtained if an appropriate cost or objective function can be reduced. The objective is a function of the design variables and quantifies the quality of the material object to be designed. The design is optimum when the cost is the lowest among all the feasible designs.

If we call $[x_1, x_2, ..., x_n]$ the design variables, $f(x)$ the objective function and Ω the feasible set, the optimization problem can be denoted

$$\text{minimize } f(x); \text{ subject to } x \in \Omega. \tag{1.1}$$

This problem is said to be a *Mathematical Program* and, the discipline that studies the numerical techniques to solve Problem (1.1), *Mathematical Programming*. Even if mathematical programs arise naturally in optimization problems related to a wide set of disciplines that employ mathematical models, several physical phenomena can be modeled by means of mathematical programs. This is the case when the "equilibrium" is attained at the minimum of an energy function.

[1]Partially written at INRIA, Institut National de Recherche en Informatique et en Automatique, Rocquencourt, France.

J. Herskovits (ed.), Advances in Structural Optimization, 71–116.
© 1995 *Kluwer Academic Publishers.*

The first stage to get an optimal design is to define the *Optimization Model*. That is, to select appropriate design variables, an objective function and the feasible set. In engineering design, we generally have

$$\Omega \equiv \{x \in R^n / g_i(x) \leq 0, i = 1, 2, ..., m; h_i(x) = 0, i = 1, 2, ..., p\},$$

where $x \equiv [x_1, x_2, ..., x_n]^t$ and g_i and h_i are the *Inequality* and *Equality Constraints* respectively. Since it involves nonlinear functions, Problem (1.1) is a *Nonlinear Program*. The optimization problem is said to be *Unconstrained* when $\Omega \equiv R^n$.

Mathematical Programming provides a general and flexible formulation for engineering design problems. Once the optimization model was established, nonlinear programming algorithms only require the computation of f, g_i, h_i and their derivatives at each iteration.

Nowadays, strong and efficient mathematical programming techniques for several kind of problems, based on solid theoretical results and extensive numerical studies, are available. Approximated functions, derivatives and optimal solutions can be employed together with optimization algorithms to reduce the computer time.

The aim of this paper is not to describe the state of the art of nonlinear programming, but to explain in a simple way most of the modern techniques applied in this discipline. With this objective, we include the corresponding algorithms in the framework of a general approach, based on Newton - like iterations for nonlinear systems. These iterations are used to find points verifying first order optimality conditions. Some favorable characteristics of optimization problem, conveniently explored, lead to strong algorithms with global convergence.

There exists a very large bibliography about nonlinear programming. We only mention the books by Bazaraa and Shetty [7], Dennis and Schnabel [10], Fletcher [14], Gill et al. [19], Hiriart - Urruty and Lemarechal [32], Luenberger [35], [36] and Minoux [39]. Several books written by engineering and/or structural designers include numerical optimization techniques in the framework of optimal design, such as Arora [1], Haftka et al. [20], Haug and Arora [22], Kirsch [34], Fox [15] and Vanderplaats [32].

We discuss some basic concepts of mathematical programming in the next section and, in the following one, optimality conditions are studied. A view of Newton - like algorithms for nonlinear systems is given in section 4. Unconstrained and equality constrained optimization techniques are studied in sections 5 and 6. Sequential Quadratic Programming method is discussed in section 7, where in addition to a classical algorithm based on this method, a feasible directions algorithm is presented. A Newton - like approach for interior points algorithms, proposed by the author [30], [31], is presented in the last section.

2. Some basic concepts

We deal with the nonlinear optimization problem

$$\left.\begin{array}{rl} \text{minimize} & f(x) \\ \text{subject to} & g_i(x) \leq 0; i = 1, 2, ..., m \\ \text{and} & h_i(x) = 0; i = 1, 2, ..., p, \end{array}\right\} \tag{2.1}$$

where f, g and h are smooth functions in R^n and at least one of these functions is nonlinear. Any inequality constraint is said to be *Active* if $g_i(x) = 0$ and *Inactive* if $g_i(x) < 0$. Denoting $g(x) = [g_1(x), g_2(x), ..., g_m(x)]^t$ and $h(x) = [h_1(x), h_2(x), ..., h_p(x)]^t$, we have

$$\left.\begin{array}{rl} \text{minimize} & f(x) \\ \text{subject to} & g(x) \leq 0 \\ \text{and} & h(x) = 0. \end{array}\right\} \tag{2.2}$$

We introduce now the auxiliary variables $\lambda \in R^m$ and $\mu \in R^p$, called *Dual Variables* or *Lagrange Multipliers* and define the *Lagrangian Function* associated with Problem (2.1) as

$$l(x, \lambda, \mu) \equiv f(x) + \lambda^t g(x) + \mu^t h(x).$$

Following, some definitions concerning Problem (2.1) and the methods to solve this problem are presented. First, the meaning of the statement of the optimization problem is discussed.

DEFINITION 2.1. A point $x^* \in \Omega$ is a *Local Minimum* (or *Relative Minimum*) of f over Ω if there exists a neighborhood $\Delta \equiv \{x \in \Omega / \parallel x - x^* \parallel \leq \delta\}$ such that $f(x) \geq f(x^*)$ for any $x \in \Delta$. If $f(x) > f(x^*)$ for any $x \in \Delta$, then x^* is a *Strict Local Minimum*.

DEFINITION 2.2. A point $x^* \in \Omega$ is a *Global Minimum* (or *Absolute Minimum*) of f over Ω if $f(x) \geq f(x^*)$ for any $x \in \Omega$.

Note that a global minimum is also a local minimum. The nature of optimization implies the search of the global minimum. Unfortunately, the global minimums can be characterized only in some particular cases, as in Convex Programming.

Usually Nonlinear Programming methods are iterative. Given an initial point x^0, a sequence of points, $\{x^k\}$, is obtained by repeated applications of an algorithmic rule. This sequence must converge to a solution x^* of the problem. The convergence is said to be asymptotic when the solution is not achieved after a finite number of iterations. Except in some particular cases, like linear or quadratic programming, such is the case in nonlinear optimization.

DEFINITION 2.3. An iterative algorithm is said to be *Globally Convergent* if for any initial point $x^0 \in R^n$ (or $x^0 \in \Omega$) it generates a sequence of points converging to a solution of the problem.

DEFINITION 2.4. An iterative algorithm is *Locally Convergent* if there exists a positive ρ such that for any initial point $x^0 \in R^n$ (or $x^0 \in \Omega$) verifying $\| x^0 - x^* \| \leq \rho$, it generates a sequence of points converging to a solution of the problem.

Modern Mathematical Programming techniques seek for globally convergent methods. Locally convergent algorithms are not useful in practice since the neighborhood of convergence is not known in advance.

The major objective of numerical techniques is to have strong methods with global convergence. Once this is obtained, engineers are worried with efficiency. In Design Optimization, evaluation of functions and derivatives generally takes more computer time than the internal computations of the algorithms itself. Then, the number of iterations gives a good idea of the computer time required to solve the problem. The following definition introduces a criterion to evaluate the speed of convergence of asymptotically convergent iterative methods.

DEFINITION 2.5. The *Order of Convergence* of a sequence $\{x^k\} \to x^*$ is the largest number p of the nonnegative numbers $\bar{p}$ satisfying

$$lim_{k \to \infty} \frac{\| x^{k+1} - x^* \|}{\| x^k - x^* \|^{\bar{p}}} = \beta < \infty.$$

When $p = 1$ we have *Linear Convergence* with *Convergence Ratio* $\beta < 1$. If $\beta = 0$ the convergence is said to be *Superlinear*. The convergence is *Quadratic* in the case when $p = 2$.

Since they involve the limit when $k \to \infty$, p and β are a measure of the asymptotic speed of convergence. Unfortunately, a sequence with a good order of convergence may be very "slow" far from the solution.

The convergence is faster when p is larger and β is smaller. Near the solution, if the convergence is linear the error is multiplied by β at each iteration, while the reduction is squared for quadratic convergence. The methods that will be studied here have rates varying between linear and quadratic.

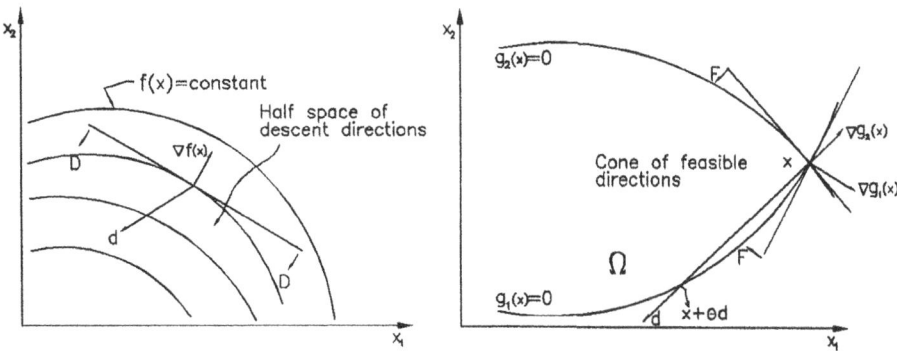

Figure 1. (a) Descent directions (left); (b) Feasible directions (right)

In general, globally convergent algorithms define at each point a search direction and look for a new point on that direction.

DEFINITION 2.6. A vector $d \in R^n$ is a *Descent Direction* of a real function f at $x \in R^n$ if there exists a $\delta > 0$ such that $f(x + td) < f(x)$ for any $t \in (0, \delta)$.

If f is differentiable at x and $d^t \nabla f(x) < 0$, it is easy to prove [7] that d is a descent direction of f.

In Figure 1.a, constant value contours of $f(x)$ are represented. We note that $f(x)$ decreases in any direction that makes an angle greater than 90 degrees with $\nabla f(x)$. The set of all descent directions constitutes the half space D.

DEFINITION 2.7. A vector $d \in R^n$ is a *Feasible Direction* of the problem (2.1), at $x \in \Omega$, if for some $\theta > 0$ we have $x + td \in \Omega$ for all $t \in [0, \theta]$.

In Figure 1.b the feasible region Ω of an inequality constrained problem is represented. The vector d is a feasible direction, since it supports a non zero segment $[x, x + \theta d]$. Any direction is feasible at an interior point. At the boundary, the feasible directions constitutes a cone F that we call *Cone of Feasible Directions*. This cone is not necessarily closed.

3. Optimality conditions

A first requirement to solve optimization problems is to characterize the solutions by conditions that are easy to verify. These conditions will be useful to identify a minimum point and, frequently, will be at the heart of numerical techniques to solve the problem.

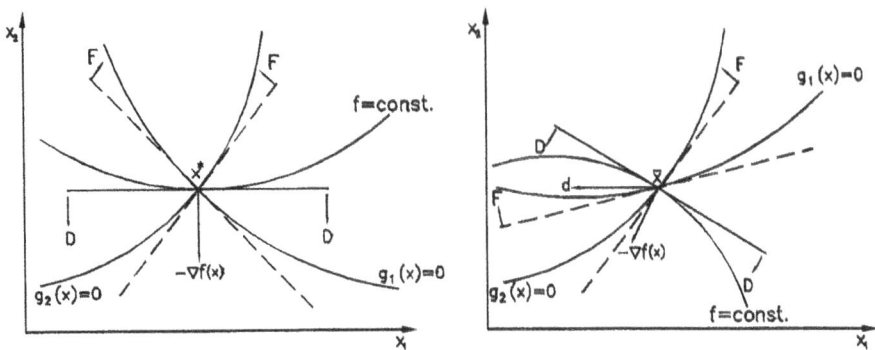

Figure 2. Illustration of the first order optimality condition

Optimality conditions are based on differential calculus, and they are said to be of first order if they involve only first derivatives. They are of second order if second derivatives are also required. In what follows we describe a series of optimality conditions for unconstrained and constrained problems based on Luenberger [36], where they are proved. Since differential calculus gives only a local information about the problem and we do not include convexity assumptions, only relative minima can be characterized.

All these conditions can be proved by considering that any feasible curve $x(t)$ passing through a local minimum x^* of problem (2.1) has a local minimum of $f[x(t)]$ at x^*. The results are obtained by applying optimality conditions of one dimensional problems to $f[x(t)]$. The following theorem gives a geometric interpretation of optimality conditions of a large class of problems.

THEOREM 3.1. *First and second order necessary conditions.* If $x^* \in \Omega$ is a local minimum of f over Ω then, for any feasible direction $d \in R^n$, it is

i) $d^t \nabla f(x^*) \geq 0$

ii) if $d^t \nabla f(x^*) = 0$, then $d^t \nabla^2 f(x^*) d \geq 0$ □

The first result means that every improving direction of f at x^* is not a feasible direction, that is $F \cap D \equiv 0$. This is the case in Figure 2, where x^* is a minimum while $\bar{x}$ is not. In fact, walking along d from $\bar{x}$, we can get a new feasible point with lower f.

In what follows, optimality conditions for unconstrained, equality constrained and inequality constrained optimization are discussed.

3.1. UNCONSTRAINED OPTIMIZATION

Since any $d \in R^n$ is a feasible direction, the well known optimality conditions for unconstrained optimization are easily derived from Theorem 3.1.

COROLLARY 3.2. First and second order necessary conditions. If x^* is a local minimum of f over R^n, then

i) $\nabla f(x^*) = 0$.

ii) for all $d \in R^n$, $d^t \nabla^2 f(x^*)d \geq 0$. That is, $\nabla^2 f(x^*)$ is positive semidefinite.
□

A sufficient local optimality condition for this problem is stated below. It requires the calculus of two derivatives of f.

THEOREM 3.3. Sufficient optimality conditions. Let f be a twice continuously differentiable scalar function in R^n and x^* such that

i) $\nabla f(x^*) = 0$

ii) $\nabla^2 f(x^*)$ is positive definite.

Then, x^* is a *strict local minimum point* of f.
□

3.2. EQUALITY CONSTRAINED OPTIMIZATION

We consider now the equality constrained problem

$$\left. \begin{array}{ll} \text{minimize} & f(x) \\ \text{subject to} & h(x) = 0 \end{array} \right\} \tag{3.1}$$

and introduce some definitions concerning the constraints of this problem.

DEFINITION 3.4. Let x be a point in Ω and consider all the continuously differentiable curves in Ω that pass through x. The collection of all the vectors tangent to these curves at x is said to be the *Tangent Set* to Ω at x.

DEFINITION 3.5. A point $x \in \Omega$ is a *Regular Point* of the constraints if the vectors $\nabla h_i(x)$, for $i = 1, 2, ..., p$, are linearly independent.

Regularity will be a requirement for most of the theoretical results and numerical methods in constrained optimization. At a regular point x, it is proved that

i) The *Tangent Set* to Ω at x constitutes a subspace , called *Tangent Space*.

ii) The tangent space at x is

$$T \equiv \{y/\nabla h^t(x)y = 0\}.$$

For a given constraint geometry, even in the case when the tangent set constitutes a subspace, a point may or may not be regular depending on how h is, defined [36]. The following optimality conditions are stated,

LEMMA 3.6. Let x^*, a regular point of the constraints h(x) = 0, be a *local minimum* of Problem (3.1). Then $\nabla f(x^*)$ is orthogonal to the tangent space. $\square$

Since $\nabla f(x^*)$ is orthogonal to all the constraints, it can be expressed as a linear combination of their gradients. Then, it gives the following result.

THEOREM 3.7. First Order Necessary Conditions. Let x^*, a regular point of the constraints $h(x) = 0$, be a *local minimum* of Problem (3.1). Then, there is a vector $\mu^* \in R^p$ such that

$$\nabla f(x^*) + \nabla h(x^*)\mu^* = 0$$
$$h(x^*) = 0$$

$\square$

This theorem is valid even in the trivial case when $p = n$. Introducing now the Lagrangian, we see that a feasible point satisfying the first order optimality conditions, can be obtained by solving the nonlinear system in (x, μ)

$$\left. \begin{array}{l} \nabla_x l(x, \mu) = 0 \\ \nabla_\mu l(x, \mu) = 0 \end{array} \right\} \tag{3.2}$$

Then, the Lagrangian plays a role similar to that of the objective function in unconstrained optimization. This is also true when we consider the following second order conditions.

THEOREM 3.8. Second Order Necessary Conditions. Let x^*, a regular point of the constraints $h(x) = 0$, be a *local minimum* of Problem (3.1).

Then there is a vector $\mu^* \in R^p$ such that the result of Theorem 3.7 is true and the matrix

$$H(x^*, \mu^*) = \nabla^2 f(x^*) + \sum_{i=1}^{p} \mu_i^* \nabla^2 h_i(x^*)$$

is positive semidefinite on the tangent space, that is, $y^t H(x^*, \mu^*) y \geq 0$ for all $y \in T$. $\qquad\square$

The matrix $H(x^*, \mu^*)$ plays a very important role in constrained optimization. When the constraints are linear, we have $H(x^*, \mu^*) = \nabla^2 f(x^*)$, and it follows from the previous theorem that $\nabla^2 f(x^*)$ is positive semidefinite on the space defined by the constraints. This is a natural result in view of Theorem 3.3. When there are nonlinear constraints, $H(x^*, \mu^*)$ takes into account their curvature.

THEOREM 3.9. Second Order Sufficiency Conditions. Let the point x^* satisfy $h(x^*) = 0$. Let $\mu^* \in R^p$ be a vector such that

$$\nabla f(x^*) + \nabla h(x^*) \mu^* = 0$$

and $H(x^*, \mu^*)$ be positive definite on the tangent space. Then x^* is a *Strict Local Minimum* of Problem 3.1. $\qquad\square$

3.3. INEQUALITY CONSTRAINED OPTIMIZATION

Let us consider the inequality constrained problem

$$\left. \begin{array}{ll} \text{minimize} & f(x) \\ \text{subject to} & g(x) \leq 0. \end{array} \right\} \tag{3.3}$$

We call $I(x) \equiv \{i / g_i(x) = 0\}$ the *Set of Active Constraints* at x and say that x is a *Regular Point* if the vectors $\nabla g_i(x)$ for $i \in I(x)$ are linearly independent. The *Number of Active Constraints* at x is $Card[I(x)]$.

It is easy to prove that, if $d^t \nabla g_i(x) < 0$ for $i \in I(x)$, then d is a feasible direction of the constraints at x.

Suppose now that x^*, a regular point, is a local minimum of Problem (3.3). It is clear that x^* is also a local minimum of $f(x)$ subject to $g_i(x) = 0$, for $i \in I(x^*)$. Then, it follows from Theorem 3.7 that there is a vector $\lambda^* \in R^m$ such that

$$\nabla f(x^*) + \nabla g(x^*) \lambda^* = 0, \tag{3.4}$$

where $\lambda_i^* = 0$ for $i \notin I(x^*)$.

The condition $\lambda_i^* = 0$ for $i \notin I(x^*)$ is called *Complementarity Condition* and can be represented by means of the following equalities

$$\lambda_i^* g_i(x^*) = 0; \text{ for } i = 1, 2, ..., m.$$

If we define $G(x) \equiv diag[g(x)]$, a diagonal matrix such that $G_{ii}(x) = g_i(x)$, the complementarity condition is expressed as

$$G(x^*)\lambda^* = 0.$$

An additional necessary condition,

$$\lambda^* \geq 0,$$

is obtained as a consequence of the first result of Theorem 3.1. In effect, let us assume that the condition is not true. Then, for some $l \in I(x^*)$, it is $\lambda_l^* < 0$. As x^* is a regular point, given $\delta_i < 0$, $i \in I(x^*)$, we can find a feasible direction d verifying $d^t \nabla g_i(x^*) = \delta_i$. It follows from (3.4) that

$$d^t \nabla f(x^*) = - \sum_{i \in I(x^*),\ i \neq l} \delta_i \lambda_i^* - \delta_l \lambda_l^*.$$

Taking now δ_i, for $i \in I(x^*)$ and $i \neq l$, small enough, we can get a feasible direction d such that $d^t \nabla f(x^*) < 0$. Then, d is a descent direction of f, but this conclusion is in contradiction with Theorem 3.1. These results constitute Karush - Kuhn - Tucker optimality conditions.

THEOREM 3.10. First Order Necessary Conditions. Let x^*, a regular point of the constraints $g(x) \leq 0$, be a *Local Minimum* of Problem (3.1). Then, there is a vector $\lambda^* \in R^m$ such that

$$\nabla f(x^*) + \nabla g(x^*)\lambda^* = 0$$
$$G(x^*)\lambda^* = 0$$
$$\lambda^* \geq 0$$
$$g(x^*) \leq 0.$$

□

In Figure 3 we have the convex cone F^*, defined by all the positive linear combinations of the gradients of the active constraints [7]. The previous theorem implies that, if x^* is a local minimum, then $-\nabla f(x^*) \in F^*$.

3.4. GENERAL CONSTRAINED OPTIMIZATION

The optimality conditions discussed above are easily generalized to optimization problem (2.1), with equality and inequality constraints.

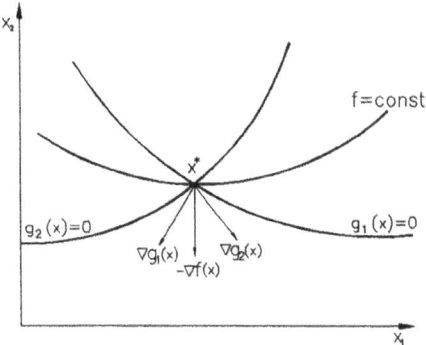

Figure 3. Illustration of Karush-Kuhn-Tucker conditions

DEFINITION 3.11. A point $x \in \Omega$ is a *Regular Point* of the constraints if the vectors $\nabla h_i(x)$, for $i = 1, 2, ..., p$, and $\nabla g_i(x)$ for $i \in I(x)$ are linearly independent. The tangent space at x is now

$$T \equiv \{y / \nabla g_i^t(x)y \text{ for } i \in I(x) \text{ and } \nabla h^t(x)y = 0\}.$$

THEOREM 3.12. *Karush - Kuhn - Tucker First Order Necessary Conditions.* Let x^*, a regular point of the constraints $g(x) \leq 0$ and $h(x) = 0$, be a *Local Minimum* of Problem (2.1). Then, there is a vector $\lambda^* \in R^m$ and a vector $\mu^* \in R^p$ such that

$$\nabla f(x^*) + \nabla g(x^*)\lambda^* + \nabla h(x^*)\mu^* = 0 \tag{3.5}$$

$$G(x^*)\lambda^* = 0 \tag{3.6}$$

$$h(x^*) = 0 \tag{3.7}$$

$$g(x^*) \leq 0 \tag{3.8}$$

$$\lambda^* \geq 0. \tag{3.9}$$

□

THEOREM 3.13. *Second Order Necessary Conditions.* Let x^*, a regular point of the constraints $g(x) \leq 0$ and $h(x) = 0$, be a *local minimum* of Problem (2.1). Then there is a vector $\lambda^* \in R^m$ and a vector $\mu^* \in R^p$ such that the result of Theorem 3.12 is true and the matrix

$$H(x^*, \lambda^*, \mu^*) = \nabla^2 f(x^*) + \sum_{i=1}^{m} \lambda_i^* \nabla^2 g_i(x^*) + \sum_{i=1}^{p} \mu_i^* \nabla^2 h_i(x^*)$$

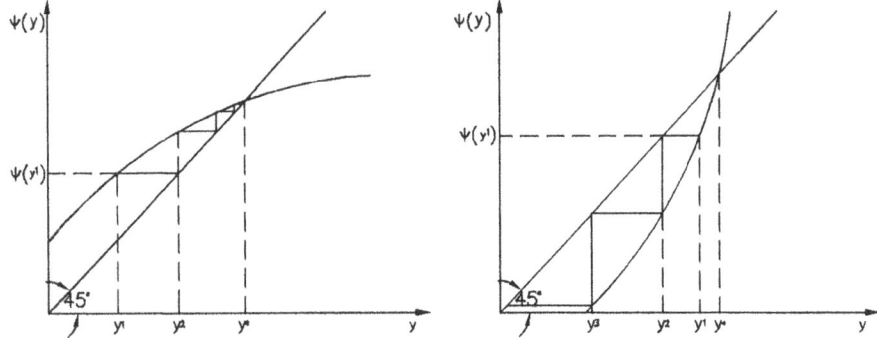

Figure 4. Iterations of successive approximations method

is positive semidefinite on the tangent space, that is, $y^t H(x^*, \lambda^*, \mu^*)y \geq 0$ for all $y \in T$. □

THEOREM 3.14. Second Order Sufficiency Conditions. Let the point x^* satisfy $g(x^*) \leq 0$ and $h(x^*) = 0$. Let there be a vector $\lambda^* \in R^m$, $\lambda^* \geq 0$ and a vector $\mu^* \in R^p$ such that

$$\nabla f(x^*) + \nabla g(x^*)\lambda^* + \nabla h(x^*)\mu^* = 0$$

and $H(x^*, \lambda^*, \mu^*)$ be positive definite on the tangent space. Then x^* is a *Strict Local Minimum* of Problem 2.1. □

4. Newton-like Algorithms for Nonlinear Systems

In this section we discuss about iterative methods for solving

$$\phi(y) = 0, \tag{4.1}$$

where $\phi : R^n \rightarrow R^n$ is continuously differentiable.

Let us write (4.1) as follows:

$$y = y - \phi(y). \tag{4.2}$$

To find a solution of (4.1) by *Successive Approximations* , we give an initial trial and make repeatedly substitutions in the right side of (4.2) obtaining the sequence

$$y^{k+1} = y^k - \phi(y^k). \tag{4.3}$$

Under appropriate conditions, this sequence converges to a solution of the system.

ALGORITHM 4.1 Successive Approximations Method

Data. Initial $y \in R^n$.

Step 1. Computation of the step d.

$$d = -\phi(y). \tag{4.4}$$

Step 2. Update.

Set

$$y := y + d.$$

Step 3. Go back to *Step 1.* □

The above algorithm is said to be a *Fixed Point Algorithm* because, if a solution is attained, then $d = 0$ and the rest of the sequence stays unchanged.

We define now the function $\psi(y) \equiv y - \phi(y)$. The theorem that follows gives conditions for global convergence and results about the speed of convergence of Algorithm 4.1. These results are proved in [35].

THEOREM 4.1. Let be $\| \nabla\psi(y) \| \leq \alpha < 1$ on R^n, then there is a unique solution y^* of (4.1) and the sequence generated by *Successive Approximations Method* converges to y^* for any initial $y^0 \in R^n$. The order of convergence is linear and the rate is equal to α. □

The assumptions of the previous theorem restrict the application of successive approximations method to a particular class of problems. In Figure 4 the process of solving a one-dimensional equation is illustrated. This is equivalent to find the point of intersection of $z = \psi(y)$ with $z = y$. In Figure 4.a the slope of $\psi(y)$ is less than the unity and the iterates converge, while in 4.b the process diverges. Since the order of convergence is linear, Algorithm 4.1 is also called *Linear Iterations Algorithm*.

Let y^k be an estimate of y^*. In a neighborhood of y^k, we have

$$\phi(y) \approx \phi(y^k) + \nabla\phi(y^k)^t(y - y^k).$$

Then, a better estimate y^{k+1} can obtained by making

$$\phi(y^k) + \nabla\phi(y^k)^t(y^{k+1} - y^k) = 0, \tag{4.5}$$

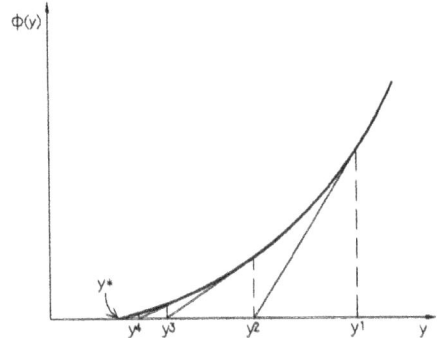

Figure 5. Newton's iterations

which defines *Newton's Method* iteration.

ALGORITHM 4.2 Newton's Method

Data. Initial $y \in R^n$

Step 1. Computation of the step d.

Solve the linear system for d

$$[\nabla \phi(y)]^t d = -\phi(y). \tag{4.6}$$

Step 2. Update.

Set

$$y := y + d.$$

Step 3. Go back to *Step 1.*   □

The process of Newton's method is illustrated in Figure 5. At a given estimate of y^* the function is approximated by its tangent. A new estimate is then taken at the point where the tangent crosses the y axis.

The theorem below gives the conditions for local convergence and studies the speed of convergence. It can be proved in a similar way as in [9], [35] and [36].

THEOREM 4.2. Let $\phi(y)$ be twice continuously differentiable and y^* a solution of $\phi(y) = 0$. Assume that $\nabla \phi(y^*)^{-1}$ exists. Then if started close enough from y^*, Algorithm 4.2 doesn't fail and it generates a sequence that converges to y^*. The convergence is at least quadratic. □

The algorithm doesn't fail if (4.6) has a unique solution. This method's major advantage comes from its speed of convergence. However it requires the evaluation of the Jacobian $\nabla\phi$ and the solution of a linear system at each iteration, which can be very expensive in terms of computer effort. Moreover, global convergence is not assured. The analytic Jacobian can be replaced by a finite - difference approximation, but this is also costly since n additional evaluations of the function per iteration are required.

With the objective of reducing computational effort, quasi - Newton method generates an approximation of the Jacobian or of its inverse. The basic idea of most quasi - Newton techniques is to try to construct an approximation of the Jacobian, or of its inverse, using information gathered as the iterates progress. Let be B^k the current approximation of $\nabla\phi(y^k)$, a new approximation B^{k+1} is obtained from

$$B^{k+1} = B^k + \Delta B^k. \tag{4.7}$$

Since

$$\phi(y^{k+1}) - \phi(y^k) \approx \nabla\phi(y^k)^t(y^{k+1} - y^k),$$

ΔB^k is defined in such a way that

$$\phi(y^{k+1}) - \phi(y^k) = [B^{k+1}]^t(y^{k+1} - y^k). \tag{4.8}$$

Substitution of (4.7) in (4.8) gives n conditions to be satisfied by ΔB^k. Since ΔB^k has n^2 elements, these conditions are not enough to determine it. Several updating rules for B^{k+1} were proposed [10], *Broyden's Rule* being the most successful,

$$B^{k+1} = B^k + (\gamma - B^k\delta)\delta^t/\delta^t\delta, \tag{4.9}$$

where $\delta = y^{k+1} - y^k$ and $\gamma = \phi(y^{k+1}) - \phi(y^k)$.

ALGORITHM 4.3 Quasi - Newton Method

Data. Initial $y \in R^n$ and $B \in R^{n \times n}$

Step 1. Computation of the step d.

Solve the linear system for d

$$B^t d = -\phi(y). \tag{4.10}$$

Step 2. Update.

(i) Set

$$y := y + d$$

and

$$B := B + \Delta B.$$

Step 3. Go back to *Step 1.*   □

In Step 2, B can be updated using (4.9) or other rules. The following theorem is proved in [10].

THEOREM 4.3. Let $\phi(y)$ be twice continuously differentiable and y^* a solution of $\phi(y) = 0$. Assume that $\nabla\phi(y^*)^{-1}$ exists. Then, if started close enough to y^* and the initial B is close enough from $\nabla\phi(y^*)$, Algorithm 4.3 doesn't fail and it generates a sequence that converges to y^*. The convergence is superlinear.   □

Although quasi - Newton methods have the advantage of avoiding the computation of $\nabla\phi(y)$, the initial B must be a good approximation of $\nabla\phi(y)$ to have local convergence.

Looking to Algorithms 4.1 to 4.3 we note that they have a similar structure. All of them define a step by the expression

$$Sd = -\phi(y),$$

where $S \equiv I$ in the linear iterations, S is an approximation of $\nabla\phi(y)$ in quasi - Newton methods or $S \equiv \nabla\phi(y)$ in Newton's method. The rate of convergence goes from linear to quadratic. We call this kind of iterations Newton - like algorithms.

5. Unconstrained Optimization

Let us consider the unconstrained optimization problem

$$\text{minimize } f(x); x \in R^n. \tag{5.1}$$

According to Corollary 3.3, a local minimum x is a solution of the system of equations

$$\nabla f(x) = 0. \tag{5.2}$$

In this section we show that the best known techniques for unconstrained optimization can be obtained by applying the Newton - like algorithms studied above to solve (5.2). Some favorable characteristics of optimization problems, conveniently explored, lead to globally convergent algorithms.

This system is generally nonlinear in x, being linear only when f is quadratic. The Jacobian $\nabla^2 f(x)$ is symmetric. As a consequence of second order optimality conditions $\nabla^2 f(x)$ is positive definite, or at least positive semidefinite, at a local minimum.

Following the techniques discussed in the previous section, (5.2) can be solved using Newton - like iterations. The change of x, called d, is given now by the expression

$$Sd = -\nabla f(x), \tag{5.3}$$

where S can be taken equal to the identity, to $\nabla^2 f(x)$ or to a quasi - Newton approximation of $\nabla^2 f(x)$.

In general this procedure is not globally convergent. To get global convergence each iterate must be nearer the solution than the previous one. In unconstrained optimization this happens if the function is reduced at each new point.

If S is positive definite, d is a descent direction of f at x. In effect, it follows from (5.3) that $d^t \nabla f(x) = -d^t S d$. Then, $d^t \nabla f(x) < 0$. This result means that d points towards the lower values of the function, but it does not imply that $f(x + d) < f(x)$.

Since d is a descent direction, we can find a new point $x + td$ in a way to have a satisfactory reduction of f. In this case, d is called the *Search Direction* and the positive number t, the *Step Length*. The procedure to find t is called the *Line Search*. Different *Line Search Criteria* can be adopted to decide whether the step length is adequate or not.

As the search direction is zero only at a solution of (5.2), the line search cannot allow step lengths that are null or that go to zero. Otherwise premature convergence to points that are not a solution would be obtained.

The algorithm that follows is based on these ideas. It is globally convergent to points satisfying first order optimality conditions.

ALGORITHM 5.1 A Basic Algorithm for Unconstrained Optimization

Data. Initialize $x \in R^n$ and $S \in R^{n \times n}$ symmetric and positive definite.

Step 1. Computation of the search direction d.

Solve the linear system for d

$$Sd = -\nabla f(x). \tag{5.4}$$

Step 2. Line search.

Find a step length t satisfying a given line search criterium.

Step 3. Update.

Set

$$x := x + td$$

and define a new $S \in R^{n \times n}$ symmetric and positive definite.

Step 4. Go back to *Step 1.*   □

Particular versions of this algorithm can be obtained by choosing S and a line search procedure. The best alternative depends on the problem to be solved, the available information about f and the desired speed of convergence. Even if far from a local minimum $\nabla^2 f(x)$ is not necessarily positive definite, to get descent search directions, S must be taken positive definite. Moreover, we make the following assumption about S.

ASSUMPTION 5.1. There exists positive numbers σ_1 and σ_2 such that

$$\sigma_1 \| d \|^2 \leq d^t S d \leq \sigma_2 \| d \|^2$$

for any $d \in R^n$.

5.1. ABOUT THE LINE SEARCH

At a given point x, once the search direction d is defined, $f(x + td)$ becomes a function of the single variable t. The first idea is to walk on d until the minimum on that direction is reached; that is, to find t that minimizes $f(x + td)$. This procedure is called *Line Search by Exact Minimization,* even though in practice the exact minimum can be rarely obtained. Exact minimization is done in an iterative way and is very costly.

Modern algorithms include *Inaccurate Line Search Criteria* that also ensure global convergence [10], [32], [36]. The line search can be completed in a few number of iterations. A very simple procedure is due to Armijo.

Armijo's Line Search

Define the step length t as the first number of the sequence $\{1, \nu, \nu^2, \nu^3, ...\}$ satisfying

$$f(x + td) \leq f(x) + t\eta_1 \nabla f^t(x)d, \tag{5.5}$$

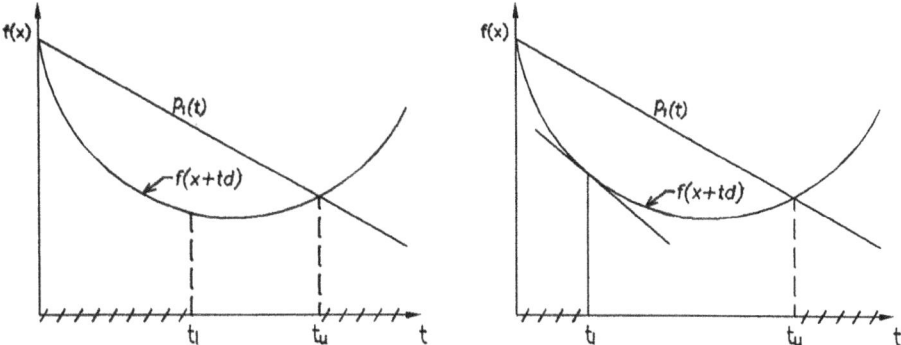

Figure 6. (a) Armijo's search (left);(b) Wolfe's criterion (right)

where $\eta_1 \in (0,1)$ and $\nu \in (0,1)$. □

Condition (5.5) on t ensures that the function is reduced at least η_1 times the reduction of a linear function, tangent to f at $t = 0$. In Figure 6.a we have $p_1(t) = f(x) + t\eta_1 \nabla f^t(x)d$. The acceptable step is in $[t_l, t_u]$. It is easy to deduce that $t_l = \inf(1, \nu t_u)$.

Wolfe's inaccurate line search criterion also establishes bounds on the step length by requiring a reduction of the function and, at the same time, a reduction of its directional derivative.

Wolfe's Criterion

Accept a step length t if (5.5) is true and

$$\nabla f^t(x + td)d \geq \eta_2 \nabla f^t(x)d \tag{5.6}$$

where $\eta_1 \in (0, 1/2)$ and $\eta_2 \in (\eta_1, 1)$. □

This criterion is illustrated in Figure 6.b, where the slope is $\eta_2 \nabla f^t(x)d$ at t_l. Condition (5.5) defines an upper bound on the step length and (5.6) a lower bound. Then, a step t is too long if (5.5) is false and too short if (5.6) is false.

A step length satisfying Wolfe's criterion can be obtained iteratively [32]. Given an initial t, if it is too short, extrapolations are done until a good or a too long step is obtained. If a too long step was already obtained, interpolations based on the longest short step and the shortest long step are done, until the criterion is satisfied. Since the function and the directional derivative are evaluated for each new t, cubic interpolations of f can be done. As the criterion of acceptance is quite wide, the process generally requires very few iterations.

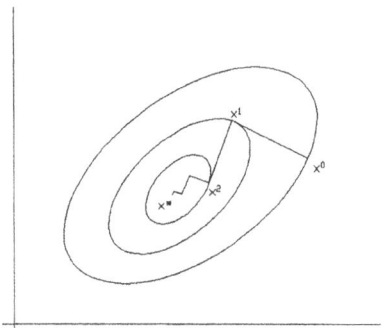

Figure 7. Steepest Descent iterates

The number of iterations required by Armijo's line search is usually greater than using Wolfe's criterion, but it is simple to code and it does not require calculation of ∇f. *Goldstein's Criterion*, described in [36], also does not require ∇f. Very efficient line search algorithms can be obtained by combining polynomial interpolations with Goldstein's criterion.

5.2. CONSIDERATIONS ABOUT GLOBAL CONVERGENCE

Global convergence of Algorithm 5.1 can be proved if Assumption 5.1 is true and any of the previously discussed line search criteria is adopted. We are not going to present the proof in this paper. In the case of Armijo's search this result is a particular case of Theorem 4.7 in [40].

5.3. FIRST ORDER ALGORITHMS

Taking $S \equiv I$ in Algorithm 5.1, we have $d = -\nabla f(x)$. That is, the search direction is opposite to the gradient. It is easy to prove that the downhill slope of f is maximum in the direction of d.

One of the most widely known methods for unconstrained optimization is the *Steepest Descent Algorithm*, that includes an exact minimization in the line search. In Figure 7 we illustrate the iterative process to solve a two dimensional problem described by some level lines $f(x) = constant$. Since the search directions are normal to the level line at the current point and $\nabla f^t(x + td)\nabla f(x) = 0$ in the line search, each direction is orthogonal to the previous one.

Modern first order algorithms include inaccurate line search procedures instead of the exact minimization. Although the number of iterations is not smaller, there is generally reduction in the overall computer time.

Let r be the *Condition Number* of $\nabla^2 f(x^*)$, defined as the ratio of the largest to the smallest eigenvalue. If the steepest descent algorithm generates a sequence $\{x^k\}$ converging to x^*, then *the sequence of objective values $f(x^k)$ converges linearly to $f(x^*)$ with a convergence ratio no greater than $\beta = \left(\frac{r-1}{r+1}\right)^2$*. This result is proved in refs. [19] and [36]. It implies that the convergence is slower as the conditioning of $\nabla^2 f(x^*)$ becomes worse.

5.4. NEWTON'S METHOD

Newton's algorithm is obtained by taking $S \equiv \nabla^2 f(x)$. To have descent search directions, S must be positive definite. This is not necessarily true at any point even though, according to Theorem 3.3, $\nabla^2 f(x)$ is positive definite at a strict local minimum.

It follows from Theorem 4.2 that, if $f(x)$ is three times continuously differentiable at x^* and there exists $K > 0$ such that $t^k = 1$ for $k > K$, then the convergence of Algorithm 5.1 with $S \equiv \nabla^2 f(x)$ is at least quadratic. When Armijo's line search or Wolfe's criterion is adopted, since $\nabla^2 f(x^*)$ is positive definite, it is easy to prove that a unit step length can be obtained near the solution. This is a requirement of quasi - Newton and Newton algorithms to have superlinear and quadratic convergence.

To make S positive definite, Newton method is modified by taking $S \equiv \nabla^2 f(x) + \epsilon I$, where $\epsilon > 0$ is large enough to satisfy Assumption 5.1 and $\epsilon \to 0$ [10], [19], [36]. Since the search direction is now a combination of steepest descent and Newton's directions, the speed of convergence is not as good as Newton's method. This approach is known as Levenberg - Marquardt method. The major difficulty is to determine ϵ, that is not too large, in a way to perturb Newton's iteration as little as possible.

5.5. QUASI-NEWTON METHOD

We define S as B, a quasi Newton approximation matrix for $\nabla^2 f(x)$ [9]. Since the Hessian is symmetric, it is reasonable to generate B symmetric. Then, given an initial symmetric B, we need symmetric ΔB. *Rank One Updating Rule* adds to B a rank one matrix of the form

$$\Delta B = \alpha z z^t,$$

where α is a number and z a vector in R^n. By imposing (4.8) the following rule is obtained [36],

$$B^{k+1} = B^k + \frac{(\gamma - B^k \delta)(\gamma - B^k \delta)^t}{\delta^t(\gamma - B^k \delta)}, \tag{5.7}$$

where now it is $\delta = x^{k+1} - x^k$ and $\gamma = \nabla f(x^{k+1}) - \nabla f(x^k)$.

To obtain descent directions in Algorithm 5.1, B is required to be positive definite. If B^k is positive definite, we need $\delta^t(\gamma - B^k \delta) > 0$ to have

B^{k+1} positive definite. Unfortunatelly this is not always true. Several *Rank Two Updating Rules* overcome this problem. This is the case of Broyden - Fletcher - Shanno - Goldfarb (BFGS) formulae

$$B^{k+1} = B^k + \frac{\gamma\gamma^t}{\delta^t\gamma} - \frac{B^k\delta\delta^t B^k}{\delta^t B^k \delta}. \tag{5.8}$$

It can be proved that, if B^k is positive definite, then

$$\delta^t\gamma > 0 \tag{5.9}$$

is a sufficient condition to have B^{k+1} positive definite. It can be easily shown that this condition is automatically satisfied if the line search in Algorithm 5.1 is an exact minimization or if Wolfe's criterion is adopted. As a consequence of Theorem 4.3, the convergence is superlinear. A unit step length near the solution is also required.

6. Equality Constrained Optimization

The techniques for unconstrained optimization studied above can be extended to the problem

$$\left.\begin{array}{l} \text{minimize} \quad f(x) \\ \text{subject to} \quad h(x) = 0. \end{array}\right\} \tag{6.1}$$

The pair (x^*, μ^*), satisfying first order optimality conditions, is obtained by solving with Newton - like iterations the nonlinear system

$$\nabla f(x) + \nabla h(x)\mu = 0 \tag{6.2}$$

$$h(x) = 0, \tag{6.3}$$

where the unknowns x and μ are called *Primal and Dual Variables* respectively. To have μ unique in (6.2), x must be a regular point of the problem. Due to this fact, the algorithms that solve first order optimality conditions, require the assumption that all the iterates x^k are regular points of the problem.

Denoting $y \equiv (x, \mu)$ and

$$\phi(y) \equiv \left[\begin{array}{c} \nabla f(x) + \nabla h(x)\mu \\ h(x) \end{array} \right], \tag{6.4}$$

we have

$$\nabla\phi(y) \equiv \left[\begin{array}{cc} H(x, \mu) & \nabla h^t(x) \\ \nabla h(x) & 0 \end{array} \right], \tag{6.5}$$

where $H(x,\mu) = \nabla^2 f(x) + \sum_{i=1}^{p}\mu_i\nabla^2 h_i(x)$ is the Hessian of the Lagrangian defined in Theorem 3.8.

A Newton's iteration that starts at (x^k,μ^k) and gives a new estimate (x^{k+1},μ^{k+1}) is then stated as follows:

$$
\begin{bmatrix} H(x^k,\mu^k) & \nabla h(x^k) \\ \nabla h^t(x^k) & 0 \end{bmatrix} \begin{bmatrix} x^{k+1}-x^k \\ \mu^{k+1}-\mu^k \end{bmatrix} = - \begin{bmatrix} \nabla f(x^k)+\nabla h(x^k)\mu^k \\ h(x^k) \end{bmatrix}.
$$
(6.6)

As the system (6.4) is linear in μ, when x^* is known Newton's method gives μ^* in one iteration. In fact, taking $x^k = x^*$ in (6.6), we have that $(x^{k+1},\mu^{k+1}) = (x^*,\mu^*)$ for any μ^k. We conclude that $\mu^{k+1} \to \mu^*$ when $\{x^k\} \to x^*$. This remark suggests that a line search concerning only the primal variables x is enough to get global convergence in (x,μ).

In a similar way as in Section 4, $\nabla\phi$ can be substituted by a quasi - Newton approximation or by the identity matrix. Since anyway, in the present problem we need ∇h to evaluate ϕ, it seems more efficient to substitute only $H(x^k,\mu^k)$ in $\nabla\phi$ by it's quasi Newton approximation or by the identity. Calling S^k this matrix and d^k the change in x, we have the linear system of equations

$$
S^k d^k + \nabla h(x^k)\mu^{k+1} = -\nabla f(x^k) \tag{6.7}
$$

$$
\nabla h^t(x^k)d^k = -h(x^k) \tag{6.8}
$$

that gives d^k, a search direction in x, and μ^{k+1}, a new estimate of μ. If x^k is a regular point and S^k is positive definite, then it can be proved that the solution of (6.7), (6.8) is unique.

In unconstrained optimization we know that a minimum is approached as the objective function decreases. This is not always true in constrained minimization since an increase of the function can be necessary to obtain feasibility. Then, an appropriate objective for the line search is needed.

With this purpose we define the auxiliary function

$$
\phi(x,r) = f(x) + \sum_{i=1}^{p}r_i\|h_i(x)\|, \tag{6.9}
$$

where r_i are positive constants. It can be shown that $\phi(x,r)$ is an Exact Penalty Function of the equality constrained problem if r_i are large enough [36]. In other words, there exists a finite $\bar{r}$ such that, for $r_i \geq \bar{r}_i$, the unconstrained minimum of $\phi(x,r)$ occurs at the solution of the problem (6.1). When compared with others penalty functions, ϕ is numerically very advantageous since it does not require penalty parameters going to infinite [35]. However, ϕ is not differentiable at points on the constraints, requiring nonsmooth optimization techniques [32].

Denoting by $SG(h)$ a diagonal matrix such that $SG_{ii}(h) \equiv sg(h_i)$, where $sg(.) = (.)/|(.)|$ and $sg(0) = 0$, we can write

$$\phi(x,r) = f(x) + r^t SG[h(x)]h(x). \tag{6.10}$$

Suppose now that x^k is a regular point, S is positive definite and

$$r_i \geq \|\mu_i^{k+1}\|; i = 1, 2, ..., p. \tag{6.11}$$

Then, d^k given by (6.7) and (6.8) is a descent direction of $\phi(x,r)$ at x^k. In effect, it exists $\tau > 0$ such that $sg[h_i(x^k + td^k)]$ doesn't change for any $t \in (0, \tau]$. Calling $\Delta\phi(x^k, td^k) \equiv \phi(x^k + td^k, r) - \phi(x^k, r)$, we have

$$\Delta\phi(x^k, td^k) = td^{kt}\nabla f(x^k) + td^{kt}\nabla h(x^k)SG[h(x^k + td^k)]r + o(t),$$

for any $t \in (0, \tau]$, where $o(t) \to 0$ faster than t. As a consequence of (6,7) and (6,8),

$$d^{kt}\nabla f(x^k) = -d^{kt}S^k d^k + h^t(x^k)\mu^{k+1}. \tag{6.12}$$

Then, considering (6.8) and (6.12), we get

$$\Delta\phi(x^k, td^k) = -td^{kt}S^k d^k + th^t(x^k)\{\mu^{k+1} - SG[h(x^k + td^k)]r\} + o(t)$$

and there exists $\delta \in (0, \tau]$ such that $\phi(x^k + td^k, r) < \phi(x^k, r)$ for any $t \in [0, \delta)$, what proves the assertion above.

The following globally convergent algorithm takes $\phi(x, r)$ as the objective of a line search along d^k.

ALGORITHM 6.1 A Basic Algorithm for Equality Constrained Optimization

Data. Initial $x \in R^n$ and $S \in R^{n \times n}$ symmetric and positive definite and $r \in R^p$, $r > 0$.

Step 1. Computation of the search direction d and an estimate of the Lagrange multipliers μ.

Solve the linear system in (d, μ)

$$Sd + \nabla h(x)\mu = -\nabla f(x) \tag{6.13}$$

$$\nabla h^t(x)d = -h(x) \tag{6.14}$$

Step 2. Line search.

i) If $r_i \leq \|\mu_i\|$, then set $r_i > \|\mu_i\|$, for $i = 1, 2, ..., p$.

ii) Find a step length t satisfying a given line search criterion on the auxiliary function

$$\phi(x, r) = f(x) + r^t SG[h(x)]h(x).$$

Step 3. Update.

Set

$$x := x + td$$

and define a new $S \in R^{n \times n}$ symmetric and positive definite.

Step 4. Go back to *Step 1.* □

Assumption 5.1 is also adopted in this algorithm. The same line search procedures as in Algorithm 5.1 for unconstrained optimization can be employed. However some precautions must be taken here since ϕ is nonsmooth [32].

6.1. FIRST ORDER METHOD

First order algorithms are obtained by taking $S \equiv I$ in Algorithm 6.1. In the case when the constraints are linear, a natural extension of gradient methods to equality constrained optimization consists on taking an initial point on the constraints and a search direction obtained by projecting $-\nabla f(x^k)$ on the constraints. This direction is known as the *Projected Gradient Direction* [36], [43] and denoted here by d_π. A new point on the constraints is then obtained.

We have that $-\nabla f(x)$ can be written as the sum of its projection on the tangent space and a vector orthogonal to all the constraints. Then, if the constraints are regular, there exists $\mu \in R^p$ such that

$$-\nabla f(x) = d_\pi + \nabla h(x)\mu \tag{6.15}$$

and, since d_π is orthogonal to the set $\{\nabla h_1, \nabla h_2, ..., \nabla h_p\}$, we have

$$\nabla h^t(x)d_\pi = 0. \tag{6.16}$$

It can be concluded that taking $S \equiv I$ in Algorithm 6.1, the search direction at points where $h(x) = 0$ becomes the Projected Gradient.

The results about the speed of convergence of gradient algorithms for unconstrained optimization are easily extended to the projected gradient method. The convergence is linear and the convergence ratio no greater than $\beta = \left(\frac{r-1}{r+1}\right)^2$, where now r is the ratio of the largest to the smallest eigenvalue of $\nabla^2 f(x^*)$ as measured on the constraints. This is not surprising since the iterates "walk" on the constraints.

The Projected Gradient Method was extended by Rosen [20], [44] to problems with nonlinear constraints. Given a point x^k on the constraints, a better point $\bar{x}^k$ on the projected gradient is obtained. Since the projected gradient no longer follows the constraints surface, $\bar{x}^k$ is generally infeasible. The new iterate x^{k+1} is the projection of $\bar{x}^k$ on the constraints. This projection is done in an iterative way, which is very costly.

When the constraints are nonlinear, is the ratio of the largest to the smallest eigenvalue of the Hessian of the Lagrangian $H(x^*, \mu^*)$ as measured on the tangent space that determines the speed of convergence [36]. The effect on the rate of convergence due to the curvature of the constraints is included in $H(x^*, \mu^*)$.

Algorithm 6.1, when applied to nonlinearly constrained problems, is much more efficient than the projected gradient method. Since the iterates are not required to be feasible, the projection stage is avoided.

6.2. NEWTON'S METHOD

Taking $S \equiv H(x^k, \mu^k)$ we have a Newton's equality constrained optimization algorithm. Even though in the majority of applications the computation of the second derivatives of f and g is very expensive, in several problems they are available or easy to obtain.

The Hessian of the Lagrangian function, $H(x, \mu)$, is not necessarily positive definite. Theorem 3.9 only ensures that $H(x^*, \mu^*)$ is positive definite on the tangent space. To have positive definite S, a procedure similar to Levenberg - Marquardt method for unconstrained optimization can be employed.

6.3. QUASI-NEWTON METHOD

In this class of algorithms, S is defined as a quasi - Newton approximation matrix of the Hessian of the Lagrangian $H(x^*, \mu^*)$, that we call B. In principle, B can be obtained using the same updating rules as in unconstrained optimization, but taking $\nabla_x l(x, \mu)$ instead of $\nabla f(x)$.

As $H(x, \mu)$ is not necessarily positive definite at (x^*, μ^*), it is not always possible to get B positive definite. To overcome this difficulty, Powell [42] proposed a modification of BFGS updating rule that takes

$$\delta = x^{k+1} - x^k$$

and

$$\gamma = \nabla_x l(x^{k+1}, \mu^k) - \nabla_x l(x^k, \mu^k).$$

If

$$\delta^t \gamma < 0.2\delta^t B\delta,$$

then it is computed

$$\phi = \frac{0.8\delta^t B\delta}{\delta^t B\delta - \delta^t \gamma}$$

and taken

$$\gamma = \phi\gamma + (1 - \phi)B\delta.$$

Finally, BFGS updating rule (5.8) is employed.

6.4. ABOUT THE SPEED OF CONVERGENCE

Since the techniques studied in this section are based on iterative algorithms for nonlinear systems, (x^k, μ^k) converges to (x^*, μ^*) at the same speed as the original algorithm. Then, the convergence of (x^k, μ^k) is quadratic, superlinear or linear, depending if a Newton's, a quasi - Newton or a first order algorithm for equality constrained optimization is employed.

In practice, we are more interested on the speed of convergence of x^k than (x^k, μ^k). Several authors studied this point, in particular Powell [41], Gabay [16], Hoyer [33], Gilbert [18]. Basically the same results are obtained, but in two steps, i. e., Powell proved that quasi - Newton algorithms are two-steps superlinearly convergent. That is

$$lim_{k \to \infty} \frac{\| x^{k+2} - x^* \|}{\| x^k - x^* \|} = 0.$$

As in unconstrained optimization, when applying Newton's or quasi - Newton algorithms, the step length must be the unity near the solution. There are examples showing that, taking $t = 1$ in Step 3) of Algorithm 6.1, it is not always possible to get a reduction of ϕ near the solution. This is known as Maratos' effect [37]. Several researches have been looking for methods to avoid Maratos' effect [18], [24], [27].

7. Sequential Quadratic Programming Method

To extend the ideas presented above in a way to solve the general nonlinear programming problem (2.1), one major difficulty has to be overcome. While in unconstrained or in equality constrained optimization, a point satisfying first order necessary optimality condition can be obtained by solving a system of equations, problems including inequality constraints require the solution of the system of equations and inequations (3.5) - (3.9). That is, a solution of the system of equation (3.5) - (3.7), that satisfies the inequalities (3.8) and (3.9) has to be found.

Sequential Quadratic Programming (SQP), that is at the moment the largest employed method for nonlinear constrained optimization, is a quasi - Newton technique based on an idea proposed by Wilson [50] in 1963 and interpreted by Beale [8] in 1967.

A *Quadratic Program* is a class of constrained optimization problems such that the objective is a convex quadratic function and the constraints are linear. Efficient techniques to solve this problem are available, even when inequality constraints are included. The exact solution is obtained after a finite number of iterations [36].

To explain Wilson's idea, we take x constant and consider the following quadratic programming problem that has $d \in R^n$ as unknown,

$$\left.\begin{array}{ll} \text{minimize} & \frac{1}{2}d^t S d + \nabla f^t(x)d \\ \text{subject to} & \nabla h^t(x)d + h(x) = 0. \end{array}\right\} \tag{7.1}$$

Since (7.1) is a convex problem, the global minimum satisfies Karush - Kuhn - Tucker optimality conditions

$$S d + \nabla f(x) + \nabla h(x)\mu = 0 \tag{7.2}$$

$$\nabla h^t(x)d + h(x) = 0, \tag{7.3}$$

where μ are the Lagrange multipliers. Then, (d, μ) in Algorithm 6.1 can be obtained by solving the quadratic program (7.1) instead of the linear system (6.13), (6.14). Based on this fact, to solve the problem

$$\left.\begin{array}{ll} \text{minimize} & f(x) \\ \text{subject to} & g(x) \leq 0 \\ \text{and} & h(x) = 0, \end{array}\right\} \tag{7.4}$$

Wilson proposed to define the search direction d and new estimates of the Lagrange multipliers λ and μ by solving at each iteration

$$\left.\begin{array}{ll} \text{minimize} & \frac{1}{2}d^t S d + \nabla f^t(x)d \\ \text{subject to} & \nabla g^t(x)d + g(x) \leq 0 \\ \text{and} & \nabla h^t(x)d + h(x) = 0. \end{array}\right\} \tag{7.5}$$

Wilson's is a Newton algorithm. Garcia Palomares and Mangasarian proposed later a quasi - Newton technique [17], Han obtained a globally convergent algorithm [21] and Powell proved superlinear convergence [41].

The exact penalty function

$$\phi(x, s, r) = f(x) + \sum_{i=1}^{m} s_i \{ \sup[0, g_i(x)] \} + \sum_{i=1}^{p} r_i \| h_i(x) \| \qquad (7.6)$$

is taken as the objective of the line search. If r satisfies (6.11) and

$$s_i \geq \lambda_i, \text{ for } i = 1, 2, ..., m, \qquad (7.7)$$

then d that solves (7.5) is a descent direction of $\phi(x, s, r)$. This result is proved in [36].

In Sequential Quadratic Programming algorithms, the matrix S is defined as a quasi - Newton approximation of the Hessian of the Lagrangian. Most of the optimizers employ BFGS rule modified by Powell, as explained in the last section. SQP algorithm can be stated as follows,

ALGORITHM 7.1 Sequential Quadratic Programming

Parameters. $r \in R^p$ and $s \in R^m$ positive.

Data. Initialize $x \in R^n$ and $B \in R^{n \times n}$ symmetric and positive definite.

Step 1. Computation of the search direction d and an estimate of the Lagrange multipliers λ and μ.

Solve the quadratic program for d,

$$\left. \begin{array}{rl} \text{minimize} & \frac{1}{2} d^t S d + \nabla f^t(x) d \\ \text{subject to} & \nabla g^t(x) d + g(x) \leq 0 \\ \text{and} & \nabla h^t(x) d + h(x) = 0 \end{array} \right\} \qquad (7.8)$$

Step 2. Line search.

i) If $r_i \leq \| \mu_i \|$, then set $r_i > \| \mu_i \|$, for $i = 1, 2, ..., p$.

ii) If $s_i \leq \lambda_i$, then set $r_i > \lambda_i$, for $i = 1, 2, ..., m$.

iii) Find a step length t satisfying a given line search criterion on the auxiliary function

$$\phi(x, s, r) = f(x) + \sum_{i=1}^{m} s_i \{ \sup[0, g_i(x)] \} + \sum_{i=1}^{p} r_i \| h_i(x) \|$$

Step 3. Updates.

Let be $\delta = td$ and $\gamma = \nabla_x l(x + td, \lambda, \mu) - \nabla_x l(x, \lambda, \mu)$.

i) If $\delta^t \gamma < 0.2 \delta^t B \delta$, then compute

$$\phi = \frac{0.8 \delta^t B \delta}{\delta^t B \delta - \delta^t \gamma}$$

and set $\gamma = \phi \gamma + (1 - \phi) B \delta$.

ii) Set

$$B := B + \frac{\gamma \gamma^t}{\delta^t \gamma} - \frac{B \delta \delta^t B}{\delta^t B \delta}$$

and

$$x := x + td$$

Step 4. Go back to *Step 1.* □

This algorithm generates sequences that are globally convergent to Karush - Kuhn - Tucker points of the problem. However it fails at points where the quadratic program has no a solution. In effect, since the constraints of the quadratic program solved in Step 1 are linear approximations of the constraints of the original problem, the feasible region may be empty.

The asymptotic speed of convergence has similar properties as quasi - Newton algorithms for equality constrained optimization and Maratos' effect can also occur.

7.1. A FEASIBLE DIRECTIONS ALGORITHM

Feasible directions algorithms are an important class of methods for solving constrained optimization problems. At each iteration, the search direction is a feasible direction of the inequality constraints and, at the same time, a descent direction of the objective or an other appropriate function. A constrained line search is then performed to obtain a satisfactory reduction of the function without loosing the feasibility.

The fact of giving feasible points makes feasible directions algorithms very efficient in engineering design, where functions evaluation is in general very expensive. Since any intermediate design can be employed, the iterations can be stopped when the cost reduction per iteration becomes small enough.

There are also several examples that deal with an objective function, or constraints, that are not defined at infeasible points. This is the case of size and shape constraints in structural optimization. When applying feasible directions algorithms to real time problems, as feasibility is maintained and cost reduced, the controls can be activated at each iteration .

In what follows we describe an SQP feasible directions algorithm based on a technique presented by Herskovits in [26] and by Herskovits and Carvalho in [27]. By solving the quadratic program (7.5), this algorithm defines first (d_0, λ_0), where d_0 is a descent direction in the primal space of $l(x, \lambda_0)$. However, d_0 is not necessarily a feasible direction since, if an inequality constraint of problem (7.4) is active and the corresponding constraint in (7.5) is also active, then d_0 is tangent to the feasible set. In a second stage, the algorithm obtains a feasible and descent search direction d by solving a modified quadratic program with equality constraints only.

To present the method, we consider the inequality constrained optimization problem

$$\left. \begin{array}{ll} \text{minimize} & f(x) \\ \text{subject to} & g(x) \leq 0, \end{array} \right\} \tag{7.9}$$

whose feasible set is $\Omega \equiv \{x \in R^n / g(x) \leq 0\}$, and introduce the following definition:

DEFINITION 7.1. A vector field $d(x)$ defined on Ω is said to be a *uniformly feasible directions field* of the problem (7.9), if there exists a step length $\tau > 0$ such that $x + td(x) \in \Omega$ for all $t \in [0, \tau]$.

This condition is much stronger than the simple feasibility of $d(x)$ for any $x \in \Omega$. When $d(x)$ constitutes a uniformly feasible directions field, it supports a feasible segment $[x, x + \theta(x)d(x)]$, such that $\theta(x)$ is bounded below in Ω by $\tau > 0$.

As a consequence of the feasibility requirement, the search directions of feasible directions algorithms must constitute a uniformly feasible directions field. Otherwise, the step length may go to zero, forcing convergence to points which are not KKT points.

We state now the algorithm

ALGORITHM 7.2 SQP Feasible Directions Algorithm

Parameters. $\alpha \in (0, 1)$ and $\varphi > 0$.

Data. Initialize $x \in R^n$ feasible and $B \in R^{n \times n}$ symmetric and positive definite.

Step 1. Computation of a descent direction d_0 and an estimate of the Lagrange multipliers λ_0.

Solve the quadratic program in d_0

$$\left. \begin{array}{cc} \text{minimize} & \frac{1}{2}d_0^t B d_0 + \nabla f^t(x)d_0 \\ \text{subject to} & \nabla g^t(x)d_0 + g(x) \leq 0 \end{array} \right\} \tag{7.10}$$

Step 2. Computation of the search direction d.

    i) Let the active set be $J(x) \equiv \{j \in 1, 2, ..., m/\lambda_{0j} > 0\}$ and $g_J(x) \equiv [g_j(x)]^t$, $j \in J(x)$.

    If $g_J^t(x)[\nabla g_J^t(x)B^{-1}\nabla g_J(x)]^{-1}e < 0$, find

$$\rho_J = \frac{(1-\alpha)\nabla l^t(x, \lambda_0)d_0}{g_J^t(x)[\nabla g_J^t(x)B^{-1}\nabla g_J(x)]^{-1}e} \tag{7.11}$$

    where the Lagrangian $l(x, \lambda) = f(x) + \lambda^t g(x)$ and $e \equiv [1, 1, ..., 1]^t$.

    ii) Let the inactive set be $\bar{J}(x) \equiv \{j \in 1, 2, ..., m/\lambda_{0j} = 0\}$.

    For each $j \in \bar{J}(x)$, if

$$\{1 - \nabla g_j^t(x)B^{-1}\nabla g_J(x)[\nabla g_J^t(x)B^{-1}\nabla g_J(x)]^{-1}e\} < 0,$$

    find

$$\rho_j = -\frac{g_j(x) + \nabla g_j^t(x)d_0}{1 - \nabla g_J^t(x)B^{-1}\nabla g_J(x)[\nabla g_J^t(x)B^{-1}\nabla g_J(x)]^{-1}e} \tag{7.12}$$

    iii) Set $\rho = \inf\{\varphi \parallel d_o \parallel^2, \rho_J; \rho_j, j \in \bar{J}(x)\}$.

    iv) Solve the quadratic program in d

$$\left. \begin{array}{cc} \text{minimize} & \frac{1}{2}d^t B d + \nabla f^t(x)d \\ \text{subject to} & \nabla g_J^t(x)d + g_J(x) = -\rho e \end{array} \right\} \tag{7.13}$$

Step 3. Line search.

Find a step length t satisfying a given constrained line search criterion on the Lagrangian function $l(x, \lambda_0)$ and such that $(x + td)$ is a feasible point of the nonlinear program (7.9).

Step 4. Updates.

Let be $\delta = td$ and $\gamma = \nabla_x l(x + td, \lambda_0) - \nabla_x l(x, \lambda_0)$.

i) If $\delta^t \gamma < 0.2 \delta^t B \delta$, then compute

$$\phi = \frac{0.8 \delta^t B \delta}{\delta^t B \delta - \delta^t \gamma}$$

and set $\gamma = \phi \gamma + (1 - \phi) B \delta$.

ii) Set

$$B := B + \frac{\gamma \gamma^t}{\delta^t \gamma} - \frac{B \delta \delta^t B}{\delta^t B \delta}$$

and

$$x := x + td$$

Step 4. Go back to *Step 1.* $\square$

As x is an admissible point of problem (7.9), the feasible region of (7.10) is non empty. In effect, $d_0 = 0$ satisfies the constraints of (7.10). Considering KKT conditions of problem (7.10), since B is positive definite, it is easy to show that d_0 is a descent direction of $l(x, \lambda_0)$. However, as discussed above, d_0 is not necessarily a feasible direction of the problem.

To obtain a feasible direction, we define the equality constrained quadratic program (7.13) for d, where $\rho > 0$. It follows from the constraints of this last problem that $\nabla g_j^t(x) d < 0$, then d is a feasible direction of the original nonlinear program. The constraints of (7.13) are obtained by moving slightly the active constraints of (7.10) to the interior of the feasible region. Since this movement is proportional to ρ, by establishing bounds on ρ we can ensure that d is a descent direction of $l(x, \lambda_0)$ and that d is also a feasible direction, even in relation to the constraints of the original problem that are not included in (7.13). These bounds are defined by (7.11) and (7.12) respectively. In fact, in [27] it is proved that

$$\nabla^t l(x, \lambda_0) d \le \alpha \nabla^t l(x, \lambda_0) d_0, \tag{7.14}$$

when (7.11) is true, and also that

$$g_j(x) + \nabla g_j^t(x) d \le -\rho, \tag{7.15}$$

if ρ satisfies (7.12). As a consequence of (7.14), to get a feasible direction we pay the price reducing the directional derivative of $l(x, \lambda_0)$ by α.

Although the present method requires a constrained line search, the objective is smooth. This is an important advantage of feasible directions algorithms.

In [27], the authors show that the search directions obtained by the present approach constitute a uniformly feasible directions field. It can be proved that if the sequence given by the algorithm converges, the limit is a KKT point of the problem. In a similar way as in [40], it can be shown that the rate of convergence of the present method is superlinear and that Maratos' effect is avoided.

7.2. THE CONSTRAINED LINE SEARCH

Feasible directions algorithms, and others methods that produce a sequence of admissible points with respect to the inequality constraints, need a constrained line search. We consider Problem (7.9) and call $\phi(x)$ the line search objective of an algorithm to solve this problem. It is assumed that $\phi(x)$ is continuously differentiable in Ω. The previous feasible directions algorithm takes $\phi(x) = l(x, \lambda_0)$. The corresponding *Constrained Line Search by Exact Minimization* consists on finding t^* that minimizes $\phi(x + td)$, subject to $g(x + td) \leq 0$, where x is the current iterate and d the search direction.

To define *Inaccurate Constrained Line Search Criteria* we must take in consideration that the exact minimum t^* may be in the interior or in the boundary. In both cases there exists t such that (5.5) is true. That is, a decrease of the line search objective can always be obtained. On the other hand, if the exact minimum is not in the interior, the existence of t such that (5.6) holds is not assured. As an example, when ϕ is linear the derivative cannot be reduced.

We propose now extensions of Armijo's and Wolfe's criteria to constrained line search.

Armijo's Constrained Line Search

Define the step length t as the first number of the sequence $\{1, \nu, \nu^2, \nu^3, ...\}$ satisfying

$$\phi(x + td) \leq \phi(x) + t\eta_1 \nabla \phi^t(x)d \tag{7.16}$$

and

$$g(x + td) \leq 0 \tag{7.17}$$

where $\eta_1 \in (0, 1)$ and $\nu \in (0, 1)$ also. $\qquad \square$

Wolfe's Constrained Line Search Criterion

Accept a step length t if (7.16) and (7.17) are true and at least one of the followings $m + 1$ conditions hold:

$$\nabla \phi^t(x + td)d \geq \eta_2 \nabla \phi^t(x)d \qquad (7.18)$$

and

$$g_i(x + td) \geq \gamma g_i(x); i = 1, 2, ..., m \qquad (7.19)$$

where now $\eta_1 \in (0, 1/2)$, $\eta_2 \in (\eta_1, 1)$ and $\gamma \in (0, 1)$.   □

Conditions (7.16) and (7.17) define upper bounds on the step length in both criteria and, in Wolfe's criterion, a lower bounds is given by one of the conditions (7.18) and (7.19). The iterative procedure to find a step length satisfying Wolfe's criterion is similar to that employed in unconstrained optimization, but interpolations of the constraints have to be also done.

8. A Newton-like Interior Point Method

In the last section, the Sequential Quadratic Programming Method was presented as a technique to extend Newton - like iterations to inequality constrained optimization. We present now a family of Newton-like algorithms to solve Karush-Kuhn-Tucker optimality conditions. To obtain these algorithms, Newton - like iterations to solve the nonlinear system in (x, λ) given by the equalities (3.5) - (3.7) in KKT conditions, are first defined. Then, these iterations are slightly modified in such a way to have the inequalities (3.8) and (3.9) satisfied at each iteration.

The algorithms based of this approach require an initial estimate of x, at the interior of the feasible region defined by the inequality constraints, and generate a sequence of points also at the interior of this set. They are feasible directions algorithms, since at each iteration they define a search direction that is a feasible direction with respect to the inequality constraints and a descent direction of the objective, or another appropriate function. When only inequality constraints are considered, the objective is reduced at each iteration. In the general problem, an increase of the objective may be necessary to have the equalities satisfied.

The present method is simple to code, strong and efficient. It does not involve penalty functions, active set strategies or quadratic programming subproblems. It only requires the solution of two linear systems with the same matrix at each iteration, followed by an inaccurate line search.

This approach, proposed by the author, is the basis of the first order algorithm and the quasi - Newton one described in [25], [26], [27] and [28]. This is also the case of the quasi - Newton algorithms described in [29] and

[40] and of a quadratic programming algorithm developed by Tits et al. [45]. A unified study of this technique is presented in [30] and [31].

Several problems in Engineering Optimization were solved using the present method. We can mention applications in structural optimization [25], fluid mechanics [4], multidisciplinary optimization with aerodynamics and electromagnetism [4], [3], and real time optimal control [11], [12].

A Newton algorithm for limit analysis of solids with nonlinear yield functions was proposed by Zouain et al. [49]. A Newton algorithm for stress analysis of linear elastic solids in contact is described in [2] and [48].

To explain the ideas behind Newton - like Interior Point Method, we consider the nonlinear inequality constrained program

$$\left. \begin{array}{ll} \text{minimize} & f(x) \\ \text{subject to} & g(x) \leq 0. \end{array} \right\} \tag{8.1}$$

and the corresponding KKT optimality conditions

$$\nabla f(x) + \nabla g(x)\lambda = 0 \tag{8.2}$$

$$G(x)\lambda = 0 \tag{8.3}$$

$$\lambda \geq 0 \tag{8.4}$$

$$g(x) \leq 0. \tag{8.5}$$

A Newton's iteration to solve the nonlinear system of equations (8.2), (8.3) in (x, λ) is stated as

$$\begin{bmatrix} H(x^k, \lambda^k) & \nabla g(x^k) \\ \Lambda^k \nabla g^t(x^k) & G(x^k) \end{bmatrix} \begin{bmatrix} x^{k+1} - x^k \\ \lambda_0^{k+1} - \lambda^k \end{bmatrix} = - \begin{bmatrix} \nabla f(x^k) + \nabla g(x^k)\lambda^k \\ G(x^k)\lambda^k \end{bmatrix} \tag{8.6}$$

where (x^k, λ^k) is the starting point of the iteration and $(x^{k+1}, \lambda_0^{k+1})$ is the new estimate, $H(x, \lambda)$ is the Hessian of the Lagrangian and Λ a diagonal matrix with $\Lambda_{ii} \equiv \lambda_i$. Proceeding in a similar way as in Section 6, we define the linear system in (d_0^k, λ_0^{k+1})

$$S^k d_0^k + \nabla g(x^k)\lambda_0^{k+1} = -\nabla f(x^k) \tag{8.7}$$

$$\Lambda^k \nabla g^t(x^k)d_0^k + G(x^k)\lambda_0^{k+1} = 0, \tag{8.8}$$

where d_0^k is a direction in the primal space, defined by $d_0^k = x^{k+1} - x^k$, and λ_0^{k+1} is a new estimate of λ. The matrix S^k is symmetric and positive definite and it can be taken as to $H(x^k, \lambda^k)$, a quasi-Newton approximation of $H(x^k, \lambda^k)$, or the identity.

It can be proved that d_0 is a descent direction of f. However, d_0 is not useful as a search direction since it is not necessarily feasible. This is due to the fact that as any constraint goes to zero, (8.8) forces d_0 to tend to a direction tangent to the feasible set. In fact, (8.8) is equivalent to

$$\lambda_i^k \nabla g_i^t(x^k)d_0^k + g_i(x^k)\lambda_{0i}^{k+1} = 0 \; ; \; 1, 2, ..., m, \tag{8.9}$$

which implies that $\nabla g_i^t(x^k)d_0^k = 0$ for i such that $g_i(x^k) = 0$.

To avoid this effect, we define the linear system in d^k and $\bar{\lambda}^{k+1}$

$$S^k d^k + \nabla g(x^k)\bar{\lambda}^{k+1} = -\nabla f(x^k) \tag{8.10}$$

$$\Lambda^k \nabla g^t(x^k)d^k + G(x^k)\bar{\lambda}^{k+1} = -\rho^k \lambda^k, \tag{8.11}$$

obtained by adding a negative vector to the right side of (8.8), where the scalar factor ρ is positive. In this case, (8.11) is equivalent to

$$\lambda_i^k \nabla g_i^t(x^k)d^k + g_i(x^k)\bar{\lambda}_i^{k+1} = -\rho^k \lambda_i^k; \; i = 1, m,$$

and then, $\nabla g_i^t(x^k)d^k < 0$ for the active constraints. Thus, d^k is a feasible direction.

The addition of a negative number in the right hand side of (8.8) produces the effect of deflecting d_0^k into the feasible region, where the deflection is proportional to ρ^k. As the deflection of d_0^k is proportional to ρ^k and d_0^k is a descent direction of f, it is possible to find bounds on ρ^k, in a way to ensure that d^k is also a descent direction. Since $d_0^{k^t} \nabla f(x^k) < 0$, we can get these bounds by imposing

$$d^{k^t} \nabla f(x^k) \leq \alpha d_0^{k^t} \nabla f(x^k), \tag{8.12}$$

which implies $d^{k^t} \nabla f(x^k) < 0$.

Condition (8.12) means that d^k is in a circular straight cone whose axis is $\nabla f(x^k)$. In general, the rate of descent of f along d^k will be smaller than along d_0^k. This is a price that we pay for obtaining a good feasible descent direction.

Let us consider the auxiliary linear system

$$S^k d_1^k + \nabla g(x^k)\lambda_1^{k+1} = 0 \tag{8.13}$$

$$\Lambda^k \nabla g^t(x^k)d_1^k + G(x^k)\lambda_1^{k+1} = -\lambda^k, \tag{8.14}$$

it is easy to deduce that

$$d^k = d_0^k + \rho d_1^k.$$

Substitution of this last expression in (8.12) gives

$$\rho \leq (\alpha - 1)d_0^{k^t} \nabla f(x^k)/d_1^{k^t} \nabla f(x^k). \tag{8.15}$$

Finally, to get a new feasible primal point and a satisfactory decrease of the objective function, an inaccurate constrained line search along d^k is done. Different updating rules can be adopted to define a positive λ^{k+1}.

In Figure 8, the search direction of an optimization problem with two design variables and one constraint is illustrated. At x^k on the boundary, d_0 is tangent to the constraint. In this example, d_1 is normal to the boundary.

ALGORITHM 8.1 *Newton-like Interior Point Algorithm for inequality con-straints.*

Parameters. $\alpha \in (0,1)$ and $\varphi > 0$.

Data. Initialize $x \in \Omega_a^0$, $\lambda > 0$ and $S \in R^{n \times n}$ symmetric and positive definite.

Step 1. Computation of the search direction d.

i) Solve the linear system for (d_0, λ_0)

$$S d_0 + \nabla g(x)\lambda_O = -\nabla f(x), \tag{8.16}$$

$$\Lambda \nabla g^t(x)d_0 + G(x)\lambda_0 = 0. \tag{8.17}$$

If $d_0 = 0$, stop.

ii) Solve the linear system for (d_1, λ_1)

$$S d_1 + \nabla g(x)\lambda_1 = 0, \tag{8.18}$$

$$\Lambda \nabla g^t(x)d_1 + G(x)\lambda_1 = -\lambda. \tag{8.19}$$

iii) If $d_1^t \nabla f(x) > 0$, set

$$\rho = inf[\varphi \parallel d_0 \parallel^2 ; (\alpha - 1)d_0^t \nabla f(x)/d_1^t \nabla f(x)]. \tag{8.20}$$

Otherwise, set

$$\rho = \varphi \parallel d_0 \parallel^2 . \tag{8.21}$$

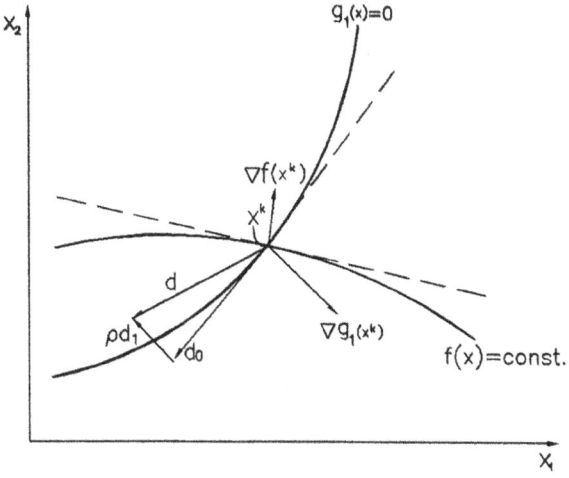

Figure 8. Search direction of Algorithm 8.1

iv) Compute the search direction

$$d = d_0 + \rho d_1, \tag{8.22}$$

and also

$$\bar{\lambda} = \lambda_0 + \rho \lambda_1. \tag{8.23}$$

Step 2. Line search.

Find a step length t satisfying a given constrained line search criterion on the objective function f and such that

$$g_i(x + td) < 0 \text{ if } \bar{\lambda}_i \geq 0, \tag{8.24}$$

or

$$g_i(x + td) \leq g_i(x) \tag{8.25}$$

otherwise.

Step 3. Updates.

i) Set
$$x = x + td$$

and define new values for $\lambda > 0$ and S symmetric and positive definite.

ii) Go back to *Step 1.*   □

In addition to (8.15), we have from (8.20) and (8.21) that ρ is bounded above by $\varphi \parallel d_0 \parallel^2$. In [31] it is shown that $\rho = O\parallel d_0 \parallel^2$; we don't need larger ρ to ensure that d constitutes a uniformly feasible directions field. The line search includes conditions (8.24) and (8.25) that force the new primal point to be in the interior. Moreover, (8.25) prevents the saturation of the constraints associated to negative dual variables, as it is needed to prove convergence to Karush-Kuhn-Tucker points [31]. Armijo's constrained line search and Wolfe's criterion, given in the previous section, can be easily modified to take into account these requirements.

Different algorithms can be obtained according to the way of updating λ and S. We assume that the updating rule for S is such that Assumption 5.1 is true and that λ satisfies the following:

Assumption 8.1. There exist positive numbers λ^I, λ^S and β such that $0 \leq \lambda \leq \lambda^S$ and $\lambda_i \geq \lambda^I$ for any i such that $g_i(x) \geq -\beta$.

The theoretical analysis in [31] includes first a proof that, if x^k is a regular point of the problem, then the solutions of the linear systems (8.16), (8.17) and (8.18), (8.19) are unique. Then, it is shown that any sequence $\{x^k\}$ generated by the algorithm converges to a Karush- Kuhn-Tucker point of the problem, for any way of updating λ and S, provided that the previous assumptions are true. We also have that (x^k, λ_0^k) converges to a Karush-Kuhn-Tucker pair (x^*, λ^*) and, depending on the way of updating λ, global convergence in the dual space can also be obtained.

The matrix S can be defined in the same way as in equality constrained algorithms discussed in Section 6. In the case of the dual variables λ, the following updating rules can be stated in Step 3:

8.1. UPDATE OF λ

Rule 8.1.

Set, for $i = 1, m$,

$$\lambda_i := \sup \, [\lambda_{0i}; \epsilon \parallel d_0 \parallel^2]. \tag{8.26}$$

If $g_i(x) \geq -\beta$ and $\lambda_i < \lambda^I$, set $\lambda_i = \lambda^I$. $\qquad \square$

The parameters ϵ, β and λ^I are taken positive. If β and λ^I are taken small enough, after a finite number of iterations λ_i becomes equal to λ_{0i}, for the active constraints. For the inactive ones, $\lambda_i \to 0$ as $x^k \to x^*$. As a consequence, $\lambda \to \lambda^*$.

If Assumption 5.1 is verified, then λ defined above satisfies Assumption 3.1. In fact, it can be shown that λ_0 and d_0 are bounded. Then λ_i has a positive upper bound λ^S.

Far from the solution it is convenient to have a search direction that is not pointing to the constraints boundary but slipping along their boundary. In this way the steps will be longer and, consequently, the efficiency of the algorithm will improve.

This effect can be obtained by increasing the dual variable as the corresponding constraint goes to zero since, as it follows from (2.9), $\nabla g_i^t(x)d_0$ becomes smaller as λ_{0i} grows. The following rule satisfies this requirement in a very simple way :

Rule 8.2.

Set, for $i = 1, m$,

$$\lambda_i := -g_i^{-1}(x). \qquad (8.27)$$

If $\lambda_i > \lambda^S$, set $\lambda_i = \lambda^S$. □

This rule is based on Dikin's algorithm for Linear Programming, presented in [13] and also studied in [46]. In the case of linear programs, since $H(x,\lambda) = 0$, we can take $S = 0$. In [31] it is shown that d_0 obtained with this rule is identical to the search direction given by Dikin's algorithm.

8.2. ASYMPTOTIC CONVERGENCE

In the case when (8.26) and BFGS rules are used for λ and B, the convergence is two-step superlinear, provided that a unit step length is obtained after a finite number of iterations. This result can be obtained in a similar way as in Theorem 4.6 in [40], by showing that the search directions of the present method and of the SQP algorithm locally differ by a term of the same order as $\| d_0 \|^2$ and then, extending to the present algorithm the proof of two-step superlinear convergence of the SQP method.

To satisfy the requirement about the step length, d must be such that Definition 7.1 holds for some $\theta > 1$. It is clear that θ increases whenever ρ grows but, unfortunately, in some problems the upper bound on ρ may be not large enough to allow the step length to be unitary. This effect, which is similar to Maratos' effect, in theory can produce rates of convergence lower than superlinear.

In the quasi-Newton algorithm studied in [24], as in Algorithm 7.2, Maratos' effect is avoided by looking at each iteration for a decrease of an estimate of the Lagrangian, instead of the objective function. An algorithm that also solves optimality conditions by means of Newton - like iterates was presented in [40]. The authors obtained global and local superlinear convergence by applying a technique presented by Mayne and Polak [38]. First, as in the present approach, a feasible descent direction d is obtained. Then, the constraints are computed at $(x + d)$ and an approximate projection, $\tilde{x}$, of $(x + d)$ on the active constraints is found. Finally, a new primal variable is determined by doing a search along a parabola which is tangent to d and contains $\tilde{x}$. In this search, a decrease of the objective and the feasibility of the new iterate are required.

8.3. INCLUDING EQUALITY CONSTRAINTS

In view of Algorithms 6.1 and 8.1, a Newton - like algorithm to solve the general nonlinear program (2.1) is easily derived, [31]. This algorithm requires an initial point in the interior of the inequality constraints. It generates a sequence $\{x^k\}$ of interior points, with respect to the inequalities, that converges to a Karush-Kuhn-Tucker point of the problem. In general, the equalities are active only at the limit. Then, to have the equalities satisfied, the objective function must be allowed to increase. The auxiliary function $\phi(x, r)$, defined in Section 6, is taken as objective in the line search. The algorithm is stated below, with d_0 as a descent direction of $\phi(x, r)$ and d obtained by deflecting d_0 with respect to the inequality constraints.

ALGORITHM 8.2 Newton-like Interior Point Algorithm for constrained optimization.

Parameters. $\alpha \in (0, 1)$, $\varphi > 0$ and $r > 0$, $r \in R^p$

Data. Initialize $x \in \Omega_a^0$, $\lambda > 0$ and $S \in R^{n \times n}$ symmetric and positive definite.

Step 1. Computation of the search direction d.

i) Solve the linear system for (d_0, λ_0, μ_0)

$$Sd_0 + \nabla g(x)\lambda_0 + \nabla h(x)\mu_0 = -\nabla f(x),$$

$$\Lambda \nabla g^t(x)d_0 + G(x)\lambda_0 = 0.$$

$$\nabla h^t(x)d_0 = -h(x)$$

If $d_0 = 0$, stop.

ii) Solve the linear system for (d_1, λ_1, μ_1)

$$Sd_1 + \nabla g(x)\lambda_1 + \nabla h(x)\mu_1 = 0,$$

$$\Lambda \nabla g^t(x)d_1 + G(x)\lambda_1 = -\lambda.$$

$$\nabla h^t(x)d_1 = 0$$

iii) If $r_i \leq \|\mu_{0i}\|$, then set $r_i > \|\mu_{0i}\|$, for $i = 1, 2, ..., p$.

iv) Let $\phi(x, r) = f(x) + r^t SG[h(x)]h(x)$. If $d_1^t \nabla \phi(x, r) > 0$, set

$$\rho = inf[\varphi \| d_0 \|^2 ; (\alpha - 1)d_0^t \nabla \phi(x, r)/d_1^t \nabla \phi(x, r)].$$

Otherwise, set

$$\rho = \varphi \| d_0 \|^2 .$$

(v) Compute the search direction

$$d = d_0 + \rho d_1,$$

and also

$$\bar{\lambda} = \lambda_0 + \rho \lambda_1.$$

Step 2. Line search.

Find a step length t satisfying a given constrained line search criterion on the auxiliary function $\phi(x, r)$ and such that

$$g_i(x + td) < 0 \text{ if } \bar{\lambda}_i \geq 0,$$

or

$$g_i(x + td) \leq g_i(x)$$

otherwise.

Step 3. Updates.

i) Set
$$x = x + td$$

and define new values for $\lambda > 0$ and S symmetric and positive definite.

ii) Go back to *Step 1.* □

Acknowledgements. I thank Prof. Jasbir S. Arora for his careful review of the manuscript and helpful comments.

References

1. Arora, J. S. "Introduction to Optimum Design", McGraw-Hill, New York, 1989.
2. Auatt, S. S. "Nonlinear Programming Algorithms for Elastic Contact Problems", (in Portuguese), MSc. Dissertation, COPPE - Federal University of Rio de Janeiro, Mech. eng. Program, Caixa Postal 68503, 21945-970, Rio de Janeiro, Brazil, 1993.
3. Baron, F. J. and Pironneau, O. "Multidisciplinary Optimal Design of a Wing Profile", Proceedings of STRUCTURAL OPTIMIZATION 93, Rio de Janeiro, Aug. 1993, J. Herskovits ed..
4. Baron, F. J. "Constrained Shape Optimization of Coupled Problems with Electromagnetic Waves and Fluid Mechanics" (in Spanish), D. Sc. Dissertation, University of Malaga, Spain, 1994.
5. Baron, F. J., Duffa, G., Carrere, F. and Le Tallec, P. "Optimisation de forme en aerodynamique", CHOCS "Revue scientifique et technique de la Direction des Applications Militaires du CEA", France. 1994.
6. Baron, F. J. "Radar Visibility Optimization under Aerodynamic Constraints for a Wing Profile", Research Report No. 2073, INRIA, BP 105, 78153 Le Chesnay CEDEX, France, 1994.
7. Bazaraa, M. S. and Shetty, C. M. Nonlinear Programming, Theory and Algorithms" John Wiley and Sons, New York, 1979.
8. Beale, E. M. L. "Numerical Methods ", in Nonlinear Programming, J. Abadie ed, North - Holland, Amsterdam, 1967.
9. Dennis, J. E. and More, J. J. "Quasi - Newton Methods, Motivation and Theory", SIAM Review, 19, pp. 46-89,1977.
10. Dennis, J. E. and Schnabel, R. "Numerical Methods for Constrained Optimization and Nonlinear Equations ", Prentice Hall, 1983.
11. Dew, M. C. "A First Order Feasible Directions Algorithm for Constrained Optimization" Technical Report No. 153, Numerical Optimization Centre, The Hatfield Polytechnic, P. O. Box 109, College Lane, Hatfield, Hertfordshire AL 10 9AB, U. K., 1985.
12. Dew, M. C. "A Feasible Directions Method for Constrained Optimization based on the Variable Metric Principle" Technical Report No. 155, Numerical Optimization Centre, The Hatfield Polytechnic, P. O. Box 109, College Lane, Hatfield, Hertfordshire AL 10 9AB, U. K., 1985.
13. Dikin, I. I. "About the Convergence of an Iterative Procedure" (in Russian), Soviet Mathematics Doklady, Vol. 8, pp. 674-675, 1967.
14. Fletcher, R. "Practical methods in Optimization", Vol. I and II, John Wiley, Chichester, 1980.
15. Fox, R. L. "Optimization Methods for Engineering Design", Addison-Wesley, Reading, 1973.
16. Gabay, D. "Reduced Quasi - Newton Methods with Feasibility Improvement for Nonlinearly Constrained Optimization", Mathematical Programming Study 16, pp. 18-44,1982.
17. Garcia Palomares, U. M. and Mangasarian, O. L., "Superlinearly Convergent Quasi-Newton Algorithms for Nonlinearly Constrained Optimization Problems", Mathematical Programming 11, pp. 1-13, 1976.
18. Gilbert, J. C. "Superlinear Convergence of a Reduced BFGS Method With Piecewise Line-search and Update Criterion", Research Report No. 103, INRIA, BP 105, 78153 Le Chesnay CEDEX, France, 1993.
19. Gill, P. E., Murray, W. and Wright, M. H. "Practical Optimization", Academic Press, New York, 1981.

20. Haftka, R. T., Gürdal, Z. and Kamat, M. P. "Elements of Structural Optimization ", Kluwer Academic Publishers, 1990.

21. Han, S. P. "A Globally Convergent Method for Nonlinear Programming", Journal for Optimization Theory and Applications, 22, pp. 297-309, 1977.

22. Haug, E. J. and Arora, J.S. "Applied Optimal Design", Wiley Interscience, New York, 1979.

23. Herskovits, J. - "A Two-Stage Feasible Direction Algorithm for Non-Linear Constrained Optimization", Research Report No. 103, INRIA, 1982.

24. Herskovits, J. "A Two-Stage Feasible Directions Algorithm Including Variable Metric Techniques for Nonlinear Optimization", Research Report No. 118, INRIA, 1982.

25. Herskovits, J. "Développement d'une méthode numérique pour l'optimisation non-linéaire", (in English), Thèse de Docteur - Ingénieur en Mathématiques de la Décision, Paris IX University, Published by INRIA, 1982.

26. Herskovits, J. "A Two-Stage Feasible Directions Algorithm for Nonlinear Constrained Optimizations Using Quadratic Programming Subproblems", 11th IFIP Conference on System Modeling and Optimization, Copenhagen, Denmark, July 1983.

27. Herskovits, J.N. and Carvalho, L.A.V. "A Successive Quadratic Programming based Feasible Directions Algorithm" Proceedings of the 17th International Conference on Analysis and Optimization of Systems, Antibes, France, Springer -Verlag, 1986.

28. Herskovits, J. "A Two-Stage Feasible Directions Algorithm for Nonlinear Constrained Optimization", Mathematical Programming, Vol. 36, pp. 19-38, 1986.

29. Herskovits, J. and Coelho, C.A.B. "An Interior Point Algorithm for Structural Optimization Problems", in Computer Aided Optimum Design of Structures: Recent Advances, Brebbia, C. A. and Hernandez, S. ed., Computational Mechanics Publications, Springer-Verlag, 1989.

30. Herskovits, J. "An Interior Points Method for Nonlinearly Constrained Optimization", in Optimization of Large Structural Systems, Rozvany, G. ed., Vol I, pp. 589-608, NATO/ASI Series, Springer - Verlag, 1991.

31. Herskovits, J. "An Interior Point Technique for Nonlinear Optimization" Research Report No. 1808, INRIA, BP 105, 78153 Le Chesnay CEDEX, France, 1992.

32. Hiriart-Urruty, J. B. and Lemaréchal, C. "Convex analysis and Minimization Algorithms ", I, II, Springer - Verlag, Berlin Heildelberg, 1993.

33. Hoyer, W. "Variants of Reduced Newton Method for Nonlinear EqualityConstrained Optimization Problems ", Optimization, 17, pp.757-774,1986.

34. Kirsch, U. "Structural Optimization, Fundamentals and Applications", Springer - Verlag, Berlin, 1993.

35. Luenberger, D. G. "Optimization By Vector Space Methods", John Wiley and Sons, New York, 1969.

36. Luenberger D.G., "Linear and Nonlinear Programming", 2nd. Edition, Addison-Wesley, Reading, 1984.

37. Maratos, N. "Exact Penalty Functions Algorithms for Finite Dimensional and control Optimization Problems", Ph. D. Dissertation, Imperial college, University of London, 1978.

38. Mayne D.Q. and Polak E., "Feasible Directions Algorithms for Optimization Problems with Equality and Inequality Constraints", Mathematical Programming 11, pp. 67-80, 1976.

39. Minoux, M. "Programation Mathématique: Théorie et Algorithmes " ,I and II, Dunod, Paris,1983.

40. Panier, E.R.,Tits A.L. and Herskovits J. "A QP-Free, Globally Convergent, Locally Superlinearly Convergent Algorithm for Inequality Constrained Optimization", SIAM Journal of Control and Optimization, Vol 26, pp 788-810, 1988.

41. Powell, M.J.D. "The Convergence of Variable Metric Methods for Nonlinearly Constrained Optimization Calculations", in Nonlinear Programming 3, edited by O.L. Mangasarian, R.R. Meyer and S.M. Robinson, Academic Press, London, 1978.

116

42. Powell M. J. D., "Variable Metric Methods for Constrained Optimization", in Mathematical Programming - The State of the Art, Edited by A. Bachem, M. Grotschet and B. Korte, Springer-Verlag, Berlin, 1983.

43. Rosen, J. "The Gradient Projection Method for Nonlinear Programming, I, Linear Constraints ", SIAM J., 8, pp.181-217,1960.

44. Rosen, J. "The Gradient Projection Method for Nonlinear Programming, II, Nonlinear Constraints ", SIAM J., 9, pp.514-532,1961.

45. Tits, L. A. and Zhou J. L. "A Simple, Quadratically Convergent Interior Point Algorithm for Linear Programming and Convex Quadratic Programming", in Large Scale Optimization: State of the Art, W. W. Hager, D. W. Hearn and P.M. Pardalos ed., Kluwer Academic Publishers B.V, 1993.

46. Vanderbei, R. J.and Lagarios, J. C., "I. I. Dikin's Convergence Result for the Affine Scaling Algorithm", Technical Report, AT&T Bell Laboratories, Murray Hill, NJ, 1988.

47. Vanderplaats, G. N. "Numerical Optimization Techniques for Engineering Design with Applications" McGraw-Hill" , New York, 1984.

48. Vautier, I., Salaun, M. and Herskovits, J. " Application of an Interior Point Algorithm to the Modeling of Unilateral Contact Between Spot-Welded Shells", Proceedings of STRUCTURAL OPTIMIZATION 93, Rio de Janeiro, Aug. 1993, J. Herskovits ed..

49. Zouain N.A., Herskovits J.N., Borges L.A. and Feijóo R.A. - "An Iterative Algorithm for Limit Analysis with Nonlinear Yield Functions", International Journal on Solids Structures, Vol. 30, n0 10, pp. 1397-1417, Gt. Britain, 1993.

50. Wilson, R. B. "A Simplicial Algorithm for Concave Programming ", Ph.D. Dissertation, Harvard University Graduate School of Business Administration, 1963.

LARGE SCALE STRUCTURAL OPTIMIZATION

G. N. Vanderplaats
VMA Engineering
225 East Cheyenne Mountain Boulevard, Suite 200B
Colorado Springs, CO 80906, USA

ABSTRACT

The purpose here is to describe efficient methods for large scale structural optimization. A brief review of historical developments shows that techniques are now available to make the structural optimization task quite efficient and reliable. Proper formulation requires that different element types, design variables and responses be treated differently. Also, because a large number of nonlinear constraints are involved, these must be efficiently handled. Examples are offered to demonstrate the range of design tasks that can now be addressed and the design efficiencies that can be expected. Using the latest methods, large scale structural optimization is possible for a wide range of practical applications.

INTRODUCTION

Structural optimization theory has advanced rapidly in recent years, and today most major finite element analysis codes include some form of optimization capability. Key to the efficiency of modern structural optimization was the development of approximation techniques in the mid 1970's. For many common design tasks, the use of formal approximations reduced the cost of optimization by more than one order of magnitude. Also, these methods could easily be incorporated into existing structural analysis programs.

During the 1980's, and continuing today, second generation approximation techniques began to evolve. These include the use of intermediate variables, force approximations for stress constraints, and Rayleigh quotient approximations for frequency constraints, as examples. These methods dramatically improve the quality of the approximations, but are more difficult to incorporate into existing analysis codes. For example, the use of intermediate variables and force approximations requires that, during the approximate optimization phase, the stress recovery routines be efficiently available to the optimization module of the pro-

J. Herskovits (ed.), Advances in Structural Optimization, 117–147.
© 1995 *Kluwer Academic Publishers.*

gram. Most existing analysis codes were not designed with this feature and are difficult to modify to accommodate it. Similarly, for frame structures, the input to the analysis is the section properties, yet the designer needs to change the physical dimensions during optimization.

The purpose here is to discuss the concepts which make large scale structural optimization possible. This discussion closely follows that of reference 1, since approximation techniques are the key ingredient to efficient optimization. It will be seen that proper formulation of the structural optimization task requires that different element types, design variables and responses must be treated differently. The proper choice of intermediate design variables and intermediate responses will have a dramatic effect on the quality of the approximations and therefore the efficiency and reliability of the design program.

Examples are offered to demonstrate the range of design tasks that can now be addressed and the design efficiencies that can be expected.

It is concluded that, using the latest methods, large scale structural optimization is possible for a wide range of practical applications. However, to make the best use of the latest methods, it is necessary to fully integrate the design/analysis process.

BASIC CONCEPTS

Regardless of the optimization method used, the structural optimization task can be stated as, find the set of design variables, X, that will

$$\text{Minimize} \quad F(X) \tag{1}$$

Subject to:

$$g_j(X) \leq 0 \qquad j=1,m \tag{2}$$

$$X_i^L \leq X_i \leq X_i^U \qquad i=1,n \tag{3}$$

The function, $F(X)$, is referred to as the objective or merit function and is dependent on the values of the design variables, X, which themselves include member dimensions or shape variables of a structure as examples. The limits on the design variables, given in Equation 3, are referred to as side constraints and are used simply to limit the region of search for the optimum. For example, it would not make sense to allow the thickness of a structural element to take on a negative value. Thus, the lower bounds would be set to a reasonable minimum gage size. If we wish to maximize $F(X)$, for example, maximize the fundamental frequency of a structure, we simply minimize the negative of $F(X)$. Typically, we can consider about 10-50 variables contained in X if no special features of the problem are taken into consideration and on the order of 100-500 or more variables with specially written programs.

The $g_j(X)$ are referred to as constraints and they provide bounds on various response quantities. The most common constraint is the limits imposed on stresses at various points within the structure. Then if $\bar{\sigma}$ is the upper bound allowed on stress, the constraint function would be written as

$$\frac{\sigma_{ijkl}}{\bar{\sigma}} - 1 \leq 0 \tag{4}$$

i = element, j = stress location, k = stress component, l = load condition

Other typical constraints include displacement, frequency, local buckling, system buckling load factor, dynamic response, aeroelastic divergence and aeroelastic flutter, as examples. Noting that most structures are required to support numerous sets of loads, and constraints such as stress limits must be imposed at hundreds or even thousands of locations within the structure, it is clear that the number, m, of such constraints can become very large. In practical design, the number of nonlinear, inequality constraints can easily exceed one million.

It is common to minimize structural weight, and this is sometimes the only option available in a given software package. However, unless special techniques, such as optimality criterion methods, are being used, there is no need to limit ourselves to minimizing weight. Any calculated response should be available to be chosen as the objective function.

There are many optimization algorithms available. However, in one form or another, most of them are based on a simple iterative formula:

$$X^q = X^{q-1} + \alpha^* S^q \tag{5}$$

where q is the iteration number, S^q is a vector search direction and α^* is called the move parameter (2). This has the physical interpretation that S is a vector direction in n-dimensional space, defining how we wish to simultaneously change the design variables, X, and α^* is the number of steps (fractional steps are allowed) in this direction. Together, the term $\alpha^* S^q$ represent the perturbation, δX, to be made in the design at this iteration. The distinction between most optimization algorithms is in how the search direction, S^q, is chosen and how the "one-dimensional" search to determine the step size, α^*, is performed. Of key importance is the fact that most modern optimization algorithms require that gradient information be available.

FINITE ELEMENT ANALYSIS AS A BASIS FOR OPTIMIZATION

The usual form of the finite element analysis method is the displacement method, where we wish to find the set of displacements, $\mathbf{u}$, that solves the system of equations;

$$[K]\,\mathbf{u} + [C]\,\dot{\mathbf{u}} + [M]\,\ddot{\mathbf{u}} = F(t) \tag{6}$$

where $\mathbf{u}$ is assumed to be a function of time. Depending on the problem at hand, this general form may be added to or parts may be deleted. For example, for aeroelastic optimization an aerodynamic coefficient matrix will be added, while for linear static, elastic analysis, Equation 6 reduces to the familiar form:

$$[K]\,\mathbf{u} = \mathbf{P} \tag{7}$$

where K is the master stiffness matrix, being the sum of elemental stiffness matrices, and $\mathbf{P}$ contains the applied loads. Vector $\mathbf{P}$ may contain multiple columns associated with multiple loading conditions. On the other hand, when considering static aeroelastic analysis, an aeroelastic influence coefficient matrix is added to the K matrix in Equation 6.

Perhaps the most significant developments of recent years then have been, not in the creation of new algorithms for solving the standard problem of Equations 1-3, but instead in the area of problem formulation. It can be argued that, by creating the proper mathematical problem, relative to structural design, the actual optimization algorithm becomes secondary. Thus, we have seen considerable effort devoted to creating approximations to the design task that are of very high quality with respect to the original problem, but which are well suited for optimization. The basic idea here is to somehow first approximate the real problem with explicit, or at least easily evaluated functions. These functions are then used for optimization, returning to the full finite element analysis only to update the approximation.

Sensitivity Analysis

Because the structural responses are implicit, nonlinear, functions of the design variables, it might be assumed that the gradients (sensitivity) of the structural response with respect to the design variables cannot be readily found. Indeed, research in structural optimization continued for approximately five years after its inception in 1960 (3) before sensitivity of displacements and stress with respect to sizing variables was analytically calculated (4). Today, we can calculate the gradient of displacement, stress, frequency, buckling load factor, and even aeroelastic quantities such as divergence and flutter speed with respect to structural design variables. However, this information is available at considerable cost in program development. That is, it is necessary to either make fundamental changes within the finite element code itself or supply some relatively complex external programming to get the needed information (5-7). To understand what is required, consider the relatively simple task of calculating the sensitivity of displacements with respect to design variables for static elastic analysis. We can implicitly differentiate Equation 7 with respect to design variable X_i to get, upon re-arranging.

$$\frac{\partial u}{\partial X_i} = [K]^{-1} \left\{ \frac{\partial P}{\partial X_i} - \left[\frac{\partial K}{\partial X_i} \right] u \right\} \tag{8}$$

where the rate of change of the master stiffness matrix, **K**, is the sum of the rates of change of the elemental stiffness matrices. Note that, once a static analysis has been performed, K^{-1} is available in decomposed form so that Equation 8 requires only some additional forward and backward substitutions for its solution, just as in the case of adding new load vectors. This is equivalent to adding a number of new right-hand sides to the basic analysis equation equal to the number of design variables times the number of load conditions for which gradients are required. Once the rates of change of displacements are calculated, the rates of change of stresses are recovered from the basic stress-displacement equations. Thus, at least mathematically, sensitivities of displacements and stresses are straightforward in static analysis. Only minor extensions are needed for sensitivity calculations of other responses such as frequencies, buckling load factors and other responses. However, experience has shown that the addition of this capability to an existing finite element program can be a major task.

Given the optimizer, finite element analysis and sensitivity analysis capabilities, it is straightforward to couple these to create the full structural synthesis capability. In fact, this was the approach used between the mid 1960's and mid 1970's, and the general program flow is shown in Figure 1.

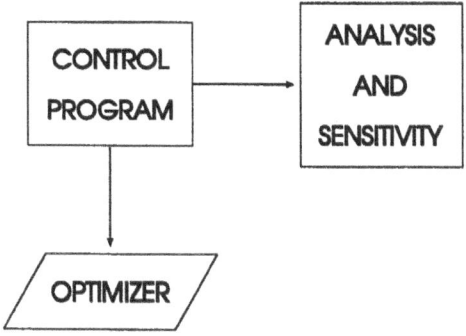

Figure 1: Pre-1974 Optimization Strategy

Whenever the objective function and constraints, or their gradients are needed, the finite element program is called via the control program. While the optimizer may be efficient as judged by optimization technology, the overall cost of optimization can be very high because the cost of a single analysis of a large, practical structure may be very high. Therefore, it is desirable to have methods that do not directly call the finite element program at every step in the optimization process.

APPROXIMATION CONCEPTS

Assuming gradients of the response quantities with respect to the design variables or some intermediate variables are available, a variety of approximation techniques are available by which this information can be manipulated in order to efficiently direct the design process. This approach has the advantages that relatively large numbers of design variables can be considered, large perturbations are allowed between full finite element analyses, and the reliability of the optimization process is improved.

Using approximation concepts, the basic program structure is shown in Figure 2.

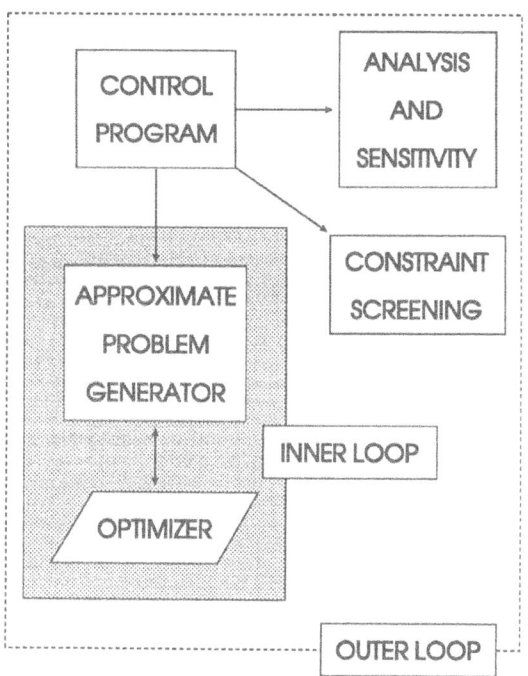

Figure 2: Post-1974 Optimization Strategy

The steps involved in the optimization process are described as follows;

1. Analyze the initial proposed design as a full finite element analysis.

2. Evaluate all constraint functions and rank them according to criticality. Retain only the critical and potentially critical constraints for further consideration during this design cycle.

3. Call the sensitivity analysis to calculate gradients of the retained set of constraints. These may be calculated as gradients of intermediate responses in terms of intermediate variables.

4. Using these gradients, construct approximations and create an optimization problem to be solved by a general purpose optimization code, and solve it. Here, the approximation could be linear, or may be modified in various ways as described in references 8 and 9. During this approximate optimization, move limits are imposed on the design variables to insure the reliability of the approximation.

5. Update the analysis data and call the analysis program again to evaluate the quality of the proposed design, If the solution has converged to an acceptable optimum, terminate. Otherwise repeat from step 2.

From these five steps, it is clear that the analysis and optimization tasks are quite closely coupled. This provides the greatest possible efficiency, but at the expense of major development costs. The overall process consists of an outer loop and an inner loop, as shown in Figure 2. The outer loop consists of analysis, constraint deletion, gradient calculations, and creation of the approximate problem. The inner loop consists of actually solving the approximate optimization problem. One cycle through the inner loop defines an optimization iteration while one cycle through the outer loop may be called a design cycle. For each design cycle, we require a full finite element analysis and gradient computations for responses that are retained for this cycle. Typically about ten design cycles are required, while perhaps 20 or more iterations are required to solve each approximate problem. Thus, the key is to create approximations of very high quality to reduce the number of design cycles (full finite element analyses) and for the approximate functions to be rapidly evaluated to reduce the cost of optimization in the inner loop. The initial concepts used here were developed in the mid 1970's (10,11). Research has continued in this area to create what may be called second generation approximation methods.

To better understand this method, three particular features require further discussion. These are the use of move limits, the concept of constraint deletion and the use of intermediate variables and responses to create the high quality approximations to the design constraints. Also, the concepts of basis vectors and synthetic functions as design aids will be briefly discussed.

Move Limits

Whenever approximations are used in optimization, it is essential to employ move limits during the approximate optimization phase to protect against unreasonably large perturbations in the design variables or the intermediate variables. This is true no matter how good the approximation is believed to be, since in production applications it is common that ill-conditioned design problems will be attempted.

If a simple linear approximation is used, a typical move limit may be ten percent of the current value of the design variables. Using advanced approximation techniques, initial move limits of fifty percent are reasonable. In the case where basis vectors are used to define the design, it is common that the design variables can have zero value and they can often change sign during the optimization process. Here, special care must be taken to define reasonable move limits because a percentage change in the design variable is not physically meaningful.

Additionally, some move limit adjustment algorithm is needed to insure overall convergence of the optimization process. This may be as simple as reducing the move limits by fifty percent whenever the projected and calculated values of the responses do not agree within a prescribed tolerance. In practice, experience has shown that move limits on the design variables should be individually adjusted, either by increasing or decreasing their value, to improve reliability of the process (12, 13). These more complicated algorithms are primarily designed to protect the quality and reliability of the overall process for those cases which include design variables that have significant differences in physical meaning so that reasonable initial move limits cannot be easily prescribed.

Constraint Deletion

During a single design cycle, we require gradients of the responses that define the objective and constraint functions. However, recognizing that calculation of thousands of gradients will be very costly, and that most constraints may be far from critical, it is reasonable to ignore many of the constraints for this design cycle. This is achieved through the concept of constraint deletion. The simplest approach would be to just ignore any constraint which is not within, say, 30% of being critical. However, in a given region of a structure, many elements may have nearly the same stress level. Therefore, we use an additional feature called regionalization. Thus, we first sort all constraints and retain all those that are within a specified tolerance of being critical. Assuming all constraints are normalized, we may retain all those with a numerical value greater than -0.3. Now we further sort the retained set and keep only a specified number (say two) in any single region of the structure, where the regions have been defined as part of the input to the program. This further reduces the number of retained constraints to a manageable set and also reduces the probability of retaining constraints that are mathematically the same. This process is shown in Figure 3, where X identifies a constraint that satisfies the threshold test of -0.3 and Y identifies the most critical constraints to be retained in each region. In general, we can use this approach to reduce the number of retained constraints to perhaps 3-5 times the number of design variables. Any constraints not retained, which become active as a result of the design changes during this cycle, will be retained on future design cycles. Constraint deletion is used simply to reduce the cost of gradient computations and otherwise should have no effect on the overall optimization process.

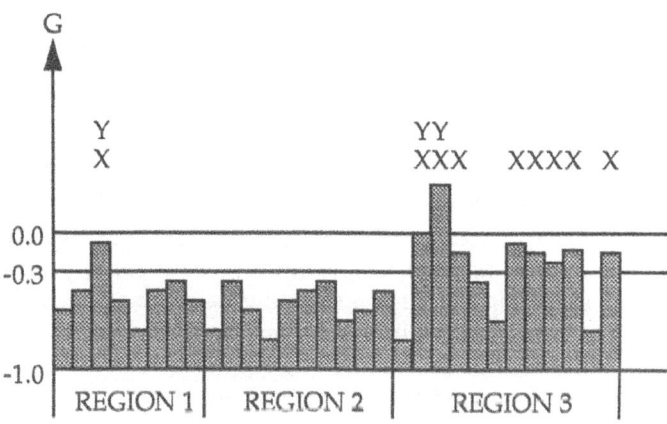

Figure 3: Constraint Deletion

Intermediate Variables and Responses

The concept of using intermediate variables and responses to create a high quality approximation is fundamental to the overall efficiency and reliability of the optimization process.

First consider the original approximation techniques presented in reference 10, where reciprocal variables were used to approximate stresses and displacements. This method may be simply explained by reference to Figure 4, which is a simple rod in tension.

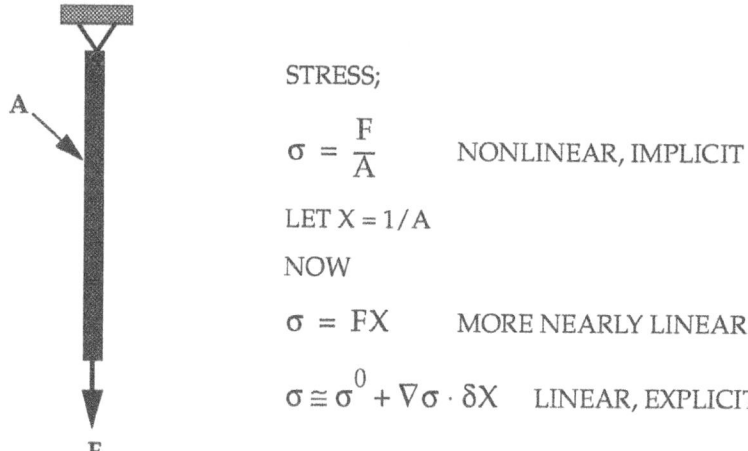

STRESS;

$$\sigma = \frac{F}{A} \qquad \text{NONLINEAR, IMPLICIT}$$

LET $X = 1/A$

NOW

$$\sigma = FX \qquad \text{MORE NEARLY LINEAR}$$

$$\sigma \cong \sigma^0 + \nabla\sigma \cdot \delta X \qquad \text{LINEAR, EXPLICIT}$$

Figure 4: Rod in Tension

It is clear that, even here, the stress is highly nonlinear with respect to the cross-sectional area. Therefore, using area as a design variable, a linear approximation would be of poor quality and we could only change it by a few percent before having to create a new approximation. However, with the simple change of variables, $X = 1/A$, we create a problem that is much more linear with respect to the new variable, and indeed is exact for a statically determinate truss. Thus in X-space, we can make large changes in the design variable before a detailed finite element analysis is needed. Of course, the original variable is easily recovered as $A = 1/X$. Also, if we are minimizing the volume of material, the objective will now be nonlinear, being $V = L/X$. However, it is easily evaluated, along with it's gradients, and so we do not need to approximate the objective. The key idea is that we are free to choose whatever intermediate variables are convenient to create a high quality approximation to the original problem.

Now, we can take this process a step further and make the intermediate variables more complicated functions of the design variables, and also use intermediate responses, rather than the original responses during gradient computations (14-19). This is best understood by considering stress constraints for beam elements. Consider a simple rectangular beam element of width B and height, H. These are the physical design variables that the engineer wishes to determine. Assume we have a simple stress constraint calculated at H/2. Then;

$$\sigma = \frac{Mc}{I} \pm \frac{P}{A} \qquad (9)$$

where c=H/2, I=BH^3/12 and A=BH are simple functions of B and H. M is the bending moment and P is the axial force. A traditional linearization would be to create a Taylor series approximation to stress as;

$$\bar{\sigma} = \sigma^0 + \nabla\sigma \bullet \delta X \tag{10}$$

where $\mathbf{X}^T = (B, H)$ and $\delta X = \mathbf{X} - \mathbf{X}^0$. However, it is clear that stress is highly non-linear in the design variables, B and H, and so very small move limits would be necessary during the solution of the approximate problem.

Now consider how we might better approximate the stress. First, we treat A and I as intermediate variables. Next, we calculate the gradients of M and P (intermediate responses) with respect to A and I;

$$\begin{bmatrix} \tilde{M} \\ \tilde{P} \end{bmatrix} = \begin{bmatrix} M^0 \\ P^0 \end{bmatrix} + \begin{bmatrix} \dfrac{\partial M}{\partial A} & \dfrac{\partial M}{\partial I} \\ \dfrac{\partial P}{\partial A} & \dfrac{\partial P}{\partial I} \end{bmatrix} \begin{bmatrix} A - A^0 \\ I - I^0 \end{bmatrix} \tag{11}$$

When the optimizer requires the value of stress, we first calculate A and I explicitly as functions of B and H. Then, we calculate the member end forces, M and P using equation 11. Finally, we recover the stress in the usual fashion.

With the use of such intermediate variables and responses, we achieve two important goals. First, we allow the engineer to treat the physical dimensions, B and H, as design variables. Second, we retain a great deal of the nonlinearity of the original problem explicitly. This allows us to make very large changes in the design variables during a given design cycle.

As another example, assume we wish to approximate the stresses in a plate element (16).

The six approximate element forces at the center of the element are first calculated using a Taylor Series expansion in terms of the intermediate design variables (see references 18 and 19 for a discussion of intermediate variables).

$$\{\tilde{M}\} = \begin{Bmatrix} \tilde{N}_x \\ \tilde{N}_y \\ \tilde{N}_{xy} \\ \tilde{M}_x \\ \tilde{M}_y \\ \tilde{M}_{xy} \end{Bmatrix} \tag{12}$$

The intermediate design variables for plate structures are the shape design variables, the plate thicknesses, t, and bending stiffnesses, D, where

$$D = \frac{t^3}{12} \tag{13}$$

The approximate surface stresses are then calculated using

$$\{\tilde{\sigma}\} = \begin{Bmatrix} \tilde{\sigma}_x \\ \tilde{\sigma}_y \\ \tilde{\tau}_{xy} \end{Bmatrix} = \begin{Bmatrix} \dfrac{\tilde{N}_x}{t} - \dfrac{\tilde{M}_x z}{D} \\ \dfrac{\tilde{N}_y}{t} - \dfrac{\tilde{M}_y z}{D} \\ \dfrac{\tilde{N}_{xy}}{t} - \dfrac{\tilde{M}_{xy} z}{D} \end{Bmatrix} \tag{14}$$

Finally the approximate principle, maximum shear, and von Mises stresses are calculated using

$$\tilde{\sigma}_I = \frac{\tilde{\sigma}_x + \tilde{\sigma}_y}{2} + \sqrt{\left(\frac{\tilde{\sigma}_x - \tilde{\sigma}_y}{2}\right)^2 + \tilde{\tau}_{xy}^2} \tag{15}$$

$$\tilde{\sigma}_{II} = \frac{\tilde{\sigma}_x + \tilde{\sigma}_y}{2} - \sqrt{\left(\frac{\tilde{\sigma}_x - \tilde{\sigma}_y}{2}\right)^2 + \tilde{\tau}_{xy}^2} \tag{16}$$

$$\tilde{\tau}_{max} = \sqrt{\left(\frac{\tilde{\sigma}_x - \tilde{\sigma}_y}{2}\right)^2 + \tilde{\tau}_{xy}^2} \tag{17}$$

$$\tilde{\sigma}_{VM} = \sqrt{\tilde{\sigma}^2_{\,x} + \tilde{\sigma}^2_{\,y} - \frac{\tilde{\sigma}_x \tilde{\sigma}_y}{2} + 3\tilde{\tau}^2_{\,xy}} \qquad (18)$$

Note that, here we first approximate the internal forces in the element. From these, we approximate the individual stress components, and finally, we calculate the combined stresses. The key again is to retain a great deal of the nonlinearity of the problem using simple, explicit expressions.

In a similar way, other responses can be approximated using our insight into the mathematical nature of the particular response. For example, eigenvalues may be approximated by what is called the Rayleigh quotient approximation;

$$\lambda_k = \frac{U}{T} = \frac{\Phi^T_k K \Phi_k}{\Phi^T_k M \Phi_k} \qquad (19)$$

We now create a linear approximation to the numerator, U, and denominator, T, independently and use the result to estimate λ_k (17). Note that we can use intermediate variables, such as section properties to create these approximations, just as for stress in the previous examples.

Conservative Approximations Of Constraints

Even though the state of the art in approximation concepts is reasonably mature, there are numerous cases where we do not yet have the necessary physical insight to create high quality approximations based on the physics of the response. In such cases, it is desirable to create an approximation that is at least conservative, relative to a simple linear approximation. This can be achieved by what is referred to as conservative approximations (8, 9).

First consider a direct (linear) approximation to a constraint. This is given as

$$g_d(X) = g(X^0) + \sum_{i=1}^{N} \left[\frac{\partial}{\partial X_i} g(X^0) \right] \left(X_i - X^0_i \right) \qquad (20)$$

Alternatively, we may use a reciprocal approximation as was presented in reference 8 for sizing problems. That is, we approximate the constraint in terms of the reciprocal of the design variables as

$$g_r(X) = g(X^0) - \sum_{i=1}^{N} \left[\frac{\partial}{\partial X_i} g(X^0) \right] \left(\frac{1}{X_i} - \frac{1}{X^0_i} \right) (X^0_i)^2 \qquad (21)$$

Neither of these approximations may be best in general. Also, when considering shape variables or using basis vectors described below, the reciprocal approximation may not valid, since the design variable may approach or cross zero. Also, in the absence of better information, we may wish for g(**X**) to be conservative relative to a linear approximation, so

$$g_c(\mathbf{X}) = g(\mathbf{X}^0) + \sum_+ \left[\frac{\partial}{\partial X_i} g(\mathbf{X}^0)\right]\left(X_i - X_i^0\right)$$
$$- \sum_- \left[\frac{\partial}{\partial X_i} g(\mathbf{X}^0)\right]\left(\frac{1}{X_i} - \frac{1}{X_i^0}\right)(X_i^0)^2 \tag{22}$$

Note that the terms contained in $\sum_+ (\ldots)$ imply the use of the direct approximation defined by equation 20, while the terms contained in $\sum_- (\ldots)$ imply the use of the reciprocal approximation defined by equation 21. In other words;

$$\text{If } X_i \frac{\partial g\left(X^0\right)}{\partial X_i} \geq 0 \qquad \text{Use direct expansion} \tag{23}$$

$$\text{If } X_i \frac{\partial g\left(X^0\right)}{\partial X_i} < 0 \qquad \text{Use reciprocal expansion} \tag{24}$$

As an example, consider the following simple function;

$$G(X) = 1 + 3X + X^2 \leq 0 \tag{25}$$

Now create local direct and reciprocal approximations at $X = -1$, $+1$ and $+3$. The corresponding information is given in Table 1.

TABLE 1: Approximations

X	-1	1	3
G(X)	-3	3	1
$\partial G/\partial X$	5	1	-3
$\tilde{G}_D(X)$	2 + 5X	2 + X	10 - 3X
$\tilde{G}_R(X)$	-8 - 5/X	4 - 1/X	-8 + 27/X

Figure 5 is a plot of the true function, as well as the direct and reciprocal approximations. Note that the approximation of choice is the one which over estimates the value of G(X). Thus, the reciprocal approximation is used if the approximate value of G(X) is greater than the linear approximation. Now, if we optimize using the conservative approximation of equation 22, and some approximate constraint is driven to zero, we would expect that the true constraint is satisfied by a small margin.

This "conservative approximation" is considered to be the best approximation in the absence of better information. Recognizing that a design variable may be either positive or negative (assuming the move limits allow this), we modify the simple decision above by always using a direct expansion if X_i is near zero. This is because the reciprocal approximation is undefined at X=0.

Finally, observe that the reciprocal approximation is conservative relative only to a linear approximation. There is no guarantee that either a direct or reciprocal approximation will be conservative relative to the true function. This is because the curvature of the true function is unknown. Therefore, it is not possible to insure that the curvature of the approximation is more positive than that of the true function.

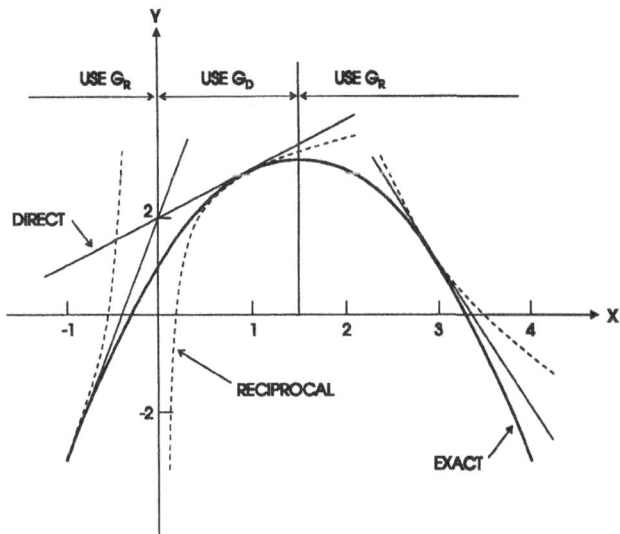

Figure 5: Conservative Approximations

The key concept here is to use the best approximation we can to a particular response. By doing this, we can create very high quality approximations which allow us to change the design variables by a large amount in any design cycle. By retaining the essential nonlinearities of the problem in explicit form, the optimization process is much less dependent on move limits. Experience has shown that these second generation approximation methods are often more efficient than previous methods and are almost always more reliable. The advantages of these "high quality approximations" are a dramatic improvement in design efficiency and reliability.

Basis Vectors

Often, the designer has several good candidate designs and just wishes to refine them to achieve an optimum, or he/she has considerable insight into what the final design should be. Also, in shape optimization, we may treat the shape as a combination of specified shapes in order to insure that the resulting optimum is reasonable. Finally, by providing a set of candidate designs, the number of independent design variables can often be dramatically reduced.

The basic concept here is that we may provide several candidate designs and then find the best linear combination of these designs to achieve the overall design objective (20). In the case of shape optimization, such "basis vectors" can be used to control internal nodes to retain a reasonable mesh in the finite element model, and thus reduce the need for remeshing the analysis model during the optimization process.

For example, let vectors Y^i define coordinates that can be changed, and X_i are the design variables. Then the resulting shape is

$$Y = X_1 Y^1 + X_2 Y^2 + \ldots + \ldots + X_N Y^N \tag{26}$$

In equation 26, the vectors Y^i represent candidate designs, any of which may define a valid design. Alternatively, we may define a nominal design with perturbations which may be added to or subtracted from this nominal. This is actually closer to traditional design, where we are seeking improvements on an existing design. Such a design would be defined as;

$$Y = Y^0 + X_1 Y^1 + X_2 Y^2 + \ldots + \ldots + X_N Y^N \tag{27}$$

A simple example of this approach for shape optimization is shown in Figure 6, which shows a 1/4th model of a plate with a hole. Note that if we treat only the positions around the hole as design variables, we are limited to very small changes in the design before it is necessary to remesh the analysis model. However, by including the positions of the interior nodes in the basis vector, we can change the external shape much more before remeshing the analysis model.

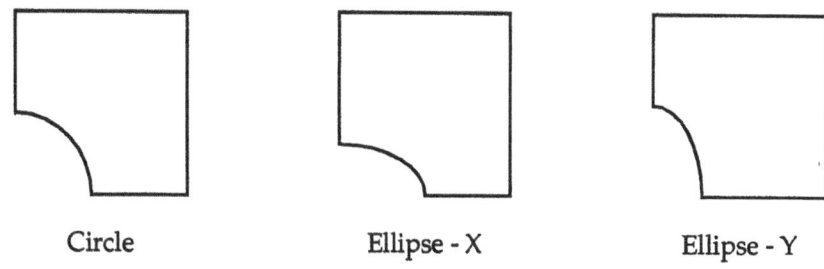

| Circle | Ellipse - X | Ellipse - Y |

Figure 6: Basis Shapes

Synthetic Functions

Finite element analysis codes provide sensitivity for specified responses in terms of specified variables. Usually, the input to the analysis is the section properties, but designers work with dimensions. Also, the analysis code may not calculate local buckling constraints or their sensitivities. Even with a specially written structural synthesis program, it is not possible to anticipate or include all possible relationships between the design variables and section properties, or special responses, such as proprietary failure criteria that the designer may require.

Synthetic functions can bridge the gap between what we have and what we want (21). The basic concept is that the user provides FORTRAN like relationships as part of the input data. The design program then processes these relationships in real time to meet the designer's needs.

Figure 7 gives examples of such relationships for the section properties of a rectangular beam and for an Euler buckling constraint on a rod element. Addition of utilities such as this can significantly enhance the generality of the optimization capability.

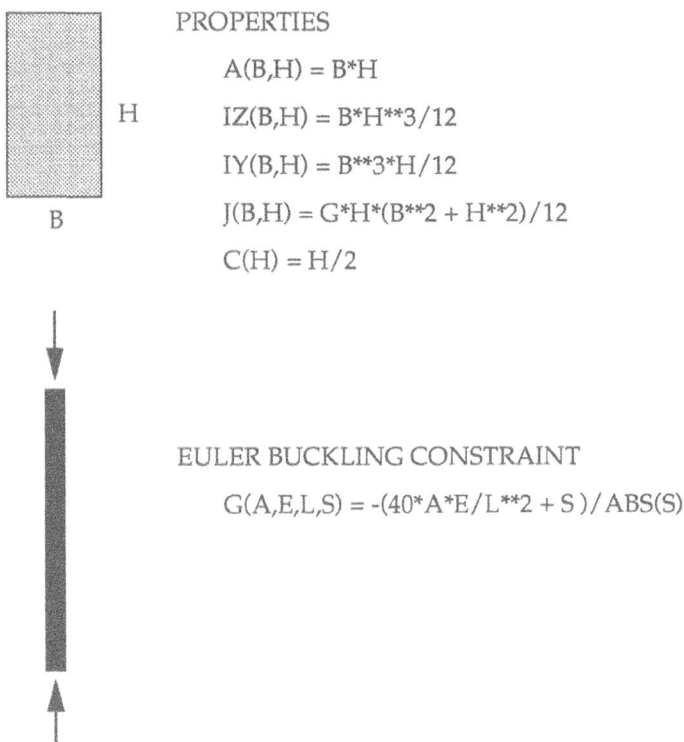

PROPERTIES

$A(B,H) = B*H$

$IZ(B,H) = B*H**3/12$

$IY(B,H) = B**3*H/12$

$J(B,H) = G*H*(B**2 + H**2)/12$

$C(H) = H/2$

EULER BUCKLING CONSTRAINT

$G(A,E,L,S) = -(40*A*E/L**2 + S)/ABS(S)$

Figure 7: Synthetic Functions

STATUS OF APPROXIMATION CONCEPTS

In addition to such features as basis vectors and synthetic functions, the key to creating an efficient structural synthesis code is in the appropriate use of intermediate variables and responses, and the approximations that are possible with such techniques. To give an indication of the present state of the art in this aspect, Table 2 summarizes the intermediate variables and responses that are presently best suited to the approximate optimization process when designing for member sizes.

TABLE 2: Intermediate Variables and Responses For Member Sizing

Element	Intermediate Response	Response	Intermediate Variable	Design Variable
Rod	Force	Stress	Area	Area
	Displacement	Displacement	Mixed	Area
	Rayleigh Quotient	Frequency	Area	Area
Beam	Force	Stress	Section Property	Dimension
	Displacement	Displacement	Section Property	Dimension
	Rayleigh Quotient	Frequency	Section Property	Dimension
Membrane	Force	Stress	Thickness	Thickness
	Displacement	Displacement	Mixed	Thickness
	Rayleigh Quotient	Frequency	Thickness	Thickness
Plate/Shell	Force	Stress	Bending Stiffness	Thickness
	Displacement	Displacement	Mixed	Thickness
	Rayleigh Quotient	Frequency	Bending Stiffness	Thickness

Table 3 summarizes the intermediate variables and responses that are presently best suited to the approximate optimization process when designing for optimum shape. In this case, member dimensions are included as design variables, since it is not meaningful to consider shape variables only when discrete elements are included.

TABLE 3: Intermediate Variables and Responses For Shape Optimization

Element	Intermediate Response	Response	Intermediate Variable	Design Variable
Rod	Force	Stress	Area	A, X, Y, Z
	Displacement	Displacement	Mixed	A, X, Y, Z
	Rayleigh Quotient	Frequency	Mixed	A, X, Y, Z
Beam	Force	Stress	Section Property	Dimension, X, Y, Z
	Displacement	Displacement	Section Property	Dimension, X, Y, Z
	Rayleigh Quotient	Frequency	Section Property	Dimension, X, Y, Z
Membrane	Force	Stress	Thickness	Thickness, X, Y, Z
	Displacement	Displacement	Mixed	Thickness, X, Y, Z
	Rayleigh Quotient	Frequency	Mixed	Thickness, X, Y, Z
Plate Shell	Force	Stress	Bending Stiffness	Thickness, X, Y, Z
	Displacement	Displacement	Mixed	Thickness, X, Y, Z
	Rayleigh Quotient	Frequency	Mixed	Thickness, X, Y, Z
Solid	Stress	Stress	Mixed	X, Y, Z
	Displacement	Displacement	Mixed	X, Y, Z
	Rayleigh Quotient	Frequency	Mixed	X, Y, Z

DESIGN EXAMPLES

Here, several examples are offered to demonstrate the efficiency and reliability of the structural synthesis process using modern approximation techniques.

Portal Frame

The portal frame shown in Figure 7 has been used in the literature to demonstrate optimization methods.

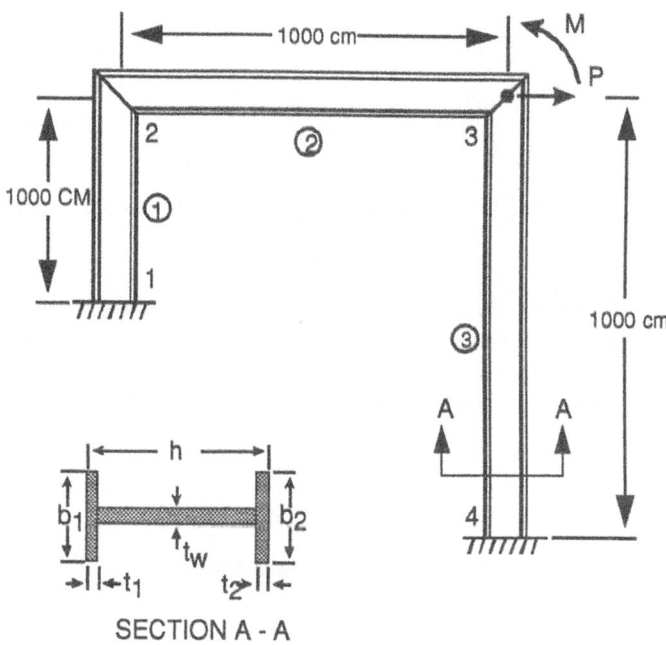

Figure 7: Portal Frame

The design task is to;

Minimize

$$V = \sum_{i=1}^{3} \text{i i} \qquad \text{Volume of material} \qquad (28)$$

Subject to;

$-12,000 \le \sigma_{ij} \le 12,000$ — Stress limits in each element for each load case (29)

$-0.4 \le \delta_{ij} \le 0.4$ — Horizontal displacement limit at each joint for each load case (30)

$-0.015 \le \theta_{ij} \le 0.015$ — Rotation limit at each joint for each load case (31)

$\sigma_{ab} = -1.4464E\left(\dfrac{t_f}{b_f}\right)^2 \le \sigma_{ij}$ — Flange buckling limit for each element and load case (32)

$\tau_{ij} = \dfrac{1.5|V|}{A} \le 11,600$ — Web shear stress limit for each element and load case (33)

Here, this problem is solved using force approximations as intermediate responses and member section properties as intermediate variables. The actual design variables are the member dimensions. The section properties are calculated using the following synthetic functions

$$A(B,H,TF,TW) = 2.0*TF*B+TW*(H-2.0*TF) \tag{34}$$

$$IZ(B,H,TF,TW) = (B*H**3-(B-TW)*(H-2.0*TF)**3)/12. \tag{35}$$

The local buckling and shear stress constraints utilize the following synthetic functions

$$G1(B,TW,E,S) = -(1.4464*E*(TW/B)**2 - S)/5000. \le 0 \tag{36}$$

$$G2(H,TW,V) = 1.5*ABS(V)/(H*TW) \le 11,600 \tag{37}$$

The iteration history is shown in Figure 8, beginning from a very feasible and a very infeasible design. A design cycle is considered one cycle through the outer loop of Figure 2, including a full finite element analysis, constraint deletion, creation and solution of the approximate optimization problem. In the worst case, only ten detailed finite element analyses were needed to reach the optimum. Here, in order to demonstrate the robustness of second generation approximation techniques, no move limits were used.

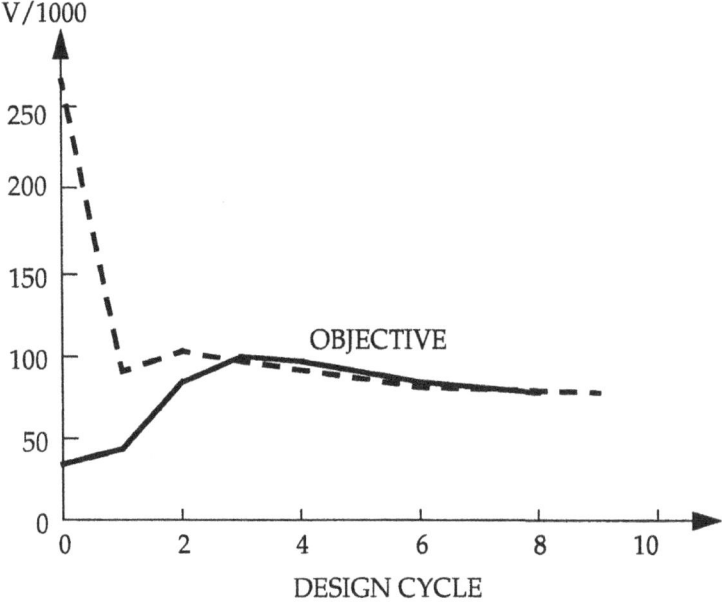

Figure 8: Iteration History

18 Bar Truss

Figure 9 shows the initial and final configuration for the weight minimization of an 18-bar truss. This structure was designed subject to stress and local buckling constraints. The details of the design problem are given in reference 22. The design variables are both member sizes and node locations. Member forces were used as intermediate responses. An optimum of 4505 was found using 8 detailed analyses. The best previous solution reported (23) was 4668, requiring 111 detailed analyses.

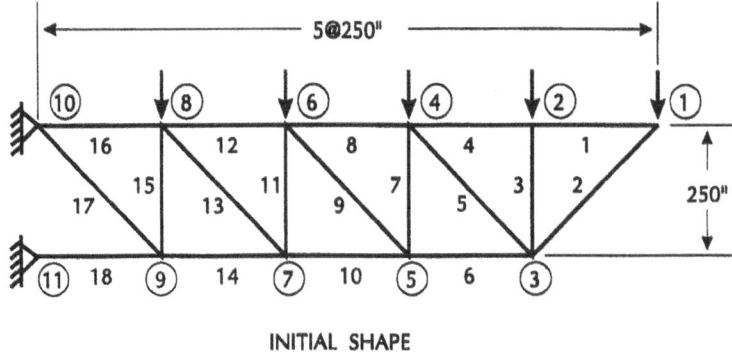

INITIAL SHAPE

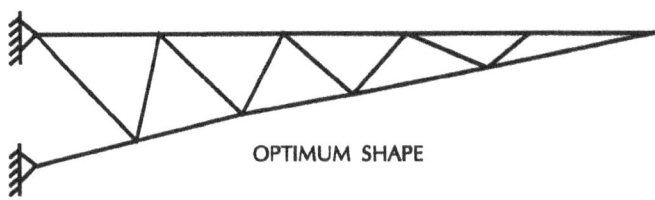

OPTIMUM SHAPE

Figure 9: Eighteen Bar Truss

Marine Gear Housing

Figure 10 shows a large marine gear housing which is designed for minimum weight, subject to stress and displacement constraints. A total 30 design variables were considered and there were 5620 inequality constraints. Six independent load conditions were considered. A one-fourth model of the doubly symmetric struc- ture was modeled with 1623 finite elements and 7239 independent degrees of free- dom. Reciprocals of the plate and rod member sizes were used as intermediate design variables. The details of this problem are given in reference (24). A sum- mary of run times for a single design cycle is given in Table 4. A total CPU time of approximately one hour was required to solve this problem. This problem was solved using first generation approximation techniques by coupling an external optimization program with the MSC/NASTRAN finite element program, Version 63 (25).

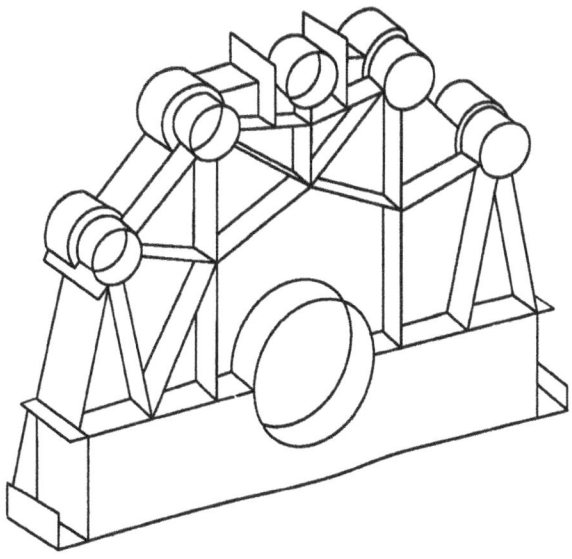

Figure 10: Marine Gear Housing

TABLE 4: Marine Gear Housing Run Time Summary (Cray 1S)

Phase	Seconds
Initial Sort	5.0
MSC/NASTRAN Analysis	243.4
Constraint Sort	13.0
MSC/NASTRAN Sensitivity	337.0
Approximate Optimization	1.5
Total Per Design Cycle	600.0

Engine Connecting Rod

The connecting rod model shown below has been used in the literature to demonstrate optimization methods. This is considered to be a simple example of problems which can be routinely solved on a workstation.

This structure was modeled using 1120 eight node elements for a one fourth model (approximately 5000 DOFs). The objective was to minimize mass, subject to stress constraints. The Basis Design approach was used and 9 shape design variables were considered. The optimization task was solved using the GENESIS program (27).

Beginning from an arbitrary design, the mass was reduced by over 30% and required nine design cycles (9 detailed FEM analyses). The design time (CPU time) on an entry level workstation was under four hours.

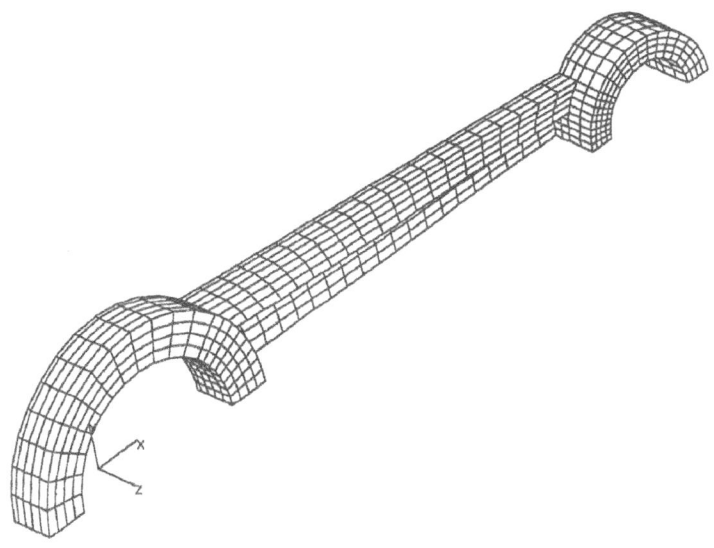

Figure 11: Engine Connecting Rod

Aircraft Wing Box

A simplified structural box of a high aspect ratio wing shown below. This structure is required to support two separate loading conditions, including a lifting load and a lift-torsion load. The structure is modeled with 3000 elements and 11,400 degrees of freedom. 125 design variables were considered, being the element thicknesses and the areas of the spar caps. A total of over 12,000 design constraints were considered. These include stress limits, displacement limits at the wing tip and a lower bound on the fundamental frequency. Beginning from an arbitrary (but reasonable) design, an optimum was achieved in 6 design cycles (7 detailed FEM analyses). The initial design violated constraints by over 10%, while the optimum design satisfied all design constraints. The weight was reduced by

20% This design problem was also solved on a workstation in under eight hours of CPU time.

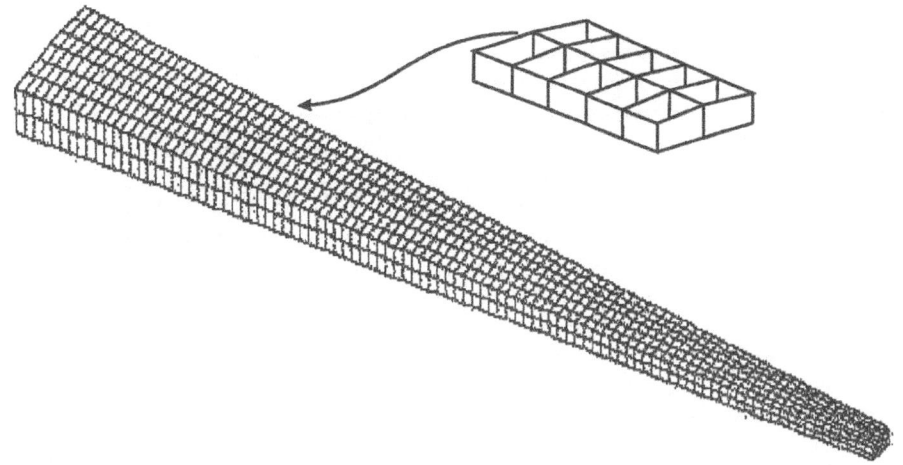

Figure 12: Wing Box

Car Body

Finally, consider the car body shown in Figure 13. Half Model Optimization was performed on a Cray Y-MP supercomputer, based on existing FEA Model. There were 20,000 Analysis Degrees of Freedom and 67 Design Variables. The optimization task was to reduce the material volume based on a total of 17,500 constraints. The optimization task required six design cycles. Beginning from an arbitrary, but slightly infeasible design, a feasible design was found while reducing the material volume by twenty percent.

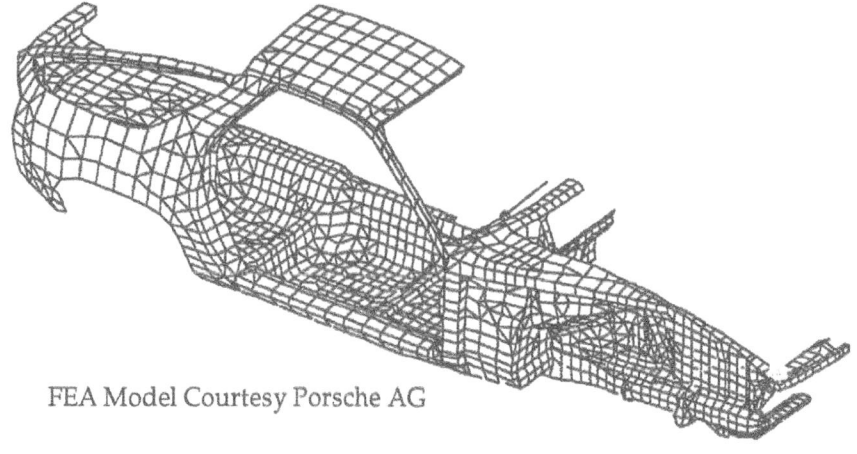

FEA Model Courtesy Porsche AG

Figure 13: Car Body

SUMMARY

Structural optimization technology is now reasonably mature and these methods have been added to most commercial finite element codes. Various methods are employed, ranging from simple coupling between the analysis and optimization to reasonably sophisticated use of approximation techniques.

Second generation finite element software, such as the GENESIS program (27), is now beginning to become available. The motivation for this is that it is quite difficult to realize the full advantages of the latest approximation methods with existing programs that are modified to use optimization.

In developing a fully integrated structural optimization program, to be a design program from the beginning, the latest methods described here and elsewhere can be efficiently used. These include providing analytical gradients in an efficient manner, use of basis vectors, constraint deletion, synthetic functions and formal approximations in an organized fashion. Most importantly, a program structure can be created that is best able to accommodate both analysis and design. Key to this is a unified data structure which can accomodate all of the necessary parts and be expandable to accomodate new methods in the future.

Finally, as seen from the summary of approximation methods, it is clear that research and development will continue to expand the realm of problems that can be solved and to continue the improve the efficiency and reliability of the overall structural synthesis process.

REFERENCES

1. Vanderplaats, G. N., "Approximation Concepts for Structural Design Optimization," S. Kodiyalam and M. Saxena, editors, Geometry and Optimization Techniques for Structural Design, Computational Mechanics Publications, Elsevier Applied Science, Southampton, 1994.

2. Vanderplaats, G.N., Numerical Optimization Techniques for Engineering Design: with Applications, McGraw-Hill, 1984.

3. Schmit, L.A., "Structural Design by Systematic Synthesis," Proc. 2nd Conference on Electronic Computation, ASCE, New York, 1960, pp. 105-122.

4. Fox, R. L., "Constraint Surface Normals for Structural Synthesis Techniques," AIAA Journal, Vol. 3(8), 1965, pp. 1517-1518.

5. Yang, R.J. and Botkin M.E., "The Relationship Between the Variational Approach and Implicit Differentiation Approach to Shape Design Sensitivities," Proc. 26th AIAA/ASME/ASCE/AHS Structures, Structural Dynamics and Materials Conference, Orlando, FL, 1985.

6. Vanderplaats, G.N. and Miura, H., "Trends in Structural Optimization: Some Considerations in Using Standard Finite Element Software, "Proc. 6th Vehicle Structural Mechanics Conference, Detroit, MI, April 1986.

7. Haug, E.J., Choi, K.K. and Komkov, V., Design Sensitivity Analysis of Structural Systems, Academic Press, 1984.

8. Starnes, J.R. Jr. and Haftka, R.T., "Preliminary Design of Composite Wings for Buckling, Stress and Displacement Constraints," Journal of Aircraft, Vol. 16, Aug. 1979, pp. 564-570.

9. Fleury, C. and Braibant, V., "Structural Optimization: A New Dual Method using Mixed Variables," Int. J. of Numerical Methods in Engineering, Vol. 23, No. 3, 1986, pp. 409-429.

10. Schmit, L. A., and Farshi, B., "Some Approximation Concepts for Structural Synthesis," AIAA J., Vol. 12, No. 5, 1974, pp. 692-699.

11. Schmit, L.A., and Miura, H., "Approximation Concepts for Efficient Structural Synthesis," NASA CR-2552, 1976.

12. Bloebaum, C. L., "Variable Move Limit Strategy for Efficient Optimization," Proceedings of the AIAA/ASME/ASCE/AHS/ASC 32nd Structures, Structural Dynamics and Materials Conference, Baltimore, Maryland, AIAA, Washington, D. C., pp. 431-437, 1991.

146

13. Thomas, H. T., Vanderplaats, G. N. and Shyy, Y-K, "A Study of Move Limit Adjustment Strategies in the Approximation Concepts Approach to Structural Synthesis," Proc. Fourth AIAA/USAF/NASA/OAI Symposium on Multidisciplinary Analysis and Optimization, Cleveland, OH, Sept. 21-23, 1992, pp. 507-512.

14. Salajegheh, E. and Vanderplaats, G.N., "An Efficient Approximation Method for Structural Synthesis with Reference to Space Structures," International Journal of Space Structures, Vol 2, No 3, 1986/87, pp. 165-175.

15. Vanderplaats, G. N. and Salajegheh, E., "A New Approximation Method for Stress Constraints in Structural Synthesis," AIAA J., Vol. 27, No. 3, March 1989, pp. 352-358.

16. Vanderplaats, G. N. and Thomas, H. L., "An Improved Approximation for Stress Constraints in Plate Structures," Structural Optimization, Vol. 6, No. 1, 1993, pp. 1-6.

17. Canfield, R. A., "An Approximation Function for Frequency Constraints in Structural Optimization," Proc. of the 2nd NASA/Air Force Symposium on Recent Advances in Multidisciplinary Analysis and Optimization, Hampton, VA, Sept. 28-30, 1988, pp. 937-953.

18. Yoshida, N. and Vanderplaats, G. N., "Structural Optimization Using Beam Elements," AIAA Journal, Vol. 26, No. 4, pp. 454-462, 1988.

19. Vanderplaats, G. N. and Salajegheh, E., "An Efficient Approximation Technique for Frequency Constraints in Frame Optimization," International Journal for Numerical Methods, Vol. 26, pp. 1057-1069, 1988.

20. Pickett, R. M., Jr., Rubinstein, M. F. and Nelson, R. B., "Automated Structural Synthesis Using a Reduced Number of Design Coordinates," AIAA Journal, Vol. 11, no. 4, 1973, pp. 489-494.

21. Vanderplaats, G.N., Miura, H., Cai, H.D. and Hansen, S.R., "Structural Optimization using Synthetic Functions," Proceedings, 1989 AIAA/ASME/ASCE/AHS 30th Structures, Structural Dynamics and Materials Conference, AIAA, Washington, DC, 1989.

22. Hansen, S.R. and Vanderplaats, G.N., "An Approximation Method for Configuration Optimization of Trusses," AIAA Journal, Vol 28, No. 1, Jan. 1990, pp. 161-172.

23. Imai, K., Configuration Optimization of Trusses by the Multiplier Method, Report No. UCLA-ENG-7842, Mechanics and Structures Department, School of Engineering and Applied Science, University of California, Los Angeles, 1978.

24. Vanderplaats, G.N., Miura, H. and Chargin, M., "Large Scale Structural Synthesis," J. Finite Elements in Analysis and Design, Vol. 1, No. 3, June 1985, pp. 117-130.

25. MSC/NASTRAN Version 63 User's Manual, The MacNeal-Schwendler Corporation, Los Angeles, CA 1983.

26. Kodiyalam, S., Vanderplaats, G. N., Miura, H., Nagendra, G. K. and Wallerstein, D. V., "Structural Shape Optimization with MSC/NASTRAN," J. Computers and Structures, (In Press).

27. GENESIS User's Manual, VMA Engineering, Santa Barbara, CA, June 1991.

WHAT IS MEANINGFUL IN TOPOLOGY DESIGN?
AN ENGINEER'S VIEWPOINT

G.I.N. ROZVANY
FB 10, Essen–University
Postfach 10 37 64, D–45117 Essen, Germany

1. Introduction

"Ingenieure beschäftigen sich mit *unbeantworteten Fragen*, Mathematiker liefern *ungefragte Antworten*" (a German mathematician, who wishes to remain anonymous, 1993).[1]

Topology design, a recently coined synonym for the traditional term "layout optimization", is rapidly becoming one of the trendiest topics in structural optimization. In an unprecedented rush for the bandwagon, diverse research groups try to outpublish and out-selfquote each other in a quest for authority. Since the nonspecialist reader is justifiably confused by conflicting claims, it is the aim of this paper to establish some basic philosophy for deciding which research activities are actually meaningful in this fascinating field.

It is certainly not the author's intention to belittle the contributions of mathematicians, for whom he has the highest respect. On the contrary, the rather ambitious aim of this paper is to bridge over the schism of comprehension that exists between the two professions mentioned above. As the highest expectation, it is hoped that our venerable colleagues from mathematics departments may

- learn as to which research aims are technologically meaningful and relevant;

---

[1] The pun is untranslatable: freely interpreted "whilst engineers deal with unanswered questions, mathematicians come up with answers nobody needs". It is based on the clever interchange of the words "Antwort" (answer) and "Frage" (question), which does not work out in English. Mathematicians may produce some technologically irrelevant results, but seem to have a better sense of humor than engineers.

J. Herskovits (ed.), Advances in Structural Optimization, 149–188.

- use the terminology of structural mechanics which has been established for some time by engineers;
- mention identical but much earlier results by their engineering colleagues (if it makes them feel better, they may call them "intuitive results by engineering methods" or "conjectures"); and possibly
- utilize concepts and methodological short-cuts of structural mechanics, after making them perhaps more respectable by "more rigorous" definitions and proofs.

The author has been somewhat surprised to find that outstanding mathematicians in this field often do not understand common terms of structural mechanics, such as "statical determinacy", "kinematic admissibility", "generalized stresses", "St. Venant's principle" or "principle of stationary mutual energy", etc.

The quest for ever-increasing rigour is commendable but also has some practical drawbacks. Beautifully concise proofs by Clerk Maxwell, A.G.M. Michell or William Prager are no longer good enough and are replaced by a ten to twenty times longer one, *without a single case where the conclusions of the former were actually found incorrect.* More importantly, the set of problems amenable to treatment of acceptable rigour is rapidly shrinking to completely artificial and technologically irrelevant ones (see Section 3).

2. Common Misnomers in Topology Optimization

"The layout of a structure is the general arrangement adopted by a designer. ... it is not commonly appreciated that the best layout is in fact subject to very strict rules. The first of these rules was stated and proved by Clerk Maxwell and ... the general theory of design was developed from Clerk Maxwell's rule by A.G.M. Michell" (Cox, 1958).

2.1. "TOPOLOGY DESIGN" VS. "LAYOUT OPTIMIZATION" AND "GENERALIZED SHAPE OPTIMIZATION"

Layout optimization was a well-defined field already in the 1960s. Considering a grid-like structure, such as a truss, for example, it involves three simultaneous operations, namely

- topology optimization (the optimal choice of the spatial sequence or configuration of members and joints);
- geometrical optimization (the determination of the best location or coordinates of the joints); and
- sizing optimization (selection of cross-sectional dimensions).

Prager and the author outlined two subfields of layout optimization (for a review, see e.g. Rozvany and Ong, 1987, p. 170). In *"classical" layout optimization*, we deal with grid-like structures consisting of very slender members. As a result of this, we can calculate the total weight (cost), by adding the weight (cost) of individual members in various directions, neglecting the effect of intersections on cost, strength or stiffness. We can also say that classical layout optimization considers structures with a *low volume fraction* (i.e. low ratio of material volume and available volume).

Considering now, for example, a *truss* bounded by two parallel planes, we can change the limiting values in the behavioural constraints progressively so that the truss members in the optimal design will become increasingly thicker. We shall then reach the stage when the effects of the intersections of members can no longer be ignored and the optimal solution will contain increasingly rounded corners at these intersections. In modelling such problems, we must consider a *perforated plate* in plane stress. Optimization of the latter requires two operations: first the optimal *microstructures* must be derived for given ratios of principal stiffnesses or forces and then the *layout* (macroscopic configuration) of such microstructures must be determined. Prager and the author termed this combined operation *"advanced layout optimization"*, and the author introduced the alternative term (e.g. Rozvany, Zhou and Sigmund, 1994) *"generalized shape optimization"* for reasons as follows. In determining the optimal solution for a perforated plate, for example, both the *shape and the topology of the internal boundaries* (holes) must be optimized simultaneously. This task is therefore more general than traditional shape optimization, in which the topology of internal boundaries is usually given (e.g. one hole of unknown shape in a plate).

The above transition is illustrated graphically in Fig. 1, in which the classical least-weight truss solution (Fig. 1a, Chan, 1960) changes gradually (Figs. 1b and c) as we increase the volume fraction. The latter two are based conceptually on Vigdergauz's (1994) optimal microstructures for a single load condition with either a stress or a compliance constraint.

In the literature, the term "topology optimization" is often used either for classical layout optimization (of trusses, for example), or for advanced layout optimization, i.e. generalized shape optimization (of perforated or composite plates, for example). This is clearly confusing, because topology optimization is a *suboperation* of the above tasks, and hence it cannot signify the combined solution process as well.

2.2. "OPTIMIZATION BY HOMOGENIZATION"

As mentioned above, in a *gridlike structure* (cf. classical layout theory) the *volume of holes* is relatively high, whilst in *perforated or porous structures* (cf. advanced layout theory or generalized shape optimization) it is rela-

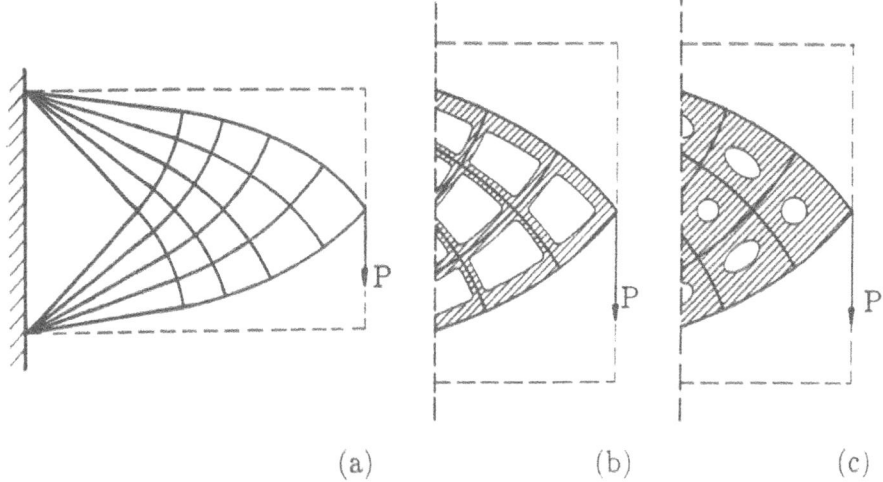

Figure 1. Example illustrating the difference between "classical" and "advanced" layout optimization at progressively increasing volume fractions.

tively low. If we fill the holes with a second material, we have a *composite structure*. It turns out, however, that if we optimize any of the above systems, we usually finish up with a so-called *periodic structure*, in which the size of the holes is infinitesimal and their shape repeats itself, with a slower variation. In *analysing* a periodic structure, we must calculate its stiffness by some averaging or "smear-out" process, also termed "homogenization". This has been done ever since the work of Michell (1904) for structures with densely spaced components. In structural mechanics, it is convenient to relate these stiffnesses to *"generalized" stresses and strains* (Prager, 1974) of a structural system (such as, for example, a twisting moment and a rate of twist in a shell grid). Working out the equivalent stiffnesses of a periodic structure either analytically or numerically is no big magic for engineers and has been done correctly for some time by using accepted principles of structural mechanics (e.g. Rozvany, Olhoff, Bendsøe *et al.* 1987). Moreover, for reinforced concrete plates, for example, engineers have used "homogenized" stiffness and strength equivalents for many decades. More recently, mathematicians defined "homogenization" in terms of elastic coefficients (E_{ijkl}) of a fictitious anisotropic elastic continuum having the same stiffnesses as the periodic structure. Apart from adding some proofs of convergence, which is to be expected and hence of little *practical* value in engineering, this approach uses 2D or 3D continua with *traditional stresses* and not generalized stresses and hence it is more restricted than the average stiffness values for *generalized stresses*. In any case, homogenization forms a part of the analysis of periodic structures and not of their optimization. Hence the term in the title makes little sense.

2.3. "OPTIMIZATION OF NEW MATERIALS"

We may optimize

- the microstructure and layout of a periodic (perforated or composite) structure, or
- the nonhomogenity of structures whose elastic properties vary.

For reasons of correctness and clarity, the author prefers the above terms to expressions like "new, optimized, artificial, advanced, high-tech, self-adaptive materials", although these may have advantages in obtaining research grants from sponsors of material sciences.

3. Technological Relevance of Various Topology Design Problems

"During the last two decades research on structural optimization became increasingly concerned with two aspects: the application of general numerical methods of optimization to structural design of complex real structures, and the analytical derivation of necessary and sufficient conditions for the optimality of broad classes of comparatively simple and more or less idealized structures. Both kinds of research are important: the first for obvious reasons; the second, because it furnishes information that is useful in testing the validity, accuracy and convergence of numerical methods and in assessing the efficiency of practical designs" (Prager and Rozvany, 1977b).

Both topology optimization problems and methods for solving them fall into two broad categories which, together with their relative importance, are shown in Table 1. The above quotation mentions two important classes of research problems, which are shown under (b) and (c). As can be seen from Table 1, there are another two, albeit without much significance.

The value judgements in Table 1 can be justified as follows:

(a) In practical, real-world problems, the boundary and loading conditions, as well as design constraints are rather complex and therefore closed form, explicit solutions are extremely rare, even with the help of special computer programs for analytical operations.

(b) Discretized solutions for real-world topology design problems are *directly* useful for obvious reasons, particularly in view of the unavailability of exact solutions for these problems.

(c) Exact analytical solutions for idealized or artificial problems are *indirectly* useful because they

- throw some light on fundamental features of optimal topologies;

TABLE 1. Basic classes of research problems and types of solutions in topology optimization

	Exact (explicit, closed form) analytical solutions	Approximate (discretized, iterative) numerical solutions
Complex, real-world, practical problems	(a) Very rare	(b) Directly useful
Simple, idealized or artificial problems	(c) Indirectly useful	(d) Not very meaningful

- can be used for testing the validity, accuracy and convergence of numerical methods used under (b); and

- provide a basis for assessing the relative economy of practical designs.

(d) Although this class of solutions occupies most of the current literature, approximate, *discretized solutions of artificial problems* cannot be used *directly* because of their technological irrelevance, nor *indirectly* because of their rather fuzzy resolution and often low accuracy for detecting fundamental features of optimal solutions (see Fig. 13 later). Moreover, analytical solutions often show that an infinite number of layouts are equally optimal in some regions. Numerical solutions pick randomly one of these nonunique solutions, thereby increasing the confusion further.

The aim of this paper is to demonstrate that, using certain optimality criteria methods,

- *discretized* optimal layouts (topologies) can be obtained for most real-world, multiload, multipurpose structures [(b) in Table 1]; and
- significant progress has been made in deriving some *exact analytical* solutions for the same problems [(c) in Table 1] for verification purposes.

4. Real-World Problems vs. Artificial Problems

"**Compliance**: action in accordance with request or command" (Concise Oxford Dictionary).

4.1. REAL-WORLD PROBLEMS

In most structural problems in practice,

- *a number of load conditions* must be considered; and
- design codes or standards *prescribe limiting values* on quantities representing structural behaviour (or response), in terms of stresses, displacements, buckling loads, natural frequencies, ultimate collapse load, service life expectancy, or alternatively, probabilities of various limit states.

Then a solution is chosen whose weight, cost (in a financial sense), some nonstructural property (e.g. an aerodynamic property for aeroplanes) or some weighted combination of the above quantities is as favourable as possible. The above *real-world problems* can be formulated as follows:

minimize an objective function (or functions, e.g. weight, cost, etc.)
subject to constraints on the behaviour (stress, displacements, etc.). (1)

4.2. EXAMPLE OF AN ARTIFICIAL PROBLEM: COMPLIANCE DESIGN

In mechanics, "compliance" is a strange term for an abstract concept, meaning the total amount of external work for a structure (which for elastic structures equals the total amount of internal work). No bridge, aeroplane, car or any other structure would be designed in a real-world situation for a compliance constraint or for minimum compliance. It has been known since the fifties (Cox, 1958) and also shown recently by mathematicians (e.g. Bendsøe *et al.*, 1994) that

- for some simple structures (e.g. trusses) and
- for a single loading condition

optimal compliance design and optimal design for equal permissible stresses in compression and tension are identical within a constant multiplier (see also Section 6.7 herein). The same conclusion was extended in the sixties (Hegemier and Prager, 1969) to optimal compliance design and optimal design for a given natural frequency. The above findings, however, do not justify extensive research into compliance design, because similar equivalences cannot be extended to

- more complicated structural elements,
- several load conditions,
- more realistic design (e.g. stress and displacement) constraints, or
- unequal permissible stresses in tension and compression (which is the case for most materials).

Compliance design is usually formulated as follows:

minimize the compliance (for one load condition)
subject to a given structural weight. (2)

For several load conditions, two types of compliance problems are used in the literature:

minimize the weighted combination of compliances
for several load conditions, (3)

or alternatively

minimize the maximum compliance out of those for several load conditions,
subject to a given structural weight. (4)

We term the above problems *inverse design problems*, because they require the minimization of some response variable subject to given weight. In real-world problems the weight (or cost) is minimized subject to constraints on the responses. The corresponding compliance problem would become:

minimize the structural weight,
subject to a given compliance value (for one load condition). (5)

Naturally, the above direct formulation gives the same results as (2) above, but this does not make compliance design more meaningful. Compliance constraints can be regarded as a special case of displacement constraints. As will be seen later, compliance problems have the advantage that the corresponding problem is

- convex and
- selfadjoint (see Section 5.1),

which makes their rigorous mathematical treatment relatively easy.

4.3. TRUSSES VS. GRILLAGES IN TOPOLOGY STUDIES

Much of the literature on exact optimal layouts and almost all literature on discretized layout solutions deals with trusses, usually ignoring the problem of member buckling of compression bars. This makes topology studies based on trusses highly unrealistic. On the other hand, grillages (or beam systems) are relatively stable systems with similar permissible stresses for positive and negative moments. This was the author's reason for devoting his first book to the optimization of flexural systems (Rozvany, 1976). It was correctly pointed out by Prager (Prager and Rozvany, 1977b) that "despite of its late start, this"(the grillage layout) "theory advanced farther than that of optimal trusses" since for grillages the exact optimal layout was already then available "for almost all loading and boundary conditions". For the few unsolved grillage problem classes, the gaps are being filled currently (e.g. Rozvany, 1994a, b; Rozvany and Liebermann, 1994).

5. Exact Analytical Solutions for Layout (Topology) Problems

"Most of the literature on structural optimization is concerned with the optimal choice of cross-sectional dimensions. When the *layout* as well as the cross-sectional dimensions are at the choice of the designer, structural optimization becomes a much more challenging problem" (Prager and Rozvany, 1977a).

5.1. HISTORICAL NOTES

The basic principles of the so-called "optimal layout theory" were outlined in a revolutionary contribution some 90 years ago by Michell (1904), which was little appreciated by his contemporaries and found response first some fifty years later (e.g. Cox, 1958). Michell's method consists of applying optimality criteria to a highly connected network of potential members, later termed "basic structure" (Prager, 1974), or "ground structure" (Dorn *et al.*, 1964) or "structural universe" (e.g. Rozvany, 1981). An important aspect of this method is that vanishing members (of zero cross-sectional area) must also fulfil some optimality criteria (usually inequalities) regarding their strain values. Michell's ingenious ideas were extended by Prager and Rozvany (e.g. 1977a) from trusses to other structures. This was a very useful but fairly obvious step requiring a good understanding of Michell's work but little inventiveness.

More recent extensions of the optimal layout theory have made use of the concept of the so-called *adjoint structure*. This is a fictitious structure which is used as a mechanical analogy for visualizing optimality criteria in a context that helps to utilize engineering intuition and insight in obtaining exact analytical solutions. In some idealized problems (e.g. trusses with a stress or a "compliance" constraint), including that of Michell (1904), the real and adjoint structures were identical, which made their treatment very simple. Such problems are termed selfadjoint, and usually also turn out to be convex. However, practical, real-world problems to be discussed subsequently are neither selfadjoint, nor convex.

5.2. WHY ALL REAL-WORLD PROBLEMS ARE NON-SELFADJOINT

As will be seen in Section 6.7, a problem with displacement constraints is selfadjoint only if (i) we have only one displacement constraint per load condition and (ii) in each displacement constraint the weighting factors for various locations are proportional to the loads at the same locations. There is no reason whatsoever, why this should be so in real-world problems! More-

158

over, problems with stress constraints for several load conditions or with a combination of stress and displacement constraints are in general non-selfadjoint. For these reasons, it is difficult to imagine any real-world structural design problem that is selfadjoint. Moreover, non-selfadjoint problems are also nonconvex, which seems to cause difficulties for the proof of existence in their rigorous treatment.

6. Derivation of Optimality Criteria for the Exact Optimal Layout of Multiload, Multipurpose Structures

"Michell structures are designed for only one loading case and depend on the appropriate specification of the strain field. Unfortunately, these structures are statically determinate and impractical, consisting of non-standard length and an infinite number of bars" (Topping, 1993).

The original theory of Michell covered, indeed, only one loading condition, although for *plastic* design of trusses, powerful methods are available for two (Nagtegaal and Prager, 1973; Spillers and Levy, 1971; and Hemp, 1973) or more (Rozvany and Hill, 1978) load conditions. Exact optimality criteria and exact solutions for multiload, multipurpose *elastic* structures were developed only recently and will be discussed in this chapter. Even for the latter, optimal truss solutions often turn out to be statically determinate, but this in itself does not make them impractical. Moreover, the exact optimal layout often contains a finite number of members (see e.g. Figs. 3, 7, 11 and 12 later). Even in the case of an infinite number of members, a very good practical design can be obtained by

- selecting a finite number of nodes in the exact layout, and
- connecting them with straight members.

This procedure is illustrated in Fig. 2 and has been found to yield structural weights within one or two percent of the exact optimal weight. Figure 2a shows a finite number of members of the exact layout (which consists of an infinite number of members) and Fig. 2b shows the corresponding simplified design. Nonstandard member lengths can be a practical disadvantage but can become more economical at very long spans or in repetitive production.

If the designer prefers a simple layout with standard lengths, then a correspondingly simple ground structure (structural universe) can be used and optimality criteria methods still have significant advantages over mathematical programming methods (see Sections 8 and 9).

Finally, the exact least-weight solution does not "depend" on the specification of a strain field; the optimal strain field for Michell trusses is usually unique and necessary for finding the exact optimal solution.

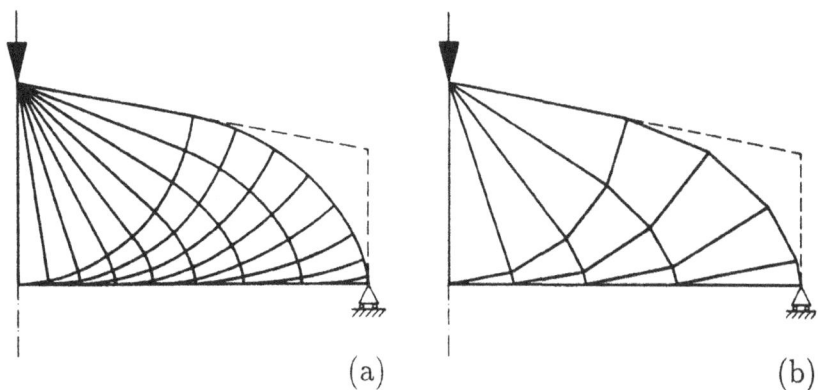

Figure 2. (a) Exact Michell layout and (b) simplified design based on it.

6.1. GENERAL PROCEDURE FOR DERIVING CONTINUUM-TYPE OPTIMALITY CRITERIA (COC AND DCOC)

The basic steps of the derivation of so-called continuum-type optimality criteria (COC, e.g. Rozvany, 1989) or its discretized equivalent (DCOC, e.g. Zhou and Rozvany, 1992/93) are outlined below.

- A relaxed problem formulation is used in which static admissibility, but no kinematic admissibility of the "real" structure is prescribed.
- The displacement constraints are expressed by means of work equations containing "real" ($\mathbf{f}$) and "virtual" ($\hat{\mathbf{f}}$) internal forces.
- The Lagrange multipliers $\overline{\mathbf{u}}$ for the equilibrium equations of the real structure are shown to represent the displacements of a fictitious structure termed *adjoint structure*, which is subject to
  - •• the sum of products of the virtual loads for displacement constraints and the corresponding Lagrangians, and
  - •• prestrains in stress-controlled members which are given by the gradients of the stress constraints.
- As an optimality condition, the adjoint displacements ($\overline{\mathbf{u}}$) and adjoint strains ($\overline{\varepsilon}$) must be kinematically admissible.
- The Lagrange multipliers ($\mathbf{u}$) of the equilibrium equations for the virtual forces are shown to represent the displacements for the real structure. As an optimality condition, the real displacements ($\mathbf{u}$) and real strains (ε) must be kinematically admissible, which was not required in the original relaxed problem formulation.

To show that $\overline{\mathbf{u}}$ and $\mathbf{u}$ are kinematically admissible displacements for the adjoint and real structures, the following forms of equilibrium conditions have been used in the literature:

- *Continua:* specific equilibrium equations (e.g. Rozvany and Zhou, 1989/90, pp. 63-64), or
  the principle of virtual displacements (e.g. Rozvany and Zhou, 1993, pp. 11-13).
- *Discretized (FE) systems:* statics matrix (e.g. Zhou and Rozvany, 1992/93, p. 14).

Note. In the above continuum formulation, small letters were used for all variables. In discretized formulation, small letters will be used at the element level and capital letters at the system level.

More recently, it was shown (Zhou and Haftka, 1994) that the same optimality conditions, as expected, are obtained by suitably modifying the traditional discretized optimality criteria (DOC, see e.g. Berke and Khot, 1987) formulation. For a detailed treatment of continuum-type optimality criteria, the reader is referred to the author's book on this subject (Rozvany, 1989).

6.2. CURRENT PROCEDURE FOR DERIVING OPTIMALITY CRITERIA FOR EXACT OPTIMAL LAYOUTS

The approach used currently by the author consists of the following steps:

(a) Derive optimality criteria for discretized systems consisting of a finite number of elements with a prescribed minimum cross-section. Such optimality criteria are now available for any combination of a great variety of design constraints, as will be seen in Sections 8 and 9.

(b) Extend these optimality criteria for members with a zero cross-sectional area (i.e. when the prescribed minimum cross-sectional area tends to zero).

(c) Extend these optimality criteria to the case when the number of elements is increased to infinity.

As will be seen in Sections 6.4 and 6.5, steps (b) and (c) above are based on engineering intuition. The latter is usually avoided in a rigorous mathematical treatment (e.g. Strang and Kohn, 1983).

6.3. ILLUSTRATION OF THE DERIVATION OF OPTIMALITY CRITERIA FOR EXACT LAYOUTS: TRUSSES WITH DISPLACEMENT AND LOWER SIDE CONSTRAINTS AND DIFFERENT STRESS CONSTRAINTS FOR TENSION AND COMPRESSION

For illustrative purposes, we consider *initially* one load condition. The considered problem can be stated as

$$\min W = \sum_{e=1}^{E} \gamma^e L^e x^e \,, \tag{6}$$

subject to

$$f^e/\sigma^{e+} - x^e \leq 0 \quad (e = 1, \ldots, E), \tag{7}$$

$$-f^e/\sigma^{e-} - x^e \leq 0 \quad (e = 1, \ldots, E), \tag{8}$$

$$\sum_{e=1}^{E} \frac{f^e \hat{f}_k^e L^e}{E^e x^e} - w_k \leq 0 \quad (k = 1, \ldots, K), \tag{9}$$

$$x_{\downarrow}^e - x^e \leq 0, \tag{10}$$

$$\mathbf{P} - \mathbf{BF} = 0, \tag{11}$$

$$\hat{\mathbf{P}}_k - \mathbf{B}\hat{\mathbf{F}}_k = 0, \tag{12}$$

where W is the truss weight, w_k a limiting displacement value, $\mathbf{P}$ and $\overline{\mathbf{P}}_k$ the real and virtual loads, $\mathbf{B}$ the statics matrix and the following symbols refer to the e-th element: γ^e is the specific weight, L^e the length, x^e the cross-sectional area (with the prescribed lower limit $x_{\downarrow}^e$), σ^{e+} and σ^{e-} the permissible stresses in tension and compression, f^e and $\hat{f}_k^e$ the real and virtual member forces (at system level $\mathbf{F}$ and $\hat{\mathbf{F}}^e$) and E^e Young's modulus. The relations (7) and (8) represent stress constraints, (9) displacement constraints, (10) lower side constraints, and (11) and (12) equilibrium of the real and virtual forces.

Adopting the Lagrange multipliers $\lambda^{e+}L^e$, $\lambda^{e-}L^e$, ν_k, $\beta_{\downarrow}^e L^e$, $\overline{\mathbf{U}}$ and $\mathbf{U}$ for the constraints (7)–(12), the Kuhn-Tucker conditions with respect to the variables stated below become:

Design Variables (x^e)

$$\gamma^e - \lambda^{e+} - \lambda^{e-} - \frac{f^e \sum_{k=1}^{K} \nu_k \hat{f}_k^e}{E^e x^{e2}} - \beta_{\downarrow}^e = 0. \tag{13}$$

Real Members Forces (f^e)

$$\lambda^{e+}/\sigma^{e+} - \lambda^{e-}/\sigma^e + \sum_{k=1}^{K} \nu_k \frac{\hat{f}_k^e}{E^e x^e} = \overline{\varepsilon}^e. \tag{14}$$

It was shown earlier (e.g. Rozvany and Zhou, 1991 for trusses or Zhou and Rozvany, 1992/1993 for any elastic structure) that $(-1/L^e)(\partial/\partial f^e)$ $(\overline{\mathbf{U}}^T \mathbf{BF}) = \overline{\varepsilon}^e$ can be interpreted as the kinematically admissible strain in the member e of a fictitious or *"adjoint" truss*.

Virtual Member Forces $(\hat{f}_k^e)$

$$\frac{f^e}{E^e x^e} = \varepsilon^e . \tag{15}$$

It was shown in the above cited papers that $(-1/L^e)\,(\partial/\partial \hat{f}_k^e)\,(\mathbf{U}^T \mathbf{B}\hat{\mathbf{F}}_k)$ $= \varepsilon^e$ respresents the kinematically admissible strain in the member e of the "real" truss.

All Lagrange multipliers are nonnegative, and nonzero only if the corresponding constraint is active.

From the above Kuhn-Tucker conditions, the following optimality criteria can be derived for various cases.

(I) Some Displacement Constraints Active
(A) Side Constraint Inactive
(a) Stress Constraint in Tension Active
$(x^e > x_\downarrow^e, \, f^e/x^e = \sigma^{e+})$

Then we have $\lambda^{e-} = \beta_\downarrow^e = 0$ and by (13) and (14)

$$\lambda^{e+} = \gamma^e - \frac{f^e \displaystyle\sum_{k=1}^{k} \nu_k \hat{f}_k^e}{E^e x^{e2}} , \tag{16}$$

$$\bar{\varepsilon}^e = \frac{\lambda^{e+}}{\sigma^{e+}} + \sum_{k=1}^{k} \frac{\nu_k \hat{f}_k^e}{E^e x^e} = \frac{\gamma^e}{\sigma^{e+}} . \tag{17}$$

(b) Stress Constraint in Compression Active
$(x^e > x_\downarrow^e, \, -f^e/x^e = \sigma^{e-})$

$$\bar{\varepsilon}^e = -\frac{\gamma^e}{\sigma^{e-}} . \tag{18}$$

(c) Stress Constraint Inactive
$(x^e > x_\downarrow^e, \, -\sigma^{e-} < f^e/x^e < \sigma^{e+})$

Then we have $\lambda^{e+} = \lambda^{e-} = \beta_\downarrow^e = 0$ and by (13) and (14)

$$x^e = \sqrt{\frac{f^e \displaystyle\sum_{k=1}^{K} \nu_k \hat{f}_k^e}{E^e \gamma^e}} , \tag{19}$$

$$\bar{\varepsilon}^e = \frac{\displaystyle\sum_{k=1}^{K} \nu_k \hat{f}_k^e}{E^e x^e} . \tag{20}$$

(B) Side Constraint Active
(a) Stress Constraint in Tension Active
$(x^e = x_\downarrow^e, \; f^e/x^e = \sigma^{e+})$
Then we have $\lambda^{e-} = 0$ and by (13) and (14)

$$\lambda^{e+} = \gamma^e - \frac{f^e \sum\limits_{k=1}^{K} \nu_k \hat{f}_k^e}{E^e x^{e2}} - \beta_\downarrow^e \geq 0, \tag{21}$$

$$\frac{\sum_{k=1}^{K} \nu_k \hat{f}_k^e}{E^e x_\downarrow^e} \leq \bar{\varepsilon}^e \leq \frac{\gamma^e}{\sigma^{e+}}. \tag{22}$$

(b) Stress Constraint in Compression Active
$(x^e = x_\downarrow^e, \; -f^e/x^e = \sigma^{e-})$

$$-\frac{\gamma^e}{\sigma^{e-}} \leq \bar{\varepsilon}^e \leq \frac{\sum_{k=1}^{K} \nu_k \hat{f}_k^e}{E^e x_\downarrow^e}. \tag{23}$$

(c) Stress Constraint Inactive
$(x^e = x_\downarrow^e, \; -\sigma^{e-} < f^e/x^e < \sigma^{e+})$
Then we have $\lambda^{e+} = \lambda^{e-} = 0$ and by (13) and (14)

$$x_\downarrow^e = \sqrt{\frac{f^e \sum\limits_{k=1}^{K} \nu_k \hat{f}_k^e}{E^e(\gamma^e - \beta^e)}}, \tag{24}$$

$$\bar{\varepsilon}^e = \frac{\sum\limits_{k=1}^{K} \nu_k \hat{f}_k^e}{E^e x_\downarrow^e}. \tag{25}$$

(II) All Displacement Constraints Inactive
For this case we have $\nu_k = 0$ (for all k).
(A) Side Constraint Inactive
(a) Stress Constraint in Tension Active
$(x^e > x_\downarrow^e, \; f^e/x^e = \sigma^{e+})$
Then we have $\lambda^{e-} = \beta_\downarrow^e = 0$, $\nu_k = 0$ (for all k) and by (13) and (14)

$$\lambda^{e+} = \gamma^e, \tag{26}$$

$$\bar{\varepsilon}^e = \frac{\gamma^e}{\sigma^{e+}}. \tag{27}$$

(b) Stress Constraint in Compression Active

$(x^e > x^e_{\downarrow}, -f^e/x^e = \sigma^{e-})$

$$\bar{\varepsilon}^e = -\frac{\gamma^e}{\sigma^{e-}}. \tag{28}$$

(B) Side Constraint Active
(a) Stress Constraint in Tension Active
$(x^e = x^e_{\downarrow}, f^e/x^e = \sigma^{e+})$
Then we have $\lambda^{e-} = 0$, $\nu_k = 0$ (for all k) and by (13) and (14)

$$\lambda^{e+} = \gamma^e - \beta^e, \tag{29}$$

$$0 \leq \bar{\varepsilon}^e \leq \frac{\gamma^e}{\sigma^{e+}}. \tag{30}$$

(b) Stress Constraint in Compression Active
$(x^e = x^e_{\downarrow}, -f^e/x^e = \sigma^{e-})$

$$-\frac{\gamma^e}{\sigma^{e-}} \leq \bar{\varepsilon}^e \leq 0. \tag{31}$$

Important Note. If all displacement constraints are inactive, then the above optimality criteria can only be used in general in plastic design, because the Kuhn-Tucker conditions do not imply kinematic admissibility of the elastic strains. However, many layout solutions for optimal plastic design are also valid for elastic design owing to statical determinacy (Sved, 1954), which always implies kinematic admissibility (see e.g. Fig. 3).

6.4. EXTENSION TO VANISHING CROSS-SECTIONS

If the prescribed minimum cross-sectional area tends to zero ($x^e_{\downarrow} \to 0$), allowing members to vanish, then the folllowing modifications are necessary.

6.4.1. Plastic Design for Different Permissible (Yield) Stress in Tension and Compression
For $x^e_{\downarrow} \to 0$, (30) and (31) combine into a single condition and hence we have the following adjoint strains:

$$(\text{for } f^e > 0) \quad \bar{\varepsilon}^e = \frac{\gamma^e}{\sigma^{e+}}, \tag{32}$$

$$(\text{for } f^e < 0) \quad \bar{\varepsilon}^e = -\frac{\gamma^e}{\sigma^{e-}}, \tag{33}$$

$$(\text{for } f^e = 0) \quad -\frac{\gamma^e}{\sigma^{e-}} \leq \bar{\varepsilon}^e \leq \frac{\gamma^e}{\sigma^{e+}}, \tag{34}$$

where $\bar{\varepsilon}^e$ must be kinematically admissible. It is to be remarked that the above optimality conditions are fully consistent with the general theory of optimal plastic design (Prager and Shield, 1967; see also Rozvany, 1989, pp. 32-58).

6.4.2. *Elastic Design for Displacement Constraints*
For this case $x_1^e \rightarrow 0$ and (19), (20), (24) and (25) with

$$\varepsilon^e = f^e/(x^e E^e) \tag{35}$$

imply

$$(\text{for } x^e > 0) \quad (E^e/\gamma^e)\varepsilon^e\bar{\varepsilon}^e = 1, \tag{36}$$

$$(\text{for } x^e = 0) \quad (E^e/\gamma^e)\varepsilon^e\bar{\varepsilon}^e \leq 1. \tag{37}$$

6.4.3. *Elastic Design for Stress and Displacement Constraints*
For stress-controlled members we have

$$(\text{for } x^e > 0) \quad \varepsilon^e = \sigma^{e+}/E, \tag{38}$$

$$(\text{for } x^e < 0) \quad \varepsilon^e = -\sigma^{e-}/E, \tag{39}$$

and then (32) and (33) reduce to (36). For $f^e = x^e = 0$ we must satisfy both inequalities (34) and (37).

6.5. EXTENSION TO AN INFINITE NUMBER OF MEMBERS

From a practical, engineering point of view, it is reasonable to assume that if (i) we consider an infinitely dense truss with potential members running at all points of a plane in all directions and (ii) the strains in such members must be kinematically admissible (compatible), then we can replace the corresponding strains with a plane strain field. This has the advantage that we can make use of known geometrical properties of plane strain for deriving *exact* optimal truss layouts. The same approach can be extended to three-dimensional (space) trusses.

As usual, the author does not believe that the above extension leads to an error in deriving exact layout solutions, but it would be nice if mathematicians made this step respectable by treating it within the framework of their theoretical concepts.

6.6. EXTENSION TO SEVERAL LOAD CONDITIONS

We consider here trusses subject to load conditions $\ell = 1, \ldots, L$ with real forces f_ℓ^e and virtual forces $\hat{f}_{\ell k}^e$ ($k = 1, \ldots, K_\ell$) for the K_ℓ displacement constraints under the load condition ℓ.

6.6.1. *Plastic Design*
For several load conditions (32)-(34) change to

$$(\text{for } f_\ell^e > 0) \quad \bar{\varepsilon}_\ell^e = \mu_\ell^e \frac{\gamma^e}{\sigma^{e+}}, \tag{40}$$

$$(\text{for } f_\ell^e < 0) \quad \bar{\varepsilon}_\ell^e = -\mu_\ell^e \frac{\gamma^e}{\sigma^{e-}}, \tag{41}$$

$$(\text{for } f_\ell^e = 0) \quad \bar{\varepsilon}_\ell^e = \mu_\ell^e \left[\alpha \frac{\gamma^e}{\sigma^{e+}} - (1-\alpha) \frac{\gamma^e}{\sigma^{e-}} \right], \tag{42}$$

$$\sum_{\ell=1}^{L} \mu_\ell^e = 1, \quad \mu_\ell^e \geq 0, \quad 1 \geq \alpha \geq 0. \tag{43}$$

Note that μ_ℓ^e can be different for each element e.

6.6.2. *Elastic Design for Displacement Constraints*
For several load conditions, (36) and (37) are replaced by

$$(\text{for } x^e > 0) \quad (E^e/\gamma^e) \sum_{\ell=1}^{L} \varepsilon_\ell^e \bar{\varepsilon}_\ell^e = 1, \tag{44}$$

$$(\text{for } x^e = 0) \quad (E^e/\gamma^e) \sum_{\ell=1}^{L} \varepsilon_\ell^e \bar{\varepsilon}_\ell^e \leq 1, \tag{45}$$

where

$$\bar{\varepsilon}_\ell = \frac{\sum_{k=1}^{K\ell} \nu_{\ell k} \hat{f}_{\ell k}}{E^e x^e}. \tag{46}$$

6.7. OPTIMALITY CRITERIA FOR SOME SELFADJOINT PROBLEMS

6.7.1. *Proportional Displacement Constraints*
Proportional displacement constraints mean that for each load condition only one displacement constraint is prescribed in the following manner.

- For a single point load, the displacement is constrained (i) at the point of application and (ii) in the direction of that point load.
- For several loads, a weighted combination of displacements, matching the location and directions of the loads, is considered, in which the weighting factors are the same as the magnitude of the corresponding loads.

This means that for proportional displacement constraints the real and virtual loads, forces, displacement and strains can be made identical for each load condition (ℓ)

$$\mathbf{P}_\ell = \hat{\mathbf{P}}_\ell, \quad \mathbf{f}_\ell = \hat{\mathbf{f}}_\ell, \quad \mathbf{u}_\ell = \hat{\mathbf{u}}_\ell, \quad \varepsilon_\ell = \hat{\varepsilon}_\ell, \tag{47}$$

and hence *for one load condition* we have

$$\bar{\varepsilon} = \nu\varepsilon. \tag{48}$$

On the basis of (48), the above problems are called "selfadjoint".

This means that (36) and (37) are replaced by

$$\text{(for } x^e > 0) \quad \nu(E^e/\gamma^e)\varepsilon e^2 = 1, \qquad \text{(for } x^e = 0) \quad \nu(E^e/\gamma^e)\varepsilon e^2 \leq 1. \tag{49}$$

Moreover, for several load conditions (44) and (45) reduce to

$$\text{(for } x^e > 0) \quad (E^e/\gamma^e)\sum_{\ell=1}^{L}\nu_\ell\varepsilon_\ell^{e^2} = 1, \tag{50}$$

$$\text{(for } x^e = 0) \quad (E^e/\gamma^e)\sum_{\ell=1}^{L}\nu_\ell\varepsilon_\ell^{e^2} \leq 1, \tag{51}$$

after replacing $\nu_{\ell k}$ with ν_ℓ in (46).

Proportional displacement constraints are equivalent, within a constant multiplier, to *compliance constraints*.

6.7.2. *Equivalence of a Proportional Displacement Constraint and Equal Stress Constraint in Tension and Compression for One Load Condition*
For $\sigma^{e+} = \sigma^{e-} = \sigma^e$, the relations (32), (33) and (34) reduce to

$$\text{(for } x^e > 0) \quad (\sigma^e/\gamma^e)|\bar{\varepsilon}^e| = 1, \tag{52}$$

$$\text{(for } x^e = 0) \quad (\sigma^e/\gamma^e)|\bar{\varepsilon}^e| \leq 1. \tag{53}$$

By squaring (52) and (53), we obtain expressions similar to (49) within a constant multiplier, which shows the equivalence of the two classes of problems given in the title of this section.

6.8. SOME GEOMETRICAL PROPERTIES OF OPTIMAL MULTILOAD PLANE TRUSS LAYOUTS FOR DISPLACEMENT CONSTRAINTS

Using the optimality conditions (44) and (45), it was shown (Rozvany, Birker and Lewiński, 1994) that irrespective of the number of load conditions and displacement constraints, *optimal members may run only in two* (usually non-orthogonal) *directions* at any given point of the plane. The only exception is the degenerate case, in which all directions are equally optimal.

Moreover, it was found that at least for some problems with a straight supporting line and the same point loaded in various directions by alternate loads, the optimal layout consists of two symmetrically located straight members (see Sections 7.3 to 7.5).

6.9. THE CONCEPT OF THE LAYOUT CRITERION FUNCTION

Most of the above criteria for layout optimality can be brought to the following form:

$$(\text{for } x > 0) \quad \phi = 1, \quad (\text{for } x = 0) \quad \phi \leq 1, \tag{54}$$

where ϕ is the so-called layout criterion functions. For several load conditions and displacement constraints, for example, by (44) and (45) the layout criterion function becomes

$$\phi = (E^e/\gamma^e) \sum_{\ell=1}^{L} \varepsilon_\ell^e \bar{\varepsilon}_\ell^e. \tag{55}$$

7. Examples of Exact Optimal Plane Truss Layouts

"While this kind of structure is not practical, it furnishes a lower bound on the structural weight of more realistic designs that is useful in the evaluation of their structural efficiency" (Rozvany and Prager, 1976, on exact optimal layout solutions).

With the increasing capability of including more and more design constraints in exact layout solutions, the first part of the above quotation is becoming somewhat obsolete.

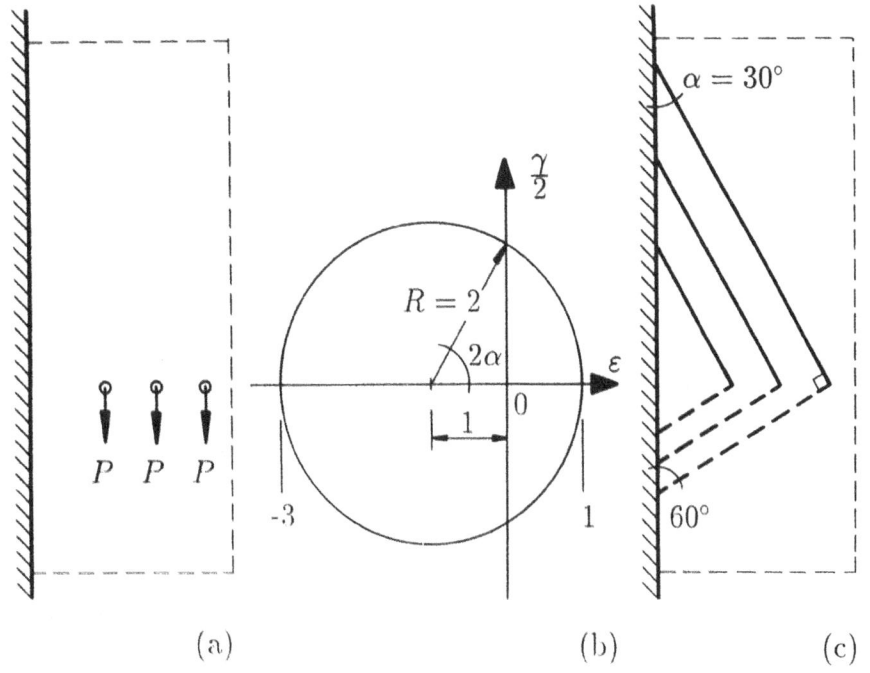

Figure 3. Optimal truss layout for different permissible stresses in tension and compression.

7.1. OPTIMAL PLANE TRUSS LAYOUT FOR DIFFERENT PERMISSIBLE STRESSES IN TENSION AND COMPRESSION, ONE LOAD CONDITION

Three vertical point loads along a horizontal line are to be transmitted to a vertical support (Fig. 3a) by a least-weight truss having the permissible stresses $\sigma^{e+} = 1$ and $\sigma^{e-} = 1/3$ for all members. The specific weight is the same, $\gamma^e = 1$, for all members. Then by (32)–(34), we have

$$(\text{for } f^e > 0) \quad \bar{\varepsilon}^e = 1, \tag{56}$$

$$(\text{for } f^e < 0) \quad \bar{\varepsilon}^e = -3, \tag{57}$$

$$(\text{for } f^e = 0) \quad -3 \leq \bar{\varepsilon}^e \leq 1. \tag{58}$$

Since nonzero forces correspond to *directional maxima and minima* of the adjoint strain field $\bar{\varepsilon}$ for a *plane* truss, optimal member directions must coincide with the *principal directions* of that strain field.

It is assumed that the direction and magnitude of principal strains is location-independent, that is, they have the same values over the entire plane. This means that in the vertical direction we must have a zero strain,

$$\bar{\varepsilon}_{\text{vert}} = 0 \,, \tag{59}$$

because along the vertical supporting line the displacements, and hence the strains, are zero. It is shown subsequently that the strain field assumed above satisfies all optimality criteria. It follows from (56)–(58) that the principal strains are

$$\bar{\varepsilon}_1 = 1 \,, \quad \bar{\varepsilon}_2 = -3 \,. \tag{60}$$

The corresponding Mohr-circle is shown in Fig. 3b. The location $\bar{\varepsilon} = 0$ on the $\bar{\varepsilon}$-axis (strains given by the point P) corresponds to the vertical direction. Assuming members in the principal directions (Fig. 3c, in which compression bars are denoted by broken lines), the angle between the vertical support and the tension bars will be denoted by α. Then by Fig. 3b we have

$$\cos(2\alpha) = 1/2 \Rightarrow \alpha = 30^\circ, \quad \alpha_1 = 60^\circ \,, \tag{61}$$

which determine the optimal directions of the bars. Since the solution is statically determinate, it is also valid for *elastic* design with stress constraints. A constant elastic real strain field given by the above solution would actually be kinematically inadmissible, because the principal strains

$$\varepsilon_1 = 1 \,, \quad \varepsilon_2 = -1/3 \tag{62}$$

would produce a nonzero vertical strain, which violates the kinematic boundary condition. This is consistent with the note at the end of Section 6.3, which explained that the underlying real elastic strain field is usually incompatible in optimal plastic layout design. Figure 3c demonstrates the fact that optimal trusses usually contain a finite number of members, if the support conditions are not too restrictive.

7.2. EXACT OPTIMAL PLANE TRUSS LAYOUTS FOR EQUAL PERMISSIBLE STRESSES IN TENSION AND COMPRESSION, ONE LOAD CONDITION

This is the classical problem of Michell (1904), for which relatively few exact solutions were known until recently. A systematic exploration of solutions for various support and load conditions was started by the author's research team in Essen around 1990. Solutions for line supports (Rozvany and Gollub, 1990), and for cantilevers and point supports (Lewiński, Zhou and Rozvany, 1994) were investigated in a comprehensive fashion, involving some rather advanced functions (e.g. Lommel functions of two variables).

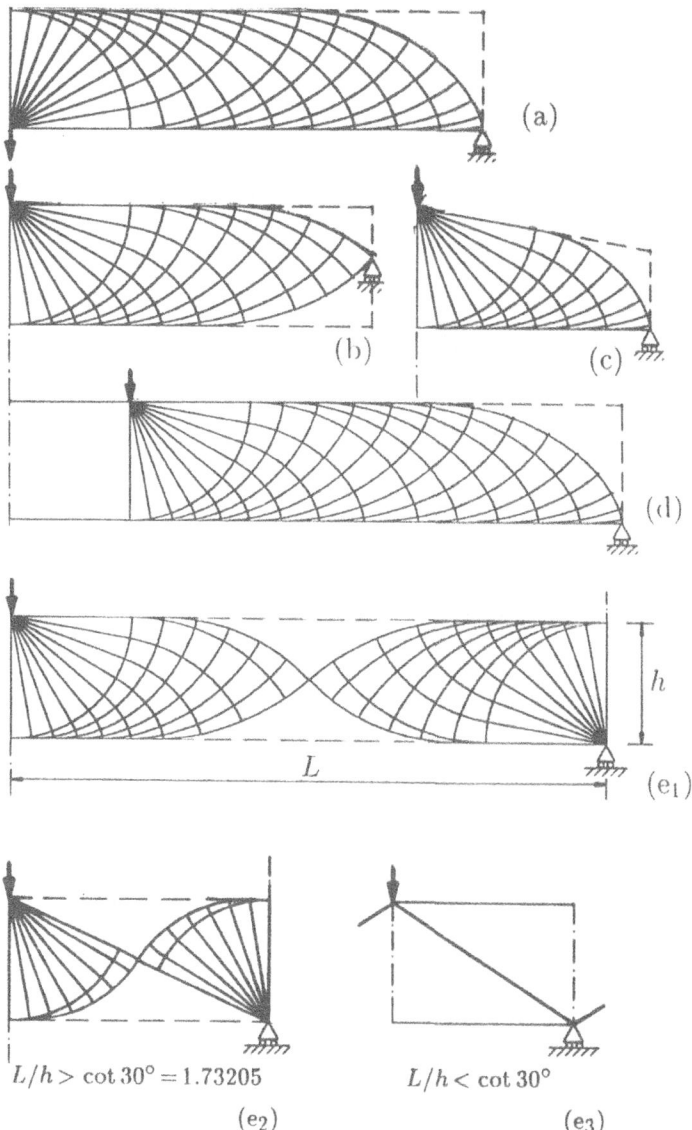

Figure 4. Exact optimal truss layouts for point supports.

Figures 4a–d show some exact layouts for simply supported trusses (axis of symmetry shown in dash-dot line) and Figs. $4e_1$–e_3 for trusses with many supports and a point load at each midspan. These solutions were plotted by computer graphics on the basis of closed form analytical expressions

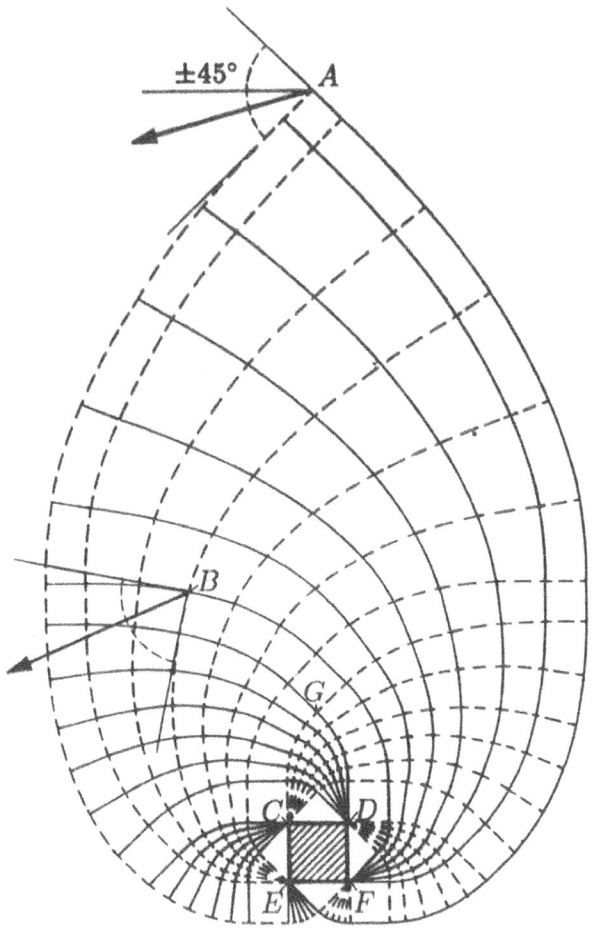

Figure 5. Optimal plane truss topology for a square support.

(Lewiński, Zhou and Rozvany, 1994). Using these results, exact optimal layouts can be constructed for plane trusses with most boundary conditions. For a cantilever truss supported along a square ($CDEF$ in Fig. 5), the optimal topology is shown in Fig. 5. This layout is valid for any number of point loads, as long as they are in the same quartal in between a compression (broken line) and tension (continuous line) bar; see, for example, the loads at the points A and B. For the area CDG the solution is identical with the classical one by Chan (1960).

In the regions j, k and ℓ of Fig. 6, both sets of members are curved, in the regions f, g, h and i one set of members is straight and in regions a, b,

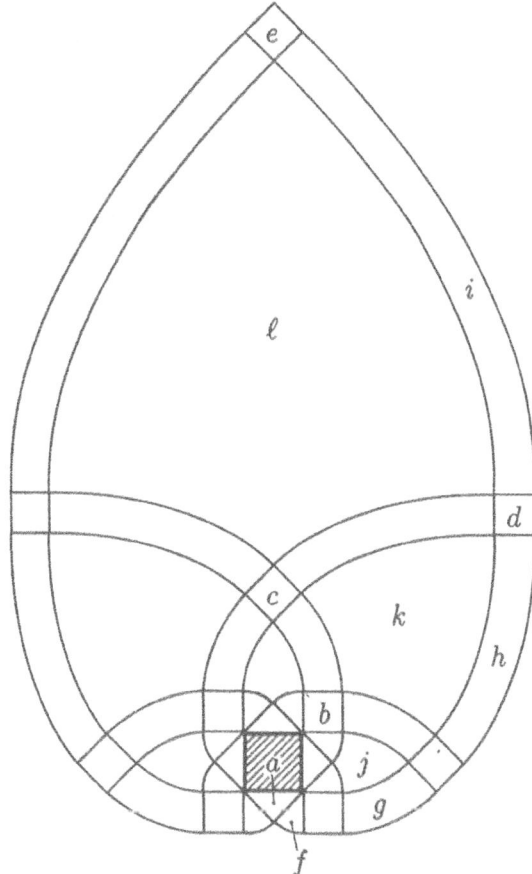

Figure 6. Regions of the optimal topology in Fig. 5.

c and d both sets of members are straight.

7.3. EXACT OPTIMAL PLANE TRUSS LAYOUTS FOR SEVERAL PROPORTIONAL DISPLACEMENT CONSTRAINTS

For two symmetrically oriented alternate point loads and a vertical support, solutions were obtained by Rozvany (1992) and Rozvany, Zhou and Birker (1993). The optimal orientation (α) of the bars to the horizontal for various orientations (β) of the load to the horizontal is given in Fig. 7a and the optimal nondimensional weights in Fig. 7b.

The layout criterion function for the above problem by (50) and (51) becomes (with $E^e = \gamma^e = 1$)

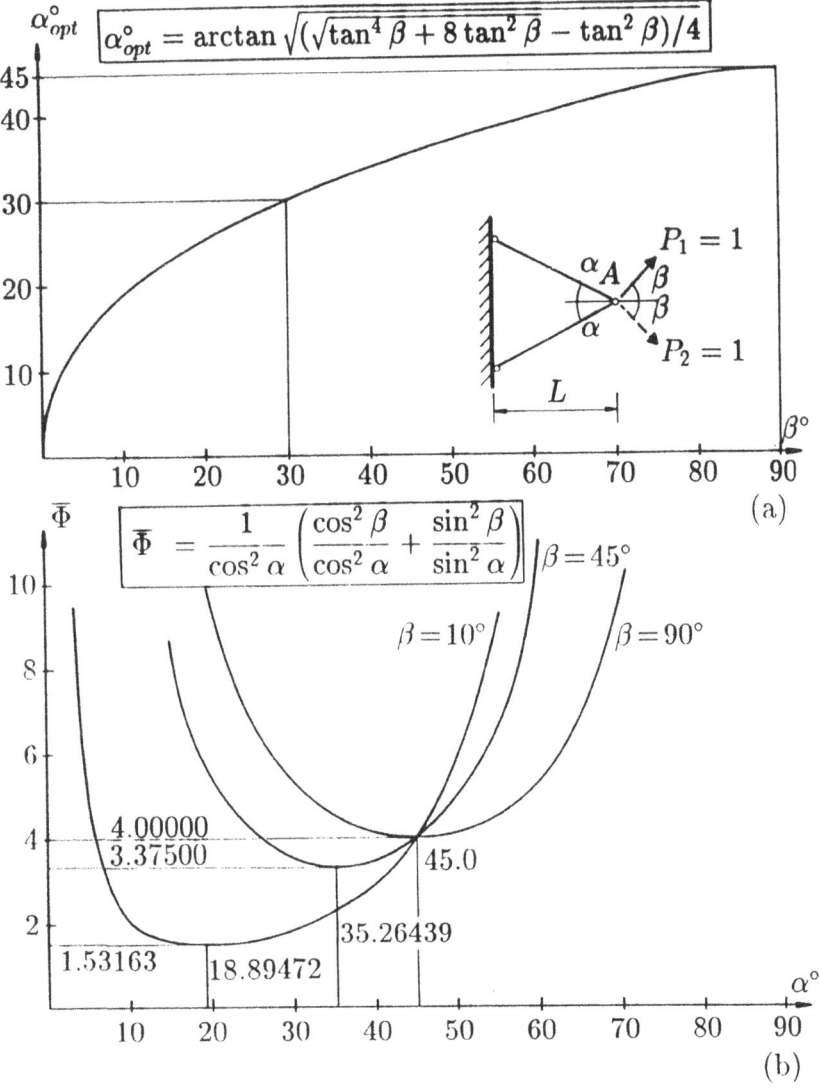

$$\alpha^\circ_{opt} = \arctan \sqrt{\left(\sqrt{\tan^4 \beta + 8 \tan^2 \beta} - \tan^2 \beta\right)/4}$$

(a)

$$\overline{\Phi} = \frac{1}{\cos^2 \alpha} \left(\frac{\cos^2 \beta}{\cos^2 \alpha} + \frac{\sin^2 \beta}{\sin^2 \alpha} \right)$$

(b)

Figure 7. Exact optimal truss layouts for two proportional displacement constraints (two load conditions).

$$\phi = \nu_1 \varepsilon_1^2 + \nu_2 \varepsilon_2^2. \tag{63}$$

Owing to symmetry, $\nu_1 = \nu_2$ and hence the variation of the *scaled* optimality criterion function

$$\phi = \varepsilon_1^2 + \varepsilon_2^2 \tag{64}$$

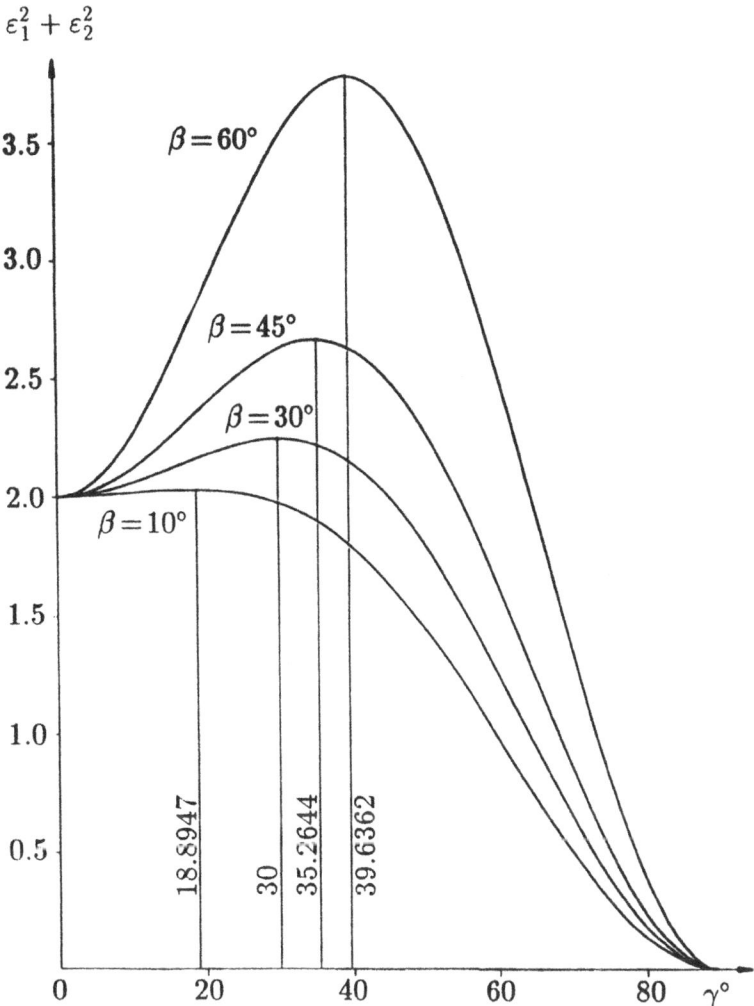

Figure 8. Variation of scaled layout criterion functions for the problems in Fig. 7.

is shown in Fig. 8. It can be seen that the maximum value of this function corresponds to the optimal bar orientations in Fig. 7. It can be checked that the values of ϕ equal 1.0 in the optimal directions, if the unscaled criterion function in (63) is used.

Further exact optimal plane truss layouts for proportional displacement constraints were obtained by Rozvany, Birker and Lewiński (1994), including three symmetrically oriented and two unsymmetrical alternative point loads.

In the above solutions, the optimal layout consists of two bars. With more restrictive support conditions, this is not the case. In order to deal

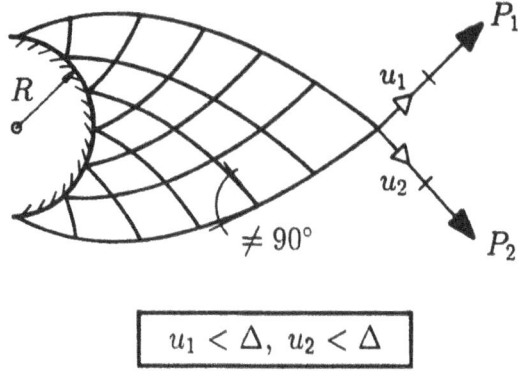

$$u_1 < \Delta, \ u_2 < \Delta$$

Figure 9. Generalized Hencky-net solution for a circular support with two load conditions and two displacement constraints.

with such problems, the theory of generalized, nonorthogonal Hencky-nets is being developed. For a circular support with two symmetric load conditions, for example, the expected type of solution is shown in Fig. 9, in which the member directions depend on the radial but not on the circumferential coordinates and intersecting members are non-orthogonal.

It is to be noted that the above problems are *selfadjoint and convex* and hence the relevant optimality criteria guarantee a global optimal solution.

7.4. EXACT OPTIMAL PLANE TRUSS LAYOUTS FOR ONE LOADING CONDITION WITH ONE NONPROPORTIONAL DISPLACEMENT CONSTRAINT (ILL-POSED NON-SELFADJOINT PROBLEMS)

The above problem was discussed by Rozvany, Sigmund *et al.* (1993). For a sloping point load and a constrained vertical displacement, for example, the variation of the weight with the member directions η is shown in Fig. 10a. For this problem, the optimality conditions in (36) and (37) are satisfied by two sets of strain fields (Fig. 10b), one giving a *local minimum* (A in Fig. 10a), but the other representing a *local maximum* (B in Fig. 10a) of the structural weight. Both local extrema correspond to a symmetrically oriented two-bar layout. More importantly, for $\eta \leq 20°$, the total weight of a truss satisfying the above displacement constraint tends to zero and the corresponding solution is non-stationary (ill-posed problem).

However, stationary solutions of finite weight can be forced by introducing

- two displacement constraints in the plane (Section 7.5), or
- one nonproportional displacement constraint and stress constraints (Section 7.6).

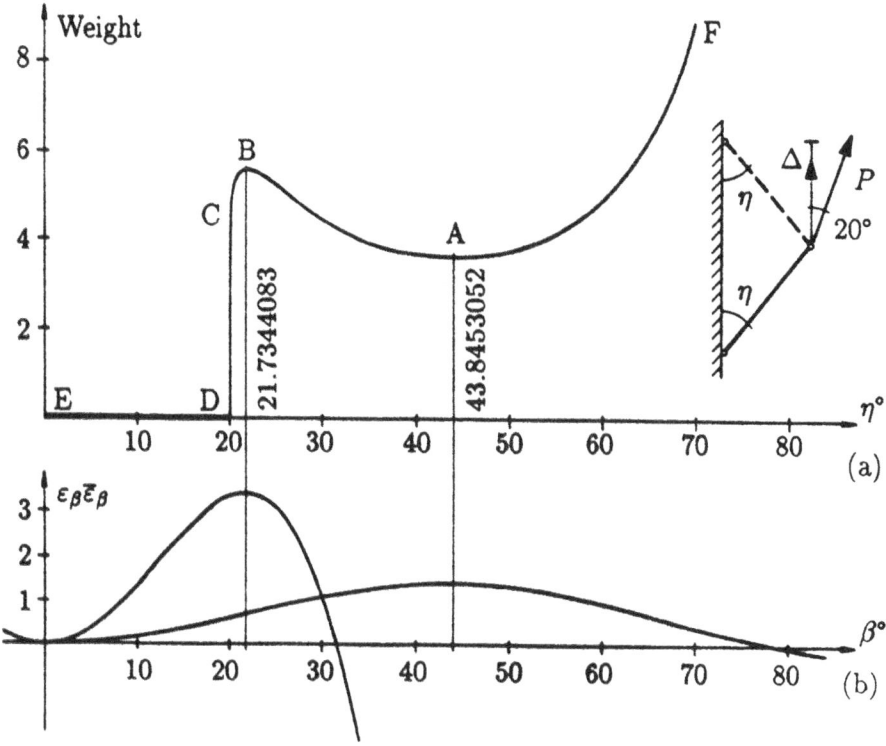

Figure 10. An ill-posed non-selfadjoint problem: plane truss with one nonproportional displacement constraint; (a) weight (w) variation with the bar angle (η); (b) variation of the layout criterion function (ϕ) with orientation (β).

7.5. EXACT OPTIMAL PLANE TRUSS LAYOUTS FOR ONE LOADING CONDITION WITH TWO DISPLACEMENT CONSTRAINTS (WELL-POSED NON-SELFADJOINT PROBLEMS)

The above class of problems was discussed in a very recent paper (Birker, Lewiński and Rozvany, 1994) in which least-weight trusses for a vertical support and a sloping point load are considered for a proportional and a nonproportional displacement constraint (Fig. 11a). It is shown that the following three types of solutions exist:

- only the proportional displacement constraint is active,
- only the nonproportional displacement constraint is active, or
- both constraints are active.

In all three cases, the solution has been shown to have a two-bar layout with symmetric bar orientations (but different cross-sectional areas). These solutions were derived independently (i) by assuming a two-bar layout and optimizing the member directions and (ii) directly from the optimality criteria in (36) and (37).

178

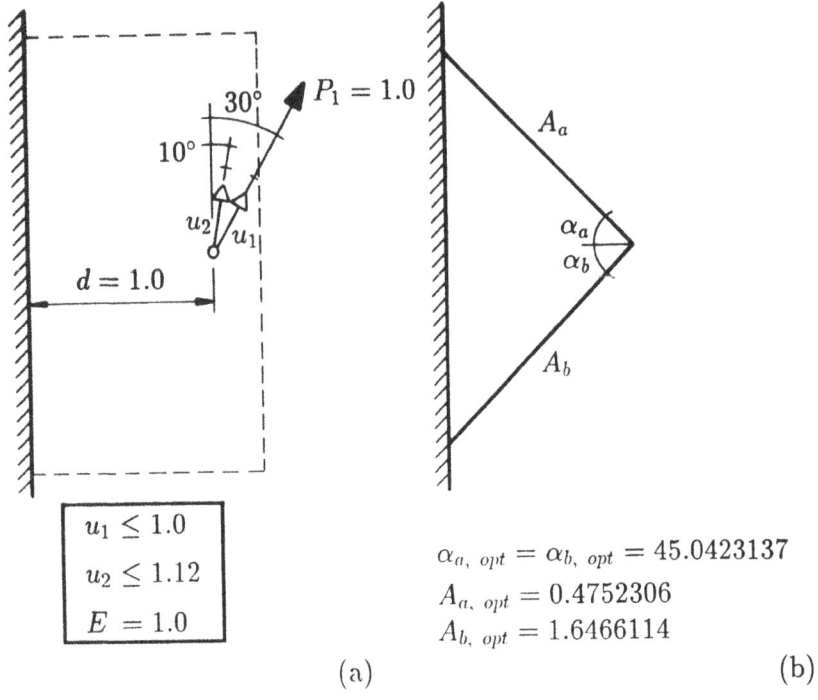

$$\alpha_{a,\ opt} = \alpha_{b,\ opt} = 45.0423137$$
$$A_{a,\ opt} = 0.4752306$$
$$A_{b,\ opt} = 1.6466114$$

(a) (b)

Figure 11. Exact optimal plane truss layout for two displacement constraints.

For the solution shown in Fig. 11, for example, both displacement constraints are active. It can be seen from Table 2 that the layout criterion function $\phi = \nu_1\varepsilon^2 + \nu_2\varepsilon\bar{\varepsilon}$, in which $\bar{\varepsilon}$ is the adjoint strain field for the nonproportional displacement constraint, does satisfy the optimality criteria (36) and (37), requiring a directional maximum of ϕ (with $\phi = 1.0$) in the optimal member directions. In this problem, the strain fields ε and $\bar{\varepsilon}$ are location-independent.

Owing to nonconvexity, the above criteria do not ensure a global optimum, but they do represent a local optimum with respect to any variation of layout (not only two-bar layouts).

7.6. EXACT OPTIMAL PLANE TRUSS LAYOUTS FOR STRESS AND DISPLACEMENT CONSTRAINTS

This class of problems is being investigated currently. An example of exact optimal plane truss layouts for a sloping point load and vertical displacement constraint as well as stress constraints is shown in Fig. 12, in which the value of the permissible stress (σ^e) is shown on the horizontal axis and the optimal bar angles (α_a, α_b) on the vertical axis. For $\sigma^e \leq 0.5$ both

TABLE 2. Values of the layout criterion function $\phi = \nu_1 \varepsilon^2 + \nu_2 \varepsilon \bar{\varepsilon}$ for various bar directions α ($\nu_1 = 2.6087381$, $\nu_2 = 0.3519804$)

α	ε	$\bar{\varepsilon}$	ϕ
-45.1	0.5864522	0.4979307	0.9999960
-45.0423137	0.5864987	0.4972206	1.0000000
-45	0.5865314	0.4966976	0.9999978
$+45$	-0.5430858	-1.2061963	0.9999979
$+45.0423137$	-0.5431172	-1.2056714	1.0000000
$+45.1$	-0.5431581	-1.2049529	0.9999960

bars are controlled by the stress constraint and for $\sigma^e \geq 0.59791$, both bars are controlled by the displacement constraint, the optimal layout being symmetrical. For $0.5 < \sigma^e < 0.59791$, one bar is controlled by the stress constraint and the other bar by the displacement constraint, and the layout is unsymmetrical.

8. Approximate, Discretized Optimal Layouts by New Optimality Criteria methods (DCOC)

"During the seventies people who worked in the field of structural optimization were divided into two distinct and somewhat belligerent camps... The mathematical programming camp decried the lack of generality in optimality criteria methods and the optimality criteria camp sneered at the inefficiency of the mathematical programming approach" (Haftka, Gürdal and Kamat, 1990, p. 283).

It is shown below that optimality criteria methods can now be used for most design constraints.

A discretized version of continuum-type optimality criteria methods (COC) was developed by Zhou (Zhou and Rozvany, 1992/1993) under the term DCOC. The above method has been extended to a wide range of constraints, including (see also Rozvany and Zhou, 1992, 1994; Rozvany, Zhou and Sigmund, 1994; Liebermann, Gerdes, Birker and Rozvany, 1994; Zhou, 1994)

- several load conditions,
- displacement constraints,
- stress constraints,
- lower and upper side constraints,
- natural frequency constraints,

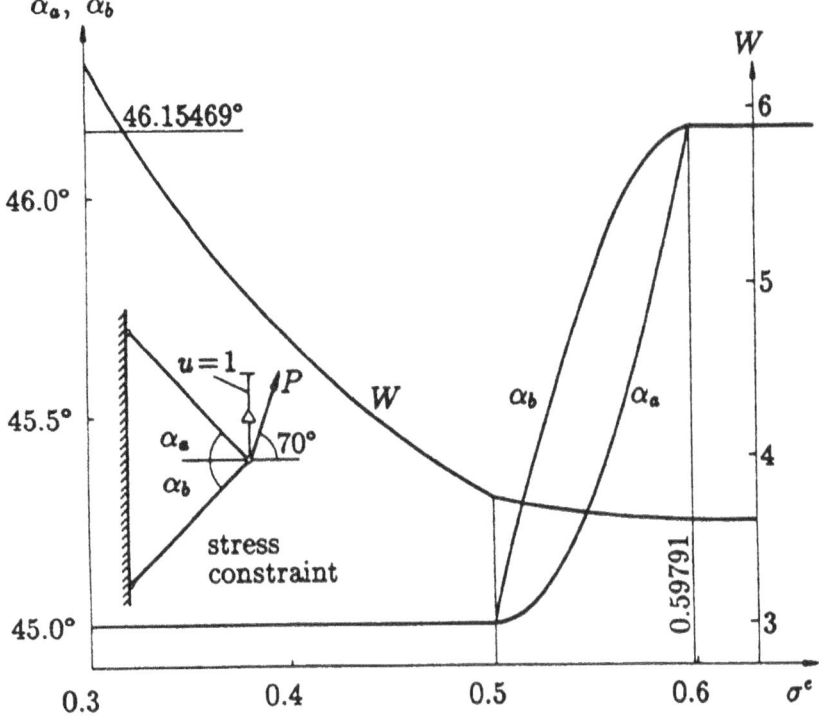

Figure 12. Optimal truss layout for one displacement constraint and stress constraints.

- local buckling constraints,
- system stability constraints,
- elastic supports,
- prestrains or temperature strains,
- selfweight,
- structural mass,
- given support settlement,
- allowance for the cost of reactions,
- passive control (variable loads, prestrains and support settlements, with possible allowance for their cost).

8.1. LAYOUT OPTIMIZATION OF DISCRETIZED GRID-LIKE STRUCTURES (LOW VOLUME FRACTION)

The main difference between discretized, approximate and exact, continuum-type layout optimization is that in the former

- a finite (but usually large) number of potential members are used in the structural universe (instead of an infinite number);
- the prescribed minimum cross-sectional area is nonzero but small (instead of zero);
- the structure is analysed by discretized (FE) methods (instead of continuum methods); and
- the solution is obtained iteratively (and not explicitly).

Discretized layout optimization methods provide only an approximation of the exact optimal layout, but this is usually of very high quality if the structural universe contains many members.

It can be seen from this section that discretized layout optimization by DCOC can readily be applied to a combination of a wide range of design constraints. The only difficulty at present is the singularity of some layout solution, which was pointed out by Sved and Ginos (1968) and discussed by Kirsch (e.g. 1990). This difficulty, however, could be eliminated through a better understanding of exact solutions (see Sections 6 and 7).

8.2. OPTIMIZATION OF DISCRETIZED PERFORATED OR COMPOSITE STRUCTURES (GENERALIZED SHAPE OPTIMIZATION OF STRUCTURES WITH A HIGH VOLUME FRACTION)

Analytical results using optimal microstructures (e.g. Rozvany, Olhoff, Bendsøe et al., 1987) and extensive numerical investigations (e.g. Bendsøe, Diaz and Kikuchi, 1993) have shown that

- optimal solutions for perforated disks, plates and shells consist of three types of regions, namely (i) *solid regions* – filled with material; (ii) *empty regions* – without material and (iii) *porous regions* – some material, with cavities of infinitesimal size;
- a *high proportion* of the optimal layout consists of *porous regions*; and
- at *low volume fractions* the optimal solution tends to that for *grid-type structures* (e.g. to least-weight trusses in the case of plane stress).

The above solutions are termed *SEP (Solid-Empty-Porous) topologies* and can be rather unpractical, although of considerable theoretical interest. From an engineering point of view, it is more practical to aim at solutions with mostly solid and empty regions at the macro-level *(SE topologies)*.

It has been demonstrated by the author's research team that for SE topologies a powerful method is the combination of DCOC with *solid isotropic microstructures* with *penalty* (SIMP) for intermediate densities. The basic idea of this approach was mentioned with some critical comments by Bendsøe (1989). Details of the SIMP-method are given elsewhere (e.g. Rozvany, Zhou and Birker, 1992). Advantages of the DCOC–SIMP method are

- *simplicity* of analysis and optimization,
- *selective suppression* of porous regions by *adjustable penalty*, and
- capability of handling a *wide variety of design conditions* (not only simple selfadjoint problems).

9. Examples of Discretized Optimal Layout Solutions

"Although we have been able to get near-optimal designs with the current update strategy, terminal convergence is slow... Another computational problem is associated with non-uniqueness of the displacement solution... parts of the structure have intermediate densities... in many practical situations we would like the final design to be comprised solely of macroscopic solid and void regions" (Jog, Haber and Bendsøe, 1993, on solutions with rank-2 microstructures).

The above quotation represents an excellent summary of some of the problems associated with rank-2 microstructures. In the next subsection, these difficulties are illustrated by comparing solutions for the same problem by various methods.

9.1. SYMMETRIC CANTILEVER TRUSS

The exact optimal layout (Lewiński, Zhou and Rozvany, 1994) for this problem is shown in Fig. 13a which is an extension of a solution by Chan (1960). The discretized truss solution given in Fig. 13b was obtained by Zhou using DCOC with a structural universe (ground structure) with 7204 truss elements and has only 0.9% higher weight than the exact optimal solution. Finally, the DCOC-SIMP solution for a perforated plate by Zhou, who used 5400 constant-strain triangular elements, is shown in Fig. 13c. The close agreement among these solutions is obvious. Further solutions for the same problem with a smaller aspect ratio, obtained by Bendsøe, Diaz and Kikuchi (1993), are given in Figs. 13d through f, which were obtained using 6400 rectangular elements with microstructures having rectangular holes, rank-2 layers and an isotropic cell, respectively. Further rank-2 solutions by Allaire and Kohn (1993), with and without penalty for intermediate densities, are given in Figs. 13g and h. It can be seen that the resolution of the latter five in terms of solid macro-regions is very limited and the layout given by them shows much bigger deviation from the exact truss layout (Fig. 13a) than the solution in Fig. 13c.

These apparent advantages of the DCOC-SIMP method can only be regarded as experimental evidence and do not necessarily apply to other examples or to improved versions of other methods.

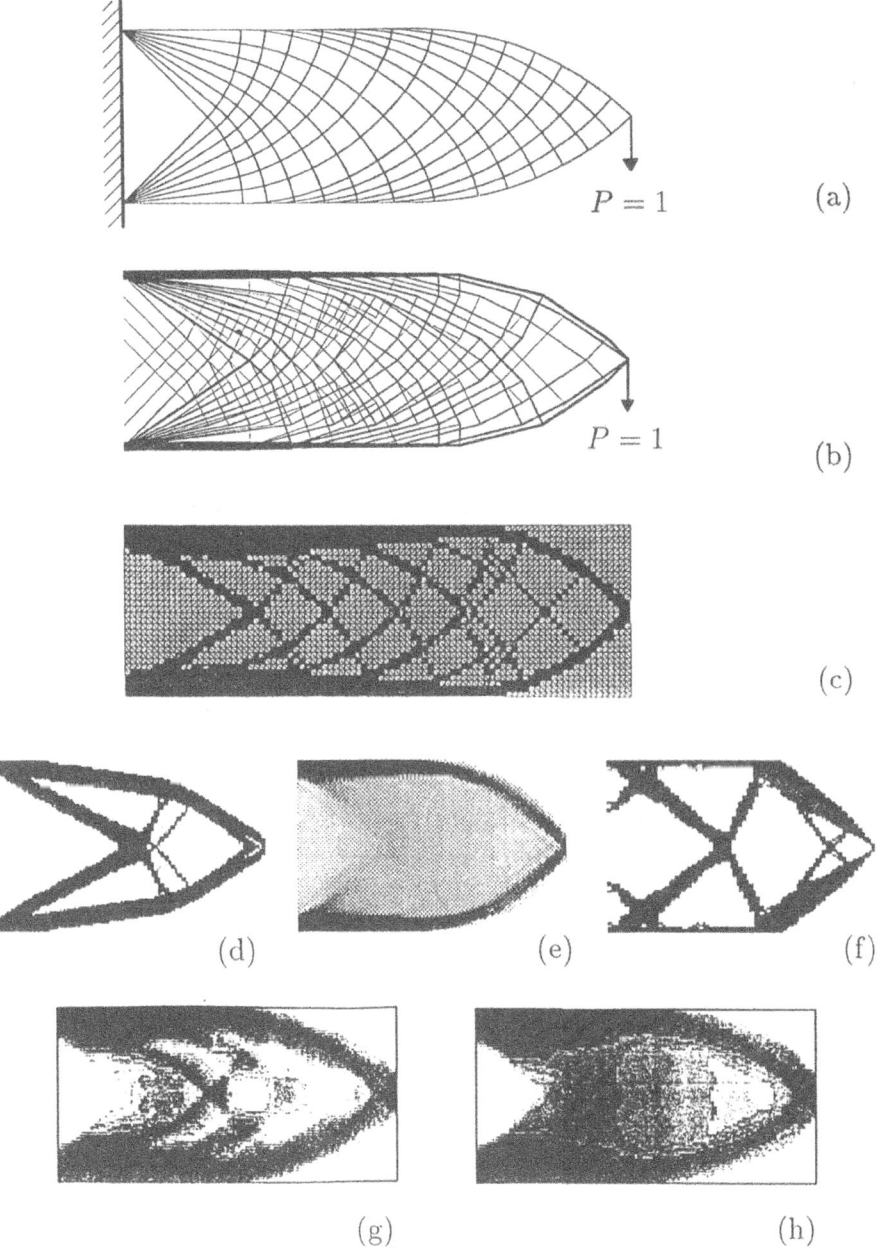

Figure 13. Comparison of results by analytical solutions, DCOC-SIMP and other methods.

184

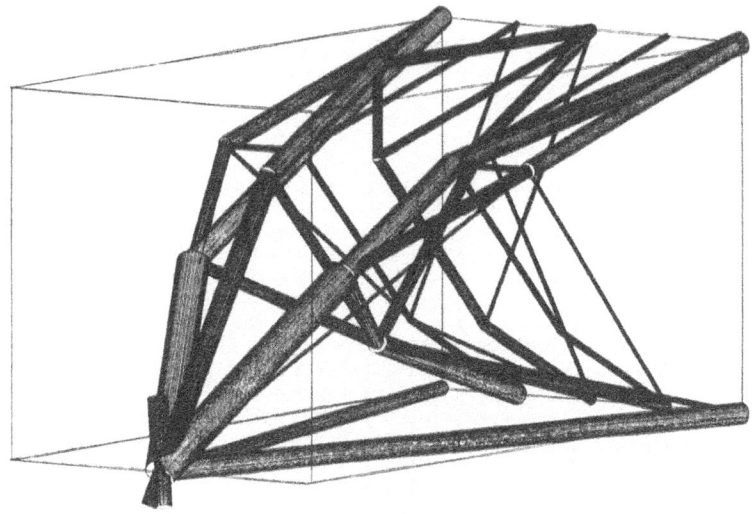

Figure 14. Optimized discretized layout of a space truss in accordance with the relevant German design standard: stress constraints and two displacement constraints.

9.2. PRACTICAL LAYOUTS BASED ON GERMAN DESIGN STANDARDS (DIN 1052 AND DIN 18800)

The DCOC algorithm has reached the stage when it is possible to derive discretized optimal layouts taking all provisions of the relevant design standards into consideration. Gerdes (1994) investigated layout optimization of trusses and rigid frames on the basis of the German design standard for timber structures (DIN 1052) and Liebermann, Rozvany *et al.* (1994) considered layout optimization of steel tubular trusses according to the standard for steel structures (DIN 18800). Both procedures include system stability and local member buckling. Figure 14, for example, shows the optimized space truss for stress and two displacement constraints.

10. Acknowledgements

The author is indebted to the German Research Foundation (DFG) for financial support, to Anne Fischer and Karin Deutscher (text processing), Elke Becker (drafting) and Susann Rozvany (editing).

11. References

Allaire, G.; Kohn, V. (1993) Topology Optimization and Optimal Shape Design Using Homogenization, in: Bendsøe, M.P.; Mota Soares, C.A. (eds.) *Topology Design of Structures*, Kluwer, Dordrecht, 207-218.

Bendsøe, M.P. (1989) Optimal Shape Design As a Material Distribution Problem, *Struct. Optim.* **1**, 193-202.

Bendsøe, M.P.; Ben-Tal, A.; Zowe, J. (1994) Optimization Methods for Truss Geometry and Topology Design, *Struct. Optim.* **7**, 141-159.

Bendsøe, M.P.; Diaz, A.; Kikuchi, N. (1993) Topology and Generalized Layout Optimization of Elastic Structures, in: Bendsøe, M.P.; Mota Soares, C.A. (eds.) *Topology Design of Structures*, Kluwer, Dordrecht, 159-206.

Berke, L.; Khot, N.S. (1987) Structural Optimization Using Optimality Criteria, in: Mota Soares, C.A. (ed.) *Computer Aided Optimal Design: Structural and Mechanical Systems* (Proc. NATO/NASA/NSF/USAF ASI, held in Troia), Springer, Berlin, 271-312.

Birker, T.; Lewiński, T.; Rozvany, G.I.N. (1994) A well posed non-selfadjoint layout problem: plane truss for one load condition and displacement constraints, *Struct. Optim.* **8** (in press).

Chan, A.S.L. (1960) The Design of Michell's Optimum Structures, *College of Aeronautics Report* **142**, Cranfield.

Cox, H.L. (1958) The Theory of Design, *Aeronaut. Res. Council Rep.* No. 19791.

Dorn, W.S.; Gomory, R.E.; Greenberg, H.J. (1964) Automatic Design of Optimal Structures, *J. de Méchanique* **3**, 25-52.

Gerdes, D. (1994) *Strukturoptimierung unter Anwendung der Optimalitätskriterien auf diskretisierte Tragwerke bei besonderer Berücksichtigung der Stabilität*, Doctoral Dissertation, Essen University.

Hegemier, G.A.; Prager, W. (1969) On Michell Trusses, *Int. J. Mech. Sci.* **11**, 209-215.

Haftka, R.T.; Gürdal, Z.; Kamat, M.P. (1990) *Elements of Structural Optimization*, Kluwer, Dordrecht.

Hemp, W.S. (1974). *Optimum Structures*, Clarendon, Oxford.

Jog, C.S.; Haber, R.B.; Bendsøe, M.P. (1993) A Displacement-Based Topology Design Method With Self-Adaptive Layered Materials, in: Bendsøe, M.P.; Mota Soares, C.A. (eds.) *Topology Design of Structures*, Kluwer, Dordrecht, 219-238.

Kirsch, U. (1990) On Singular Topologies in Optimum Structural Design, *Struct. Optim.* **2**, 133-142.

Lewiński, T.; Zhou, M.; Rozvany, G.I.N. (1994) Extended Exact Solutions for Least-Weight Truss Layouts - Part I: Cantilever with a Horizontal Axis of Symmetry; Part II. Unsymmetric Cantilevers, *Int. J. Mech. Sci.* **36**, 375-398, 399-419.

Liebermann, S.; Gerdes, D.; Birker, T.; Rozvany, G.I.N. (1994) Topology Design of Tubular Structures with a Variety of Design Constraints, *Proc. 6th Int. Symp. on Tubular Structures* (held in Melbourne, Dec. 1994), Balkema Publishers, Rotterdam.

Michell, A.G.M (1904) The Limits of Economy of Material in Frame Structures, *Phil. Mag.* **8**, 589-597.

Nagtegaal, J.C.; Prager, W. (1973) Optimal Layout of Truss for Alternative Loads, *Int. J. Mech. Sci.* **15**, 7, 583-592.

Prager, W. (1974) *Introduction to Structural Optimization*, Springer, Vienna.

Prager, W.; Rozvany, G.I.N. (1977a) Optimization of Structural Geometry, in: Bednarek, A.R.; Cesari, L. (eds.) *Dynamical Systems*, 265-293, Academic Press, New York.

Prager, W.; Rozvany, G.I.N. (1977b) Optimal Layout of Grillages, *J. Eng. Mech.* **5**, 1-18.

Prager, W.; Shield, R.T. (1967) A General Theory of Optimal Plastic Design, *J. Appl. Mech. ASME* **34**, 184-186.

Rozvany, G.I.N. (1976) *Optimal Design of Flexural Systems*, Pergamon, Oxford.

Rozvany, G.I.N. (1981) Optimality Criteria for Grids, Shells and Arches, in: Haug, E.J.; Cea, J. (eds.) *Optimization of Distributed Parameter Structures* (Proc. NATO ASI, Iowa City 1980), Sijthoff and Noordhoff, Alphen aan der Rijn, 112-151.

Rozvany, G.I.N. (1989) *Structural Design via Optimality Criteria*, Kluwer, Dordrecht.

Rozvany, G.I.N. (1992) Optimal Layout Theory: Analytical Solutions for Elastic Structures with Several Deflection Constraints and Load Conditions, *Struct. Optim.* **4**, 247-249.

Rozvany, G.I.N. (1994a) Optimal Layout of Grillages: Allowance for Cost of Supports and Optimization of Support Locations, *Mech. Struct. Mach.* **22**, 49-72.

Rozvany, G.I.N. (1994b) Topological Optimization of Grillages: Past Controversies and New Directions, *Int. J. Appl. Mech.* **36**, 495-512.

Rozvany, G.I.N.; Birker, T.; Lewiński, T. (1994) Some Unexpected Properties of Exact Least-Weight Plane Truss Layouts with Displacement Constraints for Several Alternate Loads, *Struct. Optim.* **7**, 76-86.

Rozvany, G.I.N.; Gollub, W. (1990) Michell Layouts for Various Combinations of Line Supports, *Int. J. Mech. Sci.* **32**, 1021-1043.

Rozvany, G.I.N.; Hill, R.H. (1978) Optimal Plastic Design: Superposition Principles and Bounds on the Minimum Cost, *Comp. Mech. Appl. Mech. Engrg.* **13**, 151-173.

Rozvany, G.I.N.; Liebermann, S. (1994) Exact Optimal Grillage Layouts — Part I: Combinations of Free and Simply Supported Edges, *Struct. Optim.* **7**, 260-270.

Rozvany, G.I.N.; Olhoff, N.; Bendsøe, M.P., Ong, T.G.; Sandler, R.; Szeto, W.T. (1987) Least-Weight Design of Perforated Elastic Plates, I, II, *Int. J. Solids Struct.* **23**, 521-536, 537-550.

Rozvany, G.I.N.; Ong, T.-G. (1987) Minimum Weight Plate Design via Prager's Layout Theory (Prager Memorial Lecture), in: Mota Soares, C.A. (ed.) *Computer Aided Optimal Design: Structural and Mechanical Systems* (Proc. NATO ASI, held in Troia, 1986), 165-180, Springer, Heidelberg.

Rozvany, G.I.N.; Prager, W. (1976) Optimal Design of Partially Discretized Grillages. *J. Mech. Phys. Solids* **24**, 125-136.

Rozvany, G.I.N.; Sigmund, O.; Lewiński, T.; Gerdes, D.; Birker, T. (1993) Exact Optimal Structural Layouts for Non-Self-Adjoint Problems, *Struct. Optim.* **5**, 204-206.

Rozvany, G.I.N.; Zhou, M. *et al.* (1989/1990) Continuum-Type Optimality Criteria Methods for Large Finite Element Systems with a Displacement Constraint, Part I, II, *Struct. Optim.* **1**, 47-72; **2**, 77-104.

Rozvany, G.I.N.; Zhou, M. (1991) A Note on Truss Design for Stress and Displacement Constraints by Optimality Criteria Methods, *Struct. Optim.* **3**, 45-50.

Rozvany, G.I.N.; Zhou, M. (1992) Extensions of new discretized optimality criteria methods for structures with passive control, *Proc. 4th AIAA/USAF/ NASA/OAI Symp. on Multidisciplinary Anal. and Optim.* (held in Cleveland, Sept. 1992) Part 1, 288-297.

Rozvany, G.I.N.; Zhou, M. (1993) Continuum-Based Optimality Criteria (COC) Methods: An Introduction. In: Rozvany, G.I.N. (Ed.) *Optimal Design of Large Structural Systems*, Proc. NATO/DFG ASI (held in Berchtesgaden 1991), 1-26. Kluwer, Dordrecht.

Rozvany, G.I.N.; Zhou, M. 1994: Optimality Criteria Methods by Large Discretized Systems, Chapter 2 in : H. Adeli (ed.) *Advances in Design Optimization*, Chapman & Hall, London.

Rozvany, G.I.N.; Zhou, M.; Birker, T. (1993) Why Multi-Load Topology Designs Based on Orthogonal Microstructures are in General Non-Optimal, *Struct. Optim.* **6**, 200-204.

Rozvany, G.I.N.; Zhou, M.; Sigmund, O. (1994) Optimization of Topology, Chapter 10 in : H. Adeli (ed.) *Advances in Design Optimization*, Chapman & Hall, London.

Spillers, W.R.; Lev, O. (1971) Design for Two Loading Conditions, *Int. J. Solids Struct.* **7**, 1261-1267.

Strang, G.; Kohn, R.V. (1983) Hencky-Prandtl Nets and Constrained Michell Trusses, *Comp. Meth. Appl. Mech. Engrg.* **36**, 207-222.

Sved, G. (1954) The Minimum Weight of Certain Redundent Structures, *Austral. J. Appl. Sci.* **5**, 1-8.

Sved, G.; Ginos, L. (1968) Structural Optimization under Multiple Loading, *Int. J. Mech. Sci.* **10**, 803-805.

Topping, B.H.V. (1993) Topology Design of Discrete Structures, in: Bendsøe, M.P.; Mota Soares, C.A. (eds.) *Topology Design of Structures*, Kluwer, Dordrecht, 517-534.

Vigdergauz, S. (1994) Two-Dimensional Grained Composites of Extreme Rigidity, *J. Appl. Mech. ASME* **61**, 390-394.

Zhou, M. (1994) An Efficient DCOC Algorithm Based on High Quality Approximations for Problems Including Eigenvalue Constraints, *AIAA/NASA/ USAF/ISSMO Symp. on Multidisc. Anal. Opt.* (Panama City, Florida).

Zhou, M.; Haftka, R.T. (1994) A Comparison Study of Optimality Criteria Methods for Stress and Displacement Constraints, *Proc. 35th AIAA/ASME/ ASCE/AHS/ASC SSDM Conference* (held in South Carolina, April 1994).

Zhou, M.; Rozvany, G.I.N. (1992/1993) DCOC: An Optimality Criteria Method for Large Systems. Part I: Theory, Part II: Algorithm, *Struct. Optim.* **5**, 12-25; **6**, 250-262.

Optimal Shape and Topology Design of Vibrating Structures

Noboru Kikuchi, Hsien-Chie Cheng, and Zheng-Dong Ma

Department of Mechanical Engineering and Applied Mechanics
The University of Michigan, Ann Arbor, MI 48109-2215, USA

Abstract

We shall extend the homogenization design method for the global stiffness maximization of an elastic structure to the optimization problem related to eigenvalues for free vibration such as maximization of a specified set of eigenvalues, maximization of the distance of the two specified eigenvalues, and identification of a structure that possesses a set of specified eigenvalues. To this end, the basic mathematical formulation and a solution method are proposed as well as various numerical examples of obtaining the optimum layout of both plane plate, and three-dimensional shell structures.

1. Introduction

Comparing with sizing and shape optimization, less research has been conducted on topology optimization for elastic structures. The earliest and the most significant research in the topology optimization is given by Michell (1904). His result is known as the Michell truss, and then there was a long blank period, for more than half a century to find extension of Michell's work by Hemp (1973). Then, Prager and Rozvany (1977), Rozvany (1981), Rozvany and Wang (1983) and Zhou and Rozvany (1991) have made further development of Michell's theory for more general classes of the topology related layout optimization problem. There are two review papers, Kirsch (1989) and Rozvany (1992), concerning the topology optimization, in which many other references can be found. However, not only topology optimization but also size and shape optimization, in general, could not attract much attention from practical engineers in design because existing techniques were not sufficiently straightforward to deal with rather complex problems in design practice, and more importantly, some classes of optimization problems were not quite well-posed, as pointed out by Kohn and Strang (1986). They suggested relaxation of the original design problem to be well-posed by introducing microscale perforation in some portion of the design domain to obtain the optimum solution. Similar observation was also made by Murat and Tartar (1985) for considering the shape optimization for heat conduction problems. It is, however, noted that necessity of microstructure such as very fine scale ribs on a plate was pointed out by Cheng and Olhoff (1981) for their study on the thickness optimization of a plate structure, and then mathematicians' work follows from this study. The very first response from applied mathematicians was Lurie and Cherkaev (1984), and then followed by many other researchers in French and American schools. The process of relaxation has led to the development of modern techniques for topology optimization. One of the most important techniques for topology optimization is the so-called homogenization design method introduced by Bendsøe and Kikuchi (1988). They introduced infinitely many fine scale, rectangular holes for perforation in the design domain in which a structure would be layout to relax the problem, and furthermore, derived the equivalent homogenized mechanical model of a perforated porous structure for stress analysis by applying the theory of homogenization for general composite materials, see, e.g., Sanchez-Palencia (1980) for general theory, and Guedes and Kikuchi (1990) for its computational procedure. In their work maximization of

189

J. Herskovits (ed.), Advances in Structural Optimization, 189–222.

the static global stiffness of a two-dimensional linearly elastic structure is considered, and the optimality criteria are obtained for this maximization problem by minimizing the mean compliance in the case that body forces and boundary tractions are specified. Furthermore, because of the resizing algorithm for the optimum angle of rotation of the microscale holes, the computational results in Bendsøe and Kikuchi (1988) could not show clear shape and topology of the optimum structure. Until Suzuki and Kikuchi (1991) in which the angle of rotation is identified by the direction to the principal stress direction, stable optimum solutions could not be obtained. This choice of angle was "later" proved by Pedersen (1989). It is also noted that the relaxation may yield large portion of microscale perforation, that is a porous structure, in the optimum, and it is not desirable to identify clear shape and topology of a structure. To overcome this difficulty, it was necessary to introduce some sort of "penalization" to the optimum procedure by, e.g., modifying the homogenized elasticity tensor or adding a penalty functional to make bias toward either 0 (complete void) or 1 (complete solid), see, e.g. Rozvany and Zhou (1992), Haber, Jog, and Bendsøe (1994). Multi-loading problems are also solved by Diaz and Bendsøe (1992) using the linear combination of the mean compliance of different loads, and also independently studied by Fukushima, Suzuki, and Kikuchi (1991) using an approximation of the mini-max formulation. Details of topology optimization for continuum developed in the early 90s by using the homogenization method or equivalent ones by defining nonlinear relation to the material characterization, can be found in Bendsøe, Diaz, and Kikuchi (1992) and the book edited by Bendsøe and Mota Soares(1992).

Optimum shape and topology design for dynamic problems have become increasingly important due to the wide demands in mechanical design such as design of automobile components, structures excited by engines or rotors, and others. Comparing with sizing and shape optimization, less researches have studied topology optimization of vibrating elastic structures. Despite that Olhoff and Rozvany (1982) considers an optimal grillage layout problem for a given natural frequency, there are few layout optimization study for vibrating elastic structures. All the existing work are related to the design of trusses, grillages or grid-like structures. Only very limited research has been published on this subject by using the homogenization design method, see Diaz and Kikuchi (1992) and Ma et al (1992, 1993). Thus, in this study, we shall attempt to deal with shape and topology optimization of both two-dimensional and plate/shell like structures in a dynamic system. Modeling of this requires the use of perforated plate elements. Here we shall apply the perforated laminate presented in Suzuki and Kikuchi (1992). A different plate model is also available from Soto and Diaz (1992,1993) in the homogenization design formulation. Since the model requires the material stiffness-density relation of the perforated plate, such relation is derived by using the homogenization method based on the classical laminate theory.

The major goal of this study is to improve the dynamic behavior of mechanical components and structures through optimal layout design. To achieve this goal, we consider controlling or improving the natural frequencies of structures. In the present study, three categories of eigenvalue related optimization problems are considered:

      a) Maximization of a single specified eigenvalue or a set of specified eigenvalues to obtain the desired structure.

      b) Maximization of the gaps of the specified eigenvalues from a given eigenvalue to avoid structural resonance.

      c) Identification of a structure by specifying the desired eigenvalues.

It is noted that as shown in Fig. 1 there is no idea what structure should be built in the design domain at the beginning except that this structure must be supported at the two lower corner points and must sustain three non-structural masses. No topology, shape, and size of the structure are known. The purpose of the present study is to layout the structure satisfying the design condition, for example, it has a specified set of eigenfrequencies, or it has the maximum value of the minimum eigenfrequency within a specified weight.

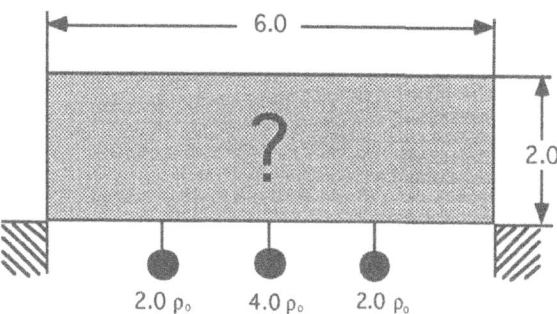

Figure 1 Design domain and work condition for a structure

In order to explore these objectives, a multimodal, multipurpose objective function, introduced by Ma et al. (1991), will be implemented. The advantages of using the multimodal formulation will also be examined by comparing the results with those using the single-modal formulation. This design problem will be solved by applying an optimality criteria method to calculate the optimal porosity of the perforated structure, and employing a generalized stress based approach to derive the optimal orientation of the perforated material. The solution to this problem can be eventually derived by integrating the platform of the homogenization design method with finite element appriximation.

2. A Relaxed Design Problem

The uniformly perforated design domain at the initial state is defined on an arbitrary surface (curved or flat) in three-dimensional space to layout a shell like structure. To simplify the problem, the curved shell domain is discretized into four-node quadrilateral elements (Q4 elements). Furthermore, the non-flat Q4 elements are assumed to be flat in the finite element approximation in order to neglect the curvature effects. This method projects the four corner nodes of the Q4 elements onto the closest plane in the mean squares sense and a local coordinate system(x, y, z) is then set up on this plane. The closest plane forms the 'x-y' plane of the local system and the 'transverse' direction, the z-direction of the local coordinate system, is defined as the thickness of the flat shell elements. Before developing the formulation for the three-dimensional perforated shell, we have to construct a perforated model for the projected flat shell and more importantly, derive the corresponding effective properties referring to the local coordinate system.

The perforated plate model introduced by Suzuki and Kikuchi (1992) is adopted in this study. This laminate plate microstructure, as shown in Fig. 2, is constructed by an isotropic base ply (h_0), sandwiched by two reinforced orthotropic plies ($h_1 - h_0$), each of which is perforated by a rectangular hole. This microstructure is symmetric with respect to the mid-plane of the base structure; therefore, the in-plane/bending coupling effect can be neglected. If h_0, the thickness of the base structure, is defined as a constant, the optimization problem can be considered as the determination of the optimal layout of stiffeners. On the other hand, it can be interpreted as the usual layout optimization of structures if h_0 is assumed to be zero. Especially, if all the applied loads are in the plane direction, this should yields layout optimization of a thin plate in plane elasticity. Variables a and b, the physical dimensions of a cavity, characterize the density of the microstructures, and θ represents the orientation of the perforated material. The effective material properties of the perforated laminate plate element with respect to the material density will be derived as follows.

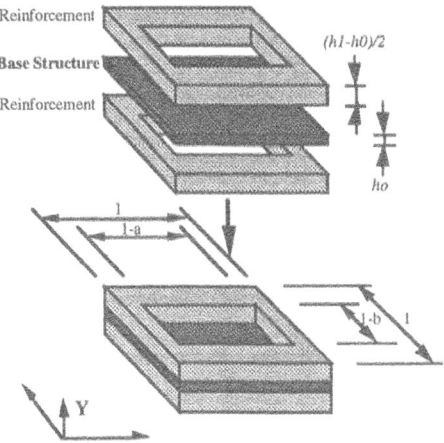

Figure 2 The Laminate Plate Microstructure

The perforated plies (the top and bottom ones in Fig. 2) can be considered as a two-dimensional microstructure under the plane stress condition. The effective properties of material for the two-dimensional, plane stress microstructure can be derived using the homogenization method based on the density and directional property of the microstructure. Solving this problem requires using a fixed, two-dimensional elastic design domain, defined in Fig. 3, in which Γ stands for the traction boundary and Γ_d represents the displacement boundary.

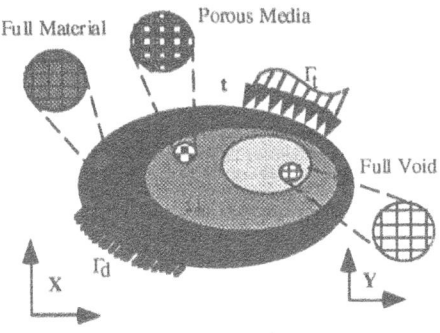

Figure 3 The Two-Dimensional Perforated Design Domain

This domain consists of an infinite number of microstructures with rapidly varying density. A cavity is embedded in the rectangular microstructure as shown in Fig. 4. In Fig. 4, parameters a, b and θ define the area density of this microstructure, where both a and b define the sizes of holes and the orientation of the rectangular hole. The perforated structure can be characterized by the rapidly varying material properties : the elasticity tensor E_{ijkl}^{ε} and the material density ρ^{ε}, which are, in general, the function of the spatial coordinate. If the microstructure is first assumed to be non-rotated (i.e. $\theta = 0$), then they can be defined as:

$$E_{ijkl}^{\varepsilon} = \begin{cases} E_{ijkl}^{0} & \text{in solid} \\ 0 & \text{in cavity} \end{cases}, \qquad \rho^{\varepsilon} = \begin{cases} \rho_0 & \text{in solid} \\ 0 & \text{in cavity} \end{cases} \tag{1}$$

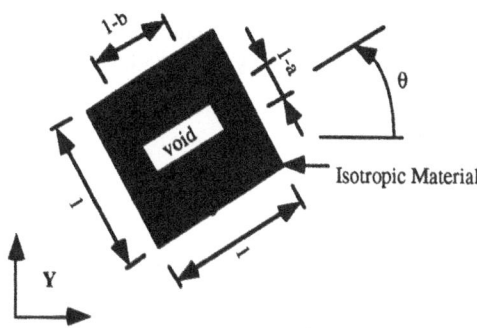

Figure 4 The Two-Dimensional Microstructure

Using the homogenization method, see, e.g., Guedes and Kikuchi (1990), the homogenized elasticity tensor E_{ijkl}^{h}, the homogenized mass density ρ^{h} can be derived respectively as:

$$E_{ijkl}^{h} = \frac{1}{|Y|} \int_{Y} (E_{ijkl}^{\varepsilon} - E_{ijpq}^{\varepsilon} \frac{\partial \chi_{p}^{kl}}{\partial y_q}) dy, \quad \rho^{h} = \frac{1}{|Y|} \int_{Y} \rho^{\varepsilon} dy \tag{2}$$

where $|Y|$ stands for the volume (area) of Y, and χ_{p}^{kl} is the solution of the microscopic problem that characterizes the micromechanical behavior of a specific microstructure :

$$\int_{Y} (E_{ijkl}^{\varepsilon} - E_{ijpq}^{\varepsilon} \frac{\partial \chi_{p}^{kl}}{\partial y_q}) \frac{\partial v_i}{\partial y_j} dy = 0 \qquad \text{for} \quad \forall v \in V_Y \tag{3}$$

Here a function space $V_Y = \{v \mid v \in H^1(Y) \text{ and } v \text{ is } Y\text{-periodic}\}$ is defined on the microstructural domain Y. When the rotation of the microstructure is also considered as shown in Fig. 4, the rotated homogenized coefficients E_{ijkl}^{H} and ρ^{H} for the plane stress problem can be obtained as:

$$E_{ijkl}^{H} = c_{iq} c_{jp} c_{ks} c_{lr} E_{qpsr}^{h}, \qquad \rho^{H} = \rho^{h} \tag{4}$$

where $c_{11} = c_{22} = \cos\theta$ and $c_{21} = -c_{12} = \sin\theta$.

The homogenized weak form (the principle of virtual displacement) of the structural system is then obtained as

$$\int_\Omega E_{ijkl}^H \frac{\partial u_i^0}{\partial x_j} \frac{\partial v_k}{\partial x_l} d\Omega + \int_\Omega \rho^H \frac{\partial^2 u_i^0}{\partial t^2} v_i d\Omega = \int_\Omega f_i^h v_i d\Omega + \int_{\Gamma_t} t_i v_i d\Gamma \quad \text{for } \forall \mathbf{v} \in V \quad (5)$$

where, u_i^0 stands for the average displacements over the unit cell, v_i the virtual displacement, f_i^h the homogenized body force, t_i the boundary traction, Ω the structural domain, Γ_t the traction boundary and Γ_d the displacement boundary. $V = \{\mathbf{v} \mid \mathbf{v} \in H^1(\Omega), \ \mathbf{v} = 0 \text{ on } \Gamma_d\}$ is the space of kinetically admissible displacement fields, where $H^1(\Omega)$ is the Sobolev space. Using a matrix expression for the rotated elastic coefficient, we can define

$$\mathbf{D}^H = \begin{bmatrix} E_{1111}^H & E_{1122}^H & E_{1112}^H \\ E_{1122}^H & E_{2222}^H & E_{2212}^H \\ E_{1112}^H & E_{2212}^H & E_{1212}^h \end{bmatrix} \quad (6)$$

then the rotated elastic matrix can be obtained as

$$\mathbf{D}^H = \mathbf{D}^0 + \mathbf{D}^1 \sin 2\theta + \mathbf{D}^2 \cos 2\theta + \mathbf{D}^3 \sin 2\theta \cos 2\theta - \mathbf{D}^4 \sin^2 2\theta \quad (7)$$

where, the coefficient matrices $\mathbf{D}^i$ ($i=0, 1, ..., 4$) are independent of the orientation variable θ, and

$$\mathbf{D}^0 = \begin{bmatrix} (E_{1111}^h + E_{2222}^h)/2 & E_{1122}^h & 0 \\ E_{1122}^h & (E_{1111}^h + E_{2222}^h)/2 & 0 \\ 0 & 0 & E_{1212}^h \end{bmatrix},$$

$$\mathbf{D}^1 = d_1 \begin{bmatrix} 0 & 0 & 1 \\ 0 & 0 & 1 \\ 1 & 1 & 0 \end{bmatrix}, \qquad \mathbf{D}^2 = 2d_1 \begin{bmatrix} 1 & 0 & 0 \\ 0 & -1 & 0 \\ 0 & 0 & 0 \end{bmatrix} \quad (8)$$

$$\mathbf{D}^3 = d_2 \begin{bmatrix} 0 & 0 & 1 \\ 0 & 0 & -1 \\ 1 & -1 & 0 \end{bmatrix}, \qquad \mathbf{D}^4 = d_2 \begin{bmatrix} 1 & -1 & 0 \\ -1 & 1 & 0 \\ 0 & 0 & -1 \end{bmatrix}$$

and

$$d_1 = \frac{1}{4}(E_{1111}^h - E_{2222}^h)$$

$$d_2 = \frac{1}{4}(E_{1111}^h + E_{2222}^h - 2E_{1122}^h - 4E_{1212}^h) \quad (9)$$

Also, the rotated homogenized compliance matrix $C^H = \left(\mathbf{D}^H\right)^{-1}$ can be written as

$$C^H = C^0 + C^1 \sin 2\theta + C^2 \cos 2\theta + C^3 \sin 2\theta \cos 2\theta - C^4 \sin^2 2\theta \qquad (10)$$

where C^k $(k=0, 1, ..., 4)$ are independent of θ,

$$C^0 = \begin{bmatrix} (C_{11}^h + C_{22}^h)/2 & C_{12}^h & 0 \\ C_{12}^h & (C_{11}^h + C_{22}^h)/2 & 0 \\ 0 & 0 & C_{33}^h \end{bmatrix},$$

$$C^1 = c_1 \begin{bmatrix} 0 & 0 & 1 \\ 0 & 0 & 1 \\ 1 & 1 & 0 \end{bmatrix}, \qquad C^2 = c_1 \begin{bmatrix} 1 & 0 & 0 \\ 0 & -1 & 0 \\ 0 & 0 & 0 \end{bmatrix}, \qquad (11)$$

$$C^3 = 2c_2 \begin{bmatrix} 0 & 0 & 1 \\ 0 & 0 & -1 \\ 1 & -1 & 0 \end{bmatrix}, \qquad C^4 = c_2 \begin{bmatrix} 1 & -1 & 0 \\ -1 & 1 & 0 \\ 0 & 0 & -4 \end{bmatrix}$$

c_1 and c_2 are the invariant material parameters:

$$c_1 = \frac{1}{2}(C_{11}^h - C_{22}^h)$$

$$c_2 = \frac{1}{4}(C_{11}^h + C_{22}^h - 2C_{12}^h - C_{33}^h) \qquad (12)$$

and

$$C_{11}^h = D_{22}^h [D_{11}^h D_{22}^h - (D_{12}^h)^2]^{-1}$$

$$C_{12}^h = -D_{12}^h [D_{11}^h D_{22}^h - (D_{12}^h)^2]^{-1}$$

$$C_{22}^h = D_{11}^h [D_{11}^h D_{22}^h - (D_{12}^h)^2]^{-1}$$

$$C_{33}^h = (D_{33}^h)^{-1} \qquad (13)$$

Using these, we can determine the optimum angle of ratation of the microstrucure in the optimization process by taking the first derivatives of the D^H or C^H matrix with respect to the anle θ. Note that in the current problem, the homogenized mass density is just an average of the mass distributed in the microstructural domain Y, and it is not a function of the orientation variable. In fact, we have a very simple formulation for the homogenized mass density in the current problem, that is

$$\rho^H = \rho_0(a + b - ab) \qquad (14)$$

It is also noted that the homogenized elasticity tensor of the microstructure with a rectangular hole behaves very similarly with the one for the so-called rank 2 material that is idealized as the strongest orthotropic material, despite that the microstructure with a rectangular hole possesses very weak shear resistance. On the other hand the rank 2 material does not have any resistance to the in-plane shear, and it only resists to the principal stresses. Figure 5(a) describes the homogenized elasticity constants for the microstructure with a rectangular hole. It is, however, noted that this choice of the homogenized elasticity tensor leads the very stable optimum layout with a possibly large portion of perforation that implies a very fine scale structure. The optimum is clearly mesh independent, and possesses convergence property with respect to the finite element approximation as Cheng and Olhoff (1981) described. Thus, in order to *idealize* the optimum structure without introducing the fine scale microstructure that is very difficult to be manufactured, we shall introduce an artificially modified (penalized) homogenized elasticity tensor by introducing a transformation nonlinear map $a \rightarrow \alpha(a)$ and $b \rightarrow \beta(b)$:

$$E_{ijkl}^{h}(a,b) \quad \rightarrow \quad E_{ijkl}^{h}(\alpha(a),\beta(b)).$$

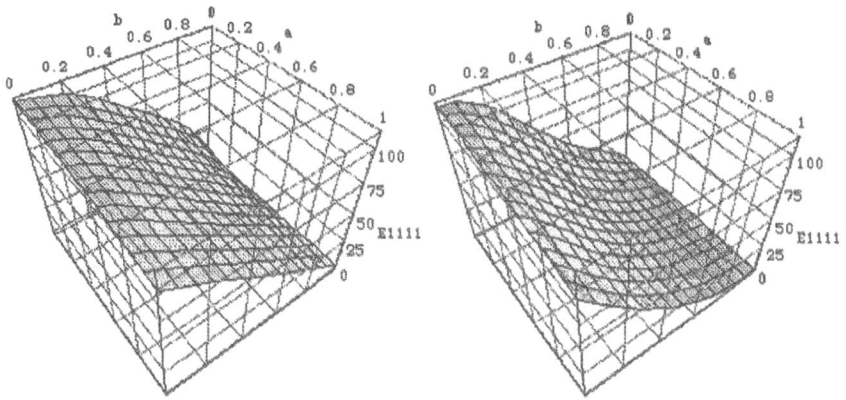

(a) Original Homogenized E_{1111} (b) Modified Constant $E_{1111}(r=1)$

Figure 5   Homogenized E_{1111} for the Microstructure with a Rectangular Hole

Here we may choose $\alpha = \beta$ such that

$$\alpha(a) = \begin{cases} a^r & \text{where} \quad 1 \le r < \infty \\ r - \sqrt{r^2 + (1-r)^2 - (a-1+r)^2} & \text{where} \quad 1 \le r < \infty \end{cases}$$

It is clear that the later choice of the nonlinear function becomes the original one when $t = +\infty$, while the nonlinearity similar to a^r can be obtained for $t = 0$, while it forms a circular arc with the radius r and passing through the two limit points $(0,0)$ and $(1,1)$. The both modified functions α are designed to approximate the characteristic function

$$\chi(a) = \begin{cases} 0 & \text{if} \quad 0 \le a < 1 \\ 1 & \text{if} \quad a = 1 \end{cases}$$

that describes the identification of the solid domain by $(a,b) = (1,1)$, and of cavity otherwise as shown in Fig. 6. Polynomial based function $\alpha = a^r$ possess zero derivatives at , while they are not $+\infty$ at $a = 1$. In other words, it provides unequal penalty at 0 and 1. On the other hand, the circular arc approximation provides equal penalty at the both limit points, and is the same to the original characteristic function.

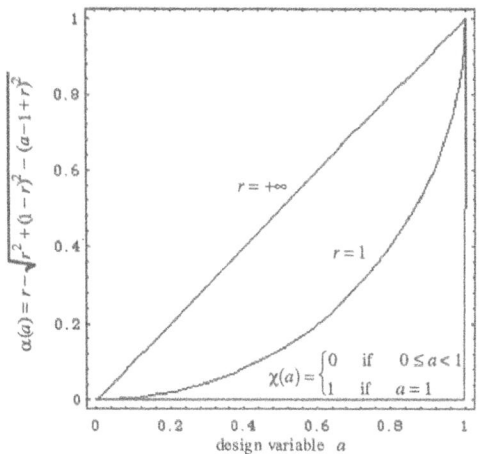

Figure 6 Approximation of the characteristic function

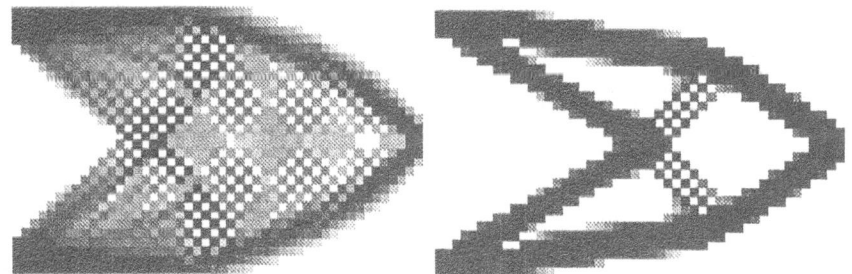

(a) Original Homogenized Tensor   (b) Transformed Homogenized Tensor

Figure 7   Optimum Layouts by Two Different Homogenized Elasticity Tensors

Clearly, there are infinitely many ways to penalize the homogenized elasticity tensor, see, e.g., Rozvany and Zhou (1993), Haber, Jog, and Bendsøe (1994), and others. This approach may be very similar to use artificially constructed elasticity tensors with respect to the density proportional to the porosity of the elastic medium using the study of porous materials. The microstructure with a rectangular hole is very close to the rank 2 material, and it can be very orthotropic in the design variables a and b, while most of other penalized elasticity tensors or constitutive relations for porous materials are isotropic with respect to the density. Since intermediate structures toward the optimum implies very different values of the maximum and minimum principal stresses, we should have orthotropic nature in the elasticity tensor to respond to this stress field. If the isotropic elasticity tensor is assumed, the structure would respond only isotropic way to the stress field, and it is clearly far from the optimum. Figure 5(b) shows the profile of the modified elasticity constants. Figure 7 describes the difference of the optimums by the original and modified elasticity tensors to find the layout of a structure for a static problem.

Based on these homogenized material properties for the two-dimensional microstructure, the effective properties for the perforated laminate element in a dynamic system can be obtained using the classical lamination theory. Assume that the displacement field of an arbitrary point P in the plate element Ω_e is approximated as (u, v, w) :

$$
\begin{aligned}
u(x,y,z,t) &= u(x,y,t) + z\theta_y(x,y,t) \\
v(x,y,z,t) &= v(x,y,t) - z\theta_x(x,y,t) \\
w(x,y,z,t) &= w(x,y,t)
\end{aligned}
\tag{15}
$$

where the italic alphabetic (u,v,w) represents the local displacement field of the middle plane at the point P' that is the projection of P, and θ_y and θ_x are the rotations of point P about the x and y axes respectively. The strain field $\boldsymbol{\varepsilon}^T = \{\varepsilon_{xx}, \varepsilon_{yy}, \varepsilon_{zz}, \gamma_{xy}, \gamma_{yz}, \gamma_{zx}\}$ is derived by taking the first derivatives of this displacement field with respect to the local coordinate system (x, y, z). The strain energy for the e-th shell element associated with the local coordinate system can then be written as:

$$
U_e = \frac{1}{2} \int_{\Omega_e} \int_{-h_1/2}^{h_1/2} \boldsymbol{\varepsilon}^T D^H \boldsymbol{\varepsilon} \, dz \, d\Omega
\tag{16}
$$

where D^H is the homogenized rotated elasticity matrix derived under the plane stress assumption. Using these strain vectors and the classical laminate theory, the strain energy of a shell element Ω_e can be yielded in a concise form:

$$
U_e = \frac{1}{2} \int_{\Omega_e} \{\boldsymbol{\varepsilon}_m^T D_o^H \boldsymbol{\varepsilon}_m + \boldsymbol{\kappa}_b^T D_2^H \boldsymbol{\kappa}_b + \boldsymbol{\gamma}_s^T D_{ts} \boldsymbol{\gamma}_s\} d\Omega
\tag{17}
$$

where $\boldsymbol{\varepsilon}_m$ is the membrane strain, $\boldsymbol{\kappa}_b$ is the curvature, and $\boldsymbol{\gamma}_s$ is the transverse shear strain, i.e.

$$
\boldsymbol{\varepsilon}_m{}^T = \left\{ \frac{\partial u}{\partial x} \quad \frac{\partial u}{\partial y} \quad \frac{\partial v}{\partial y}, + \frac{\partial v}{\partial x} \right\}
$$

$$
\boldsymbol{\kappa}_b{}^T = \left\{ \frac{\partial \theta_y}{\partial x} \quad -\frac{\partial \theta_x}{\partial y} \quad \frac{\partial \theta_y}{\partial y} - \frac{\partial \theta_x}{\partial x} \right\}
\tag{18}
$$

$$
\boldsymbol{\gamma}_s{}^T = \left\{ -\theta_x + \frac{\partial w}{\partial y} \quad \frac{\partial w}{\partial x} + \theta_y \right\}
$$

D_o^H is the homogenized membrane stiffness, D_2^H stands for the homogenized bending rigidity, and D_{ts} represents the transverse shear stiffness, i.e.

$$D_o^H = \int_{-h_1/2}^{-h_o/2} \left[D^H(a,b,\theta)\right]dz + \int_{-ho/2}^{h_o/2} \left[D^H(0,0,0)\right]dz + \int_{ho/2}^{h_1/2} \left[D^H(a,b,\theta)\right]dz$$

$$D_2^H = \int_{-h_1/2}^{-h_o/2} \left[D^H(a,b,\theta)\right]z^2dz + \int_{-ho/2}^{h_o/2} \left[D^H(0,0,0)\right]z^2dz + \int_{ho/2}^{h_1/2} \left[D^H(a,b,\theta)\right]z^2dz$$

$$D_{ts} = \int_{-h_1/2}^{-h_o/2} \left[D_s(a,b,\theta)\right]dz + \int_{-ho/2}^{h_o/2} \left[D_s(0,0,0)\right]dz + \int_{ho/2}^{h_1/2} \left[D_s(a,b,\theta)\right]dz \qquad (19)$$

Here, D^H is the rotated homogenized elastic matrix and D_s the shear modules defined as:

$$D_s = \beta \begin{bmatrix} E_{2323} & 0 \\ 0 & E_{3131} \end{bmatrix} \qquad (20)$$

where β is so-called the shear correction factor. It is noted that E_{2323} and E_{3131} must be obtained by the homogenization method for the three-dimensional unit cell with plane stress assumption only by $\sigma_{zz} = 0$, and that they are not well-defined at this moment.

Similarly, the kinetic energy of the e-th shell element associated with the local coordinate system is of the form:

$$T_e = \frac{1}{2} \int_{\Omega_e} \int_{-h_1/2}^{h_1/2} \rho^H \left[\left(\frac{\partial u}{\partial t}\right)^2 + \left(\frac{\partial v}{\partial t}\right)^2 + \left(\frac{\partial w}{\partial t}\right)^2\right] dz d\Omega$$

$$+ \frac{1}{2} \int_{\Omega_e} \int_{-h_1/2}^{h_1/2} \rho^H \left[\left(\frac{\partial \theta_x}{\partial t}\right)^2 + \left(\frac{\partial \theta_y}{\partial t}\right)^2\right] z^2 dz d\Omega \qquad (21)$$

If we assume $u = \{u,v,w\}$ and $\vartheta = \{\theta_x,\theta_y\}$, then the corresponding homogenized material density $\rho_{u\vartheta}^H = \{\rho_u^H, \rho_\vartheta^H\}$ of this perforated element can be derived as:

$$\rho_u^H = \int_{-h_1/2}^{-h_o/2} \left[\rho^H(a,b)\right]dz + \int_{-ho/2}^{h_o/2} \left[\rho^H(0,0)\right]dz + \int_{ho/2}^{h_1/2} \left[\rho^H(a,b)\right]dz$$

$$\rho_\vartheta^H = \int_{-h_1/2}^{-h_o/2} \left[\rho^H(a,b)\right]z^2dz + \int_{-ho/2}^{h_o/2} \left[\rho^H(0,0)\right]z^2dz + \int_{ho/2}^{h_1/2} \left[\rho^H(a,b)\right]z^2dz \qquad (22)$$

There are several important problems in dealing with vibrating structures which are very different from static problems. First, the nature of the objective function for a dynamic problem may not be as good as that for a static problem. In other words, in an eigenfrequency optimization problem, the objective function represented by a single eigenvalue may not be smooth because some eigenmodes switch their orders in the optimization process as well as they shear the same eigenvalue, as shown in Ma et al (1993). Second, the sensitivity of the objective function may be discontinuous. Third, the approximated optimization problem may become non-convex in a dynamic problem, and then the updating rule used in the static problem may become invalid in the dynamic problem. Thus, we need to develop a modification of the optimality criteria method to solve an optimization problem of a vibrating structure. Major extension we have made for dynamic problems are as follows:

1) We propose a new objective function corresponding to multi-eigenfrequency optimization problems. This new objective function improves the convergence of the optimization process, and it may be sufficiently general in dealing with several different eigenfrequency optimization problems.

2) We consider a frequency response optimization problem for reducing the dynamic response of a vibration structure.

3) We introduce a modified optimality criteria method for updating design variables of dynamic systems, which can improve the convergence and can overcome the oscillation and divergence of the original algorithm.

4) We also introduce a method for determining the optimal orientation of the microstructure in the optimization process.

3. Objective functions for the Optimal Design Problem

3.1. Eigenfrequency optimization problems

Maximizing a chosen eigenvalue of the system is usually used in an eigenfrequency optimization problem :

$$\text{Maximize } \lambda_n \tag{23}$$

where λ_n is a chosen eigenvalue which satisfies the eigenvalue problem:

$$\int_{\Omega} E_{rspq}^{H} \frac{\partial \phi_r^n}{\partial x_s} \frac{\partial v_p}{\partial x_q} d\Omega - \lambda_n \int_{\Omega} \rho^H \phi_r^n v_r d\Omega = 0 \quad (\text{ for } \forall \mathbf{v} \in V) \tag{24}$$

under the assumption that this eigenvalue is not isolated. Here ϕ_r^n stands for the component of the eigenvector corresponding to the eigenvalue λ_n. However, Eq. (23) may results in a non smooth optimization problem because eigenmodes may switch their orders during the optimization process. Indeed we shall consider an example that maximizes the second eigenfrequency under a given volume constraint for the rectangular design domain fixed in the left edge, and added a non-structural mass at the center of the right edge. Noting that details of this example can be found in Ma et al (1993), we shall see Fig. 8 that shows the iteration history of the three (second to fourth) eigenfrequencies in the optimization process using the optimality criteria method similar to the stiffness maximization problem for static structures. While we maximize the second eigenfrequency, the third and fourth eigenfrequencies at the original stage may fall down to the lower values, and the eigenfrequencies switch their orders very frequently during the optimization process. Thus, the sensitivity of the objective function becomes discontinuous, and large oscillation can be observed in the objective function, and then this may lead a non-convergent result to the optimization problem.

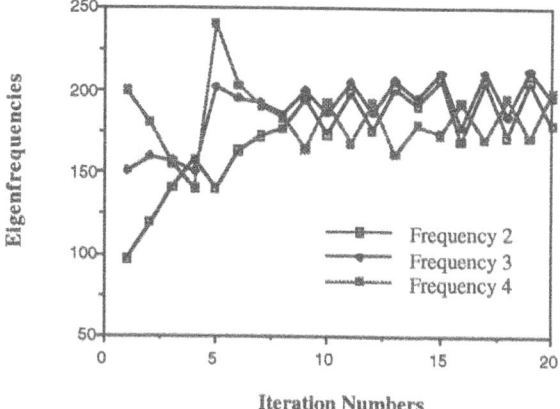

Figure 8 Eigenfrequencies switch their orders in the optimization process

In order to overcome this problem, a mean-eigenvalue Λ, which is a combination of multiple eigenvalues, is suggested as the objective function:

$$\Lambda = \begin{cases} \Lambda_0 + \left(\dfrac{1}{\alpha}\sum_{i=1}^{m} w_i(\lambda_{n_i} - \lambda_{0_i})^n\right)^{\frac{1}{n}} & (\text{ for } n = \pm 1, \ \pm 2, \ \cdots; \ n \neq 0) \\[4mm] \Lambda_0 + \exp\left(\dfrac{1}{\alpha}\sum_{i=1}^{m} w_i \ln|\lambda_{n_i} - \lambda_{0_i}|\right) & (\text{ for } n = 0) \end{cases} \tag{25}$$

where λ_{n_i} $(i = 1,2,\cdots,m)$ are specified eigenvalues, $n_i(i = 1,2,\cdots,m)$ stand for order numbers of the specified eigenvalues. Also, $w_i(i = 1,2,\cdots,m)$ are given weighting coefficients, $\lambda_{0_i}(i = 1,2,\cdots,m)$ are given parameters, and n is a given power. Λ_0 and α are arbitrary constants which are used only for making some physical meaning of the objective function and adjusting the dimension of the objective function. For example, we can use $\Lambda_0=0$ and

$$\alpha = \sum_{i=1}^{m} w_i.$$

In most of problems, Eq. (25) defines a much smoother objective function. When two modes, whose eigenvalues are in the definition of the mean-eigenvalue Λ, exchange their orders during the optimization process, the change in the objective function will be smooth because the contributions of these modes have already been accounted for in the objective function. Thus, as shown in Fig. 9, significant improvement is obtained by using the new objective function in the iteration process by using the OC method. Here, $n=-1$, and $\Lambda_0=\lambda_{0_i} = 0(i = 1,2,\cdots,m)$ are assumed. It is noted that the third and fourth eigenfrequencies practically coincides, although they might be slightly different in the finite element approximation.

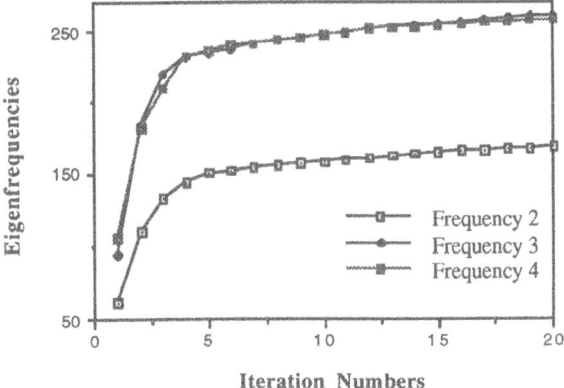

Figure 9 Convergent process obtained using the new objective function

Equation (25) also defines a general multiple eigenfrequency optimization problem. By choosing parameters λ_{0_i} ($i = 1, 2, \cdots, m$) and power n properly, three kinds of eigenfrequency optimization problems can be obtained. First, if choose power n as an odd number, then the optimization problem

Maximize Λ

becomes to maximize the mean-average of specified eigenvalues, where parameters λ_{0_i} ($i = 1, 2, \cdots, m$) can be chosen as any value. For example, if choose $n = -1$, and $\Lambda_0 = \lambda_{0_i} = \lambda_0 (i = 1, 2, \cdots, m)$, the optimization problem becomes

$$\text{Maximize } \Lambda = \lambda_0 + \alpha \left(\sum_{i=1}^{m} \frac{w_i}{\lambda_{n_i} - \lambda_0} \right)^{-1} \tag{26}$$

In the case of Eq. (26), if assume all the weighting coefficients w_i ($i = 1, 2, \ldots, m$) to be the same, the eigenvalue which is the closest to the given parameter λ_0 will experience the largest increase in the optimization process, because this eigenvalue has the largest contribution to the objective function. Furthermore, if let $\lambda_0 = 0$, then the lowest eigenvalue λ_{n_i} in Λ would have the largest increase in the optimization process. Note that, in the general case, weighting coefficients w_i ($i = 1, 2, \ldots, m$) can be chosen as different values for adjusting the contributions of the specified eigenvalues to make a desired optimization problem.

Second, if choose power n as an even number, λ_{0_i} ($i = 1, 2, \cdots, m$) as the given eigenvalues, and define the optimization problem by

Maximize Λ

then the problem becomes to maximize the mean-average of distances between specified eigenvalues and the given eigenvalues. For example, if choose $n=-2$ and $\Lambda_0 = 0$, the optimization problem becomes

$$\text{Maximize } \Lambda = \left[\frac{1}{\alpha}\sum_{i=1}^{m}\frac{w_i}{(\lambda_{n_i} - \lambda_{0_i})^2}\right]^{-\frac{1}{2}} \tag{27}$$

In the case of Eq. (27), the specified eigenvalues $\lambda_{n_i} (i=1,2,\cdots,m)$ will turn far away from their corresponding given eigenvalues $\lambda_{0_i} (i=1,2,\cdots,m)$ respectively, and the eigenvalue which is the closest to a corresponding given eigenvalue will be the fastest away from this eigenvalue assuming the all weighting coefficients are the same.

Third, if choose power n as an even number, $\lambda_{0_i} (i=1,2,\cdots,m)$ as the desired eigenvalues, and define the optimization problem as

Minimize Λ

then the optimization problem becomes to minimize the mean-average distance between specified eigenvalues and their desired values. For example, if choose $n=2$, $\Lambda_0 = 0$ and weighting coefficients as $w_i = \lambda_{0_i}^{-2} (i=1,2,\cdots,m)$, the optimization problem becomes

$$\text{Minimize } \Lambda = \left[\alpha\sum_{i=1}^{m}\frac{1}{\lambda_{0_i}^2}(\lambda_{n_i} - \lambda_{0_i})^2\right]^{\frac{1}{2}} \tag{28}$$

In the case of Eq. (28), all the specified eigenvalues $\lambda_{n_i} (i=1,2,\cdots,m)$ will approach to their desired values $\lambda_{0_i} (i=1,2,\cdots,m)$, respectively, and the frequency which is the farthest from its desired value will the fastest approach to its desired value.

3.2 Frequency response optimization problem

For the frequency response problem, assuming that $u_i^0 = U_i e^{j\omega t}$, $f_i^h = F_i e^{j\omega t}$ and $t_i = T_i e^{j\omega t}$, where $U_i = U_i(\omega)$ stands for the frequency response, ω the exciting frequency, F_i and T_i the amplitudes of f_i^h and t_i, then the equation of motion, Eq. (7), can be reduced to

$$\int_\Omega E_{ijkl}^H \frac{\partial U_i}{\partial x_j}\frac{\partial v_k}{\partial x_l} d\Omega - \omega^2 \int_\Omega \rho^H U_i v_i d\Omega = \int_\Omega F_i v_i d\Omega + \int_{\Gamma_t} T_i v_i d\Gamma \quad \text{for } \forall v \in V \tag{29}$$

Here, Eq. (29) gives the state equation for the frequency response optimization problem. Only the second term of the left hand of Eq. (29) is an additional dynamic term comparing with the state equation of a static structure. Extending the objective function used in dealing with a static structure, i. e., the "mean-compliance," to the current dynamic problem, we propose a frequency response optimization problem:

$$\text{Minimize} \quad \sigma = \int_{\Omega} U_i F_i \, d\Omega + \int_{\Gamma_i} U_i T_i \, d\Gamma \tag{30}$$

where objective function σ is called "dynamic compliance."

Note that if using a standard mode-superposition technique, the frequency response U_i can be expressed as

$$U_i = \sum_{n=1}^{N} \phi_i^n Q_n \tag{31}$$

where

$$Q_n = \frac{f_n}{\omega_n^2 - \omega^2} \quad \text{and} \quad f_n = \int_{\Omega} \phi_i^n F_i \, d\Omega + \int_{\Gamma_i} \phi_i^n T_i \, d\Gamma \tag{32}$$

Then the objective function σ can be simplified by using the mode-superposition technique:

$$\sigma = \sum_{n=1}^{N} \frac{f_n^2}{\omega_n^2 - \omega^2} \tag{33}$$

as shown in Ma and Hagiwara (1991).

4. Updating Rule : Modified Optimality Criteria Method

4.1. The traditional optimality criteria

In abstract form, the optimization problem described in this work can be represented by

$$\underset{X,\,\Theta}{\text{Minimize}} \qquad f(X, \Theta) \tag{34}$$
$$\begin{array}{c} \text{subject to} \\ h(X) \leq 0 \\ \underline{x}_i \leq x_i \leq \overline{x}_i \ , \ i=1,2,\cdots,N \end{array}$$

Where, f stands for the objective function, h the total volume (mass, weight) constraint function, x_i the i-th sizing design variable which corresponds to a size of the microstructure for a finite element model, X is the collection of x_i, N is the twice of the total number of finite elements n_{el}, θ_i the i-th orientation design variable which corresponds to the rotation of the microstructure, Θ is the collection of θ_i. $\underline{x}_i$ and $\overline{x}_i$ are the upper bound and lower bound of design variable x_i. The optimality criteria for the optimization problem can be written as

$$\frac{\partial f}{\partial x_i} + \lambda \frac{\partial h}{\partial x_i} = \alpha_{-i} - \alpha_{+i} \ , \qquad (i = 1,2,\cdots,N) \tag{35}$$

$$\frac{\partial f}{\partial \theta_i} = 0 \qquad (i = 1,2,\cdots,n_{el}) \tag{36}$$

where, λ stands for the Lagrange multiplier corresponding to the volume constraint, α_{-i} and α_{+i} the Lagrange multipliers corresponding to the side constraints while $\alpha_{-i} = \alpha_{+i} = 0$ (for $\underline{x}_i < x_i < \overline{x}_i$). Equation (35) can also be rewritten as following for a design variable x_i :

$$D_i = \frac{1}{\lambda}\left(-\frac{\partial f}{\partial x_i}\bigg/\frac{\partial h}{\partial x_i}\right) = 1 \quad (\text{for } \underline{x}_i < x_i < \overline{x}_i)$$

where D_i is called "cost function", and if the structure is not in the optimum, D_is won't be equal to one. A standard updating rule based on the Optimality Criteria (OC) is given by

$$x_i^{k+1} = (D_i^k)^\eta x_i^k \quad (\text{for } \underline{x}_i < x_i^{k+1} < \overline{x}_i) \tag{37}$$

where, x_i^k stands for the design variable in the k-th iteration, x_i^{k+1} the updated design variable, $D_i^k = D_i\big|_{x=x^k, \theta=\theta^k}$. η is a given parameter. Equation (37) implies that if $\exists i: D_i^k \neq 1$ then the design variable x_i will be updated, while if $D_i^k > 1$ then x_i is increased, if $D_i^k < 1$ then x_i is reduced. This updating rule was employed by Bendsøe and Kikuchi (1988) and others for solving static stiffness maximization problems. The optimization algorithm based on updating rule Eq. (37) is very efficient in computation and converges for the static problem. However, it does not work well in the dynamic case. The reason is that in the static problem, we almost always have

$$D_i^k > 0 \quad (\forall i, \forall k)$$

It means that the sensitivity of the objective function divided by the sensitivity of the constraint function is negative for any design variable (see Eq.(35)). Thus the design variables can be updated as positive real number using Eq. (37). But in a dynamic problem, we may have $D_i^k < 0$ for some design variables. In this case, the design variable x_i^{k+1} updated by Eq. (37) may be a negative or complex number. However, the design variable could not be a negative nor complex number by its physical meaning! Even though one can solve this problem by an intuitive treatment -- letting the design variable x_i^{k+1} to be a moving limit of the design variable, but it could cause a discontinuous jump in the design variable, and there is no guarantee that the optimization problem is going to converge to the optimal solution.

4.2. A relaxed form of the optimality criteria

To overcome the difficulty mentioned in above, a relaxed form of the optimality criteria is proposed. The basic idea is to use a shift parameter which is corresponding to the Lagrange multiplier. First, Eq. (35) can be rewritten as following form:

$$\left(\frac{\partial f}{\partial x_i} - \mu\frac{\partial h}{\partial x_i}\right) + \lambda^*\frac{\partial h}{\partial x_i} = 0 \quad (\text{for } \underline{x}_i < x_i < \overline{x}_i) \tag{38}$$

here, μ is the shift parameter, λ^* ($\lambda^* = \lambda + \mu$) the shifted Lagrange multiplier. Equation (38) is nothing than identifying to Eq. (35) for $\underline{x}_i < x_i < \overline{x}_i$. Then, the relaxed form of the Optimality Criteria using Eq. (38) can be written as

$$\overline{D}_i = \frac{1}{\lambda^*}\left(\mu - \frac{\partial f}{\partial x_i}\Big/\frac{\partial h}{\partial x_i}\right) = 1 \quad (\text{for } \underline{x}_i < x_i < \overline{x}_i) \tag{39}$$

where, $\overline{D}_i$ stands for the modified cost function. Note that $\overline{D}_i$ can have different value from D_i when the structure is not at the optimal stage. By choosing μ properly, i. e.,

$$\mu \geq \max_{1 \leq i \leq N}\left\{\frac{\partial f}{\partial x_i}\Big/\frac{\partial h}{\partial x_i}\right\} \tag{40}$$

we can have $\overline{D}_i$ always to be positive, therefore the improved updating rule based on the relaxed optimality criteria is

$$x_i^{k+1} = (\overline{D}_i^k)^{\eta} x_i^k \quad (\text{for } \underline{x}_i < x_i^{k+1} < \overline{x}_i) \tag{41}$$

then now the updated design variable x_i^{k+1} can always be positive. It should be noted that this updating rule can also be obtained by a Mathematical Programming (MP) approach -- using a convex approximation and the dual method [see Ma et al (1992) and (1993)].

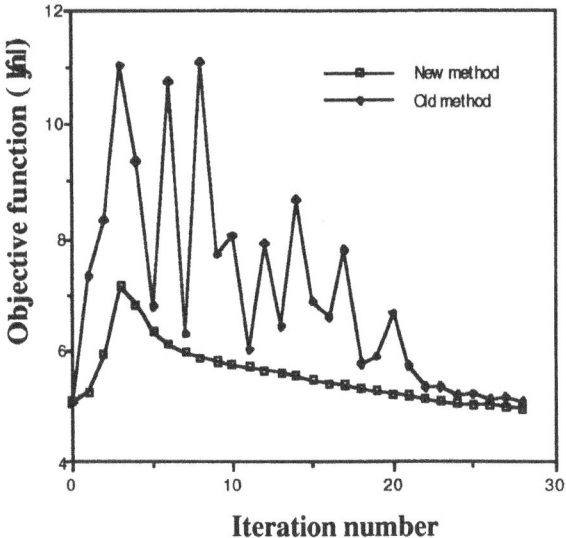

Figure 10 A comparison of "with and without shift parameter μ"

To give a comparison of the results obtained by using the improved method (with the shift parameter μ) and the previous method (without the shift parameter μ), a frequency response optimization problem is calculated (the definition of the problem can be found in the later section, Section 5.1). As shown in Fig. 10, a serious oscillation is occurred in the objective function when the shift parameter μ is not applied, while a significant improvement can been obtained by using the shift parameter μ.

4.3. Updating orientation variables

Using Eq. (10), the optimality criteria Eq. (36) corresponding to a discrete orientation variable θ_e can be obtained as

$$g_e^1 \cos 2\theta_e - g_e^2 \sin 2\theta_e + g_e^3 \cos 4\theta_e - g_e^4 \sin 4\theta_e = 0 \tag{42}$$

where $-\pi/2 \le \theta_e \le \pi/2$ and $e = 1,2,...,n_{el}$. If coefficients $g_e^k (k = 1,2,\cdots,4)$ are assumed to be constants as mentioned before, then Eq. (42) can be approximately solved separately for each individual discrete orientation variable. In fact, assuming $z = \tan \theta_e$, then Eq.(42) can be transformed into a 4th-order polynomial equation as:

$$f_4 z^4 + f_3 z^3 + f_2 z^2 + f_1 z + f_0 = 0 \tag{43}$$

where

$$
\begin{aligned}
f_4 &= -g_e^1 + g_e^3 \\
f_3 &= -2g_e^2 + 4g_e^4 \\
f_2 &= -6g_e^3 \\
f_1 &= -2g_e^2 - 4g_e^4 \\
f_0 &= g_e^1 + g_e^3
\end{aligned}
\tag{44}
$$

Solving Eq. (40), we may obtain several solutions for variable z. In order to determine a unique solution which is the optimum for the problem, first we only consider the solutions of Eq. (40) which are real number, second we defined an evaluation function, see Cheng et al (1993):

$$\psi_e(\theta_e) = g_e^1 \sin 2\theta_e + g_e^2 \cos 2\theta_e + g_e^3 \sin 2\theta_e \cos 2\theta_e - g_e^4 \sin^2 2\theta_e \tag{45}$$

By comparing the values of the evaluation function ψ_e with respect to all real solutions of Eq. (42) and $\theta_e = \pm\pi/2$, we can determine the optimal solution which gives the largest value (for the maximization problem) of the evaluation function ψ_e (note that, for the minimization problem, the optimal solution gives the smallest value of ψ_e). If there is no real solution for Eq. (42), then we just need to compare the values of the evaluation function at $\theta_e=-\pi/2$ and $\theta_e=\pi/2$ and choose one as the optimal solution.

5. Examples of Optimum Layout Design for Vibration

5.1 Frequency Response Problem for a Short Beam

The first example is the topological design for a frequency response problem. As shown in Fig. 11, the design domain is specified as a 8.0cm by 5.0cm rectangle with two fixed support boundaries at the end of the left hand side. A shear-like, periodic exciting load is assumed to act on the center of the right hand side of the design domain. Young's modulus is assumed to be $E=100$Kg/cm^2, Poisson's ratio $\nu=0.3$, mass density $\rho_0=10^{-6}$Kg/cm^2. The total volume of the material is considered as $V_0=14.4$cm^2 which is 36% of the volume of the whole design domain. The optimization process is started from the initial values $a_i=b_i=0.20$ ($i=1, 2, \cdots, N$) which satisfy the volume constraint. Two finite element models are applied to solve the problem. The first one is a coarse mesh using 32×20 finite elements, while the second case is a fine mesh using 64×40 finite elements. Figure 12 shows the convergence history of the optimal topology using the rough mesh model. Where, the objective function is the dynamic compliance, k is the iteration number, and the excitation frequency is considered 60Hz. Figure 13 gives the results with respect to the various excitation frequencies, 0Hz,

15Hz, 30Hz, 45Hz, 60Hz, 75Hz, respectively, using the fine mesh model. Where, σ^* stands for the objective function of the optimal structure. Here, when the excitation frequency is 0Hz, the result is reduced to the one obtained in the static optimization problem. It can be seen that the optimal topologies are very different when the excitation frequency is changed, and the optimal structure obtained in the static problem is not the optimal one for the dynamic problem, especially when the excitation frequency is high.

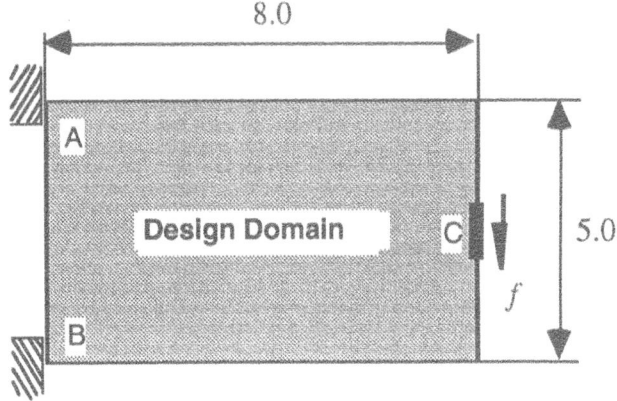

Figure 11 Design Domain for a Frequency Response Problem

5.2 Maximization of the Lowest Frequency of a Frame Structure

This example demonstrates the benefit of using the topology optimization technique with respect to the eigenfrequency optimization problem. As shown in Fig. 14(a), the design domain is specified as a 14.0cm by 2.0cm rectangle with a concentrated mass at the center of the design domain. Here, Young's modulus is assumed to be $E=100$Kg/cm^2, Poisson's ratio $\nu=0.3$, mass density $\rho_0=10^{-6}$Kg/cm^2. We shall maximize the fundamental eigenfrequency of the structure. Figure 14(b) shows the optimal structure obtained using the topological optimization technique with respect to the total mass constraint 1.4×10^{-5}Kg which includes the concentrated mass $M_0=5.0\times10^{-6}$ Kg. The fundamental eigenfrequency of the initial structure that is uniformly perforated, is 16.3Hz, with the initial values of design variables are $a_e = b_e = 0.36$ ($e=1, 2, \cdots, 2,800$) satisfying the total mass constraint. The optimized fundamental eigenfrequency by the topological change is 63.2Hz, that is, a three-fold improvement is obtained by using the topological optimization technique.

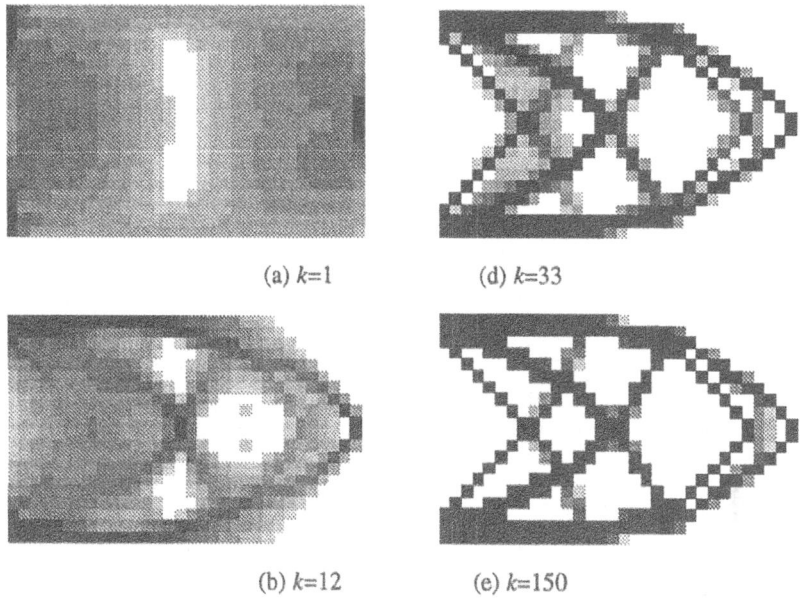

(a) $k=1$ (d) $k=33$

(b) $k=12$ (e) $k=150$

Figure 12 Generation history of the optimal topology (k: iteration number)

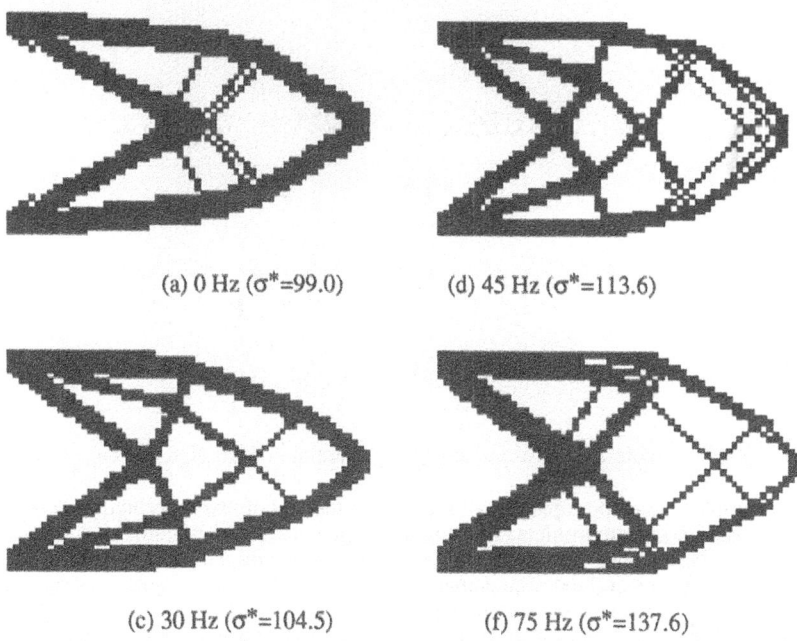

(a) 0 Hz ($\sigma^*=99.0$) (d) 45 Hz ($\sigma^*=113.6$)

(c) 30 Hz ($\sigma^*=104.5$) (f) 75 Hz ($\sigma^*=137.6$)

Figure 13 Optimal topologies w.r.t. different excitation frequencies

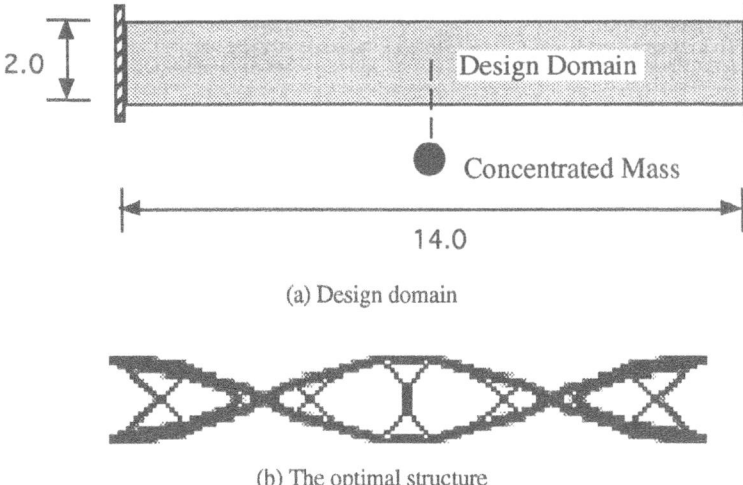

(a) Design domain

(b) The optimal structure

Figure 14 Topology optimization of a frame for the fundamental frequency

5.3 Topology Optimization with a Constraint in the Design Domain

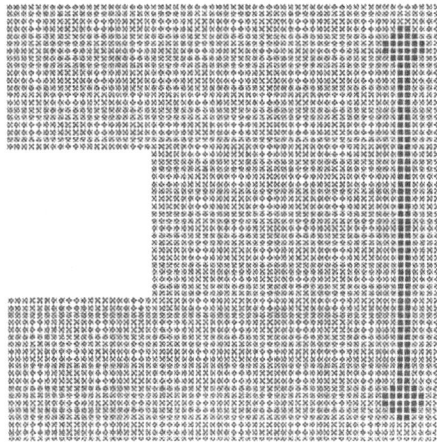

Figure 15 An example with a constraint in the design domain

This example shows the capability of the topological optimization technique to deal with a rather complex design problem. As shown in Fig. 15, the design domain is specified as a 3.0cm by 3.0cm square area with a 1.0cm by 1.0cm square non-design domain in the left side. The left edges of the design domain are fixed, and two concentrated masses which are considered as the loads of the structure are imposed on the right side of the design domain as shown in Fig. 15. The concentrated masses are connected by a bar structure, which is considered as a non-design domain (i. e., it is filled by the full material and will not change within the optimization process). For the structural analysis, the design domain is divided to 3,200 finite elements with 3,341 nodes. Here, Young's modulus is assumed to be $E=100\text{Kg/cm}^2$, Poisson's ratio $v=0.3$, mass density $\rho_0=10^{-6}\text{Kg/cm}^2$. Each concentrated mass is 3.2×10^{-6} Kg. The total mass constraint for the design problem is 5.0×10^{-6} Kg excluded the concentrated masses but included the mass of the non-design domain.

Two cases are considered for this design problem: (a) maximizing the first eigenfrequency, (b) maximizing the second eigenfrequency. The objective function Eq. (26) is utilized for the optimization problems, where it is assumed that $\lambda_0=0$ and all weights in Eq. (26) are the same. For the case (a), three modes, which are 1st, 2nd and 3rd modes, are employed for the objective function, while for the case (b), 2nd, 3rd and 4th modes are employed. Figure 16 shows the optimal structures obtained by using the method presented in this paper for the two cases respectively. The iteration number of the optimization process was about 50 for both cases, where, each iteration elapses CPU about 300 seconds for this example, using an IBM RS6000-350 machine. Table 1 shows the lowest four eigenfrequencies obtained in both cases. As shown in Table 1, both of the first eigenfrequency and the second eigenfrequency are significantly improved in the two cases respectively.

Table 1. The lowest four eigenfrequencies obtained for Case a and Case b

frequencies	Uniform Design	Case (a)	Case (b)
ω_1	62.652	158.320	80.644
ω_2	196.521	316.752	381.987
ω_3	198.013	371.195	392.676
ω_4	345.277	381.587	498.854

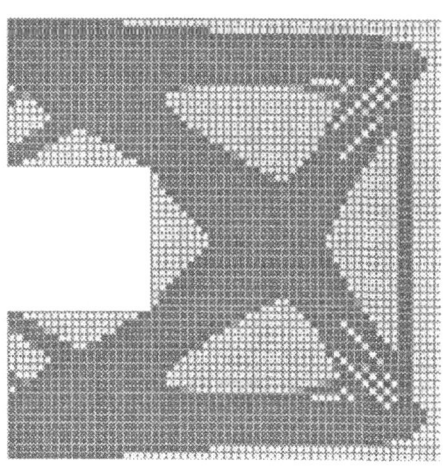

a) The optimal structure for maximizing the first eigenfrequency

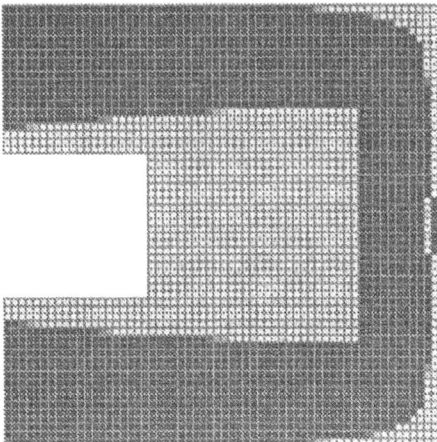

b) The optimal structure for maximizing the second eigenfrequency

Figure 16 The optimal structures for the eigenfrequency optimization problems

5.4   Maximization of the 5th Eigenmode of a Square Plate

We attempt to obtain the optimum reinforcements of a simply-supported square plate by maximizing the fifth mode of the eigensystem. The formulation of the first eigenvalue optimization problem as Eq. (26) will be applied to substantiate this goal. We apply two formulations for this example, which are a single eigenvalue formulation (a single eigenvalue optimization problem) and a multi-eigenvalue formulation (a multi-eigenvalue optimization problem). The single eigenvalue formulation simply uses the fifth eigenmode as the objective in the mean eigenvalue ($m=1$), while the multi-eigenvalue formulation employs the fifth, sixth, seventh, and eighth eigenmode as the objective ($m=4$). The mean eigenvalue adopts an exponent $n = 1$ in order to maximize the lowest mode applied in the objective function, which is the fifth eigenvalue for this problem. It is also known that both the fifth and sixth eigenvalues and the seventh and eighth eigenvalues of this system are "repeated", and more importantly, these two pairs are very close to each other. Thus, we will show that the preference for this case is the multimodal objective formulation.

The design domain that is to be reinforced is a square, simply-supported, base plate. This structure is discretized into 30x30 four-node quadrilateral Q4 finite elements with a base thickness of 0.1 (h_0), the Young's modulus of 100, and the Poison's ratio of 0.3. The maximum reinforced thickness will be up to 0.4 ($h_1 - h_0$). Using the aforementioned technique, we can eventually derive the optimum layouts of reinforcements for these two problems under two volume constraints (V.C.) as shown in Fig. 17. The volume constraints considered for each problem are 170 and 270 of the total volume (450) in the design domain. Fig. 17-(1) shows the results from the single modal objective formulation (the fifth mode), and Fig. 17-(2) presents the results from the multimodal objective formulation.

The comparison of Fig. 17-(1) and Fig. 17-(2) illustrates the advantages of using the multimodal formulation rather than the single modal formulation when the specified eigenmode that is to be optimized is very "close" to the upper mode, or the eigensystem is "repeated". In addition, as shown in Fig. 18, the convergence history of the fifth, sixth, seventh, and eighth eigenvalues using the multimodal formulation appears to be more stable and smooth than that using the single modal formulation.

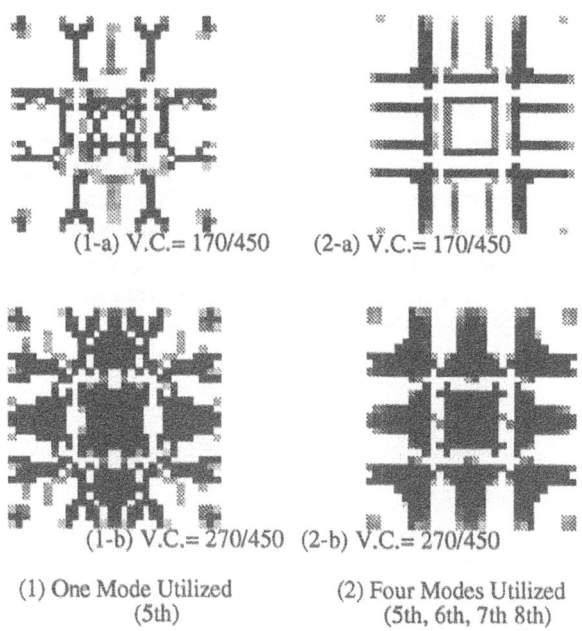

(1-a) V.C.= 170/450   (2-a) V.C.= 170/450

(1-b) V.C.= 270/450   (2-b) V.C.= 270/450

(1) One Mode Utilized   (2) Four Modes Utilized
(5th)   (5th, 6th, 7th 8th)

Figure 17   The Optimum Reinforcements for a Simply-Supported Plate

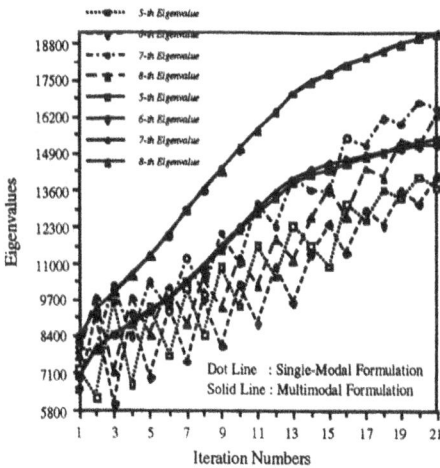

Figure 18   The Convergence History of the 5th, 6th, 7th, and 8th Eigenvalues

5.5 Maximization of the Gap Between the Third and Fourth Eigenmodes of a Simply Supported Shell

One of the primary goals of this paper is to design a structure so that the natural frequency of this structure can avoid a resonance from any external or internal source of exciting frequencies. The structure considered for this example is a square simply-supported core shell structure, which is defined by

$$z = z_{max} \sin(x\pi / x_{max}) \sin(y\pi / y_{max})$$

$z_{max}=2.5$, $x_{max}=30$, and $y_{max}=3$, where y_{max} and y_{max} stand for the length and width of the square shell, as shown in Fig. 19. This core shell is discretized into 30x30 Q4 finite elements with a base thickness of 0.1 (h_0), the Young's modulus of 100, and the Poison's ratio of 0.3. The maximum reinforced thickness can be up to 0.4 ($h_1 - h_0$). We consider reinforcing this core shell such that the eigensystem of this shell structure holds a maximal gap between the third and fourth modes.

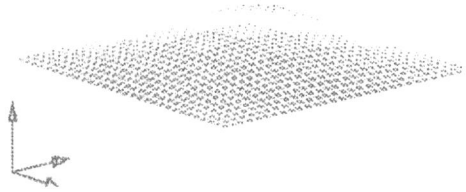

Figure 19 The Simply-Supported Core Shell

In order to achieve this goal, we attempt to apply two different optimization approaches. The first approach employs the second formulation of the mean eigenvalue shown in Eq. (27) with an exponent $n=2$. The objective function (i.e. the mean eigenvalue) is formed by a multimodal formulation (i.e. $m=4$), which includes the third, fourth, fifth, and sixth eigenmodes. Based on this multimodal objective formulation, we illustrate three sub-problems, each of which consists of a various volume constraint (V.C.) and a different exciting frequency ω_{0_i} as shown in Table 2. The second approach applies the first formulation of the mean eigenvalue expressed in Eq. (26) with an exponent $n=1$. Likewise, this approach also utilizes a multimodal formulation (i.e. $m=3$) in the objective function, which contains the fourth, fifth, and sixth eigenmodes. The fourth eigenvalue (i.e. the first mode employed in the objective function) is mainly maximized such that the gap between the third and fourth eigenvalues can be accordingly magnified. Additionally, three sub-examples are also defined with various volume constraints.

Table 2 The Eigenfrequencies of FINAL Designs

V.C.	Objective	ω_{0_i}	3rd Mode	4th Mode	5th Mode	6th Mode
170	1st Approach	24	18.68	26.30	26.32	26.40
170	2nd Approach		24.53	28.94	29.35	29.64
270	1st Approach	30	23.06	32.98	33.07	33.17
270	2nd Approach		25.97	33.10	35.07	36.86

The final optimal layout designs of reinforcements of the shell structure are shown in Fig. 20, where the results derived from the first approach are displayed in Part (1) and the results derived from the second approach are put in Part (2). Furthermore, the corresponding eigenfrequencies of these optimum designs are listed in Table 2. As shown in Fig. 20, both approaches yield the similar optimal layout design of reinforcements of the core shell structure. By comparing the data in Table 2, the gap between the third and fourth mode obtained from the first approach is apparently larger than that derived from the second approach. However, the results derived from the first approach appear to be less solid than those from the second approach. This can be attributed to two factors: 1)the third eigenmode tends to be minimized in the first approach and 2)the fourth eigenmode can not be optimized up to the value derived from the second approach. From the stiffness point of view, these "weaker" structures appear to be the disadvantage of the first approach, but they turn out to be "stiffer" in the sense of avoiding the structural resonance.

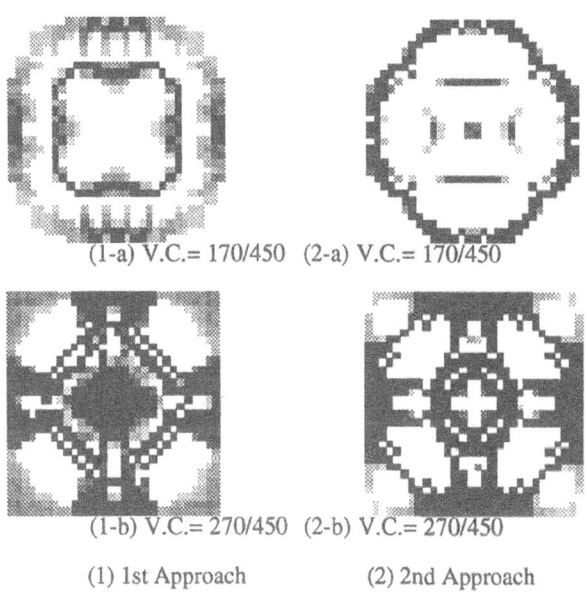

(1-a) V.C.= 170/450   (2-a) V.C.= 170/450

(1-b) V.C.= 270/450   (2-b) V.C.= 270/450

(1) 1st Approach   (2) 2nd Approach

Figure 20 The Optimal layouts of Reinforcements of the Square Shell

5.6 Identifying a Structure Possessing Specified Eigenfrequencies

Now we consider the problem stated in Introduction that yields a structure with a set of specified eigenfrequencies in the design domain shown in Fig. 1. As shown in Fig. 1, the design domain is specified as a 6.0cm by 2.0cm rectangle and a structure must be supported at both the left and right hand sides of the bottom of the design domain. Young's modulus is assumed to be $E=100Kg/cm^2$, Poisson's ratio $\nu=0.3$, mass density $\rho_0=10^{-6}Kg/cm^2$, and the total volume constraint 3.5cm^2 (34% of the volume of the whole design domain). The loading is assumed by three concentrated masses, 2.0×10^{-6}Kg, 4.0×10^{-6}Kg and 2.0×10^{-6}Kg on the bottom of the bridge. As shown in Table 3, the initial values of the lowest three eigenfrequencies are 21.5Hz, 34.7Hz and 49.9Hz for the uniformly perforated structure.

Two designs are considered for the lowest three eigenfrequencies. In the Design I, the desired eigenfrequencies are given as 80Hz, 170Hz and 200Hz. In the Design II, the desired eigenfrequencies are given as 110Hz, 130Hz and 170Hz. There is a significant difference between the requests of the two designs.

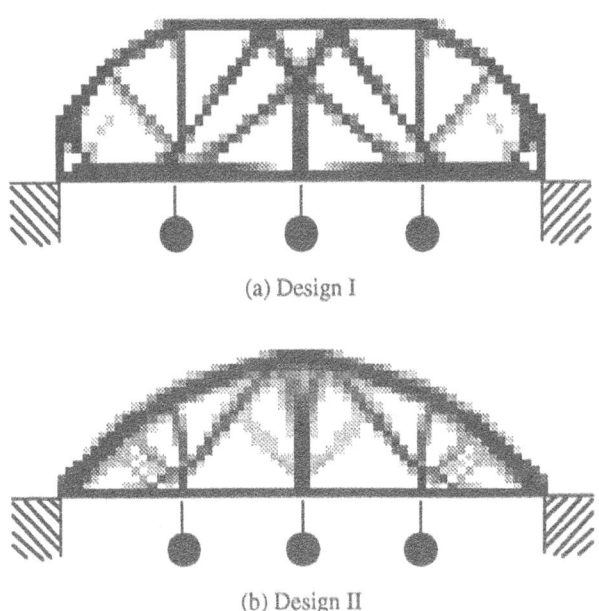

(a) Design I

(b) Design II

Figure 21 Optimal topologies obtained for desired eigenfrequencies

Table 3 Desired and obtained frequencies for the bridge-like structure

Hz	Initial	Design I		Design II	
		Desired	Obtained	Desired	Obtained
ω_1	21.5	80.0	79.5	110.0	108.7
ω_2	34.7	170.0	169.2	130.0	128.4
ω_3	49.9	200.0	199.9	170.0	170.9

Using the technique presented here, we can obtained the optimal structures for both cases (Fig. 21). As shown in Table 3, for the design I, the obtained eigenfrequencies are 79.5Hz, 169.2Hz and 199.9Hz. For the design II, the obtained eigenfrequencies are 108.7Hz, 128.4Hz and 170.9Hz. In both cases, the largest error is less than 1%. It shows that the use of the topological optimization technique makes it possible to design the optimal structure for a wide range of design purposes some of which have never been considered before.

5.7 Identification of an Unknown Shell Structure That Possesses Prescribed Eigenfrequencies

We shall extend the eigenvalue identification problem to a shell structure using a core structure shown in Fig. 22. The objective of this design is to identify an unknown structure that possesses a set of prescribed frequencies. Here, we consider reinforcement of the core structure with a given structural mass, 0.11588 (m^3), such that the eigenvalues of the prescribed modes of the final design can be the same with the prescribed eigenvalues. In engineering application, this core box structure, formed by isotropic plates with a uniform thickness of 2.0 (mm), is considered as a simplified model of a truck cabin. This box is discretized into 1600 Q4 finite elements with the Young's modulus of 2.100E+04 (Pa), the Poison's ratio of 0.3, and mass density of 0.801E-06 (kg/cm^3), and will be topologically reinforced up to a maximum thickness of 6.0 (mm). The objective function is made dimensionless corresponding to the desired eigenvalues λ_{0_i} as $(\lambda_{v_i} - \lambda_{0_i})^v / \lambda_{0_i}^v$ in order to improve the history of convergence toward the final solution. We adopts an exponent $n = -2$ in the mean eigenvalue, in which this implies that the eigenmode farthest from the corresponding prescribed eigenvalue yields the largest contribution upon the objective function.

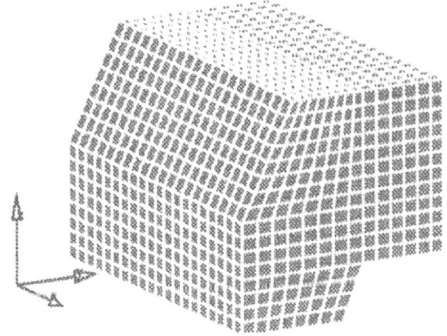

Figure 22 The Cabin Structure

Three design problems, Case (a), (b) and (c), will be considered with different sets of eigenfrequencies. We shall identify the 1st, 2nd, and 3rd eigenvalues for Case (a), and the 1st, 2nd, 3rd and 4th eigenvalues for Case (b) and Case (c). The prescribed eigenfrequencies ω_{0_i} and the corresponding weighting functions w_i for these cases are listed in Table 4. The identified structures in terms of the prescribed frequencies are shown in Fig. 23, and the corresponding computed eigenfrequencies ω_{v_i} are listed in Table 4. As shown in Table 4 these identified eigenfrequencies appear to be considerably close to the prescribed frequencies. In other word, the method presented here can identify a structure that possesses a set of eigenfrequencies prescribed.

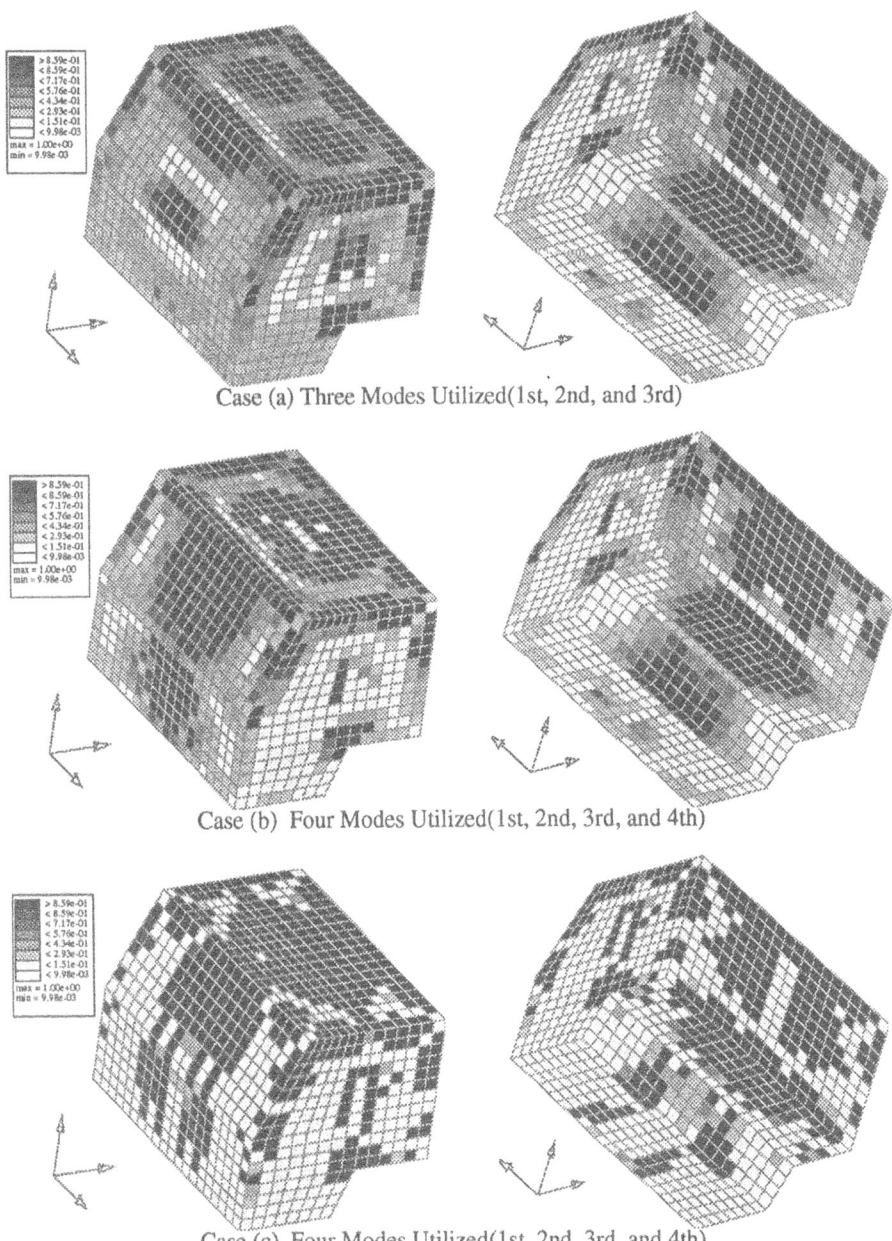

Case (a) Three Modes Utilized(1st, 2nd, and 3rd)

Case (b) Four Modes Utilized(1st, 2nd, 3rd, and 4th)

Case (c) Four Modes Utilized(1st, 2nd, 3rd, and 4th)

Figure 23 The Final Identified Structures Using Three Sets of Prescribed Frequencies

Table 4 The Desired and Derived frequencies

		1st Mode	2nd Mode	3rd Mode	4th Mode
	ω_{0_i}	6.50	8.00	11.00	---
Case (a)	w_i	200	1	50	---
	ω_{n_i}	6.60	8.28	10.83	---
	ω_{0_i}	8.00	10.00	10.00	10.00
Case (b)	w_i	250	70	1	1
	ω_{n_i}	8.08	9.84	10.02	10.12
	ω_{0_i}	9.50	12.00	13.00	13.00
Case (c)	w_i	1	1	1	1
	ω_{n_i}	9.23	12.01	12.89	13.26

6 Conclusion

We have been described an extension of the concept of the homogenization design method for maximizing static stiffness to the one we can consider optimal design of topology and reinforcement of vibrating structures. More specifically, we have introduced

1) a mean-eigenvalue is proposed to solve multiple eigenvalue optimization problems, and is applied to solve maximization of the eigenvalues of the targeted eigenmodes, maximizing the distance of two eigenfrequencies, and identifying a structure that possesses a set of prescribed eigenfrequencies,

2) a frequency response optimization problem is solved by modifying the original optimality criteria method,

3) relaxed optimality criteria method is introduced in order to improve the updating algorithm , and

4) a solution method for determining the optimal orientation of the microstructure is also discussed for the problem related to eigenvalues.

Here we have also shown that the homogenization design method makes it possible to obtain the optimal structure for various objectives for vibrating structures using many computational results.

Acknowledgment

The authors are supported by NSF, DDM-9300326, US Army TACOM, DAAE07-93-C-R125, and NAVY N00014-94-1-0022, and are grateful to these support.

References

Bendsøe, M. P., Diaz, A., and Kikuchi, N. (1992), Topology and generalized layout optimization of elastic structures, in: Bendsøe, M. P. and Soates, C. A. M. ed., Topology design of structures, NATO ASI Series, (Kluwer Academic Publishers) pp.159-205.

Bendsøe, M. P. and Kikuchi, N. (1988), Generating optimal topologies in structural design using a homogenization method, Comput. Methods Appl. Mech. Energ. 71, pp.197-24.

Bendsøe, M. P. and Soates, C. A. M. (ed), Topology Design of Structures, NATO ASI Series (Kluwer Academic Publishers)

Berke, L. and Venkayya, V.B. (1974), Review of optimality criteria approaches to structural optimization, in: Schmit, L. A., ed., Structural optimization symposium, (New York) pp.23-34.

Cheng, H.-C., Kikuchi, N., and Ma, Z.-D. (1993), An improved approach for determining the optimal orientation of the orthotropic material, Structural Optimization, to appear.

Cheng, K.T. and Olhoff, N. (1981), An investigation concerning optimal design of solid elastic plates, Int. J. Solids Structures, 17, pp.305-323.

Diaz, A. and Bendsøe, M. P., (1992), "Shape Optimization of Structures for Multiple Loading Conditions Using a Homogenization Method", Structural Optimization, 4, pp. 17-22.

Diaz, A. and Kikuchi, N. (1992), Solutions to shape and topology eigenvalue optimization problems using a homogenization method. preprint, Internat. J. Numer. Methods Engrg. 35 , pp.1487-1502.

Guedes, J. M. and Kikuchi, N., (1990), Preprocessing and Postprocessing for Materials Based on the Homogenization Method with Adaptive Finite Element Methods, Comp. Meth. Appl. Mechs. Engng., 83, pp. 143-198

Fukushima, J., Suzuki, K., and Kikuchi, N. (1991), Applications to car bodies : generalized layout design of three-dimensional shells, in Optimization of Large Structual Systems, Ed., G.I.N. Rozvany, Kluwer Academic Publishers, Dordrecht, pp. 177-192

Habor, R.B., Jog, C.S., and Bendsøe, M.P. (1994), Variable Topology Shape Optimization with a Control on Perimeter, in Advances in Design Automation 1994, DE-Vol.69-2, eds. Gilmore, B.J., Hoeltzel, D.A., Dutta, D., and Eschenauer, H.A., ASME, New York, pp. 261-272

Hemp, W. S. (1973), Optimum Structures. Clarendon (Oxford).

Kikuchi, N., Suzuki, K., and Fukushima, J. (1991), Layout optimization using the homogenization method: generalized layout design of three-dimensional shells for car bodies, in: Rozvany, G. I. N., ed., Optimization of Large Structural Systems, NATO-ASI Series (Berchtesgaden) 3, pp.110-126.

Kirsch, U. (1989), Optimum topologies of structures, Appl. Mech. Rev., 42 , pp.223-239.

Kohn, R. V. and Strang, G. (1986), Optimal design and relaxation of variational problem, Comm. Pure. Appl. Math., 39, pp.1-25, 139-182, 353-377.

Ma, Z.-D., Cheng, H.-C., Kikuchi, N., and Hagiwara, I. (1992), Topology and shape optimization technique for structural dynamic problems, Recent Advances in Structural Problems, PVP-248/NE-10 ,pp.133-143.

Ma, Z. D., Cheng, H. C., and Kikuchi, N., (1993), "Structural Design for Obtaining Desired Frequencies by Using the Topology and Shape Optimization Method", Computing Systems in Engineering, Vol. 5, No.1, pp. 77-89.

Ma, Z.-D. and Hagiwara, I. (1991), Improved Mode-Superposition Technique for Modal Frequency Response Analysis of Coupled Acoustic-Structural Systems, AIAA Journal, 29 (10), pp.1720-1726.

Ma, Z.-D., Kikuchi, N., and Hagiwara, I. (1993), Structural topology and shape optimization for a frequency response problem. Computational Mechanics, 13 (3), pp.157-174.

Maxwell, G., On reciprocal figures, frames, and diagrams of forces, Sci. Papers II, Cambridge Univ. Press, (1890) 175-177.

Michell, A. G. M.(1904), The limits of economy in frame structures. Philo Mag Sect 6 (8) , pp.589-597.

Murat, F., and Tartar, L. (1985), Optimality conditions and homogenization, in Nonlinear Variational Problems, eds., Marino A., et al, Pitman Advanced Publishing Program, Boston, pp.108

Li, Xing-Si (1991), An aggregate function method for nonlinear programming, Science in China (series A), 34 (12), pp.1467-1473.

Luire, K.A., and Cherkaev, A.V. (1984), G-Closure of some particular sets of admissible material characteristics of the problem of bending of thin plates, J. Optim. Theory Appl., 42, pp.305-315

Olhoff, N. and Rozvany, G. I. N. (1982), Optimal Grillage layout for given natural frequency, J Struc Mech ASCE 108, pp.971-974.

Olhoff, N., Bendsøe, M. P. and Rasmussen, J. (1991), On CAD-integrated structural topology and design optimization, Comput. Methods Appl. Mech. Energ., 89, pp. 259-279.

Prager, W. and Rozvany, G. I. N. (1977), Optimal layout of grillages, J Struct Mech 5, pp.1-18.

Pedersen, P., (1988), On optimal orientation of orthotropic materials, Structural Optimization, 1, pp.101-106

Rozvany, G. I. N. (1981), Optimality Criteria for grids, shells and arches, in: Haug, E. J. and Cea, J., ed., Optimization of distributed parameter structures (Sijthoff & Noordhoff) 1, pp.112-151.

Rozvany, G. I. N. (1992), Layout theory for grid-type structures. In: Bendsøe, M. P. and Soates, C. A. M. (ed): Topology design of structures, NATO ASI Series (Kluwer Academic Publishers), pp.251-272.

Rozvany, G. I. N. and Wang, C. M. (1983), Extensions of Prager's layout theory, in: Eschenauer, H. and Olhoff, N., ed., Optimization in structural design (Wissenschafsverlay, Mannheim), pp.103-110.

Rozvany, G.I.N., Zhou, M., Birker, T., and Sigmund, O., Topology optimization using iterative continuum-type optimality criteria methods for discretized systems, in Bendsøe, M.

P. and Soates, C. A. M. (ed): Topology design of structures, NATO ASI Series (Kluwer Academic Publishers), pp.273-286.

Sanchez-Palencia, E., (1980), Non-homogeneous Media and Vibration Theory, Lecture Notes in Physics, #127, Springer-Verlag, Berlin

Soto, C. and Diaz, A. (1992), On the modeling of ribbed plates for shape optimization. Technical Report, CDL-92-2, Computational Design Laboratory, Michigan State University, East Lansing, Michigan.

Soto, C. and Diaz, A., (1993), "Layout of Plate Structures for Improved Dynamic Response using a Homogenization Method", Design Automation Conference, ASME, New Mexico.

Suzuki, K. and Kikuchi, N. (1991), A homogenization method for shape and topology optimization, Comput. Methods Appl. Mech. Energ., 93 , pp 291-318.

Suzuki, K. and Kikuchi, N. (1992), Generalized layout optimization of three-dimensional shell structures, D. A. Komkov, V. (Eds.), Geometric Aspects of Industrial Design, SIAM, Philadelphia, pp.62-88.

Zhou, M. and Rozvany, G. I. N. (1991), The COC algorithm, part II: Topology, geometrical and generalized shape and optimization, Comput. Methods Appl. Mech. Energ. 89, pp.309-336.

MATERIAL OPTIMIZATIONS – AN ENGINEERING VIEW

Pauli Pedersen
Department of Solid Mechanics
Technical University of Denmark, Lyngby, Denmark

ABSTRACT – The very broad scope of material optimization is shown from an engineering point of view, i.e. with eyes that favours the energy interpretations. The mathematical point of view as well as the optimization procedure point of view are given less priority. Concentrating mainly on the results of personal research, it fails as a review paper.

With these reservations it covers linear as well as non–linear power–law materials in the general anisotropic description. Orientational optimization for minimum stiffness and with strength constraints plays a central role, and this optimization is combined with the optimal design of thickness distributions. Shape optimal design with these new materials are also described in details.

Recent results on direct design of constitutive parameter are explained as a natural extension of previous research. From this then follows the need for solving the inverse homogenization problem, where interesting results are already available. All in all the paper tries to give **the state of the art**, but from a subjective point of view.

J. Herskovits (ed.), Advances in Structural Optimization, 223–261.
© 1995 *Kluwer Academic Publishers.*

1. Introduction

New materials such as fibre–reinforced laminates and ceramics have increased the need for optimal design. Two aspects of optimization are then of practical interest, viz. the influence on size, shape and topology design **with these materials**, and the more or less detailed design **of the material** itself for a specific purpose.

A general characteristic of the new materials is that they are anisotropic. However, material anisotropy is seldom treated in great depth and even in the well–known reference of Lekhnitskii (1981), we cannot find all the results needed for optimal design based on anisotropic behaviour. Similarly, the classic books on elasticity, give only little guidance with respect to the most simple non–linear elasticity. Thus the second section will be a self–contained introduction to and **classification of non–linear, anisotropic elasticity.**

Research related to material design has focused also on the very basic sensitivity analysis related to elastic strain energy. Many important results are not well known and as it is possible to prove these results at the energy level we shall do so. This analysis holds for one–, two– and three–dimensional problems, holds for different models, and holds for analytical as well as different numerical solutions. The most important result is the **localized determination of sensitivities** that we often encounter. In section three this is presented also from a self-contained point of view.

With this basic knowledge we then, in section four, treat **optimal material orientation** for plates and discs. Here too, a number of rather general results are available, and we shall focus on these rather than on the specific problems. We may say that two– dimensional problems are to a large extent solved, but the three–dimensional problems are still a subject of intensive research. An interesting result is the **co–alignment of principal strains and stresses**. Present activities relates to orientational design with **strength constraints**. After the general formulation of this problem, we introduce a **global design description** which seems to be a good tool for future research.

Size optimization, such as distribution of plate thickness, is naturally influenced by the use of anisotropic material but in general the methods of isotropic design are applicable. Therefore, in section five, we focus on the **concurrent design of orientation and thickness**. The important result that the optimal thickness distribution is **independent of the power** in a power–law material is derived, and the examples focus on those from non–linear elasticity.

A short account of the detailed material design, in terms of **design of constitutive matrices**, is given in section six. This very active research area already has a number of important results available, and we go through two specific examples. The idea here is to see this as the third step after having first solved the orientational problem and then the total density (thickness) problem. A very recent thesis by Bendsøe (1994) includes an extended biography on these problems, and in great details this thesis covers the important integration of **homogenization–material design–topology optimization**. Active research on identifying materials with prescribed constitutive matrices are also commented in this section. These new problems may be termed as **inverse homogenization problems**.

Finally, in section seven, the shape optimization based on anisotropic materials are discussed and recent solutions shown. The **influence from numerical modelling** on the optimal shape is focused on, i.e. the number of elements and the element type in a finite element model.

2. Non–linear, Anisotropic Elasticity

We shall deal only with elasticity, i.e. with materials that give reversible response. However, we shall cover not only linear but also non–linear elasticity, and thus include problems where non–elastic behaviour is modelled by non–linear elasticity. For anisotropic elasticity rotational transformations are of vital importance, and the presentation will be based on contracted notations for stresses, strains, modulus etc. that transforms by orthonormal matrices. As these simple transformations are not well known we shall here give the results, and for a detailed discussion refer to Pedersen (1994).

2.1 ROTATIONAL TRANSFORMATIONS

The tensor description of the 2D–constitutive behaviour is given by

$$\sigma_{ij} = C_{ijk\ell}\varepsilon_{k\ell} = C\alpha_{ijk\ell}\varepsilon_{k\ell} \tag{2.1}$$

where indices run through 1 and 2 only and where $\alpha_{ijk\ell}$ is a fourth order **non–dimensional** tensor. We have separated the possible non–linear behaviour to the dimensional factor C and the load independent anisotropic behaviour to the tensor $\alpha_{ijk\ell}$, the quantities of which we shall use also in our matrix formulation.

Orthonormal transformations of stresses and strains are obtained by the definitions

$$\{\sigma\}^T := \left\{\sigma_{11},\ \sigma_{22},\ \sqrt{2}\sigma_{12}\right\},\quad \{\varepsilon\}^T := \left\{\varepsilon_{11},\ \varepsilon_{22},\ \sqrt{2}\varepsilon_{12}\right\} \tag{2.2}$$

which with (2.1) written out gives the matrix formulation of the constitutive behaviour by

$$\{\sigma\} = C\begin{bmatrix} \alpha_{1111} & , & \alpha_{1122} & , & \sqrt{2}\alpha_{1112} \\ \alpha_{1122} & , & \alpha_{2222} & , & \sqrt{2}\alpha_{2212} \\ \sqrt{2}\alpha_{1112} & , & \sqrt{2}\alpha_{2212} & , & 2\alpha_{1212} \end{bmatrix}\{\varepsilon\} = C[\alpha]\{\varepsilon\} \tag{2.3}$$

The orthonormal matrix $[T]$ that transforms between the Cartesian coordinate systems from x to y as shown in fig. 2.1 is

$$[T] := \frac{1}{2}\begin{bmatrix} 1 + c_2 & , & 1 - c_2 & , & \sqrt{2}s_2 \\ 1 - c_2 & , & 1 + c_2 & , & -\sqrt{2}s_2 \\ -\sqrt{2}s_2 & , & \sqrt{2}s_2 & , & 2c_2 \end{bmatrix} \tag{2.4}$$

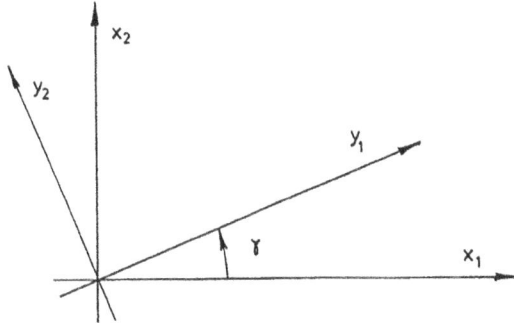

Fig. 2.1: The two Cartesian coordinate systems with the definition of the angle γ.

with the short notations

$$c_2 := \cos(2\gamma) \, , \quad s_2 := \sin(2\gamma)$$

$$c_4 := \cos(4\gamma) \, , \quad s_4 := \sin(4\gamma) \tag{2.5}$$

The transformation of the constitutive forms are often given by

$$\begin{matrix} (\alpha_{1111})_y \\ (\alpha_{2222})_y \end{matrix} = \frac{1}{2}(\alpha_{1111} + \alpha_{2222})_x \pm \alpha_2 c_2 - \alpha_3(1 - c_4) \pm \alpha_6 2 s_2 + \alpha_7 s_4$$

$$\begin{matrix} (\alpha_{1122})_y \\ (\alpha_{1212})_y \end{matrix} = \begin{matrix} (\alpha_{1122})_x \\ (\alpha_{1212})_x \end{matrix} + \alpha_3(1 - c_4) - \alpha_7 s_4 \tag{2.6}$$

$$\begin{matrix} (\alpha_{1112})_y \\ (\alpha_{2212})_y \end{matrix} = -\frac{1}{2}\alpha_2 s_2 \mp \alpha_3 s_4 + \alpha_6 c_2 \pm \alpha_7 c_4$$

with the definition of the practical (not invariants) parameters $\alpha_2, \alpha_3, \alpha_6, \alpha_7$ by

$$\alpha_2 := \frac{1}{2}(\alpha_{1111} - \alpha_{2222})_x$$

$$\alpha_3 := \frac{1}{8}(\alpha_{1111} + \alpha_{2222} - 2(\alpha_{1112} + 2\alpha_{1212}))_x$$

$$\alpha_6 := \frac{1}{2}(\alpha_{1112} + \alpha_{2212})_x \tag{2.7}$$

$$\alpha_7 := \frac{1}{2}(\alpha_{1112} - \alpha_{2212})_x$$

A valuable but not well known alternative to the formulas (2.6)–(2.7) is obtained by also representing the $[\alpha]$ matrix with $\sqrt{2}$ factors for the off–diagonal elements. The orthonormal matrix $[R]$ $\left([R]^{-1} = [R]^T\right)$ for this transformation is

$$[R] = \frac{1}{8} \cdot \tag{2.8}$$

$$\begin{bmatrix}
3 + 4c_2 + c_4 \, , & 3 - 4c_2 + c_4 \, , & 2 - 2c_4 & , & \sqrt{2} - \sqrt{2}c_4 & , & 4s_2 + 2s_4 & , & 4s_2 - 2s_4 \\
3 - 4c_2 + c_4 \, , & 3 + 4c_2 + c_4 \, , & 2 - 2c_4 & , & \sqrt{2} - \sqrt{2}c_4 & , & -4s_2 + 2s_4 \, , & -4s_2 - 2s_4 \\
2 - 2c_4 & , & 2 - 2c_4 & , & 4 + 4c_4 & , & -2\sqrt{2} + 2\sqrt{2}c_4 \, , & -4s_4 & , & 4s_4 \\
\sqrt{2} - \sqrt{2}c_4 \, , & \sqrt{2} - \sqrt{2}c_4 \, , & -2\sqrt{2} + 2\sqrt{2}c_4 \, , & 6 + 2c_4 & , & -2\sqrt{2}s_4 & , & 2\sqrt{2}s_4 \\
-4s_2 - 2s_4 \, , & 4s_2 - 2s_4 & , & 4s_4 & , & 2\sqrt{2}s_4 & , & 4c_2 + 4c_4 \, , & 4c_2 - 4c_4 \\
-4s_2 + 2s_4 \, , & 4s_2 + 2s_4 & , & -4s_4 & , & -2\sqrt{2}s_4 & , & 4c_2 - 4c_4 \, , & 4c_2 + 4c_4
\end{bmatrix}$$

and then with the contracted notation $[\alpha] \rightarrow \{\alpha\}$

$$\{\alpha\}^T := \left\{\alpha_{1111}, \alpha_{2222}, 2\alpha_{1212}, \sqrt{2}\alpha_{1122}, 2\alpha_{1112}, 2\alpha_{2212}\right\} \tag{2.9}$$

we can write the transformations simply by

$$\{\alpha\}_y = [R]\{\alpha\}_x \, , \quad \{\alpha\}_x = [R]^T\{\alpha\}_y \tag{2.10}$$

2.2 CLASSIFICATION OF THE ANISOTROPY

In the contracted notations chosen here the constitutive matrix is **symmetric** as shown in (2.3), and furthermore the matrix must be **positive definite**, i.e.

$$\alpha_{1111} > 0, \ \alpha_{2222} > 0, \ \alpha_{1212} > 0$$

$$\left(\alpha_{1111}\alpha_{2222} - \alpha_{1122}^2\right) > 0, \ 2\left(\alpha_{2222}\alpha_{1212} - \alpha_{2212}^2\right) > 0 \tag{2.11}$$

$$2\left(\alpha_{1111}\alpha_{1212} - \alpha_{1112}^2\right) > 0, \ \det[\alpha] > 0$$

From this follows that the trace tr is positive

$$\text{tr} = \alpha_{1111} + \alpha_{2222} + 2\alpha_{1212} > 0 \tag{2.12}$$

and also the squared Frobenius norm $(\text{Fr})^2$ is positive

$$(\text{Fr})^2 = \left(\alpha_{1111}\alpha_{2222} - \alpha_{1122}^2\right) + 2\left(\alpha_{2222}\alpha_{1212} - \alpha_{2212}^2\right)$$

$$+ 2\left(\alpha_{1111}\alpha_{1212} - \alpha_{1112}^2\right) > 0 \tag{2.13}$$

Note that with the representation (2.2) are these norms **invariants**, which can be seen directly from (2.6).

For **orthotropic** materials we have

$$\alpha_6 = \alpha_7 = 0 \tag{2.14}$$

However, the orthotropic directions may be unknown and then we need a criterion to **test for existence of orthotropic directions**. This is derived in Pedersen (1990b) and is expressed using the parameters $\alpha_2, \alpha_3, \alpha_6$ and α_7 (based on any reference axis) as

$$\alpha_7\alpha_2^2 - 1\alpha_7\alpha_6^2 - 4\alpha_6\alpha_3\alpha_2 = 0 \tag{2.15}$$

If (2.15) is satisfied we choose the direction of orthotropy for which $\alpha_2 > 0$. If (2.15) is not satisfied we have a non–orthotropic constitutive matrix, but still need a specific reference axis for comparison of constitutive matrices which may in reality just be mutually rotated. No clear tradition for selection of this direction seems to be available in the literature.

In our studies we have chosen the direction which maximizes α_{1111}, i.e.

$$\text{Max}\left(\alpha_{1111} = \alpha_{1111}(\gamma)\right) \ \text{for} \ 0 \leq \gamma \leq \pi \tag{2.16}$$

and in most (but not all) practical cases this is uniquely determined by

$$\left(\alpha_{1112}\right)_x = 0, \ \text{i.e.} \ \left(\alpha_6\right)_x = -\left(\alpha_7\right)_x \tag{2.17}$$

from which follows that a **general constitutive matrix** can be described by

$$[C] = C\begin{bmatrix} \alpha_{1111} , & \alpha_{1122} , & 0 \\ \alpha_{1122} , & \alpha_{2222} , & \sqrt{2}\,\alpha_{2212} \\ 0 , & \sqrt{2}\,\alpha_{2212} , & 2\alpha_{1212} \end{bmatrix}_{\substack{\text{reference} \\ \text{coordinate system}}}$$

$$\alpha_{1111} \geq \alpha_{2222} \tag{2.18}$$

For **isotropic materials** we have

$$\alpha_2 = \alpha_3 = \alpha_6 = \alpha_7 = 0 \qquad (2.19)$$

and thus from (2.7)

$$\alpha_{1212} = \frac{1}{2}(\alpha_{1111} - \alpha_{1122}) \qquad (2.20)$$

Our studies on optimal design are based on orthotropic material with the two parameters α_2 and α_3. **The difference in modulus** between the two orthotropic directions is described by α_2 which by choosing a specific orthotropic direction always is non–negative, i.e. $(\alpha_{1111} - \alpha_{2222}) \geq 0$. The physical interpretation of the parameter α_3 is as **relative shear modulus**, and we define the not well known classification by

Low shear modulus for

$$\alpha_3 > 0, \text{ i.e. } 0 < 4\alpha_{1212} < \alpha_{1111} + \alpha_{2222} - 2\alpha_{1112}$$

$$(2.21)$$

High shear modulus for

$$\alpha_3 < 0, \text{ i.e. } 4\alpha_{1212} > \alpha_{1111} + \alpha_{2222} - 2\alpha_{1112} > 0$$

In fig. 2.2 on the following page we have shown schematically the classification of the constitutive behaviour for 2D–problems.

| Any reference axis | A conveniently chosen, specific reference axis |

[α] symmetric =

$$\begin{bmatrix} \alpha_{1111} & \alpha_{1122} & \sqrt{2}\,\alpha_{1112} \\ \alpha_{1122} & \alpha_{2222} & \sqrt{2}\,\alpha_{2212} \\ \sqrt{2}\,\alpha_{1112} & \sqrt{2}\,\alpha_{2212} & 2\alpha_{1212} \end{bmatrix}$$

$\Rightarrow$ 6 parameters

[α] positive definite

$$[\alpha] = \begin{bmatrix} \alpha_{1111} & \alpha_{1122} & 0 \\ \alpha_{1122} & \alpha_{2222} & \sqrt{2}\,\alpha_{2212} \\ 0 & \sqrt{2}\,\alpha_{2212} & 2\alpha_{1212} \end{bmatrix} \Rightarrow 5 \text{ parameters } (+ \text{ direction of axis})$$

axis by $\alpha_{1111} > \alpha_{2222}$ and $\alpha_{1112} = 0$,
i.e. $\alpha_2 > 0$ and $\alpha_6 = -\alpha_7$

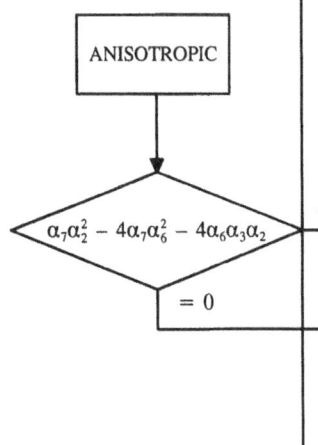

ANISOTROPIC

$\alpha_7 \alpha_2^2 - 4\alpha_7 \alpha_6^2 - 4\alpha_6 \alpha_3 \alpha_2$

$= 0$

$\neq 0$

NON-ORTHOTROPIC positive definite
$$\begin{cases} 0 < \alpha_{2222} \;;\; 0 < \alpha_{1212} \\ |\alpha_{1122}| < \sqrt{\alpha_{1111}\alpha_{2222}} \\ |\alpha_{2212}| < \sqrt{\alpha_{2222}\alpha_{1212} - \dfrac{\alpha_{1122}^2 \alpha_{1212}}{\alpha_{1111}}} \end{cases}$$

$$[\alpha] = \begin{bmatrix} \alpha_{1111} & \alpha_{1122} & 0 \\ \alpha_{1122} & \alpha_{2222} & 0 \\ 0 & 0 & 2\alpha_{1212} \end{bmatrix} \Rightarrow 4 \text{ parameters } (+ \text{ direction of axis})$$

axis by $\alpha_{1111} > \alpha_{2222}$ and
$\alpha_{1112} = \alpha_{2212} = 0$,
i.e. $\alpha_2 > 0$ and $\alpha_6 = \alpha_7 = 0$

ORTHOTROPIC positive definite
$$\begin{cases} 0 < \alpha_{2222} \;;\; 0 < \alpha_{1212} \\ |\alpha_{1122}| < \sqrt{\alpha_{1111}\alpha_{2222}} \;\left(< \tfrac{1}{2}(\alpha_{1111} + \alpha_{2222}) \right) \end{cases}$$

α_2, α_3 $\neq 0,0$ α_3 < 0 **HIGH SHEAR MODULUS**

≥ 0 **LOW SHEAR MODULUS**

$= 0,0$

Practical parameters:

$\alpha_2 := \dfrac{1}{2}(\alpha_{1111} - \alpha_{2222})$

$\alpha_3 := \dfrac{1}{8}\big((\alpha_{1111} + \alpha_{2222})$
$\qquad -2(\alpha_{1122} + 2\alpha_{1212})\big)$

$\alpha_6 := \dfrac{1}{2}(\alpha_{1112} + \alpha_{2212})$

$\alpha_7 := \dfrac{1}{2}(\alpha_{1112} - \alpha_{2212})$

ISOTROPIC → PLANE STRESS
→ PLANE STRAIN
→ OTHERS

Fig. 2.2: Overview

$$[\alpha] = \begin{bmatrix} \alpha_{1111} & \alpha_{1122} & 0 \\ \alpha_{1122} & \alpha_{1111} & 0 \\ 0 & 0 & (\alpha_{1111} - \alpha_{1122}) \end{bmatrix} \Rightarrow 2 \text{ parameters}$$
positive definite $0 \leq |\alpha_{1122}| < \alpha_{1111}$

2.3 ENERGY DENSITIES IN NON–LINEAR ELASTICITY

To relate 2D and 3D problems to 1D physical experiments, we introduce the effective strain scalar ε_e and the effective stress scalar σ_e . We define them so that the differential strain energy density du and the differential stress energy density du^C are found by

$$du = \{\sigma\}^T\{d\varepsilon\} = \sigma_e d\varepsilon_e$$

$$du^C = \{\varepsilon\}^T\{d\sigma\} = \varepsilon_e d\sigma_e \tag{2.22}$$

and thus as illustrated in fig. 2.3 the energy densities are obtained by

$$u = \int_0^{\varepsilon_e} du \ ; \ u^C = \int_0^{\sigma_e} du^C \ ; \ u + u^C = \{\sigma\}^T\{\varepsilon\} = \sigma_e \varepsilon_e \tag{2.23}$$

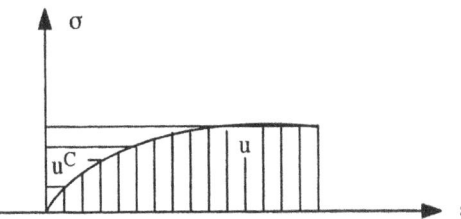

Fig. 2.3: Graphical illustration of the strain energy density u , and the stress energy density u^C.

In the 1D modelling of a **power law non–linear elasticity** we have

$$\sigma_e = E\varepsilon_e^p \rightarrow \sigma_e/\varepsilon_e = E\varepsilon_e^{p-1} \ , \ \underset{\text{(tangent)}}{d\sigma_e/d\varepsilon_e = pE\varepsilon_e^{p-1}} \tag{2.24}$$

$\underset{\text{(secant)}}{}$

or in terms of stresses

$$\varepsilon_e = \left(\sigma_e/E\right)^n \text{ with } n = 1/p \tag{2.25}$$

(The material parameters E , p are obtained from experimental data). From the definitions (2.23)–(2.24) follows directly

$$u = \frac{1}{p+1}E\varepsilon_e^{p+1} \ , \ u + u^C = E\varepsilon_e^{p+1}$$

$$\text{i.e. } u^C = \frac{p}{p+1}E\varepsilon_e^{p+1} = pu \tag{2.26}$$

For the 2D and 3D modelling the constitutive secant modulus by

$$\{\sigma\} = E\varepsilon_e^{p-1}[\alpha]\{\varepsilon\} \tag{2.27}$$

with the energy based definition of the effective strain by

$$\varepsilon_e^2 := \{\varepsilon\}^T[\alpha]\{\varepsilon\} \rightarrow 2\varepsilon_e d\varepsilon_e = 2\{\varepsilon\}^T[\alpha]\{d\varepsilon\} \tag{2.28}$$

gives the energy density by

$$du = \{\sigma\}^T\{d\epsilon\} = E\epsilon_e^{p-1}\{\epsilon\}^T[\alpha]\{d\epsilon\} = E\epsilon_e^p d\epsilon_e \tag{2.29}$$

as for the 1D–case.

Similar the effective stress σ_e need to have the energy related definition

$$\sigma_e^2 := \{\sigma\}^T[\alpha]^{-1}\{\sigma\} \rightarrow 2\sigma_e d\sigma_e = 2\{\sigma\}^T[\alpha]^{-1}\{d\sigma\} \tag{2.30}$$

and therefore a **von Mises stress can only be used for the isotropic, incompressible case**.

In relation to our optimization studies and sensitivity analysis, the main result now proven for 1D–, 2D– and 3D–problems is

$$\boxed{u^C = pu} \tag{2.31}$$

a not too well known result, that is not even graphically clear from fig. 2.3. In words it says: **the ratio between stress energy density and strain energy density is the power p and thus independent of the actual density level**.

2.4 SUMMARY

The main conclusions of this section are:

- Matrix notation is preferred, and the clarity from tensor notation is kept by the $\sqrt{2}$–contracted notations (2.2) and (2.9).

- A new transformation matrix (2.8) is presented.

- Test for orthotropy (2.15) is given.

- The classification in low and high relative shear stiffness by (2.21) is important.

- Energy based definition of effective strain and stress are necessary, i.e. (2.28) and (2.30).

- The anisotropy described by the non–dimensional matrix $[\alpha]$ is assumed constant in the non–linear description by power–law.

- The major result of this section is $u^C = pu$, which as we shall see will imply major simplifications in the following sensitivity analysis.

3. Localized Determination of Sensitivity Results

The results from sensitivity analysis have strong relations to research on optimal design but are in reality of much wider importance and applicability. Often the results are derived for specific models, and the generality is lost or at least not visible.

When the quantity for which we seek the sensitivity is related to a global energy quantity, we have results that can be **calculated locally**. These results will be derived here without reference to a specific model and are thus valid for one-, two- and three-dimensional models, for analytical calculation, and for numerical model and are valid independent of the numerical method chosen, say finite difference, finite element, or more global Galerkin approaches.

3.1 VARIATIONS WITH FIXED STRAIN OR FIXED STRESS FIELDS

Changing a design, the strain field, the stress field and the energy density field are changed. Therefore, our final results which may be evaluated relative to a fixed strain field or a fixed stress field can easily cause misunderstandings.

It is important to point out that the localized determination relative to the fixed fields does not imply that only local changes are involved, neither that the non-local effects totally are zero. The reality is that the physical total change can be determined by a factor times **a local change, which is only part of the physical local change**.

Although dynamic problems (eigenfrequencies) are more simple, we shall concentrate on static problems for which the sensitivity results are less intuitive. In general this is due to the different nature of the work of the external forces compared with the kinetic energy. Let us start with the **work equation**

$$W + W^C = U + U^C \tag{3.1}$$

where W, W^C are physical and complementary work of the external forces, $-\left(W + W^C\right)$ is the external potential, and U, U^C are physical and complementary elastic energy, also named strain and stress energy, respectively. The work equation (3.1) holds for any design h and therefore for the total differential quotient with respect to h

$$\frac{dW}{dh} + \frac{dW^C}{dh} = \frac{dU}{dh} + \frac{dU^C}{dh} \tag{3.2}$$

Now in the same way as h represents the design field generally, ε represents the total strain field and σ represents the total stress field. Remembering that as a function of h, ε we have W, U, while the complementary quantities W^C, U^C are functions of h, σ, then we get (3.2) in greater detail by means of

$$\frac{\partial W}{\partial h} + \frac{\partial W}{\partial \varepsilon}\frac{\partial \varepsilon}{\partial h} + \frac{\partial W^C}{\partial h} + \frac{\partial W^C}{\partial \sigma}\frac{\partial \sigma}{\partial h} = \frac{\partial U}{\partial h} + \frac{\partial U}{\partial \varepsilon}\frac{\partial \varepsilon}{\partial h} + \frac{\partial U^C}{\partial h} + \frac{\partial U^C}{\partial \sigma}\frac{\partial \sigma}{\partial h} \tag{3.3}$$

The **principles of virtual work**, which hold for solids/structures in equilibrium, are

$$\frac{\partial W}{\partial \varepsilon} = \frac{\partial U}{\partial \varepsilon} \tag{3.4}$$

for the physical quantities with strain variation, and for the complementary quantities with stress variation we have

$$\frac{\partial W^C}{\partial \sigma} = \frac{\partial U^C}{\partial \sigma} \tag{3.5}$$

Inserting (3.4) and (3.5) in (3.3) we get

$$\frac{\partial U^C}{\partial h} - \frac{\partial W^C}{\partial h} = - \left(\frac{\partial U}{\partial h} - \frac{\partial W}{\partial h} \right) \tag{3.6}$$

and for design–independent loads

$$\boxed{\left(\frac{\partial U^C}{\partial h} \right)_{\text{fixed stresses}} = - \left(\frac{\partial U}{\partial h} \right)_{\text{fixed strains}}} \tag{3.7}$$

as stated by Masur (1970). Note that the only assumption behind this is the design–independent loads $\partial W / \partial h = 0$, $\partial W^C / \partial h = 0$.

3.2 RESULTS FOR LINEAR ELASTICITY

To get further into a **physical interpretation** of $(\partial U / \partial h)_{\text{fixed strains}}$ (and by (3.7) of $\left(\partial U^C / \partial h \right)_{\text{fixed stresses}}$) we need the relation between external work W and strain energy U . For linear elasticity and dead load this relation is given by

$$W = 2U \tag{3.8}$$

Parallel to the analysis from (3.1) to (3.3) and on the basis of (3.8), we obtain

$$\frac{\partial W}{\partial h} + \frac{\partial W}{\partial \varepsilon} \frac{\partial \varepsilon}{\partial h} = 2 \frac{\partial U}{\partial h} + 2 \frac{\partial U}{\partial \varepsilon} \frac{\partial \varepsilon}{\partial h} \tag{3.9}$$

which for design–independent loads $\partial W / \partial h = 0$ with virtual work (3.4), gives

$$\frac{\partial W}{\partial \varepsilon} \frac{\partial \varepsilon}{\partial h} = \frac{\partial U}{\partial \varepsilon} \frac{\partial \varepsilon}{\partial h} = - 2 \frac{\partial U}{\partial h} \tag{3.10}$$

and thereby

$$\boxed{\frac{dU}{dh} = \left(\frac{\partial U}{\partial h} \right)_{\text{fixed strains}} + \frac{\partial U}{\partial \varepsilon} \frac{\partial \varepsilon}{\partial h} = - \left(\frac{\partial U}{\partial h} \right)_{\text{fixed strains}}} \tag{3.11}$$

Now let us discuss the **localized determination** of sensitivities as given generally by (3.11). The strain energy is summed over all domains, and the design parameter h_j is local. Then we get

$$\frac{dU}{dh_j} = \frac{dU_j}{dh_j} + \sum_{i \neq j} \frac{dU_i}{dh_j} = \frac{\partial U_j}{\partial h_j} + \frac{\partial U_j}{\partial \varepsilon_j} \frac{\partial \varepsilon_j}{\partial h_j} + \sum_{i \neq j} \frac{\partial U_i}{\partial \varepsilon_i} \frac{\partial \varepsilon_i}{\partial h_j} \tag{3.12}$$

where ε_j describes the strain field of domain j . The results (3.11) tells that we need not calculate all the terms $\partial U_i / \partial \varepsilon_j$, because

$$\frac{dU}{dh_j} = - \left(\frac{\partial U_j}{\partial h_j} \right)_{\text{fixed strains}} \tag{3.13}$$

and from (3.12)–(3.13) we can again determine the "indirect" effect

$$\frac{\partial U_j}{\partial \varepsilon_j} \frac{\partial \varepsilon_j}{\partial h_j} + \sum_{i \neq j} \frac{\partial U_i}{\partial \varepsilon_i} \frac{\partial \varepsilon_i}{\partial h_j} = \sum_{\text{all } i} \frac{\partial U_i}{\partial \varepsilon_i} \frac{\partial \varepsilon_i}{\partial h_j} = - 2 \frac{\partial U_j}{\partial h_j} \tag{3.14}$$

which includes a local effect as well as non–local effects.

We will often determine the element strain energy U_j by the mean strain energy density $\bar{u}_j$ in domain j and the domain volume V_j , i.e.

$$U_j = \bar{u}_j V_j \tag{3.15}$$

We then naturally treat two groups of design parameters, i.e. the ones without influence on V_j and the ones without **explicit** influence on $\bar{u}_j$. For the first group, say h_j **is a parameter of the constitutive matrix**, the sensitivity determination (3.13) simplifies to

$$\frac{dU}{dh_j} = -V_j \left(\frac{\partial \bar{u}_j}{\partial h_j} \right)_{\text{fixed strains}} \tag{3.16}$$

and for the second group, say h_j **is a thickness or area parameter**, the sensitivity determination (3.13) simplifies to

$$\frac{dU}{dh_j} = -\bar{u}_j \frac{\partial V_j}{\partial h_j} \tag{3.17}$$

In the following we shall apply these general sensitivity results to a number of specific problems.

In order to make the important result (3.11) more familiar, we shall prove it also relative to a finite element model with the equilibrium

$$[S]\{D\} = \{A\} \tag{3.18}$$

that with constant external loads $\{A\}$ gives the displacement variations $\{\delta D\}$ by

$$[S]\{\delta D\} = -[\delta S]\{D\} \tag{3.19}$$

The total strain energy is determined by

$$U = \frac{1}{2} \{D\}^T [S]\{D\} \tag{3.20}$$

and thus with a symmetric stiffness matrix $[S]$ the variation is determined by

$$\delta U = \frac{1}{2} \{D\}^T [\delta S]\{D\} + \{D\}^T [S]\{\delta D\} \tag{3.21}$$

which by (3.19) gives

$$\delta U = -\frac{1}{2} \{D\}[\delta S]\{D\} \tag{3.22}$$

in agreement with (3.11), and we see that the variation can be calculated with a fixed displacement field.

3.3 RESULTS FOR NON–LINEAR ELASTICITY

Our goal is again to determine dU/dh, where U is the **total strain energy** and h is some design parameter. Even when the design parameter h_j is a **local design parameter**, related to design domain j, strain energy outside this domain is changing and thus one would expect an accumulative determination of dU/dh_j to be necessary. However, for the class of non–linear constitutive models described in section 2.3 we can prove that a localized calculation is still possible.

Using the result (2.31) of constant ratio p between strain and stress energy densities the total strain energy U and complementary strain energy U^C satisfy

$$U + U^C = (1 + p)U \tag{3.23}$$

Then for a dead load system where the external potential is $-W$ we have

$$W = (1 + p)U \tag{3.24}$$

Note that the argument is made without reference to a specific model and is thus valid for one–, two– and three–dimensional models, for analytical calculations and for numerical modelling.

Now, as (3.24) is valid also for the changed design it follows that

$$\frac{dW}{dh} = (1 + p)\frac{dU}{dh} \tag{3.25}$$

which can be written with the "direct" and "indirect" terms separated as

$$\frac{\partial W}{\partial h} + \frac{\partial W}{\partial \varepsilon}\frac{d\varepsilon}{dh} = (1 + p)\left(\frac{\partial U}{\partial h} + \frac{\partial U}{\partial \varepsilon}\frac{d\varepsilon}{dh}\right) \tag{3.26}$$

The strain symbol ε in (3.26) represents again the **strain field in total**. Then using from the virtual work principle

$$\frac{\partial W}{\partial \varepsilon} = \frac{\partial U}{\partial \varepsilon} \tag{3.27}$$

and assuming **design independent loads** $\partial W/\partial h = 0$, we can obtain from (3.26)

$$\frac{\partial U}{\partial \varepsilon}\frac{d\varepsilon}{dh} = \frac{-(1 + p)}{p}\frac{\partial U}{\partial h} \tag{3.28}$$

Using this result to eliminate the "indirect" effect, the derivative dU/dh is expressed as

$$\boxed{\frac{dU}{dh} = -\frac{1}{p}\left(\frac{\partial U}{\partial h}\right)_{\text{fixed strains}}} \tag{3.29}$$

which for linear elasticity (i.e., $p = 1$) reduces to the well–known result $dU/dh = -\left(\partial U/\partial h\right)_{\text{fixed strains}}$, see Pedersen (1990a).

With h_j being a local design parameter we also for this non–linear model have the localized calculation

$$\frac{dU}{dh_j} = -\frac{1}{p}\left(\frac{\partial U_j}{\partial h_j}\right)_{\text{fixed strains}} \tag{3.30}$$

It is **important to note** that the **local physical change** in strain energy is not easily determined

$$\frac{dU_j}{dh_j} \neq -\frac{1}{p}\left(\frac{\partial U_j}{\partial h_j}\right)_{\text{fixed strains}} \tag{3.31}$$

even though the change in total strain energy is available via (3.30).

3.4 SUMMARY

The main conclusions of this section are:

- Change in total elastic energy can be calculated with the strain field fixed – (3.11).
- This result also holds for non–linear power–law elasticity – (3.29).
- With local design variables this implies localized determination – (3.13) and (3.30).
- Changes with fixed strain field are directly related to changes with fixed stress field – (3.7).
- With the proofs given in terms of energy, no specific model is assumed.

4. Optimal Material Orientation

In this section our general design parameter h is taken to be the material orientation θ_j in the domain j . With power–law elasticity and dead loads we have directly from (3.29)

$$\frac{dU}{d\theta_j} = -\frac{1}{p} V_j \left(\frac{\partial \bar{u}_j}{\partial \theta_j}\right)_{\text{fixed strains}} \tag{4.1}$$

where like in (3.16), $\bar{u}_j$ is the mean strain energy density in domain j with volume V_j .

4.1 GENERAL RESULTS FOR ELASTIC ENERGY OPTIMIZATION

For **coupled plate/disc problems** using traditional symbols from laminate analysis the energy density per plate area $u_j t_j$ is for linear elasticity (p = 1) given by

$$u_j t_j = \frac{1}{2}\{\varepsilon^0\}^T\{N\} + \frac{1}{2}\{\varkappa\}^T\{M\}$$

$$\{N\} = [A]\{\varepsilon^0\} + [B]\{\varkappa\} \; ; \; \{M\} = [B]\{\varepsilon^0\} + [D]\{\varkappa\} \tag{4.2}$$

and the combined result is

$$u_j t_j = \frac{1}{2}\{\varepsilon^0\}^T[A]\{\varepsilon^0\} + \{\varepsilon^0\}^T[B]\{\varkappa\} + \frac{1}{2}\{\varkappa\}^T[D]\{\varkappa\} \tag{4.3}$$

with t_j for plate thickness, $\{\varepsilon^0\}, [A]$ for midsurface strains and extensional stiffnesses; and $\{\varkappa\}, [D], [B]$ for curvatures, bending stiffnesses and coupling stiffnesses. Applying the result (4.1) we get

$$\frac{dU}{d\theta_j} = -a_j \left[\frac{1}{2}\{\varepsilon^0\}^T\left[\frac{\partial A}{\partial \theta_j}\right]\{\varepsilon^0\} + \{\varepsilon^0\}^T\left[\frac{\partial B}{\partial \theta_j}\right]\{\varkappa\} + \frac{1}{2}\{\varkappa\}^T\left[\frac{\partial D}{\partial \theta_j}\right]\{\varkappa\}\right] \tag{4.4}$$

with a_j for domain area.

Even for the **fully coupled problems** this result can be written

$$\frac{dU}{d\theta_j} = U_1 \sin 2\theta_j + U_2 \cos 2\theta_j + U_3 \sin 4\theta_j + U_4 \cos 4\theta_j \tag{4.5}$$

This follows from the fact that all the matrices [A], [B], and [D] originate from the constitutive matrix [C] , which, as seen from (2.6), contains only the trigonometric functions of eq. (4.5).

Before treating the specific and simplified problems, it should be appreciated that according to (4.5) we, in the general case, can find **at most four different** solutions to $dU/d\theta_j = 0$. This follows from rewriting $dU/d\theta_j$ as a fourth order polynomial. However, analytical solutions for this general case are too complicated to be shown here.

For **orthotropic materials** and models where only the cosine terms are involved, analytical solutions to $dU/d\theta_j$ are obtainable. Keeping in eq. (4.5) only the sine terms, we have for these (specially orthotropic/balanced) models

$$\frac{dU}{d\theta_j} = 2U_3 \sin 2\theta_j \left(\frac{U_1}{2U_3} + \cos 2\theta_j\right) \tag{4.6}$$

Stationarity is then obtained for

$$\theta_j = 0 \; ; \; \theta_j = \pi/2 \; ; \; \theta_j = \pm \frac{1}{2}\arccos\left(-\frac{U_1}{2U_3}\right) \tag{4.7}$$

and, furthermore, supplementary angles will return the same energy density $u(\pi - \theta) = u(\theta)$ when only $\cos 2\theta$ and $\cos 4\theta$ appear in the stiffness expressions. Thus, for this case the orientational dependence is described completely by the interval $0 \leq \theta \leq \pi/2$.

4.2 SIMPLY SUPPORTED PLATES IN BENDING

For simply supported plates in pure bending we introduce a mode parameter η as the ratio between the two actual halfwave lengths of the deformation. Then the optimal angle is given by fig. 4.1. These results can be found in Bert (1977), Pedersen (1987) or Muc (1988).

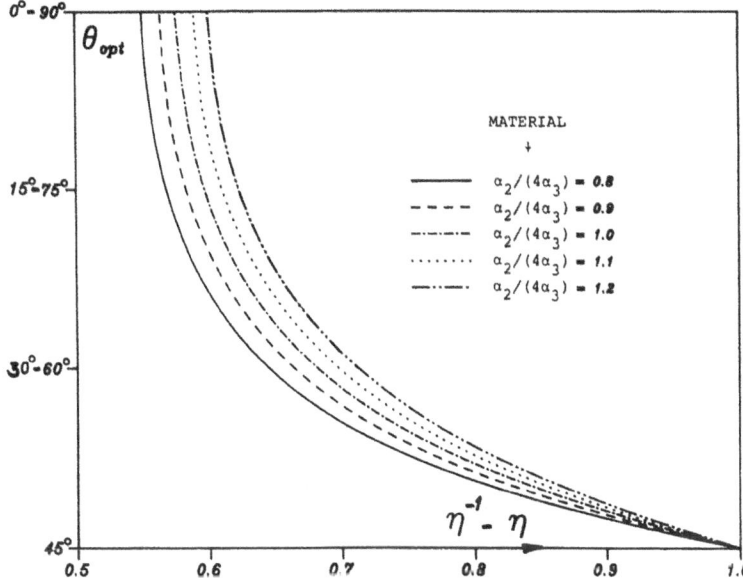

Fig. 4.1: Optimal orientation as a function of the mode parameter η for plate bending.

From an engineering point of view the main conclusions to be drawn from fig. 4.1 are:

- The optimal orientation depends mainly on the mode parameter η . Thus if the deformation pattern is known, the optimal fibredirection can be estimated directly.
- Cases of inverse mode parameters $\eta_1 = \eta_2^{-1}$ have complementary solutions $\theta_2 = \pi/2 - \theta_1$.
- For "extreme" values of η the optimal orientation is perpendicular to the long wavelength.
- The change of optimal fibredirection to a skew direction is very sensitive to the mode parameter.
- The optimal orientation is rather insensitive to the material parameters.
- The optimal orientation is independent of the position of the ply in the laminate, and thus the same for all plies.
- It is not seen in the figure but local optima exist.

By simple numerical analysis, problems with combined modes, can also be optimized; Pedersen (1986) shows such solutions.

4.3 THE MEMBRANE PROBLEM

We shall then move on to the disc problem and specifically treat optimal orientation for **maximum extensional stiffness**. For details see Pedersen (1989), and (1990a). The description is limited to **orthotropic material**.

In agreement with the general result (4.7) the analytical results are very familiar with our plate results. We get

$$\psi_{opt} = 0 \quad \text{or} \quad \psi_{opt} = \pm \, \pi/2 \quad \text{or} \quad \psi_{opt} = \pm \frac{1}{2} \arccos(- \gamma)$$

$$\gamma := \frac{\alpha_2}{4\alpha_3} \frac{1 + \varepsilon_{II}/\varepsilon_I}{1 - \varepsilon_{II}/\varepsilon_I}$$

(4.8)

where ψ is the angle from the larger principal strain direction $\left(|\varepsilon_I| > |\varepsilon_{II}|\right)$ to the larger modulus direction. The fact that $|\varepsilon_{II}/\varepsilon_I| < 1$ and $\alpha_2 > 0$ means that the **optimization parameter** γ has the same sign as the material parameter α_3. (Low shear modulus for $\alpha_3 > 0$ and high shear modulus for $\alpha_3 < 0$.

The second order derivate of the energy density in the fixed strain field is

$$\left(\frac{\partial^2 u_j}{\partial \psi_j^2}\right)_{\text{fixed strains}} = - \, 8\alpha_3 (\varepsilon_I - \varepsilon_{II})^2 \left(\cos 2\psi(\gamma + \cos 2\psi) - \sin^2 2\psi\right)$$

(4.9)

and with this we obtain the results given in table 4.1 for solutions of global maximum or global minimum. Note that although $dU/dh = - \left(\partial U/\partial h\right)_{\text{fixed strains}}$ we have $d^2U/dh^2 \neq - \left(\partial^2 U/\partial h^2\right)_{\text{fixed strains}}$ and thus a more extended analysis is necessary. In a paper by Cheng and Pedersen (1994), this analysis and the conclusions will be given.

Angle ψ of stationarity	Low shear modulus material $\alpha_3 > 0$		High shear modulus material $\alpha_3 < 0$	
	$0 < \gamma < 1$	$\gamma < 1$	$\gamma < -1$	$-1 < \gamma < 0$
$0°$	Global min.	Global min.	Global min.	Local max.
$\pm \, 90°$	Local min.	Global max.	Global max.	Global max.
$\cos 2\psi = - \gamma$	Global max.			Global min.

Table 4.1: Table for selection of the optimal orientation wrt. global minimum or global maximum of total strain energy.

The numerical procedure for solving a specific problem is, briefly stated, as follows:

1) For a given design a finite element analysis gives the actual strain field, i.e. the principal strains with the ε_I direction in each element.

2) For each element the optimization parameter γ_j is evaluated by (4.8).

3) Based on table 4.1 the new material angle (relative to the ε_I direction) is determined.

4) If actual changes are not within a given convergence criterion, return to 1) for a new analysis.

The total number of necessary iterations is normally about 5–10 , provided that an extreme convergence criterion is not specified. Convergence in terms of total energy is much faster than in terms of design variables.

It is shown in Pedersen (1990a) that for the optimal design the principal strain directions are aligned with the principal stress directions. Numerically it is found more efficient to redefine the material angle relative to the larger principal stress direction as an alternative to the larger principal strain direction. In a specific paper, Pedersen (1994), these different iteration possibilities are discussed. A number of different examples can be found in Pedersen (1990a), (1991) and shall not be repeated here.

For **non–orthotropic** materials, analytical solutions are difficult and extremum are found numerically, say with Newton–Raphson iterations. A more extended analysis than can be shown here, cf. Poulsen (1990) offers information about appropriate starting points for such iterations.

4.4 STRENGTH OPTIMAL DESIGN

On a more or less heuristic basis designs for optimal strength have been taken to be designs where principal stress direction is coaligned with material orthotropy direction. With a non–symmetric strength criterion like the Tsai–Wu criterion, such solutions can not be the optimal ones, because optimal stiffness design will be different from optimal strength design.

The strength problem is a local problem and therefore simple results from sensitivity analysis can not be expected. To solve the optimal design problem in a proper way, we have to use mathematical programming. We shall in this section formulate the problem and show a result from Hammer (1994).

The optimization problem formulated in words is: maximize a common load factor, subject to given strength constraints. Multiple layers, multiple loads, and multiple strength constraints are included in the formulation. Let $\{A\}_\ell$ be the load distribution vector corresponding to the load case ℓ, then from finite element analysis the resulting nodal displacement vector $\{D\}_\ell$ is found by

$$[S]\{D\}_\ell = \lambda\{A\}_\ell \quad \ell = 1, 2, \ldots, L \tag{4.10}$$

where $[S]$ is the stiffness matrix of the actual design (finally the optimal design) and λ is a load factor, common to all load cases L.

From $\{D\}_\ell$ strains and stresses in every layer k of every element j follows directly

$$\{D\}_\ell \Rightarrow \{\varepsilon\}_{jk\ell} , \ \{\sigma\}_{jk\ell} \tag{4.11}$$

$$\text{for } j = 1, 2, \ldots, J ; \ k = 1, 2, \ldots, K ; \ \ell = 1, 2, \ldots, L$$

Now the load strength F_n corresponding to a given strength criterion n can be determined for each of these $'jk\ell'$ and compared to the strength limit $(F_0)_n$. Formulating our problem in relation to first ply failure (FPF) we thus have the constraints

$$(F/F_0)_{jk\ell n} \leq 1 \tag{4.12}$$

$$\text{for } j = 1, 2, \ldots, J ; \ k = 1, 2, \ldots, K ; \ \ell = 1, 2, \ldots, L ; \ n = 1, 2, \ldots, N$$

The objective of our optimization is by means of orientational design to

$$\text{Maximize } \lambda \tag{4.13}$$

subject to (4.12).

Analytical sensitivity analysis, as shown in Hammer (1994), can be carried through and then the main problem is related to the existence of a large number of local optima. In a way this problem was expected. Different solutions to this problem includes strategies for choosing the initial design before mathematical programming is applied. A valuable alternative is to describe the design with global design parameter. This new approach will be described before an example is shown.

4.5 GLOBAL DESIGN PARAMETERS

The idea of a global design description used frequently in shape optimization, is extended from curve to surface parametrization. The design is given as a linear combination of orthogonal functions. Application is here to orientational design of laminates for strength optimization, but the technique is directly applicable to design also of thickness distribution. The advantages by this design description are many, including control on smoothness, control on slopes, and control on connections at the design boundaries. This simplified parametrization of the design makes it possible to work with only few design variables, say 25 for a whole design domain.

Let us first present the design description from Pedersen, Tobiesen and Jensen (1992) in a form that makes our extension from curve to surface more clear. In fig. 4.1 we show the reference domain $-1 \leq \xi < 1$, and the base curve C_0 with the mapping illustrated. (In section seven we return to this problem).

Fig. 4.2: Illustration of the mapping from reference line to the base curve.

The design curve C is then described by

$$C = C_0 + \sum_{n=1}^{N} z_n \phi_n(\xi) \qquad (4.14)$$

where ϕ_n are given functions and the linear combination factors z_n are the design variables. In section seven the mutual orthogonal functions are chosen as the first vibration modes to a clamped/clamped beam, because the positions and slopes at both ends of the curve C should be identical to those of C_0. Naturally, this selection of functions has no relation to vibration at all – it is just a convenient way of obtaining well known orthogonal functions, and surely other possibilities exist. Note that other conditions at the ends of C can be obtained from other boundary conditions of the related beam problem.

Now for our 2D problem we show in fig. 4.3 the reference domain $\left(-1 \leq \xi_1 < 1, -1 \leq \xi_2 < 1\right)$ and a finite element division put onto it. If a constant design (thickness and/or orientation) is wanted in each element, this may be controlled by the values in the center of the

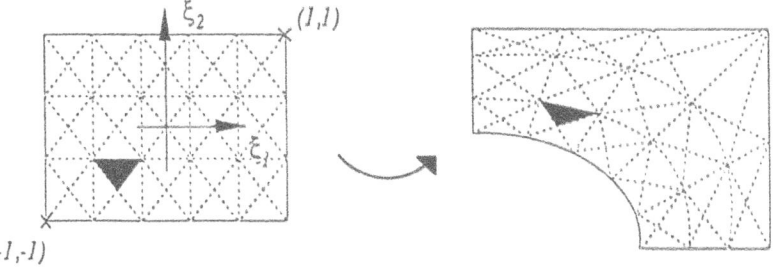

Fig. 4.3: Illustration of the mapping from reference domain to the actual finite element model.

corresponding element in the reference domain. Mapping is from element in reference model to corresponding element in actual model, and from triangle center to corresponding triangle center.

The material orientation γ in the reference domain is given by

$$\gamma(\xi_1, \xi_2) = \sum_{m,n}^{N} z_{mn} \phi_m(\xi_1) \phi_n(\xi_2) \qquad m, n = 0, 1, 2, ..., N \qquad (4.15)$$

where the ϕ functions are given mutual orthogonal functions, and thus the z_{mn} are the design parameters. In this specific case we choose the ϕ functions to be

$$\phi_0(\xi) = 1 \; ; \; \phi_1(\xi) = \xi$$

$$\phi_m(\xi) = \cosh(k_m\xi) + \mu_m \cos(k_m\xi) \qquad m = 2, 4, ...$$

$$\phi_m(\xi) = \sinh(k_m\xi) + \mu_m \sin(k_m\xi) \qquad m = 3, 5, ...$$

$$(4.16)$$

where k_m, μ_m are constants determined by

$$\tan(k_m) + (-1)^m \tanh(k_m) = 0 \qquad m = 2, 3, 4, ...$$

$$\mu_m = \frac{\cosh(k_m)}{\cos(k_m)} \quad m = 2, 4, ... \quad , \quad \mu_m = \frac{\sinh(k_m)}{\sin(k_m)} \quad m = 3, 5, ... \qquad (4.17)$$

With this design description, i.e. using the z_{mn}'s in (4.15) as design variables, a solution from Hammer (1994) will illustrate the possibilities. In relation to a two layer, two load cases problem with Tsai–Wu as well as maximum strain criteria. The problem and the quarter model is shown in fig. 4.4 and the optimal orientational design in fig. 4.5, with the optimized load factor for increasing number of design variables.

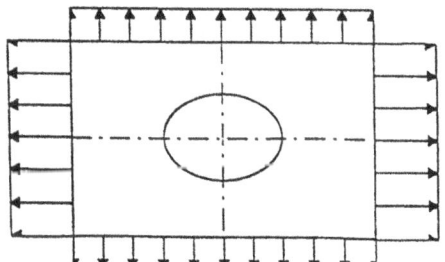 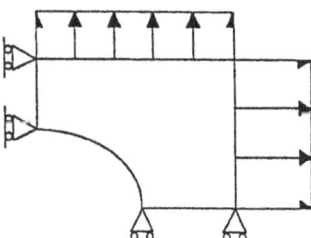

Fig. 4.4: Problem with an elliptic hole where symmetry is assumed, and the two uniform loads act independently.

4.6 SUMMARY

The main conclusions of this section are:

- Even for the most coupled problems the gradient of elastic energy will be a simple function of the orientation – (4.5).
- Simple supported rectangular plates in bending can be optimized analytical – fig. 4.1.
- For the membrane problem knowledge on global/local optima are available – table 4.1.
- Redesign according to the direction of numerically larger principal stress is a practical procedure.
- Optimal design with strength constraints is formulated and solved using a new global design description – (4.15).

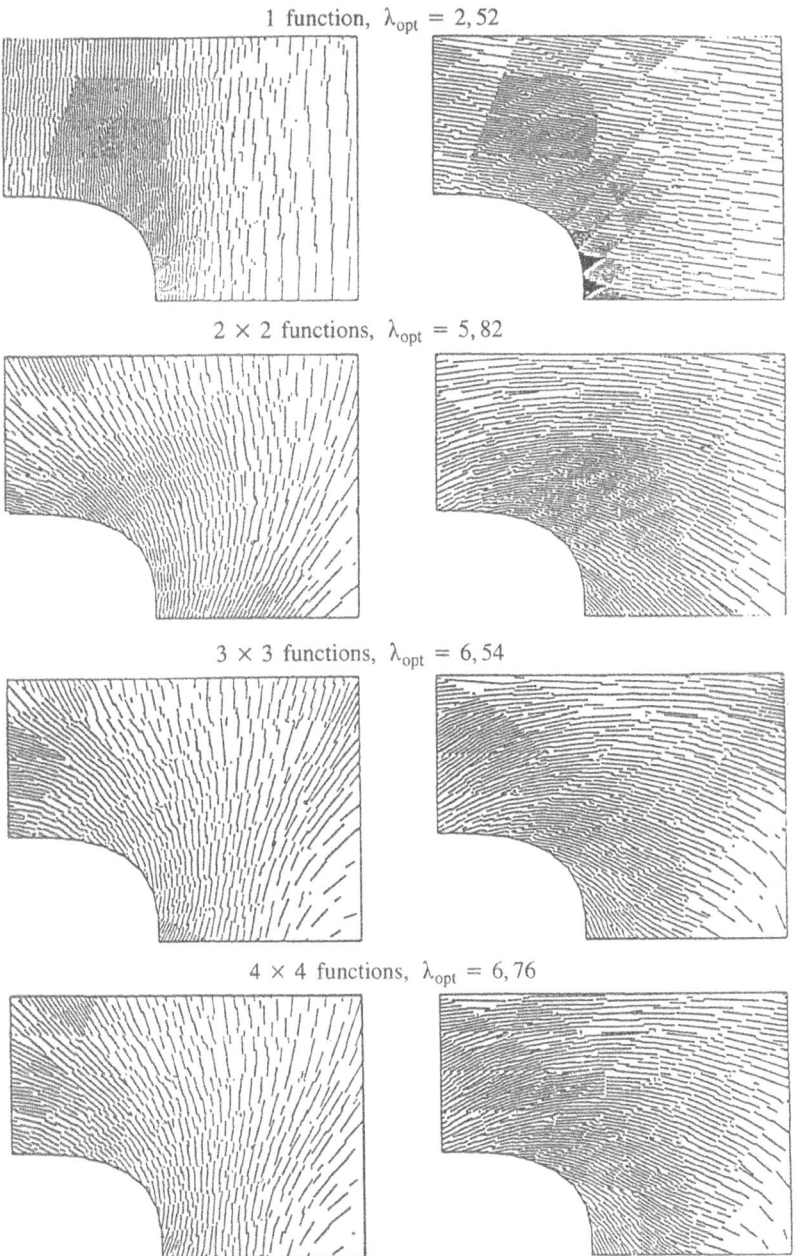

Fig. 4.5: Optimal orientation with increasing number of design variables. Upper layer orientations to the left and lower layer orientations to the right. From Hammer (1994).

5. Optimal Thickness Distribution

The sizing problem of optimal design for a thickness distribution has a long history, and for the most simple formulations the optimality criteria methods plays a dominating role.

We shall concentrate the present description on the non–linear elastic problems, but primarily give the important result based on linear elasticity.

5.1 GENERAL CRITERION WITH ONLY ONE CONSTRAINT

For the unconstrained optimization problem of orientational design, the necessary optimality criterion is zero gradients. For optimization problem with only a single constraint, an almost as simple optimality criterion is given by **constant ratio of the gradients**. Let us show this in relation to the problem of

$$\text{Minimizing } U$$

$$\text{for given } V = \overline{V} \quad \left(\overline{V} \text{ fixed}\right) \tag{5.1}$$

Now with t_j being design parameters then the differentials of U and of $\left(V - \overline{V}\right)$ are

$$dU = \sum \frac{\partial U}{\partial t_j} dt_j$$

$$d\left(V - \overline{V}\right) = dV = \sum \frac{\partial V}{\partial t_j} dt_j \tag{5.2}$$

From this follow that with

$$\boxed{\frac{\partial U}{\partial t_j} = A\frac{\partial V}{\partial t_j} \quad \text{for all } j} \tag{5.3}$$

where A is the constant ratio of gradients we get by the constraint condition $dV = 0$ also $dU = 0$
This follows by inserting (5.3) in (5.2)

$$dU = \sum A\frac{\partial V}{\partial t_j} dt_j = A\sum \frac{\partial V}{\partial t_j} dt_j = AdV = 0 \tag{5.4}$$

5.2 OPTIMALITY CRITERION FOR LINEAR ELASTICITY

As we have indicated in this more general presentation, we are interested in the design problem where in (5.3) U is elastic energy, V is volume and t_j is thickness in design domain j. Using the sensitivity result (3.17), the general optimality condition (5.3) is

$$\frac{\partial U}{\partial t_j} = -\overline{u}_j\frac{\partial V_j}{\partial t_j} = A\frac{\partial V_j}{\partial t_j} \tag{5.5}$$

from which we get directly the well–known and intuitive criterion of **constant energy density**, equal to the mean energy density $\overline{u}$

$$\boxed{\overline{u}_j = \overline{u} \quad \text{for all } j} \tag{5.6}$$

This optimality criterion is first derived by Wasiutynski (1960).

5.3 OPTIMALITY CRITERION FOR NON–LINEAR ELASTICITY

We shall immediately see that the sensitivity result (3.29) will again result in the optimality criterion (5.6). From (3.29) follows with t_j as design parameter

$$\frac{\partial U}{\partial t_j} = -\frac{1}{p}\left(\frac{\partial U}{\partial t_j}\right)_{fixed\ strain} = -\frac{1}{p}\left(\frac{\partial U_j}{\partial t_j}\right)_{fixed\ strain} = -\frac{\bar{u}_j}{p}\left(\frac{\partial V_j}{\partial t_j}\right) \qquad (5.7)$$

and therefore the condition of constant ratio of the gradients give as in (5.5)

$$-\frac{\bar{u}_j}{p}\frac{\partial V_j}{\partial t_j} = A\frac{\partial V_j}{\partial t_j} \ , \ \bar{u}_j = \bar{u} \ \ \text{for all} \ \ j \qquad (5.8)$$

With uniform energy density (5.8), we get uniform effective strain, and then by (2.27) the same constitutive secant matrix in all domains (all points). Thus **the optimal thickness distribution is independent of the power p of the constitutive matrix**. To state it in other terms: the optimal structure is equally loaded (in terms of strain energy density) at all points. We thus "bridges" the classical solutions of linear elasticity by Wasiutynski (1960) and of ideal plasticity by Prager and Shield (1967).

5.4 EXAMPLES

An example taken from Pedersen (1991) shall now be optimized first with respect to only thickness distribution and then simultaneously with respect to thickness and orientation.

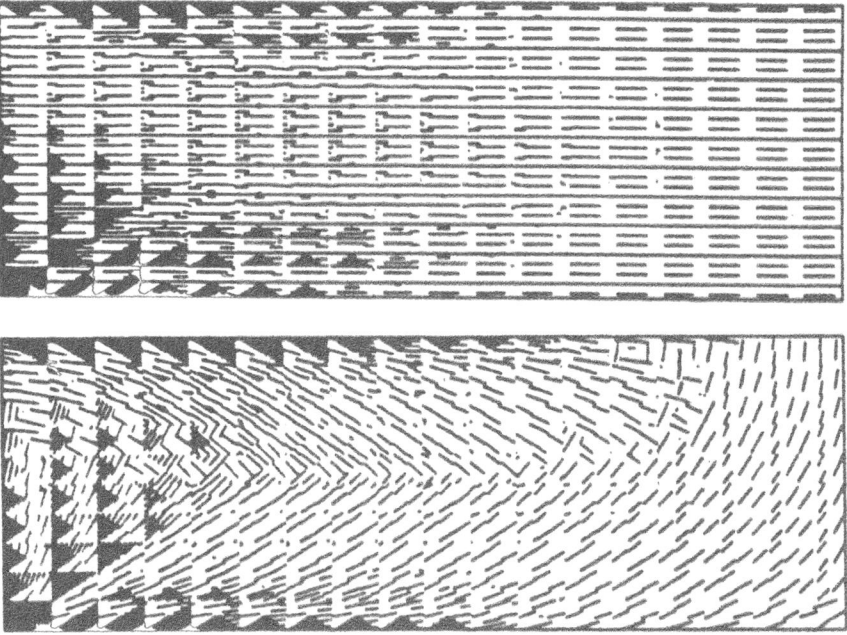

Fig. 5.1: Above: the optimal thickness distribution for a uniformly loaded cantilever. Below: the combined thickness and orientational optimization.

In all the figures we have used the same way of visualizing of the results. The **design is character-ized** by thickness and orientation, which are shown by hatching the triangular finite elements in the direction of the larger modulus direction and with the hatch density proportional to the thickness. Dark areas are therefore areas with large thicknesses. The **cantilever example** shown in fig. 5.1 is based on a 720 element model, with constant thickness and orientation in each element. For the uniform cantilever the mean and the maximum values of energy density are 787, 32919 (relative measures of stiffness and stress concentration). Only thickness optimization gives 414, 450, which means stiffness improved by a factor 1.9 and almost no energy concentration. With simultaneous thickness, orientational optimization we obtain 181, 199, i.e. a stiffness improved by a factor of 4.3.

We shall also show results based on non–linear elasticity. Results from iterative analysis based on the constitutive relation (2.27) of section 2, and optimization based on the optimality criterion (5.8) are presented for the three cases shown in fig. 5.2, taken from Pedersen and Taylor (1993).

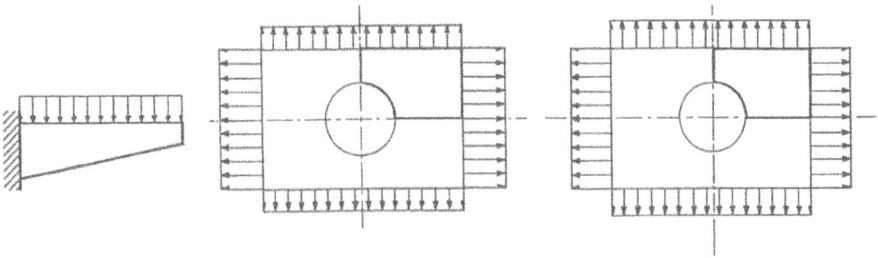

Fig. 5.2: The three example cases: a) uniformly loaded cantilever of isotropic material. b) circular hole loaded biaxially 2:1, isotropic material. c) optimal designed hole, loaded biaxially 3:2, ortho-tropic material.

The main result is that the optimal thickness distribution is independent of the power p . Thus the results for the problems of fig. 5.2 mainly illustrates the influence from the power p on two given designs, i.e. the uniform thickness design and the optimal thickness design as obtained from linear elasticity.

In terms of relative values table 5.1 gives the strain energy densities. Min., mean and max. values are related to the elements of the finite element models. The table clearly shows the different results from uniform and optimal thickness distribution. It is well known from optimization based on linear elasticity that certain areas in a model cannot be fully stresses (too little energy density). Thus the min. values are of minor interest, and the agreements of the max. values with the mean values better show the fulfillment of the optimality criterion (5.6).

The values of the objective function (work = compliance = (1+p) × strain energy) are also given by the relative mean values in table 5.1. The factor between energy in uniform design and energy in optimal design is almost constant for the three cases, with a weak tendency to be more important with increasing non–linearity (decreasing p) . The stronger effect for the cantilever problem reflects the initial less uniform/strain distribution. Also the actual stress/strain level (higher for the cantilever problem) will have an influence, and thus the three cases should be read as individual cases. More detailed information are obtainable in stress/displacement/strain graphs, omitted in this paper.

	Strain energy densities in % of reference energy density								
	cantilever, isotropic			circ. hole, isotropic			opt. hole, orthotropic		
	min	mean	max	min	mean	max	min	mean	max
$U_{1\cdot0}$	0.5	100	577	4	100	677	47	100	387
$O_{1\cdot0}$	14	50	50	81	87	89	59	90	92
$U_{0\cdot9}$	0.7	163	935	4	140	951	60	138	528
$O_{0\cdot9}$	23	75	76	112	120	122	63	123	138
$U_{0\cdot8}$	0.9	280	1599	3	202	1391	79	196	737
$O_{0\cdot8}$	39	120	121	159	170	173	69	173	221
$U_{0\cdot7}$	1	512	2916	2	304	2127	108	288	1060
$O_{0\cdot7}$	68	201	202	236	251	256	80	253	369
$U_{0\cdot6}$	2	1005	5730	0.3	482	3430	153	445	1581
$O_{0\cdot6}$	126	357	359	365	387	394	90	385	641
$U_{0\cdot5}$	2	2151	12310	0.1	808	5891	227	722	2464
$O_{0\cdot5}$	251	683	686	595	630	640	107	618	1169

Table 5.1: Table of relative results with uniform thickness U and with optimal thickness distribution O for linear elasticity $(p = 1)$ and for five models of non–linear elasticity modelled by the power $p < 1$. The three independent cases are shown in fig. 5.2.

We note that the small difference between mean and maximum density (90 and 92) for the ortho-tropic case has a strong influence for the non–linear solutions (at $p = 0.5$ the values are 618 and 1169).

5.5 SUMMARY

The main conclusions of this section are:

- The optimality criteria of uniform energy density, also holds for power–law non–linear elas-ticity – (5.8).
- The optimal thickness distribution is independent of the power p.
- Combined orientational design and thickness design should be performed.
- For the idealized problems treated, the stress concentration was eliminated by thickness design.
- Orientational design improves stiffness, but cannot eliminate stress (energy) concentration.

6. Detailed Material Design

The orientational design in section four and the thickness design in section five, in fact show how to use given materials. From these studies extends a natural wish to design the material itself. A very up to date account for this active research area is given in the thesis by Bendsøe (1994). Here we shall discuss a two parameter model and a full parameter model. Finally, we shall give a short introduction to solution of inverse homogenization problems, as in the papers by Sigmund (1993 and 1994).

6.1 A TWO PARAMETER MODEL

Let us focus only on the objective of maximize stiffness (minimize strain energy) based on ortho-tropic materials classified as low shear modulus materials. In reality we shall go directly to the material model from Bendsøe (1991), as illustrated in fig. 6.1.

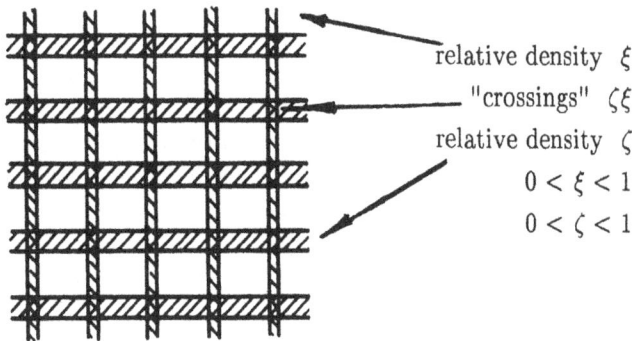

relative density ξ

"crossings" $\zeta\xi$

relative density ζ

$$0 < \xi < 1$$

$$0 < \zeta < 1$$

Fig. 6.1: A material model with two directional densities.

The total, relative volume densities ϱ_j give the total volume V by

$$V = \sum_j V_j = \sum_j \tilde{V}_j \varrho_j \quad , \quad \frac{\partial V}{\partial \varrho_j} = \frac{\partial V_j}{\partial \varrho_j} = \tilde{V}_j = \frac{V_j}{\varrho_j} \tag{6.1}$$

where $\tilde{V}_j$ is the maximum volume of domain j. In principle, ϱ_j acts like the thickness in section five. Often we shall omit the domain index j, and the relations of the directional relative densities ζ, ξ are

$$\varrho = \zeta + \xi - \zeta\xi$$

$$\frac{\partial \varrho}{\partial \zeta} = 1 - \xi \; ; \; \frac{\partial \varrho}{\partial \xi} = 1 - \zeta \tag{6.2}$$

From Bendsøe (1991) we take the orthotropic constitutive matrix, expressed in known modulus E and Poisson's ratio v and in the directional parameters ζ, ξ, as

$$[C] = \begin{bmatrix} [\tilde{C}] & ; & \begin{matrix} 0 \\ 0 \end{matrix} \\ 0 ; 0 & ; & \text{not actual} \end{bmatrix}$$

$$[\tilde{C}] = \frac{E\xi}{\alpha - \beta^2}\begin{bmatrix} 1 & ; & \beta \\ \beta & ; & \alpha \end{bmatrix} ; \quad [\tilde{C}]^{-1} = \frac{1}{E\xi}\begin{bmatrix} \alpha & ; & -\beta \\ -\beta & ; & 1 \end{bmatrix}$$

(6.3)

$$\alpha = \xi^2 + \xi(1 - \xi)/\zeta ; \quad \beta = \xi\nu$$

and we assume the major principal axis chosen by

$$1 > \alpha \Rightarrow \zeta > \xi/(1 + \xi) \tag{6.4}$$

The conditions of positive definite and limited $[\tilde{C}]$ are satisfied for

$$0 < \zeta < 1 ; \quad 0 < \xi < 1 ; \quad -1 < \nu < 1 \tag{6.5}$$

With the assumptions given, we know **from the orientational results** that the larger principal modulus direction should optimally be the same as the direction of the numerically larger strain ε_I $(|\varepsilon_I| > |\varepsilon_{II}|)$. Therefore with **strain ratio** $-1 < \eta := \varepsilon_{II}/\varepsilon_I < 1$ the strain energy density is given by

$$u = \frac{1}{2}\varepsilon_I\{1; \eta\}[\tilde{C}]\begin{Bmatrix} 1 \\ \eta \end{Bmatrix}\varepsilon_I \quad \text{or}$$

(6.6)

$$u = \frac{1}{2}\varepsilon_I^2\frac{E\xi}{(\alpha - \beta^2)}(1 + 2\beta\eta + \alpha\eta^2)$$

We know that principal stresses are generally also in the same direction, and for the present case it is proved that the numerically larger principal stress σ_I $(|\sigma_I| > |\sigma_{II}|)$ match with ε_I. Thus expressed in stresses and **stress ratio** $-1 < \mu := \sigma_{II}/\sigma_I < 1$ we have from (6.3)

$$u = u^C = \frac{1}{2}\sigma_I^2\frac{1}{E\xi}(\alpha - 2\beta\mu + \mu^2) \tag{6.7}$$

This is the results from the orientational optimization. Next, in the same way as in the thickness optimization, we have

$$U = \sum_j U_j = \sum \bar{u}_j\tilde{V}_j\varrho_j$$

(6.8)

$$\frac{\partial U}{\partial \varrho_j} = -\left(\frac{\partial U_j}{\partial \varrho_j}\right)_{\text{fixed strains}} = -\frac{\bar{u}_jV_j}{\varrho_j}$$

and thus, with (6.1), again the result of **uniform energy density** equal to mean energy density $\bar{u}$

$$\bar{u}_j = \bar{u} \tag{6.9}$$

Solution by optimality criterion iterations can thus give us ϱ_j for $j = 1, 2, ..., J$.

Lastly, we come to **the detailed design**. Knowing ϱ_j , then how to optimize ζ_j, ξ_j ? We again have a single constraint problem (omitting index j)

$$\text{Minimize}_{\text{over } \zeta, \xi} \ U = u\tilde{V}\varrho$$

with constraint $\varrho = \zeta + \xi - \zeta\xi$ (6.10)

Derivatives with respect to the constraint are stated in (6.2) and derivatives with respect to u can be found from (6.6) to be:

$$\partial u/\partial \zeta = (1 - \xi)(1 + \xi v\eta)^2/N$$

$$N = \left(1 - \xi + \zeta\xi(1 - v^2)\right)^2$$

(6.11)

$$\partial u/\partial \xi = \left(\zeta\left(1 - \zeta(1 - v^2)\right) + 2\zeta v\eta\right.$$

$$\left. + \eta^2\left((1 - \xi)^2 + \zeta\xi(2(1 - \xi) + \xi v^2) + \zeta^2\xi^2(1 - v^2)\right)\right)/N$$

Then, by means of (6.11) and (6.2), the optimality criterion of constant ratios between objective and constraint gradients

$$\frac{\partial u/\partial \zeta}{\partial \varrho/\partial \zeta} = \frac{\partial u/\partial \xi}{\partial \varrho/\partial \xi}$$

(6.12)

gives the equation of optimality

$$(1 - \zeta)(1 + \xi v\eta)^2 = \zeta\left(1 - \zeta(1 - v^2)\right) + 2\zeta v\eta$$

$$+ \eta^2\left((1 - \xi)^2 + \zeta\xi(2(1 - \xi) + \xi v^2) + \zeta^2\xi^2(1 - v^2)\right)$$

(6.13)

With the help of the Mathematica programme, Wolfram (1991), it is possible to find that (6.13) correspond to the product of two linear equations

$$((1 - \zeta) + v\zeta + \eta(1 + \xi(\zeta - 1)(1 + v))) \times$$

$$((1 - \zeta) - v\zeta - \eta(1 + \xi(\zeta - 1)(1 - v))) = 0$$

(6.14)

and thus we have with (6.2) and (6.4)–(6.5) an analytical solution, see Jog, Haber and Bendsøe (1992).

In the stress formulation (6.7) the results are more simple to derive, and we get the final results after little algebra:

$$\zeta = \frac{\varrho}{1 + |\mu|} \ ; \ \xi = \frac{|\mu|\varrho}{1 + |\mu| - \varrho}$$

(6.15)

in terms of the total relative density ϱ and stress ratio $\mu := \sigma_{II}/\sigma_I$.

In the paper by Thomsen (1991) concurrent density and orientational design is solved with constraints on a cost defined functional.

6.2 THE FULL PARAMETER MODEL

We now want to deal with material properties that are represented in **the most general form** possible for a linear elastic continuum, namely the unrestricted set of elements of a positive semi–definite constitutive matrix. The results to be presented are taken from Bendsøe, Guedes, Haber, Pedersen and Taylor (1994), and in this paper also the three dimensional problem is solved and the appropriate references are given.

In the previous formulations, the total volume of material, defined at the micro–level, provides a natural cost function for the optimization problem formulation. There is no such natural cost function for the general material design formulation we consider here. Instead, we use certain invariants of the constitutive matrix as the measure of cost, thus ensuring that the optimal design solutions are not influenced by the choice of reference frame. Moreover, we can then express cost and energy in any frame that is suitable for our formulation, a feature that is crucial for the developments below. We shall here restrict us to the trace of the constitutive matrix as a measure of cost. Stated differently, we put constraints on the sum of the eigenvalues of the constitutive matrix. The trace is given in (2.12) and the conditions for a position definite constitutive matrix is listed in (2.11).

We now make use of our frame–independent descriptions of cost (trace) and energy to express the strain energy density in the frame of the principal strains ε_I and ε_{II}

$$u = \frac{1}{2}\left(\alpha_{1111}\varepsilon_I^2 + \alpha_{2222}\varepsilon_{II}^2 + 2\alpha_{1122}\varepsilon_I\varepsilon_{II}\right) \tag{6.16}$$

Our goal is to minimize the total strain energy U, and thus according to (3.16) to maximize locally the strain energy density u as stated by (6.16). This function does not depend on the value of α_{1212}, and as this non–negative parameter enters in the cost function (2.12), we can conclude that the parameter should have value zero. Moreover, it then follows from the conditions (2.11) that

$$\alpha_{1212} = \alpha_{1112} = \alpha_{2212} = 0 \tag{6.17}$$

This implies that the optimal material at each point is an orthotropic material, which is co–aligned with the principal strain and principal stress axes and which has zero shear stiffness. We thus have

$$\begin{Bmatrix} \sigma_I \\ \sigma_{II} \\ 0 \end{Bmatrix} = C\begin{bmatrix} \alpha_{1111} & \alpha_{1122} & 0 \\ \alpha_{1122} & \alpha_{2222} & 0 \\ 0 & 0 & 0 \end{bmatrix}\begin{Bmatrix} \varepsilon_I \\ \varepsilon_{II} \\ 0 \end{Bmatrix} \tag{6.18}$$

$$\alpha_{1111} \geq 0 , \ \alpha_{2222} \geq 0 , \ \alpha_{1111}\alpha_{2222} - \alpha_{1122}^2 \geq 0$$

The parameter α_{1122} have no influence on the trace (2.12), so to maximize the strain energy density this parameter should be chosen as numerically large as possible and with the same sign as $\left(\varepsilon_I\varepsilon_{II}\right)$. Thus the constraint with α_{1122} in (6.18) must be active. Accordingly, the strain energy density u and the trace ϱ are reduced to

$$u = \frac{1}{2}\left(\alpha_{1111}\varepsilon_I^2 + \alpha_{2222}\varepsilon_{II}^2 + 2\sqrt{\alpha_{1111}\alpha_{2222}}\,|\varepsilon_I\varepsilon_{II}|\right)C$$

$$= \frac{1}{2}\left(\sqrt{\alpha_{1111}}\,|\varepsilon_I| + \sqrt{\alpha_{2222}}\,|\varepsilon_{II}|\right)^2 C \tag{6.19}$$

$$\varrho = \alpha_{1111} + \alpha_{2222}$$

The final problem of finding the vector $\left(\sqrt{\alpha_{1111}}, \sqrt{\alpha_{2222}}\right)$ that maximizes u for given ϱ, is the problem of finding a positive vector of length $\sqrt{\varrho}$ for which the inner product with the vector $\left(|\varepsilon_I|, |\varepsilon_{II}|\right)$ is maximal. The solution is unique and is the obvious one, namely

$$\alpha_{1111} = \varrho\frac{\varepsilon_I^2}{\varepsilon_I^2 + \varepsilon_{II}^2} , \ \alpha_{2222} = \varrho\frac{\varepsilon_{II}^2}{\varepsilon_I^2 + \varepsilon_{II}^2}$$

$$\alpha_{1122} = \varrho\frac{\varepsilon_I\varepsilon_{II}}{\varepsilon_I^2 + \varepsilon_{II}^2} \tag{6.20}$$

and thus the optimal strain energy density is

$$u = \frac{1}{2}\varrho\left(\varepsilon_I^2 + \varepsilon_{II}^2\right)C = \frac{1}{2}\varrho\left(\varepsilon_{11}^2 + \varepsilon_{22}^2 + 2\varepsilon_{12}^2\right)C \qquad (6.21)$$

Before ending this analysis let us point out that according to (6.21) the optimal energy density is the same as the one from a zero Poisson–ratio, isotropic material

$$\left\{\begin{matrix} \sigma_{11} \\ \sigma_{22} \\ \sqrt{2}\,\sigma_{12} \end{matrix}\right\} = C\varrho \begin{bmatrix} 1 & 0 & 0 \\ 0 & 1 & 0 \\ 0 & 0 & 1 \end{bmatrix} \left\{\begin{matrix} \varepsilon_{11} \\ \varepsilon_{22} \\ \sqrt{2}\,\varepsilon_{12} \end{matrix}\right\} \qquad (6.22)$$

which however has a cost that is three times (trace = 3ϱ) greater than the optimized material.

6.3 THE INVERSE HOMOGENIZATION PROBLEM

The first reaction to the simple result (6.20) is the immediate question about the existence of such a material. In fact this question goes beyond the present optimization and are similar to questions like: Can we find a material with negative Poisson's ratio? or generally: Can we find a material corresponding to a given constitutive matrix?

The answers to all these questions are yes and the corresponding problem is here termed the inverse homogenization problem or the material identification problem. Like with most identification problems we can not expect a unique solution, but in fact this is an advantage because then it is possible to put priorities and/or penalties.

The direct homogenization problem is a natural part of the solution, just as analysis is a natural part of any optimal design approach. Shortly we may state that the elastic strain energy of a base cell represented by its homogenized constitutive matrix $\left[C^H\right]$ subjected to any test strain field, should be the same as the elastic energy in the microstructure subjected to the same test strain field. For details see Sigmund (1993), (1994).

The inverse homogenization problem is solved by an optimality criterion based algorithm and in fig. 6.2. we show the solution to a specific problem.

6.4 SUMMARY

The main conclusions of this section are:

- Material design is a very active research area.

- The full parameter design problem has a surprisingly simple solution (6.20), which is also available for 3D problems.

- Constrained parameter formulations are more complicated. However, the shown two parameter problem also has an analytical solution (6.15).

- Formulations in strain energy density as well as in stress energy density are possible.

- The formulation of inverse homogenization problems open a new very interesting area of research.

- It seems that we are able to find a micro–mechanical model for a material with any wanted constitutive matrix.

252

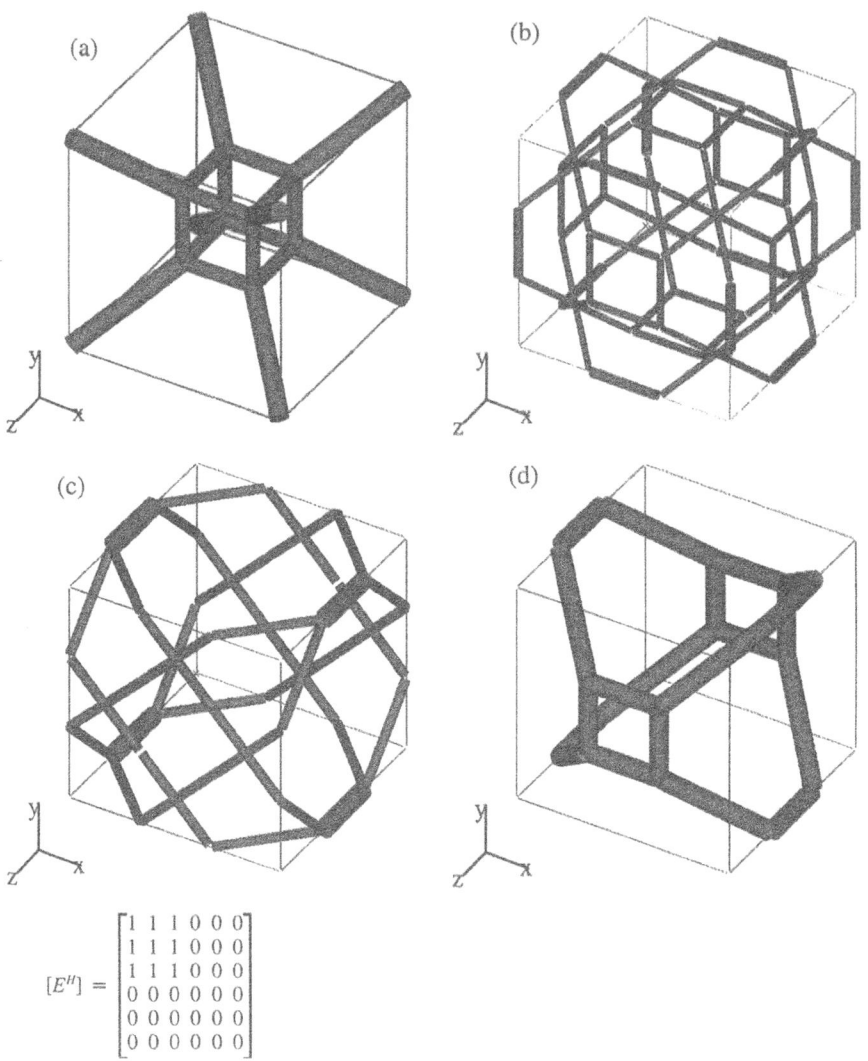

$$[E^H] = \begin{bmatrix} 1 & 1 & 1 & 0 & 0 & 0 \\ 1 & 1 & 1 & 0 & 0 & 0 \\ 1 & 1 & 1 & 0 & 0 & 0 \\ 0 & 0 & 0 & 0 & 0 & 0 \\ 0 & 0 & 0 & 0 & 0 & 0 \\ 0 & 0 & 0 & 0 & 0 & 0 \end{bmatrix}$$

Fig. 6.2: Microstructures in a 3D–truss model for an optimal material corresponding to $\varepsilon_I = \varepsilon_{II} = \varepsilon_{III}$. (The four different solutions have the same mass). From Sigmund (1994).

7. Shapes for Minimum Energy Concentration

The problem of shape design for minimum energy concentration is highly nonlinear, and must be solved iteratively. Thus, we can immediately convert the problem into a sequence of problems of **optimal redesign**, i.e. how do we change a given shape into a better "neighbouring" shape? The solution to this involves three steps: finite element strain **analysis** for the given shape – **sensitivity** analysis with respect to the parameters z_n describing the design – and **optimal decision** of redesign.

7.1 PROBLEM FORMULATION

In mathematical terms the objective of our shape design is to

$$\begin{array}{cc} \text{Minimize} & \text{Maximum} \quad u \\ \text{\small (over feasible shapes)} & \text{\small (over the structural space (x) and load cases)} \end{array} \qquad (7.1)$$

where u is the **strain energy density**. Choosing u as our objective can naturally be questioned, but alternative objectives can be treated in a similar way. In section 4.4 we described shortly the orientational design for optimal strength, which also involves local design constraints. For these problems, which were solved later than the present shape optimizations, the envelope of a number of different criteria were formulated including the Tsai–Wu criterion and the maximum strain criteria.

Converting (7.1) to a Min–problem, concentrating on redesign by design parameters Δz_n , and including – for computational reasons – an area (volume) constraint $A = \overline{A}$, we have the actual **redesign formulation**

$$\begin{array}{cc} \text{Minimize} & u_{max} \quad \text{subject to} \\ \text{\small (within move-limits on Δz_n)} & \end{array}$$

$$\tag{7.2}$$

$$u(x) + \sum_i \frac{\partial u(x)}{\partial z_n} \Delta z_n - u_{max} \leq 0 \ , \ A + \sum_i \frac{\partial A}{\partial z_n} \Delta z_n - \overline{A} = 0$$

where u_{max} is a further unknown to be determined.

Specifically, we shall concentrate in this section on the two dimensional problem of a biaxially loaded hole. The extensions relative to earlier works by Kristensen and Madsen (1976), Pedersen and Laursen (1982–83), Dybbro and Holm (1986) are to study the influence from orthotropic material behaviour, although still assuming linear elasticity. The studies by Lee, Kikuchi and Scott (1989) and by Bäcklund and Isby (1988) are also related to shape optimization with anisotropic materials, but in all very few papers are published on the subject. We also shall study the influence of finite element modelling and compare the solutions based on elements with constant and linear strain assumptions, respectively. Extensive details can be found in the M.Sc. thesis by Tobiesen and Jensen (1990), written in Danish, and the present section is closely related to the paper Pedersen, Tobiesen and Jensen (1992).

The techniques used for the design description, for the finite element modelling, and for the sensitivity analysis, have proved accurate, robust and effective. Therefore, a number of detailed comments on the methods are included.

7.2 DESCRIPTION OF THE DESIGN

The success of shape optimization depends to a large extent on the chosen design parameters, i.e. on the design parametrization. Many possibilities exist and we use a **global** description that **enforces smoothness and desired connections** to neighbouring shapes, as exemplified by symmetric requirements. The description is illustrated in fig. 4.2 and the design parameters are the factors z_n in the description (4.14).

254

The finite element modelling is by no means straightforward, either. A good finite element model is strongly related to the actual problem, which constitutes the design as well as the load case. We need to choose elements and a mesh and have to be prepared to perform mesh adaption for the individual designs as well as for the changed designs. Automatic mesh generation is a necessity.

In general, our analysis is based on a "deformed" rectangular mesh with each quadrangle divided into four triangular basic elements, illustrated in fig. 7.1. The node concentrations can be controlled by a few parameters (exponential powers), and a simple scheme for automatic mesh adaption (uniform element total energy) has also been worked out. Convergence tests confirm the accuracy, and we shall present a comparative study between the three–node constant strain triangles and the six–node linear strain triangles.

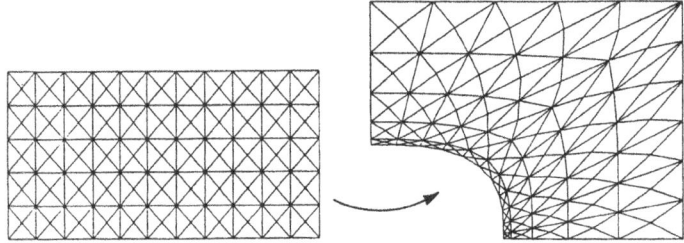

Fig. 7.1: Illustration of the mesh generations.

In fig. 7.2 we show the total model, the quarter part for analysis, and a representative finite element model.

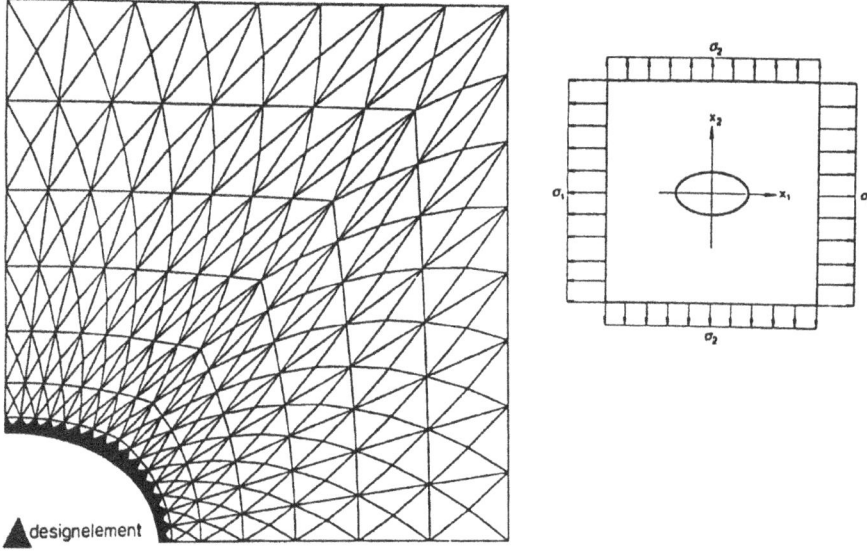

Fig. 7.2: A non–redefined finite element model with pointers on the actual design elements, i.e. elements influenced directly by boundary design. The total model with loads shown on the right.

7.3 OPTIMIZATION METHOD

The key information needed for the optimal design process is how the strain energy–density u changes with the parameters of the shape design, here symbolically denoted by z . With the constitutive matrix [C] from $\{\sigma\} = [C]\{\varepsilon\}$ and assuming linear elasticity, the strain energy density at a given point is

$$u = \frac{1}{2}\{\varepsilon\}^{T}[C]\{\varepsilon\} \tag{7.3}$$

We emphasize here that orthotropy enters only through the matrix [C] and that the following derivations are completely analogous to the analyses needed for isotropic materials. The derivations are repeated for the sake of completeness of presentation. From (7.3) follows

$$\frac{\partial u}{\partial z} = \{\varepsilon\}^{T}[C]\frac{\partial\{\varepsilon\}}{\partial z} \tag{7.4}$$

with [C] symmetric and independent of z .

From the strain/displacement relation $\{\varepsilon\} = [B]\{D_j\}$ follows

$$\frac{\partial\{\varepsilon\}}{\partial z} = \frac{\partial[B]}{\partial z}\{D_j\} + [B]\frac{\partial\{D_j\}}{\partial z} \tag{7.5}$$

The partial derivative $\partial[B]/\partial z$ is local, meaning that this term will only be non–zero if z changes the nodal positions of the actual element. The partial derivative $\partial\{D_j\}/\partial z$ is global and we traditionally determine it by the pseudo load approach, i.e. from the system equilibrium $[S]\{D\} = \{A\}$, which, with design–independent loads $\partial\{A\}/\partial z = \{0\}$, gives

$$[S]\frac{\partial\{D\}}{\partial z} = -\frac{\partial[S]}{\partial z}\{D\} \tag{7.6}$$

Then we get the gradient of the design element nodal displacements $\partial\{D_j\}/\partial z$, directly contained in $\partial\{D\}/\partial z$.

Very simply, we apply the semi–analytic approach, see Haftka, Gürdal and Kamat (1990),

$$\frac{\partial[S]}{\partial z} \approx \frac{[S(z + \Delta z)] - [S(z)]}{\Delta z} \tag{7.7}$$

This technique is a powerful tool that does not require the details of the analytical evaluation of $\partial[S]/\partial z$ while it does not require the extreme computational efforts of an overall finite difference approach by $\partial\{D\}/\partial z \approx (\{D(z + \Delta z)\} - \{D(z)\})/\Delta z$.

We also use the difference technique to obtain the gradient of the strain/displacement matrix

$$\frac{\partial[B]}{\partial z} \approx \frac{[B(z + \Delta z)] - [B(z)]}{\Delta z} \tag{7.8}$$

In many cases, central differences are preferred to (7.7) and/or (7.8). The relative changes of z are of the order 10^{-3} .

Only at the critical points, i.e. points of large u , do we perform the analysis to determine $\partial u/\partial z$. It should also be remembered that only elements connected to the shape to be designed are included in the semi–analytical analyses (7.7) and (7.8), see fig. 7.2. So what may at first seem overwhelming is in fact a fast and simple computer analysis.

Lastly, in the sensitivity analysis we need the gradient of model area $\partial A/\partial z$. Again the difference approach is applied

$$\frac{\partial A}{\partial z} \approx \frac{A(z + \Delta z) - A(z)}{\Delta z} = \frac{1}{\Delta z}\sum_{j}\left(A_j(z + \Delta z) - A_j(z)\right) \tag{7.9}$$

with element summation only for connections to the shape of design.

The optimization procedure used for each design improvement is **linear programming with move limits**. The general technique of converting the Min–max problem into a pure min problem is to introduce the further unknown u_{max} and then constraint the strain energy density everywhere to be less than u_{max}. Details are given in Kristensen and Madsen (1976) and in many other papers, so we will omit them here.

7.4 EXAMPLES

The Model Problem. A number of interesting parameter studies can be performed with the programs developed. We shall concentrate on the **biaxial single–load case** illustrated in fig. 7.1. External stress ratios σ_1/σ_2 of 3/2 and 3/1 are taken as examples. Material with the x_1, x_2 **axes as orthotropic axes** are assumed, and we shall then vary the degree of orthotropy. From the law of mixture (see Jones (1975)) the material parameters in table 7.1 are taken as examples.

Material #	$E_1/10^{11}$ Pa	$E_2/10^{11}$ Pa	ν_{12}	$G_{12}/10^{11}$ Pa	E_1/E_2
I, Isotropic	1.0	1.0	0.3	0.3846	1.0
II, 5% fiber	3.450	1.052	0.3	0.4044	3.281
III, 10% fiber	5.900	1.109	0.3	0.4264	5.322
IV, 20% fiber	10.80	1.244	0.3	0.4784	8.683
V, 30% fiber	15.70	1.416	0.3	0.5448	11.08
VI, Extreme	60.914	1.4503	0.3	0.13778	42.00

Table 7.1: Applied material data.

Dependence on Element Model. It is well known and should always be borne in mind that we are optimizing the finite element model. Thus, not only the analysis result but also the resulting optimal shape could be expected to change with the model. We therefore performed a study of the influence of model refinement, in terms of "better" mesh and/or "better" elements. A better mesh is obtained with more elements, i.e. more degrees of freedom and more computer time. Better elements relative to the simple three–node, constant strain triangles are the six–node, linear strain triangles.

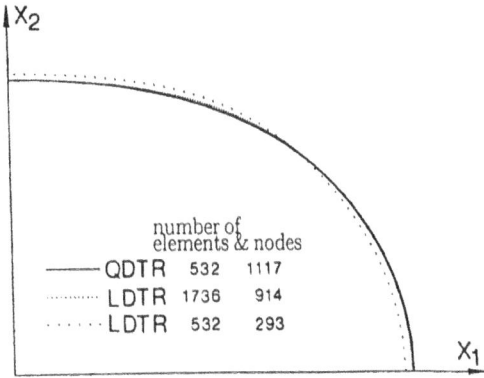

Fig. 7.3: Resulting optimal shapes with three different finite element models.

No unexpected results were obtained and the shape information from simple models was not changed in principle. Thus, refined models are not needed, at least not before the final iterations. More detailed optimal shapes were obtained with six–node triangles, but compared at equal computer time, the three–node triangles seems to be "the winner". In fig. 7.3 we show the results of comparative studies based on the strongly orthotropic material VI in table 7.1. Note that even with the same number of nodes, the computer time for the model with six–node elements will be about four times greater.

The Design History. For all the examples the history of iteration is very much the same. The necessary number of iterations is between 5 and 10, naturally depending on the convergence test.

In fig. 7.4 we show the design history with a large decrease in the maximum strain energy density especially in the first iterations. In fig. 7.5 the distribution of strain energy density along the shape of design is shown, and we notice that for this geometrically unconstrained problem the optimal design have uniform strain energy density.

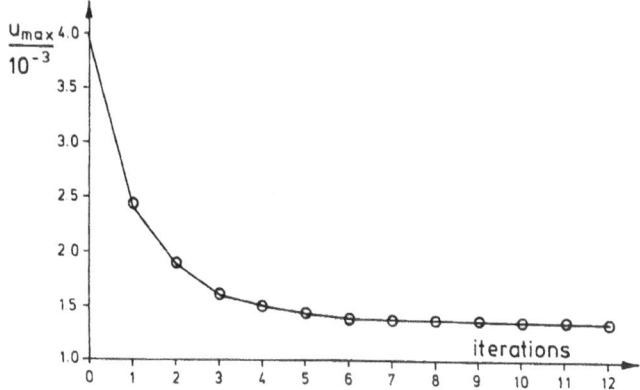

Fig. 7.4: Design history for the objective, i.e. maximum strain energy density.

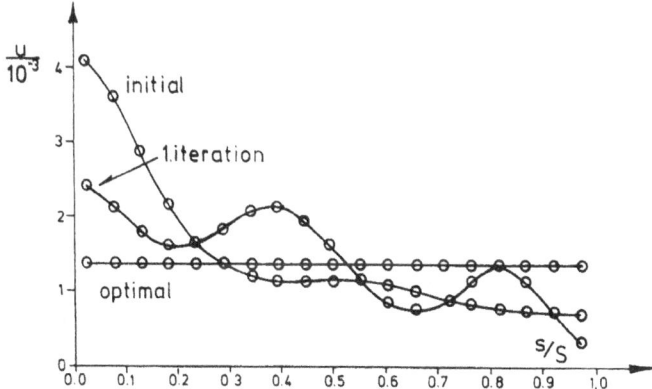

Fig. 7.5: Distribution of strain energy density along the shape of design for three different designs.

Influence of Degree of Orthotropy. In fig. 7.6, we show the optimal shape designs corresponding to the materials I–V in table 7.1. The external stress ratio σ_1/σ_2 is $3/2$.

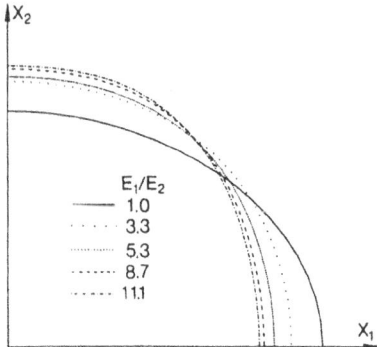

Fig. 7.6: Optimal boundary shapes for different "degrees" of orthotropy.

The corresponding values of the design parameters are given in table 7.2. These values relate to the modelling with linear strain elements, 532 elements and 1117 nodes. Refined models confirm the results. (For these results the power parameter n could in fact be fixed to 2, i.e. a normal elliptic basic curve). In table 7.3 we compare the optimal shape design with the initial designs, which for all cases were the isotropic optimal shape design.

Through the present study we have seen that the earlier methods for optimal shape design with isotropic materials also work effectively when the design is based on the use of orthotropic materials. The recommended methods are – global design description – finite element analysis – semianalytical sensitivity analysis – linear programming.

Mate-rial	C_0–ellipse parameters			Design parameters					$\dfrac{\Delta u_{SHAPE}}{u_{max}} \cdot 100\%$
	a	b	n	$c_1/10^{-5}$	$c_2/10^{-5}$	$c_3/10^{-5}$	$c_4/10^{-5}$	$c_5/10^{-5}$	
I	3.0092	2.0415	1.9908	0	0	0	0	0	
II	2.7141	2.2999	1.9987	4721.2	331.24	–231.46	–31.108	–2.8422	0.194
III	2.5529	2.3420	2.0306	5349.2	–212.04	–433.04	32.934	–7.7518	0.335
IV	2.4563	2.4095	1.9688	8115.8	–870.03	–379.46	146.15	0	1.403
V	2.4109	2.4387	2.0895	5324.7	–1326.5	–451.93	223.33	–68.25	0.376

Table 7.2: Resulting optimal design parameters and non uniformity of energy at the shape.

E_1/E_2	3.3	5.3	8.7	11.1
material	II	III	IV	V
$\dfrac{u_{max\ initial}}{u_{max\ optimal}}$	1.8	1.6	1.5	1.2

Table 7.3: Increase in maximum energy density if isotropic optimal shape were applied.

Comparisons related to the refinement of the analysis show that even a course finite element analysis gives a rather accurate optimal design. The design description with a few global design parameters is the basis for this possibility, and is generally of major importance.

The objective of minimum–maximum energy density could easily be substituted by an alternative objective, say an experimentally verified strength criterion. Problem formulations with multiple load cases and with constraints on displacements as well as on geometry could also be incorporated, but the results would then naturally be of only specific interest.

It should be noted that the resulting uniform energy density along the optimal shape is not enforced in the formulation and cannot be expected with additional load cases and/or additional constraints.

7.5 SUMMARY

The main conclusions of this section are:

- With localized design constraints (as in section 4.4), the tools of mathematical programming are necessary. We have successfully applied sequential linear programming.

- Of major importance is the design description, and the global description (4.14) has again proven to be efficient.

- The optimal shape is not very sensitive to the finite element modelling – fig. 7.3.

- Iteration history is very stable – fig. 7.4.

260

8. Conclusion

A subjective engineering view on results obtained with material optimization is given. The more general aspects have been focussed on, and as always when optimizations are performed, we learn a lot in relation also to the analysis and the sensitivity analysis.

Our goal is to design concurrently structural shape, pointwise density or thickness, material orientation, and the detailed constitutive behaviour. However, certain aspects, such as material orientation and material density, are given by optimality criteria that hold generally. Thus, advantageous decouplings are possible and should be utilized.

The present research focus on strength constraints, on the design parametrization, and on the inverse homogenization problem. Some preliminary results from these studies are presented, and a number of further results are expected in the near future.

All in all it would be fair to say that the use of advanced materials has put optimal design into a higher level of application, and has strengthened the need of the involved optimization techniques.

References

Bäcklund, J. and Isby, R. (1988) 'Shape optimization of holes in composite shear panels', Proc. IUTAM Symposium on Structural Optimization, Melbourne, Australia, 9–16.

Bendsøe, M.P. (1991) 'Optimal topology design and homogenization', to appear in Blanc, Raous, Suquet (eds.), Mechanics, Numerical Modelling and Dynamics of Materials, CNRS.

Bendsøe, M.P. (1994) 'Methods for the optimization of structural topology, shape and material', Mathematical Institute, Techn. Univ. of Denmark, 299 p.

Bendsøe, M.P., Guedes, J.M., Haber, R.B., Pedersen, P. and Taylor, J.E. (1994) 'An analytical model to predict optimal material properties in the context of optimal structural design', J. Appl. Mech. (to appear).

Bert, C.W. (1977) 'Optimal design of a composite–material plate to maximize its fundamental frequency', J. Sound and Vibration 50, 229–239.

Cheng, G. and Pedersen, P. (1994) 'Sufficiency conditions for optimal design based on extremum principles of mechanics' (being reported).

Dybbro, J.D. and Holm, N.C. (1986) 'On minimization of stress concentration for three–dimensional models', Computers & Structures 4, 637–643.

Haftka, R.T., Gürdal, Z. and Kamat, M.P. (1990) Elements of structural optimization, 2nd ed., Kluwer, 396 p.

Hammer, V.B. (1994) 'Strength optimization by fibre orientation', Solid Mechanics, Techn. Univ. of Denmark.

Jog, C.S., Haber, R.B. and Bendsøe, M.P. (1992) 'Topology design using a material with self–optimizing microstructure', IUTAM Symposium on Optimal Design with Advanced Materials, Lyngby, Denmark.

Jones, R.M. (1975) Mechanics of Composite Materials, 2, McGraw–Hill, New York, N.Y., 355 p.

Kristensen, E.A. and Madsen, N.F. (1976) 'On the optimum shape of fillets in plates subjected to multiple in–plane loading cases', Int. J. Numer. Meth. Engng. 10, 1007–1019.

Lee, M.S., Kikuchi, N. and Scott, R.A. (1989) 'Shape optimization in laminated composite plates', Comp. Meth. in Appl. Mech. and Eng. 72, 29–55.

Lekhnitskii, S.G. (1981) Theory of elasticity of an anisotropic body, Mir Publishers, Moscow, 430 p.

Masur, E.F. (1970) 'Optimum stiffness and strength of elastic structures', J. of the Engineering Mechanics Div., ASCE, EM5, 621–649.

Muc, A. (1988) 'Optimal fibre orientation for simply–supported, angle–ply plates under biaxial compression', Composite Structures 9, 161–172.

Pedersen, P. (1986) 'Minimum flexibility of non–harmonic loaded laminated plates', in R. Pyrz (ed.), Mechanical Characterization of Fibre Composite Materials, AUC, Denmark, pp. 182–196.

Pedersen, P. (1987) 'On sensitivity analysis and optimal design of specially orthotropic laminates', Eng. Opt. 11, 305–316.

Pedersen, P. (1988) 'Design for minimum stress concentration – some practical aspects', in G.I.N. Rozvany and B.L. Karihaloo (eds.), Structural Optimization, Kluwer, pp. 225–232.

Pedersen, P. (1989) 'On optimal orientation of orthotropic materials', Structural Optimization 1, 101–106.

Pedersen, P. (1990a) 'Bounds on elastic energy in solids of orthotropic materials', Structural Optimization 2, 55–63.

Pedersen, P. (1990b) 'Combining material and element rotation in one formula', Comm. in Appl. Num. Meth. 6, 549–555.

Pedersen, P. (1991) 'On thickness and orientational design with orthotropic materials', Structural Optimization 3, 69–78.

Pedersen, P. (Ed.) (1993) 'Optimal design with advanced materials', Proc. IUTAM Symp., Elsevier, 515 p.

Pedersen, P. (1994) 'Simple transformations by proper contracted forms', Solid Mechanics, Techn. Univ. of Denmark, 11 p.

Pedersen, P. and Hammer, V.B. (1994) 'On global design description for orientational strength optimization', Solid Mechanics, Techn. Univ. of Denmark, 9 p.

Pedersen, P. and Laursen, C.L. (1982–83) 'Design for minimum stress concentration by finite elements and linear programming', J. Struct. Mech. 10, 243–271.

Pedersen, P. and Taylor, J.E. (1993) 'Optimal design based on power–law non–linear elasticity', in P. Pedersen (ed.) loc. cit., 51–66.

Pedersen, P., Tobiesen, L. and Jensen, S.H. (1992) 'Shapes of orthotropic plates for minimum energy concentration', Mechanics of Structures and Machines 20, 4.

Poulsen, H.H. (1990) 'Optimal material–orientation', Thesis for M.Sc., Solid Mechanics, DTH (in Danish).

Prager, W. and Shield, R.T. (1967) 'A general theory of optimal plastic design', J. Appl. Mech. 34, 184–186.

Sigmund, Ole (1993) 'Materials with prescribed constitutive parameters: An inverse homogenization problem', Int. J. Solids Structures (to appear).

Sigmund, Ole (1994) 'Tayloring materials with prescribed elastic properties', DCAMM, 23 p.

Thomsen, Jan (1991) 'Optimization of composite discs', Structural Optimization 3, 89–98.

Tobiesen, L. and Jensen, S.H. (1990) 'Optimal hole shape in orthotropic plate', Solid Mechanics, DTH (in Danish).

Wasiutynski, Z. (1960) 'On the congruency of the forming according to the minimum potential energy with that according to equal strength', Bull. de L'Academie Polonaise des Sciences, Serie des sciences technique, VIII, 6, 259–268.

Wolfram, S. (1991) 'Mathematica: a system for doing mathematics by computer', Addison–Wesley Publ. Co., 961 p.

INTEGRATED OPTIMIZATION OF INTELLIGENT STRUCTURES

Abdon E. Sepulveda
Mechanical, Aerospace, and Nuclear Engineering Department
University of California, Los Angeles
Los Angeles, CA 90024

1. INTRODUCTION

An intelligent or smart structure is a system that consists of a structural system, sensors, actuators and some sort of signal processing to allow the structure to respond in a desired manner to stimulus. These active systems have the ability to adapt to different design conditions, thus providing a greater design flexibility, as opposed to a pure passive structure.

To perform design and analysis in this field requires detailed knowledge of deformable solids, structural dynamics, materials, electronics and control theory. This cross disciplinary aspect makes the study and design of intelligent structures a very stimulating field. The main objective of this work is to present the design of the control/structural system as an integrated problem in which optimization techniques can be used to obtain an optimal design.

1.1 The Concept of an Active Structure

The main components of an active structural system are: the structure, a system of sensors, and a system of actuators. These three components are shown schematically in Figure 1, for the case of a cantilever beam with piezoelectric sensors and actuators. The usual way of implementing the control system is to use a feedback control law, that is the signal measured by the sensor (e.g., displacements, velocities) is processed using electronic devices and then fedback to the actuators, which in turn induce loads in the structure according to a prescribed design criterion.

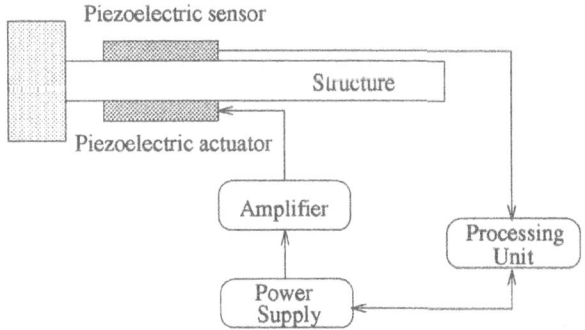

Schematics of a Feedback Control System

From the previous schematic, the main differences between an intelligent structure and a traditional passive structure are easily seen. For the passive case, the structure must meet, by itself, all design criteria, and therefore must be designed for the worst possible conditions (safety factors), which usually leads to over-designed, and redundant structures. In the active case, on the other hand, the structural behavior of the passive structure is coupled with a corrective active control effect, which means that the structure does not have to satisfy all design requirements by itself. Moreover, with a feedback control law, even

J. Herskovits (ed.), Advances in Structural Optimization, 263–315.
© 1995 *Kluwer Academic Publishers.*

unknown conditions can be corrected within the useful range of the controllers. The main drawback is that the system is dependent on electronic devices (hardware) which require an external source of energy and therefore reliability becomes a crucial design issue.

1.2 Design of Active Structures - Background

Conventionally structural design is independent of controller design, namely, the structure is designed first satisfying structural constraints and then a control law is found for the fixed structure minimizing some kind of control performance index. The structure is designed with constraints on allowable stresses, displacements at given nodes, natural frequencies, etc. When the structure or plant, i.e. the geometry, sizing and materials have been selected, the control system is designed with constraints on closed-loop eigenvalues, control effort, location of sensors and actuators, reliability, etc. This sequential approach generates a final design that cannot ensure the best performance of the overall system since the dynamic interaction between the two systems is not directly considered in the design process. Several works (see for example Refs. 1 and 10) have shown that slight structural modification can lead to a considerable improvement in the control system performance. The natural approach is then to integrate the structural design and the control design into a single design process. There has been a considerable effort to integrate the design optimization of structures and control systems in order to achieve a better performance and directly handle cross coupling effects and dynamic interactions between the two systems. During the last ten years a number of papers have been written in this area. If what follows a brief summary of the main results for the case of linear control laws based on output feedback or state feedback is given.

For the case of output feedback, several studies have been made where the structural variables and the control gains are treated simultaneously as strictly independent design variables in the optimization (Refs. 1-9). On the other hand, in the case of full state feedback control, a sequential approach is usually adopted in which the control gains are determined by solving Riccati equations corresponding to the changing structural system during design iterations (Refs. 10-16). When the gain variables are determined by solving Riccati equations for a fixed plant, they implicitly become dependent design variables and the resulting design optimization is constrained to a subspace where the optimality conditions of a control subproblem are satisfied. The tendency to subordinate gains to a dependent variable status can be attributed to the fact that for system models with a large number of degrees of freedom, the gain matrix $[H]$ contains prohibitively large numbers of independent design variables (i.e., $n_a \times 2n$ control design variables, where n_a is the number of actuators and n is the number of degrees of freedom in the structural model).

In Refs. 1-9, both structural and control variables are treated as independent in the optimization. Output feedback is adopted as a control law such that the number of elements in the feedback gain matrix is relatively small. As a result, the gain elements along with structural design variables can be directly treated as independent design variables. In Ref. 1 collocated direct velocity output feedback, which is similar to a viscous dashpot, is used as a controller. The viscous damping coefficient is minimized, with constraints on the closed-loop eigenvalues, by allowing small changes of the structural dimensions.

In Refs. 2-4, in addition to structural dimensions and control gains, sensor/actuator locations are also used as design variables. Homotopy and sequential linear programming algorithms are used to optimize either structural mass, robustness, or eigenvalue sensitivity. Also in Ref. 2, the state feedback control case is cast into a similar simultaneous optimization form by using elements of weighting matrices of the LQR (linear quadratic regulator) as independent design variables, but only the output feedback case is illustrated by giving a numerical example.

In Refs. 5 and 6, collocated output feedback is chosen as the control law to optimize a structural/control system. In addition to the structural sizing variables and control feedback gains, lumped nonstructural masses are treated simultaneously as independent design variables. Harmonic dynamic loadings are applied and a variety of constraints are considered including natural frequencies, static displacement and stress, dynamic displacements, and actuator forces. Due to the characteristics of the collocated sensor/actuator pairs, system stability can be ensured by imposing side constraints on the control gains. In Refs. 7 and 8, noncollocated output feedback is chosen instead of collocated sensor/actuator pairs, and constraints on the stability (real parts of closed-loop eigenvalues) are included. In Ref. 9 several output feedback control laws are used in the case of stochastic disturbances with constraints on the allowable mean square deflection or control effort.

In Ref. 10, LQR theory is used for the case of state feedback control. In the LQR problem, once the weighting matrices in the quadratic performance index are chosen, the control law (or all the closed-loop characteristics) is determined by solving a nonlinear matrix Riccati equation. Then, the choice of the weighting matrices in the LQR problem is very important and two problems arise. The first is how to select meaningful weighting matrices, and the second is the solution of the Riccati equation for problems involving a large number of degrees of freedom. In Ref. 10 the sum of the mass of the structure and the quadratic control effort was minimized using structural parameters as design variables. A different approach was presented in Ref. 11 in which feedback controllers for maneuvers and vibration control of flexible structures were designed using a eigenspace optimization approach.

In Ref. 12, the structural mass is minimized using structural design variables while satisfying open-loop frequency constraints and then the LQR problem is solved for the fixed structure with given weighting matrices. Here weighting matrices are chosen such that the quadratic performance index represents the absolute weighted sum of kinetic, strain, and potential energies, and the effect of the relative weighting of these energy terms is discussed. In Refs. 13 and 14, structural variables are optimized with constraints on the closed-loop eigenvalues and modal damping ratios, then the LQR problem is solved for the fixed structure with given weighting matrices (identity matrices in this case). In Refs. 15-18, the Frobenious norm of the gain matrix is introduced as either an objective or a constraint.

Reference 17 points out the difficulties of simultaneous structural/control design and suggests optimization of the closed-loop system using only structural tailoring. In this case, the objective of structural tailoring is to maximize modal stiffness in order to minimize control effort. The control law is determined by solving the Riccati equation and the weighting matrices for the LQR problem are similar to those in Ref. 12 except that only two independent weighting coefficients are used instead of three.

In Refs. 20-21, the weighted sum of the structural mass and control system performance is minimized. In Ref. 22, in addition to the structural design variables, the coefficients of the weighting matrices and orientation of an actuator are also used as design variables and results are shown for a two bar truss example. In Ref. 23, a nested optimization method is presented for finding the state feedback control which minimizes the total equivalent mass of the system (structural mass plus a mass effect of the control effort). Structural dimensions and coefficient of the control effort are optimized simultaneously to minimize the objective with a constraint on the mean square of the response. The control law is determined by solving the Riccati equation with a new set of weighting matrices (since the coefficient of the control effort is optimized, the performance index is updated at each iteration).

The effect of different objective functions on the optimum structural and control design was studied in Ref. 24. The multiobjective case was studied in Ref. 25 using concepts of game theory and other techniques are given in Ref. 26. The effect of the passive damping on closed loop eigenvalues at the optimum design was studied in Ref. 27.

An integrated approach with transient response constraints is given in Refs. 28-34. In Ref. 28, the feedback is implemented using nonlinear controllers. In Refs. 30-31 a control gain linking scheme is introduced that fully integrates the control and structural optimization for the case of state feedback.

In Refs. 29 and 33, locations of actuators and sensors are treated in terms of (0,1) discrete variables. A utopian multiobjective function containing structural mass, control effort and number of actuators is minimized by treating structural variables, (0,1) actuator/sensor location variables and open-loop gains as independent design variables in the optimization.

2. STRUCTURAL/CONTROL SYSTEM DESCRIPTION

The design of actively controlled structures requires a detailed model that describes the structural behavior as well as the control-structure interaction. The modelling of the structure is usually based on a finite element formulation, thus allowing the modelling of general structures with complicated shapes. In this section, the basic concepts necessary for structural/control design will be presented.

2.1 Equations of Motion

Using a finite element representation for the structural system, the dynamic equations of motion for the control/structural system can be written as

$$[M]\{\ddot{q}(t)\} + [C]\{\dot{q}(t)\} + [K]\{q(t)\} = [b]\{U(t)\} + \{f(t)\} \tag{1}$$

where $[M]$ is the system mass matrix, $[C]$ is the structural (passive) viscous damping matrix, $[K]$ is the structural stiffness matrix, $\{f(t)\}$ is the vector of nodal external forces, $[b]$ is the matrix that places the control forces (actuator forces) at nodal degrees of freedom, $\{U(t)\}$ is the vector of actuator forces, and $\{q(t)\}$ is the vector of nodal displacements and rotations. The number of degrees of freedom for the model will be denoted by n and the number of actuators by n_a. With this notation, the matrices $[M]$, $[C]$ and $[K]$ are $n \times n$, $[b]$ is an $n \times n_a$ matrix, and $\{U(t)\}$ is a vector of dimension n_a.

The system mass matrix $[M]$ includes the mass of the structure and the mass of actuators and sensors in the system, usually modelled as lumped non-structural masses. The pre-assigned damping inherent to the structure (matrix $[C]$) is usually assumed to be a proportional damping matrix or it is completely neglected.

The second order equation (1) is transformed to a first order differential equation in the state space, by defining the state space variables as

$$\{X(t)\} = \begin{Bmatrix} \{X_1(t)\} \\ \{X_2(t)\} \end{Bmatrix} = \begin{Bmatrix} \{\dot{q}(t)\} \\ \{q(t)\} \end{Bmatrix} \tag{2}$$

Equation (1) can be written as

$$[M]\{\dot{X}_1(t)\} + [C]\{\dot{X}_2(t)\} + [K]\{X_2(t)\} = [b]\{U(t)\} + \{f(t)\} \tag{3}$$

Appending to this equation the identity

$$[M]\{\dot{X}_2(t)\} = [M]\{X_1(t)\} \tag{4}$$

the first order state space equations of motion can be written as

$$[M^*]\{\dot{X}(t)\} + [K^*]\{X(t)\} = [B]\{U(t)\} + \{F(t)\} \tag{5}$$

where

$$[M^*] = \begin{bmatrix} [0] & [M] \\ [M] & [C] \end{bmatrix} \tag{6a}$$

$$[K^*] = \begin{bmatrix} -[M] & [0] \\ [0] & [K] \end{bmatrix} \tag{6b}$$

$$[B] = \begin{bmatrix} [0] \\ [b] \end{bmatrix} \tag{6c}$$

and

$$\{F(t)\} = \begin{Bmatrix} \{0\} \\ \{f(t)\} \end{Bmatrix} \tag{6d}$$

In classical control theory, Eq. (5) is written as

$$\{\dot{X}(t)\} = [A]\{X(t)\} + [B_C]\{U(t)\} + \{F_C(t)\} \tag{7}$$

where

$$[A] = -[M^*]^{-1}[K^*] = \begin{bmatrix} -[M]^{-1}[C] & -[M]^{-1}[K] \\ [I] & [0] \end{bmatrix} \tag{8a}$$

$$[B_C] = [M^*]^{-1}[B] = \begin{bmatrix} [M]^{-1}[b] \\ [0] \end{bmatrix} \tag{8b}$$

and

$$\{F_C(t)\} = [M^*]^{-1}\{F(t)\} = \begin{Bmatrix} [M]^{-1}\{f(t)\} \\ \{0\} \end{Bmatrix} \tag{8c}$$

Representation (8) is more appropriate when control theory concepts such as observability and controllability of the system are to be analyzed. On the other hand, representation (5) has the advantage that the matrix $[M^*]$ and $[K^*]$ are explicitly known, in terms of structural parameters so they provide a better basis for constructing approximations, which will be used extensively during the optimization process.

2.2 Sensors

Sensors are devices that allow direct or indirect measurements of displacement and/or velocity at given degrees of freedom. The output of the sensors is denoted by $\{Y(t)\}$ such that

$$\{Y_P(t)\} = [C_P]\{q(t)\} \tag{9a}$$

denotes the measured outputs for displacements, and

$$\{Y_v(t)\} = [C_V]\{\dot{q}(t)\} \tag{9b}$$

denotes the measured outputs for velocities. The matrix $[C_P]$ locates the displacement sensors at nodal degrees of freedom and $[C_V]$ is a matrix locating the velocity sensors at nodal degrees of freedom. In state space form, the sensors outputs can be expressed as

$$\{Y(t)\} = [C_C]\{X(t)\} \tag{10}$$

where

$$[C_C] = \begin{bmatrix} [0] & [C_P] \\ [C_V] & [0] \end{bmatrix} \tag{11}$$

The number of velocity sensors is denoted by n_v, the number of position sensors by n_p and the total number of sensors by n_s. With this notation $[C_V]$ is an $n_v \times n$ matrix and $[C_P]$ and $n_p \times n$ matrix.

2.3 Model Reduction

It is normal in practice to reduce the structural model given by Eq. (1) in order to consider only the dominant structural modes. For this purpose, the following coordinate transformation is introduced

$$\{q(t)\} = [\phi]\{\eta(t)\} \tag{12}$$

where $\{\eta(t)\}$ is the vector of modal participation coefficients and $[\phi]$ is the $n \times n$ modal matrix (natural modes). Assuming proportional damping, Eq. (1) can be decoupled as

$$[m]\{\ddot{\eta}\} + [c]\{\dot{\eta}\} + [k]\{\eta\} = [\phi]^T[b]\{U\} + \{\phi\}^T\{f\} \tag{13}$$

where

$$[m] = [\phi]^T[M][\phi] = [I] \tag{14}$$

assuming that the modes are normalized with respect to the mass matrix,

$$[c] = [\phi]^T[C][\phi] = [2\xi\omega] \tag{15}$$

and

$$[k] = [\phi]^T[K][\phi] = [\omega^2] \tag{16}$$

The matrices $[m]$, $[c]$, and $[k]$ are diagonal square matrices, ω_i, $i = 1,...,n$ are the natural structural frequencies and ξ_i, $i = 1,...,n$ are the modal damping ratios (passive) for the structure.

Equation (13) represents a system of n uncoupled equations, one for each structural mode. For a reduced model, only N_R modes are retained and the equations associated to these modes are

$$\ddot{\eta}_i + 2\xi_i\omega_i\dot{\eta}_i + \omega_i^2\eta_i = \{\phi\}_i^T [b]\{U\} + \{\phi\}_i^T \{f\} \quad i = 1,...,N_R \tag{17}$$

These equations have the same structure as Eq. (1) and can be transformed to state space representation using the techniques described in the previous section.

A model reduction has the advantage of reducing the dimension of the gain matrices for feedback control but only the retained modes can be controlled and stabilized as will be discussed in subsequent sections. The modes not retained in the model are still excited by the control forces and therefore they can become unstable due to the spillover effect. An approach to account for the unmodeled dynamics in order to avoid spillover effect will be discussed later in this chapter.

2.4 Types of Feedback Control Laws

The control laws discussed in this section apply to the full model or the reduced model. For simpler notation they will be written for the full model case.

There are two basic approaches to feedback control for linear systems. The first and simplest approach is direct output feedback in which the vector of control forces is considered proportional to the measured outputs, i.e.

$$\{U(t)\} = -[H]\{Y(t)\} \tag{18}$$

The matrix $[H]$ is the gain matrix, whose dimension is $n_a \times n_s$, where n_a is the number of actuators and n_s is the number of sensors. From Eq. (10), the vector of control forces can be related to the state vector $\{X(t)\}$ by

$$\{U(t)\} = -[H][C_c]\{X(t)\} \tag{19}$$

which shows that only some components of the nodal displacements and velocities are fed back into the system. The second type of feedback control is full state feedback, in which the complete state is fed back

$$\{U(t)\} = -[H]\{X(t)\} \tag{20}$$

Clearly this control law incorporates all the information available for the system and therefore, allows more design flexibility. The gain matrix has $n_a \times 2n$ elements, where n is the number of degrees of freedom. For a practical implementation, there are only a limited number of sensors, therefore, it is not possible to measure all the components of the state vector. In order to implement a state feedback control law, the state vector has to be estimated from the measured output using an observer, which is in itself a dynamic system. The construction of such an observer will be discussed later in this chapter.

Relations (20) can be written, in general terms, separating position and velocity components of the state vector as

$$\{U(t)\} = -[H_P]\{q(t)\} - [H_V]\{\dot{q}(t)\} \tag{21}$$

where $[H_P]$ and $[H_V]$ denote the position and velocity gain matrices, respectively. Using Eq. (21), the closed loop equations of motion in second order form become

$$[M]\{\ddot{q}\} + [C + b\, H_V]\{\dot{q}\} + [K + b\, H_P]\{q\} = \{f\} \tag{22}$$

from the previous relation, it is observed that the matrix $[H_V]$ provides damping and the matrix $[H_P]$ provides stiffness to the structure.

For future reference, the following notation is used

$$[C_A] = [C] + [b]\,[H_V] \quad \text{(augmented damping matrix)} \tag{23a}$$

and

$$[K_A] = [K] + [b]\,[H_P] \quad \text{(augmented stiffness matrix)} \tag{23b}$$

depending on the type of control law adopted, the matrices $[C_A]$ and $[K_A]$ may not be symmetric. Relation (19) gives similar expressions as Eqs. (21-23) with the gain matrices replaced by $[H_P][C_P]$ for the position gain matrix and $[H_V][C_V]$ for the velocity gain matrix.

From the control point of view, the goal is to design the matrices $[H_P]$ and $[H_V]$ according to a certain objective. The three most common approaches are; (1) pole placement, in which the gains are determined such that a preassigned stability of the system is achieved, (2) optimal control theory, which is mostly used for state feedback and (3) direct optimization in which the gains are used directly as design variables.

2.5 Stability

The closed loop equations (Eq. (22)) in state space form, with no external disturbances, are given by

$$\{\dot{X}(t)\} = [A_C]\{X(t)\} \tag{24}$$

where (see Eq. (8a))

$$[A_C] = \begin{bmatrix} -[M]^{-1}[C_A] & -[M]^{-1}[K_A] \\ [I] & [0] \end{bmatrix} \tag{25}$$

The solution to Eq. (24) is given by

$$\{X(t)\} = e^{[A_C]t}\{X_0\} \tag{26}$$

where $\{X_0\}$ denotes the state of the system at $t = 0$.

Consider the following eigenvalue problems

$$[A_C]\{\phi\}_i = \lambda_i\{\phi\}_i \quad i = 1, ..., 2n \quad \text{right eigenproblem} \tag{27a}$$

$$\{\chi\}_i^T[A_C] = \lambda_i\{\chi\}_i^T \quad i = 1, ..., 2n \quad \text{left eigenproblem} \tag{27b}$$

The left and right eigenvectors satisfy the orthogonality condition $\{\chi\}_i^T\{\phi\}_j = 0$ if $i \neq j$. Using this property, the solution of Eq. (24) can be expressed in modal expansion as

$$\{X(t)\} = \sum_{i=1}^{2n} \{\chi\}_i^T\{X_0\}\, e^{\lambda_i t}\{\phi\}_i \tag{28}$$

Since the eigenvectors and eigenvalues appear in complex conjugate pairs, the right hand side of Eq. (28) is a real number. For complex eigenvalues (Eqs. (27)), the following notation is adopted

$$\lambda_i = \sigma_i + j\,\omega_{di} \tag{29}$$

where the real part σ_i represents damping and the imaginary part ω_{di} is the damped frequency for the complex (closed loop) mode i ($j = \sqrt{-1}$).

The system is asymtotically stable if $\{X(t)\} \to 0$ as $t \to \infty$, which is equivalent to saying that $[A_C]$ is a stable matrix, which, in terms of the complex eigenvalues is equivalent to the condition that $\sigma_i \leq 0$ for all modes (see Eq. (28)). For a single mode i the following conditions apply

$\sigma_i > 0$ mode i is unstable

$\sigma_i < 0$ mode i is strictly stable

$\sigma_i = 0$ mode i is marginally stable

The gain matrix $[H]$ has then to be designed such that the system is stable, i.e. $\sigma_i \leq 0$, $i = 1, \ldots, 2n$. Usually the lower modes (lower damped frequencies) are constrained to be strictly stable and the higher modes to be marginally stable. The most commonly used measure of stability for a given mode is the damping ratio for the closed loop system, denoted by ξ_i and defined as

$$\xi_i = \frac{-\sigma_i}{\sqrt{\sigma_i^2 + \omega_{di}^2}} \tag{30}$$

Clearly, from the previous definitions the following conditions apply

$\xi_i < 0$ mode i is unstable

$\xi_i = 0$ mode i is marginally stable

$\xi_i > 0$ mode i is strictly stable

$\xi_i = 1$ mode i is critically damped (no vibration)

2.6 Controllability and Observability

Consider a generic dynamic system described by the following equations (see Eqs. 7 and 10)

$$\{\dot{X}\} = [A]\{X\} + [B_C]\{U\} + \{F_C\} \tag{31a}$$

$$\{Y\} = [C_C]\{X\} \tag{31b}$$

where the definitions of $[A]$, $[B_C]$ and $\{F_C\}$ are given in Eqs. (8). Clearly $\{X\} \in \mathbf{R}^{2n}$, $\{U\} \in \mathbf{R}^{n_a}$, $\{Y\} \in \mathbf{R}^{n_s}$.

If the external disturbance $\{f\}$ is such that

$$\{f(t)\} \in Q \subseteq \mathbf{R}^n \quad \forall\, t \geq 0 \tag{32}$$

under the assumption that Q is compact (i.e. closed and bounded), which is in general the case for most practical situations, the controllability and observability of the perturbed system (31) is equivalent to the controllability and observability of the disturbance-free system (Refs. 35 and 36)

$$\{\dot{X}\} = [A]\{X\} + [B_C]\{U\} \tag{33a}$$

$$\{Y\} = [C_C]\{X\} \tag{33b}$$

Controllability

A state $\{X(t)\}$ of system (33) is said to be controllable if there exists a control vector $\left[U(t')\big|_{0 \leq t' \leq t} \right]$ such that $\{X(t)\}$ can be steered from some initial state $\{X_0\}$. The system is said to be completely controllable if all the states can be reached from $\{X_0\}$.

According to this definition, complete controllability for (33) is equivalently characterized by any one of the following conditions (Refs. 36-38)

(1) $M_{\mu c}(A, B_C) = \left\{ \{X\} \mid [B_C]^T ([A]^T)^k \{X\} = \{0\}, k = 0, 1, \ldots, 2n - 1 \right\} = \{0\}$

   $M_{\mu c}(A, B_C)$ is called the uncontrollable subspace

(2)   Rank $[L_c] = 2n$, where $[L_c] = [[B_C], [A][B_C], \ldots, [A]^{2n-1}[B_C]]$

(3) $\det [L_c][L_c]^T \neq 0$

(4) $W_c(t) = \int_0^t [e^{[A]s}][B_C][B_C]^T [e^{[A]^T s}] ds, \quad t > 0$ is non singular.

In general, condition (3) is the most suitable for computational procedures if controllability constraints must be imposed. The matrix $[L_c][L_c]^T$ is symmetric and semi-positive definite for any matrix $[A]$ and $[B_C]$, therefore condition (3) can be replaced by

$$\mu_c = \lambda_{\min}([L_c][L_c]^T) > 0 \tag{34}$$

where $\lambda_{\min}$ is the smallest eigenvalue of $[L_c][L_c]^T$. In fact, μ_c can be used as a measure of controllability for the system (Ref. 38). The system is more controllable, the larger μ_c is. This measure may be interpreted as the minimum distance from the origin to the quadratic form related to $[L_c][L_c]^T$ and can be called the degree of controllability by analogy to the definition of degree of stability. The associated eigenvector of μ_c is the most controllable direction of the system.

Unfortunately, for large systems the solution of the eigenvalue problem associated to $[L_c][L_c]^T$ can be computationally very expensive, particularly if more than one eigenvalue must be constrained due to possible mode switching. An alternative approach is to compute the uncontrollable subspace directly using a modal representation of the system. For this purpose, consider the eigenvalue problem

$$[A]\{\chi\} = \alpha\{\chi\} \tag{35}$$

giving the eigenvalues $\alpha_1, ..., \alpha_n$ and the associated eigenvectors $\{\chi\}_1, ..., \{\chi\}_n$. Defining $[\Lambda]$ as the diagonal matrix formed by the eigenvalues and $[\chi]$ the modal matrix, the matrix $[A]$ is given by

$$[A] = [\chi][\Lambda][\chi]^{-1} \tag{36}$$

and

$$\{\dot{X}\} = [\chi][\Lambda][\chi]^{-1}\{X\} + [B_c]\{U\} \tag{37}$$

defining $\{\eta\} = [\chi]^{-1}\{X\}$ and $[\bar{B}] = [\chi]^{-1}[B]$ equation (37) is transformed to

$$\{\dot{\eta}\} = [\Lambda]\{\eta\} + [\bar{B}]\{U\} \tag{38}$$

Therefore, it is observed that mode k is controllable if and only if row k associated to the matrix $[\bar{B}]$ is not zero. A controllability index associated to mode k is then defined as

$$\mu_k = |\{\chi_k^{-1}\}^T [B_c]|^2 \tag{39}$$

where $\{\chi_k^{-1}\}^T$ is the k-th row of the matrix $[\chi]^{-1}$, and complete controllability of the system is given by the condition $\mu_c = \underset{k}{\text{Min }} \mu_k > 0$

For structural/control systems in which the uncontrolled structure has negligible damping ($[C] = [0]$), definition (39) gives a simple physical insight to the meaning of controllability. Recalling that matrices $[A]$ and $[B_c]$ are given by (see Eqs. 8a and 8b)

$$[A] = \begin{bmatrix} [0] & -[M]^{-1}[K] \\ [I] & [0] \end{bmatrix} \tag{40a}$$

$$[B_c] = \begin{bmatrix} [M]^{-1}[b] \\ [0] \end{bmatrix} \tag{40b}$$

it is easily seen that

$$\alpha_k = \pm j\,\omega_k \tag{41a}$$

$$\{\chi\}_k = \begin{Bmatrix} \mp\omega_k\{\phi\}_k \\ \{\phi\}_k \end{Bmatrix} \tag{41b}$$

where ω_k, $\{\phi\}_k$ are the corresponding frequency and mode shape for natural mode k. Equation (39) defines therefore, a controllability index for the k-th mode given by

$$\mu_k = \frac{\{\phi\}_k^T [b][b]^T \{\phi\}_k}{\{\phi\}_k^T [M]\{\phi\}_k} \tag{42}$$

From the previous expression, it is seen that $\mu_k = 0$ if $[b]^T\{\phi\}_k = 0$, which can only happen if all the actuators are located at the nodes of mode k. Therefore, a mode is uncontrollable if all the actuators are placed at the nodes of the mode.

Since higher modes are poorly modeled by the finite element model of the structure, controllability constraints for these modes should be relaxed or omitted, thus, for computational implementation, only some modes should be retained depending on the accuracy of the model. In summary, it is seen that the controllability depends on the modal behavior of the undamped structure and the placement of the actuators described by the matrix $[b]$.

Observability

Two states $\{X_1\}$ and $\{X_2\}$ in the state space X are called equivalent if for every $t > 0$, $[C_C][e^{[A]t}]\{X_1\} = [C_C][e^{[A]t}]\{X_2\}$. A state space X is called observable, if and only if, for any $\{X_1\}$ and $\{X_2\}$ in X, $\{X_1\}$ equivalent to $\{X_2\}$ implies that $\{X_1\} = \{X_2\}$.

Complete observability is characterized equivalently by any one of the following conditions

(1) $M_{\mu 0}(C_C, A) = \{\{X\} \mid [C_C][A]^k\{X\} = \{0\}, k = 0, 1, ..., 2n - 1\} = \{0\}$

   $M_{\mu 0}$ is called the unobservable subspace

(2) Rank $[L_0] = 2n$, where $[L_0] = \left[[C_C]^T, [A]^T[C_C]^T, ..., ([A]^T)^{2n-1}[C_C]^T\right]$

(3) det $[L_0][L_0]^T \neq 0$

(4) $W_0(t) = \int_0^t [e^{[A]^T s}][C_C]^T[C_C][e^{[A]s}]ds$, $t > 0$ is non singular

In analogy with controllability, the degree of observability of the system is defined as

$$\mu_0 = \lambda_{min}([L_0][L_0]^T) \tag{43}$$

Again, the computation of the degree of observability can be computationally burdensome, so it is necessary to use a modal representation of the system. From (31b) and the definition of $\{\eta\} = \{\chi\}^{-1}\{X\}$, the measured outputs are given by

$$\{Y\} = [C_C][\chi]\{\eta\} \tag{44}$$

So, mode k is observable if column k of $[C_C][\chi]$, i.e. $[C_C]\{\chi\}_k$, is not zero.

Considering that for a an actively controlled structure, the matrix $[C_C]$ defining the placement of the sensors is given by Eq. (11) and $\{\chi\}_k$ is given by Eq. (42b), the observability index for natural mode k is defined as

$$\mu_k^0 = \{\phi\}_k^T [\omega_k^2[C_V]^T[C_V] + [C_P]^T[C_P]]\{\phi\}_k \tag{45}$$

and the system is completely observable if $\mu_k^0 > 0$ $\forall k$.

and again only a reduced set of modes can be retained in this condition. In analogy with controllability, Eq. (45) shows that a mode is unobservable if all the sensors are placed at the nodes of the mode. In summary, it is seen that the observability of the system depends on the modal behavior of the undamped structure and the placement of the displacement and velocity sensors described by the matrices $[C_P]$ and $[C_V]$.

3. SIMULTANEOUS STRUCTURES/CONTROL DESIGN

Qualitatively, the simultaneous structure/control synthesis problem can be stated as follows: seek a design which minimizes some measure of system performance subject to the condition that all appropriate measures of the system behavior and all design variables remain within prescribes bounds. The foregoing problem statement can be quantified by formulating it as a general non-linear mathematical programming problem of the form

$$\text{Min} \quad f(y) \tag{46}$$

$$\text{s.t.} \quad g_j(y) \leq 0 \quad j = 1, ..., NCON$$

$$y_i^L \leq y_i \leq y_i^U \quad i = 1, ..., NDV$$

where NDV is the total number of design variables, $\{y\}^T = [y_1, y_2, ..., y_{NDV}]^T$, $NCON$ is the total number of constraints, g_j is the j-th behavior constraint, and y_i^L, y_i^U are lower and upper bounds for the i-th design variable, respectively.

In this problem, there are two types of design variables. The first group contains structural design variables such as cross-sectional dimensions for sizing and geometric variables for shape. The second group contains the control design variables which determine the gain matrices $[H_P]$ and $[H_V]$.

The set of constraints may include static stresses and displacement, dynamic (steady state, transient) displacements, accelerations and control forces, stability (damping ratio or real part of complex eigenvalues), natural frequencies, damped frequencies, control effort (energy), system mass, system reliability, controllability, observability, etc. For aerospace applications, the objective function is usually taken as the total system mass or weight, but any other performance measure can be easily considered with this formulation.

The set of variables that describe the control feedback law (i.e., matrices $[H_P]$ and $[H_V]$ and constraints depend on the type of feedback control law chosen. The next two sections discuss in detail the direct output feedback and state feedback cases.

3.1 Direct Output Feedback

For direct output feedback, as discussed in section 2.4, the actuator control forces are directly proportional to sensed displacements and velocities. The constants of proportionality are called the displacement gains for the sensed displacements and the velocity gains for the sensed velocities. If a sensor and actuator pair are at the same degree of freedom in the structure, the controller is called collocated. If the sensor and actuator are at different degrees of freedom then the controller is said to be noncollocated. While a control system composed entirely of collocated controllers is inherently stable, when the gains are constrained to be positive, the introduction of noncollocated controllers opens up the possibility of dynamic instability. There are two types of output feedback controller models used. The first is the axial controller which is modeled as a truss-spring type element. The axial controller is always collocated by definition. The second type of control element is the general controller which can implement in a collocated or noncollocated manner.

In general, for direct output feedback, the vector of control forces is given by

$$\{U(t)\} = -[H_P]\{Y_P(t)\} - [H_V]\{Y_V(t)\} \tag{47}$$

The dimension of $[H_P]$ is $n_a \times n_p$, where n_a is the number of actuators and n_p is the number of position sensors. The dimension of $[H_V]$ is $n_a \times n_v$, where n_v is the number of velocity sensors. Since the number of actuators and sensors is usually small, the elements of the gain matrices $[H_P]$ and $[H_V]$ can be used directly as design variables.

The collocated axial controller is implemented in the same fashion as a truss type element for the position gain, but in parallel with an axial viscous damper type element for the velocity gain. The element level control gain matrices are generated in element coordinates as

$$[H_P]^e = \begin{bmatrix} h_P & -h_P \\ -h_P & h_P \end{bmatrix} \tag{48a}$$

$$[H_V]^e = \begin{bmatrix} h_V & -h_V \\ -h_V & h_V \end{bmatrix} \tag{48b}$$

The element level gain matrices are then transformed into global coordinates by the transformation:

$$[H]^g = [T]^T [H]^e [T] \tag{49}$$

where $[T]$ contains the direction cosines for the element. The element matrices in global coordinates are then assembled into the global gain matrices $[H_V]$ and $[H_P]$. In practice the element level gain matrices, in global coordinates, are assembled directly into the augmented matrices $[C_A]$ and $[K_A]$. If each end of the collocated controller is attached to a node point in the structure then the forces are applied equally at each end and they are proportional to the relative differences of the positions and velocities at each end (along the length of the controller element).

The general controller is defined to act in the global coordinate system. The actuator and sensor degrees of freedom are selected by the designer and the gains are then assembled directly into the augmented matrices. The gain value is added to the row of the actuator degree of freedom in the column of the sensor degree of freedom (see Fig. 2). For a positive gain, the value that is added is positive. This choice of sign convention is logical for the case when the forcing and sensing are done at the same translational degree of freedom because it gives the same result as a grounded axial controller. In other cases involving sensing, a displacement or rotation at one degree of freedom and applying a force or moment at another, the definition of positive gain is less physical and it simply becomes a sign convention.

Since the measured outputs are fed back directly into the system, observability is not relevant for this case, and observability constraints are not necessary when the control law is implemented using the full model. On the other hand, if a reduced model is used, only the retained modes must be fed back and the sensed signal must be filtered, using the techniques discussed in Section 3.3. For direct output feedback, the stabilization of a given mode can be accomplished if this mode is controllable. For this purpose, the placement of the actuators has to be done such that the controllability conditions given in section 2.5 are satisfied.

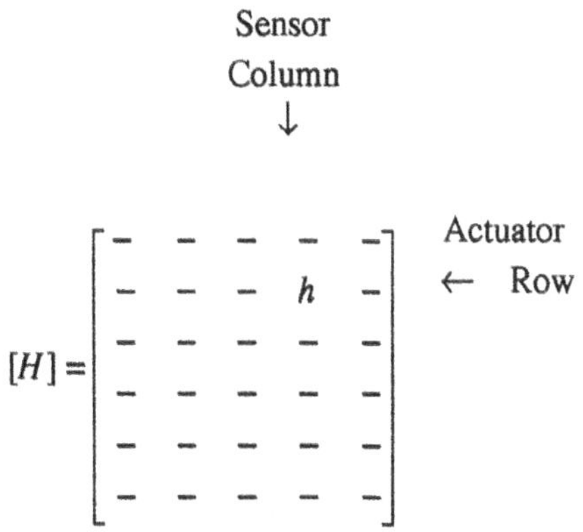

Sensor
Column
↓

$$[H] = \begin{bmatrix} - & - & - & - & - \\ - & - & - & h & - \\ - & - & - & - & - \\ - & - & - & - & - \\ - & - & - & - & - \\ - & - & - & - & - \end{bmatrix} \begin{matrix} \text{Actuator} \\ \leftarrow \quad \text{Row} \end{matrix}$$

FIGURE 2
General Control Element Gains

3.2 State Feedback

In this case the vector of control forces is proportional to the complete state vector and is given by

$$\{U(t)\} = -[H_P]\{q(t)\} - [H_V]\{\dot{q}(t)\} \tag{50}$$

and the dimensions of the matrices $[H_P]$ and $[H_V]$ is $n_a \times n$, where n is the number of degrees of freedom for the model. Since for real applications the number of degrees of freedom is very large, the elements of the gain matrices cannot be considered as independent design variables for models with large number of degrees of freedom.

The two options mostly used for integrated optimization are: (1) use control theory to determine the gains as a function of the structural variables and (2) use some kind of linking scheme to reduce the number of independent elements in the matrices $[H_P]$ and $[H_V]$. We will now discuss both approaches and a numerical comparison will be given in the numerical examples section.

3.2.1 Control Theory Approach

Using optimal control theory, the common practice is to design the controller using a linear quadratic regulator (see Ref. 37 for basic theory). The following performance index is defined

$$J = \int_0^\infty (\{X(t)\}^T [Q]\{X(t)\} + \{U(t)\}^T [R]\{U(t)\})dt \tag{51}$$

where $[Q]$ and $[R]$ are state and control weighting matrices. The matrix $[Q]$ is considered to be positive semi-definite and $[R]$ is assumed to be positive definite. The performance index J represents a combination of the state vector norm and the control forces norm or control effort (control energy). Thus, the selection of the elements of $[Q]$ and $[R]$ determines the closed-loop damping, which is directly related to the time required to control the disturbances and the energy required by the controllers.

The optimal solution for $\{U(t)\}$ is obtained by minimizing the performance index J subject to the system dynamics (see Eq. 7). Assuming a state feedback control law of the form

$$\{U(t)\} = -[H]\{X(t)\} \tag{52}$$

the optimal gain matrix $[H]$ is given by (see Ref. 37)

$$[H] = [R]^{-1} [B_C]^T [P] \tag{53}$$

where the matrix $[P]$ satisfies the following non-linear algebraic matrix equation, called the Riccati equation

$$[P]^T [A] + [A]^T [P] - [P][B_C][R]^{-1}[B_C]^T [P] + [Q] = [0] \tag{54}$$

The closed loop system (with no external disturbances) is then given by

$$\{\dot{X}\} = [A_C]\{X\} \tag{55}$$

where

$$[A_C] = [A] - [B_C][R]^{-1}[B_C]^T [P] \tag{56}$$

If the system is controllable and $[R]$ is positive definite then there is a unique solution $[P]$ for the Riccati equation, this solution is positive definite, and the closed loop system is stable (Ref. 37). The assumption that the system is controllable, allows the system to be stabilized to any degree, that is to say, one can find a feedback such that the eigenvalues of the closed loop system can be made to have real parts less than or equals to an arbitrary negative constant. For this purpose the Riccati equation can be modified as follows

$$[P_\alpha]^T ([A] + \alpha[I]) + ([A] + \alpha[I])^T [P_\alpha] - [P_\alpha]^T [B_C][R]^{-1}[B_C]^T [P_\alpha] + [Q] = [0] \tag{57}$$

where $\alpha \geq 0$ and the closed loop matrix

$$[A_C] = [A] - [B_C][R]^{-1}[B_C]^T [P_\alpha] \tag{58}$$

has eigenvalues $(\sigma + j\omega)$ such that $\sigma \leq -\alpha$.

By solving the Riccati equation for the feedback gains, the optimality conditions for the control subproblem are automatically satisfied and the gains become implicit functions of the structural variables (Eqs. (54) or (57)). Thus, design space integration is not completely achieved since the control variables (gains) are not treated as strictly independent design variables. However, the freedom to use elements of the matrices $[Q]$ and $[R]$ as design variables in order to tailor the solution of the Riccati equation, does exist.

The dimension of the matrices $[Q]$ and $[R]$ depends on the size of the vectors $\{X\}$ and $\{U\}$ respectively. Several alternatives have been studied on how to consider these matrices as design variables. The simplest and most commonly used approach is to assume that these matrices are formed by multiplying a single design variable by a specified constant matrix that does not get modified during the optimization process (Ref. 43).

3.2.2 Control Variable Linking

The idea of gain linking was introduced in Refs. 30 and 31, and the main concept is to express the gain matrix as a combination of independent basis matrices.

$$[H] = \sum_i \alpha_i \, [H^{(i)}] \tag{59}$$

where the constant matrices $[H^{(i)}]$ are the basis matrices, and the participation coefficients α_i are the design variables.

Linking can be imposed on the complete gain matrix or independently on the position and velocity parts. In the latter case, the gain matrices are written as

$$[H] = [[H_P] \, [H_V]] \tag{60a}$$

$$[H_P] = \sum_i \alpha_i \, [H_P^{(i)}] \tag{60b}$$

$$[H_V] = \sum_i \beta_i \, [H_V^{(i)}] \tag{60c}$$

and α_i, β_i are the design variables.

The main ideas underlying the creation of alternative control design variable linking schemes are: (1) separation of velocity and position parts of the gain matrix; (2) various row and column schemes corresponding to actuator and sensor degree of freedom linking; and (3) linking schemes based on only allowing changes in various sets of velocity gains (damping). Combining the foregoing ideas leads to numerous linking schemes with distinct sets and various numbers of independent control system design variables (CDV), ranging from 1 to $n_a \times 2n$ (see Table 1).

For example, consider option number 5 in Table 1. The feedback gain matrix can be written as follows:

$$
[H] =
\begin{bmatrix}
[H_P]_1 & [H_V]_1 \\
[H_P]_2 & [H_V]_2 \\
\cdot & \cdot \\
\cdot & \cdot \\
\cdot & \cdot \\
[H_P]_{n_a} & [H_V]_{n_a}
\end{bmatrix}
= [H] =
\begin{bmatrix}
\alpha_1[H_P^o]_1 & \alpha_{n_a+1}[H_V^o]_1 \\
\alpha_2[H_P^o]_2 & \alpha_{n_a+2}[H_V^o]_2 \\
\cdot & \cdot \\
\cdot & \cdot \\
\cdot & \cdot \\
\alpha_{n_a}[H_P^o]_{n_a} & \alpha_{2n_a}[H_V^o]_{n_a}
\end{bmatrix}
\tag{61}
$$

The left hand side represents the $n_a \times 2n$ feedback gain matrix in partitioned row-wise form ($[H_P]_j$, $[H_V]_j$ represent j-th rows of $[H_P]$ and $[H_V]$, respectively), and the right hand side has participation coefficients (α_i's) in front of the partitioned rows of the original matrix (superscript 0 denotes the original matrix). During optimization the α_i's are treated as independent design variables and as they change, $[H]$ is modified in accordance with the fixed ratios established in the rows of the original matrix.

TABLE 1
Control Design Variable Linking Options

Option	Description	Design Variables	No. of CDV's
1	totally unlinked	elements of $[H]$	$n_a \times 2n$
2	$[H_P]$ fixed, $[H_V]$ unlinked	elements of $[H_V]$	$n_a \times n$
3	columns of $[H]$ linked	coefficients of columns of $[H]$	$2n$
4	$[H_P]$ fixed, columns of $[H_V]$ linked	coefficients of columns of $[H_V]$	n
5	rows of $[H_P]$ and $[H_V]$ linked	coefficients of rows of $[H_P]$ and $[H_V]$	$2n_a$
6	rows of $[H]$ linked	coefficients of rows of $[H]$	n_a
7	$[H_P]$ fixed, rows of $[H_V]$ linked	coefficients of rows of $[H_V]$	n_a
8	$[H_P]$, $[H_V]$ linked	coefficients of $[H_P]$, $[H_V]$	2
9	$[H_P]$ fixed, $[H_V]$ linked	coefficient of $[H_V]$	1
10	$[H]$ linked	coefficient of $[H]$	1

When a specific control design variable linking scheme is used, it is important to choose an acceptable set of initial feedback gains, because the relative values of certain elements in the feedback gain matrix remain frozen throughout the design process according to the linking scheme selected. Three different methods for generating initial control design variable values are suggested in Ref. 30 and 31.

The first initializing method sets the feedback gains arbitrarily (e.g. $[H] = [0]$) and then carries out a few design iterations without any control design variable linking. This allows all the gains as well as the structural design variables to change freely for a few iterations in order to find a reasonable initial design prior to imposing some linking on the set of $n_a \times 2n$ control design variables. Even though the unlinked option is used for only a few iterations, this can still be a serious restriction, limiting the application of the method to small problems.

The second initializing method is to solve the $2n \times 2n$ nonlinear matrix Riccati equation once in order to find the linear optimal control law corresponding to the initial structural design. The initial gain values obtained from the matrix Riccati equation solution are then used to establish fixed ratios between the gains that are assumed to hold throughout the design optimization process.

The third approach is the decoupled Riccati equation method, which gives an approximate solution to the full order Riccati equation. This method uses normal modes to diagonalize the original equations of motion, and then for each second order scalar equation a linear quadratic regulator problem is solved neglecting the coupling effects.

The second and third initializing methods are described further in what follows.

Full Order Riccati Equation

Consider the equations of motion with the external disturbance terms set to zero (i.e. $\{f\} = \{0\}$):

$$[M]\{\ddot{q}\} + [C]\{\dot{q}\} + [K]\{q\} = [b]\{U\} \tag{62}$$

or

$$\{\dot{X}\} = [A_0]\{X\} + [B_C]\{U\} \tag{63}$$

The optimal control law that minimizes the performance index given by Eq. (51) where $[Q]$ and $[R]$ are $2n \times 2n$ positive semi definite and $n_a \times n_a$ positive definite weighting matrices for states and control forces respectively, is determined as

$$\{U\} = -[H]\{X\} = -[R]^{-1}[B_C]^T [P]\{X\} \tag{64}$$

where the $2n \times 2n$ positive semi definite symmetric matrix $[P]$ satisfies the following matrix Riccati equation

$$[P][A_0]^T + [A_0]^T[P] - [P][B_C][R]^{-1}[B_C]^T[P] + [Q] = [0] \tag{65}$$

where the index 0 for the matrix $[A]$ indicates the initial design for the structure.

Decoupled Riccati Equation Solution

An alternative method which bypasses solution of the full order Riccati equation is based on modal decomposition. The natural frequencies and normal modes of the uncontrolled disturbance free system

$$[M]\{\ddot{q}\} + [K]\{q\} = \{0\} \tag{66}$$

are given by the solution of the standard eigenproblem

$$\omega_i^2[M]\{\phi\}_i = [K]\{\phi\}_i \quad i = 1, 2, ..., r \tag{67}$$

The modes $\{\phi\}_i$ are normalized so that

$$\{\phi\}_i^T [M]\{\phi\}_j = \delta_{ij}, \quad i, j = 1, 2, ..., r \tag{68}$$

where r is the number of normal modes retained ($r \leq n$) and δ_{ij} is the Kroneocher delta. Considering the transformation

$$\{q\} = [\phi]\{z\} \tag{69}$$

where the i-th column of the $n \times r$ normal mode matrix $[\phi]$ is the i-th normal mode $\{\phi\}_i$ and $\{z\} = [z_1, z_2, ..., z_r]^T$ is the normal coordinate vector. Substituting Eq. (69) into Eq. (62) and pre-multiplying $[\phi]^T$ results in r sets of scalar equations as follows

$$\ddot{z}_i + c_i \dot{z}_i + \omega_i^2 z_i = \{\phi\}_i^T [b]\{U\}$$

$$= \{\phi\}_i^T [b]\left(\{U\}^{(i)} + \sum_{k \neq i}^r \{U\}^{(k)}\right) \quad i = 1, 2, ..., r \tag{70}$$

where

$$\{U\}^{(i)} = -[\{\tilde{H}_P\}^{(i)} \{\tilde{H}_V\}^{(i)}] \begin{Bmatrix} z_i \\ \dot{z}_i \end{Bmatrix} \quad i = 1, 2, \ldots, r \tag{71}$$

The vector $\{U\}^{(i)}$ is an $n_a \times 1$ control vector which contains only i-th normal mode components (z_i and $\dot{z}_i$). Furthermore, $\{\tilde{H}_P\}^{(i)}$, $\{\tilde{H}_V\}^{(i)}$ are $n_a \times 1$ feedback gain vectors which relate $\{U\}^{(i)}$ with z_i, $\dot{z}_i$ respectively.

Assuming that the i-th control vector $\{U\}^{(i)}$ can be calculated independently neglecting the coupling term (second term on the right hand side of Eq. (70)) and that the resulting $\{U\}$ is the sum of all the $\{U\}^{(i)}$, i.e.,

$$\{U\} = \sum_{i=1}^{r} \{U\}^{(i)} = -[[\tilde{H}_P][\tilde{H}_V]] \begin{Bmatrix} z \\ \dot{z} \end{Bmatrix} \tag{72}$$

where $[\tilde{H}_P]$ and $[\tilde{H}_V]$ denote $n_a \times r$ position and velocity gain matrices in normal coordinates the columns of which are $\{\tilde{H}_P\}^{(i)}$ and $\{\tilde{H}_V\}^{(i)}$, respectively. Using the relation $\{z\} = [\phi]^T [M] \{q\}$, the feedback gain matrix in physical coordinates can be recovered as follows

$$[H_P] = [\tilde{H}_P][\phi]^T [M] \tag{73a}$$

$$[H_V] = [\tilde{H}_V][\phi]^T [M] \tag{73b}$$

The remaining problem consists of finding r sets of $\{U\}^{(i)}$, i.e. $[\{\tilde{H}_P\}^{(i)}, [\tilde{H}_V]^{(i)}]$. Neglecting the coupling term in Eq. (70) yields

$$\ddot{z}_i + c_i \dot{z}_i + \omega_i^2 z_i \cong \{\phi\}_i^T [b] \{U\}^{(i)} \tag{74}$$

which can be transformed into the standard first order state space form as

$$\{\dot{w}_i\} = [A_i] \{w_i\} + [B_i] \{U\}^{(i)} \tag{75}$$

where

$$[A_i] = \begin{bmatrix} 0 & 1 \\ -\omega_i^2 & -c_i \end{bmatrix} \tag{76a}$$

$$[B_i] = \begin{bmatrix} [0] \\ \{\phi\}_i^T [b] \end{bmatrix} \tag{76b}$$

where $\{w_i\} = [z_i \ \dot{z}_i]^T$ is the state vector for node i, $[A_i]$ is the 2×2 system open-loop matrix and $[B_i]$ is the $2 \times n_a$ system input matrix for the i-th modal equation.

The performance index for the i-th mode (J_i) is chosen as

$$J_i = \int_0^{\infty} (\{w_i\}^T [Q_i] \{w_i\} + \{U\}^{(i)T} [R_i] \{U\}^{(i)}) \, dt \tag{77}$$

where $[Q_i] = \text{Diag}(Q_{11}^i, Q_{22}^i)$, $(Q_{11}^i, Q_{22}^i \geq 0)$ is a 2×2 diagonal weighting matrix for the i-th state and $[R_i] = \gamma_i[I]$ (where $[I]$ is an $n_a \times n_a$ identity matrix, $\gamma_i > 0$) is a weighting matrix for the i-th modal control force vector. The i-th component of the control $\{U\}^{(i)}$ can then be determined from

$$\{U\}^{(i)} = -[R_i]^{-1}[B_i]^T[P_i]\{w_i\} \tag{78}$$

where the 2×2 positive semi definite symmetric matrix

$$[P_i] = \begin{bmatrix} p_{11}^i & p_{12}^i \\ p_{12}^i & p_{22}^i \end{bmatrix} \tag{79}$$

satisfies the 2×2 Riccati equation

$$[P_i][A_i] + [A_i]^T[P_i] - [P_i][B_i][R_i]^{-1}[B_i]^T[P_i] + [Q_i] = [0] \tag{80}$$

Equation (80) can be solved in closed form and the solution is given by (Ref. 30)

$$p_{12}^i = p_{21}^i = \frac{-\omega_i^2 + \sqrt{\omega_i^4 + W_i Q_{11}^i}}{W_i} \tag{81a}$$

$$p_{22}^i = \frac{-c_i + \sqrt{c_i^2 + W_i Q_{22}^i - 2\omega_i^2 + 2\sqrt{\omega_i^4 + W_i Q_{11}^i}}}{W_i} \tag{81b}$$

$$p_{11}^i = c_i p_{12}^i + \omega_i^2 p_{22}^i + p_{12}^i p_{22}^i W_i \tag{81c}$$

where

$$W_i = \{\phi\}_i^T[b][R_i]^{-1}[b]^T\{\phi\}_i$$

$$= \frac{1}{\gamma_i}\{\phi\}_i^T[b][b]^T\{\phi\}_i \tag{81d}$$

By comparing Eq. (71) and (78) the i-th feedback gain vector in normal coordinates can be obtained as follows

$$\{\tilde{H}_P\}^{(i)} = \frac{p_{12}^i}{\gamma_i}[b]^T\{\phi\}_i \tag{82a}$$

$$\{\tilde{H}_V\}^{(i)} = \frac{p_{22}^i}{\gamma_i}[b]^T\{\phi\}_i \tag{82b}$$

Subsituting Eqs. (82) into Eq. (72) and (74), yields, after some manipulation

$$\ddot{z}_i + (c_i + W_i p_{22}^i)\dot{z}_i + (\omega_i^2 + W_i p_{12}^i)z_i \cong 0 \tag{83}$$

The initial feedback gain matrix obtained by solving r sets of 2×2 Riccati equations (Eq. 80) can be rewritten in the following form

$$[H^o] = [[H_P^o][H_V^o]] = \sum_{i=1}^{r} \left[[H_P^o]^{(i)} \ [H_V^o]^{(i)} \right] = \sum_{i=1}^{r} [H^o]^{(i)} \tag{84}$$

$$[H_P^o] = \sum_{i=1}^{r} \{\tilde{H}_P\}^{(i)} \{\phi\}_i^T [M] = \sum_{i=1}^{r} [H_P^o]^{(i)} \tag{85a}$$

$$[H_V^o] = \sum_{i=1}^{r} \{\tilde{H}_V\}^{(i)} \{\phi\}_i^T [M] = \sum_{i=1}^{r} [H_V^o]^{(i)} \tag{85b}$$

where superscripts (i) indicate that these quantities correspond to the i-th Riccati equation. The foregoing equations imply that the $[H_P^o]^{(i)}$'s and $[H_V^o]^{(i)}$'s or the $[H^o]^{(i)}$'s may be interpreted as basis matrices which can be used to generate the initial gain matrix. This suggests that the actual feedback gain matrix can be well approximated as a linear combination of these basis matrices, namely

$$[H] = \sum_{i=1}^{r} \alpha_i [H^o]^{(i)} \tag{86}$$

or

$$[H] = \sum_{i=1}^{r} \left[\alpha_i [H_P^o]^{(i)} \ \alpha_{i+r} [H_V^o]^{(i)} \right] \tag{87}$$

The whole feedback gain matrix can be linked, or the position and velocity parts of the gain matrix can be block linked separately. During the optimization the participation coefficients α_i's are treated as independent design variables. It should also be noted that these participation coefficients can be further linked with each other.

To summarize, the original full order Riccati equation is replaced by r sets of 2×2 Riccati equations which have explicit closed form solutions. Then the feedback gain matrix $[\tilde{H}] = [[\tilde{H}_P][\tilde{H}_V]]$ in normal coordinates is transformed to $[H]$ in the original coordinate system by using normal mode information.

The method presented in this section has a considerable advantage over the full order Riccati equation solution approach. It is in fact explicit and efficient enough to permit periodic updating of the initial fixed ratios between elements of the gain matrix $[H]$. When the fixed ratios of the feedback gain matrix are updated, special attention has to be given to preserve continuity of the real part of the closed-loop eigenvalues or closed loop damping ratios, so that the dynamic behavior remains relatively smooth between the updating stages.

3.3 About Spill Over Effect

When only a reduced set of modes are controlled and stabilized, some other higher modes may become unstable, due to observation or control spillover effects. Using a modal analysis, the vector of modal degrees of freedom, $\{q\}$, can be written without any truncation as follows

$$\{q\} = [\phi_c]\{z_c\} + [\phi_u]\{z_u\} \tag{88}$$

where $\{z_c\}$ is the $r \times 1$ controlled normal coordinate vector, $[\phi_c]$ is the $n \times r$ eigenmatrix made up from the normal eigenvectors of the r controlled modes $\{z_u\}$ is the $(n-r) \times 1$ uncontrolled normal coordinate vector, and $[\phi_u]$ is the $n \times (n - r)$ eigenmatrix corresponding to the uncontrolled modes $\{z_u\}$.

Eigenmatrices $[\phi_c]$ and $[\phi_u]$ in Eq. (88) are normalized such that

$$[\phi_c]^T [M] [\phi_c] = [I] \tag{89a}$$

$$[\phi_u]^T [M] [\phi_u] = [I] \tag{89b}$$

$$[\phi_c]^T [M] [\phi_u] = [0] \tag{89c}$$

and also

$$[\phi_c]^T [K] [\phi_c] = [\Delta_c] \tag{90a}$$

$$[\phi_u]^T [K] [\phi_u] = [\Delta_u] \tag{90b}$$

$$[\phi_c]^T [C] [\phi_c] = [C_c] \tag{90c}$$

$$[\phi_u]^T [C] [\phi_u] = [C_u] \tag{90d}$$

Substituting Eq. (88) into the equations of motion and pre-multiplying by $[\phi_c]^T$ and $[\phi_u]^T$ results in

$$\{\ddot{z}_c\} + [C_c]\{\dot{z}_c\} + [\Delta_c]\{z_c\} = [\phi_c]^T [b] \{U\} \tag{91a}$$

$$\{\ddot{z}_u\} + [C_u]\{\dot{z}_u\} + [\Delta_u]\{z_u\} = [\phi_u]^T [b] \{U\} \tag{91b}$$

On the right hand side of Eq. (91b) $[\phi_u]^T [b] \neq [0]$, which means there exists control spillover which means that the control vector will excite the modes neglected from the model and can make them unstable.

Substituting Eq. (88) into the expression for the control forces leads to

$$\{U\} = -([H_P][\phi_c]\{z_c\} + [H_V][\phi_c]\{\dot{z}_c\}) - ([H_P][\phi_u]\{z_u\} + [H_V][\phi_u]\{\dot{z}_u\}) \tag{92}$$

On the right hand side of Eq. (92) those matrices in front of the uncontrolled modes are not zero (i.e., $[H_P][\phi_u] \neq [0]$, $[H_V][\phi_u] \neq [0]$) which can be interpreted as observation spillover.

As can be seen from Eqs. (91b) and (92), there exist both control and observation spillover. According to Ref. 44, the closed-loop system has potential instability when the system has both observation and control spillover and it is more important to eliminate observation spillover in order to prevent instability. Therefore, the desired control input, $\{U^*\}$, should contain only controlled modes, $\{z_c\}$, namely

$$\{U^*\} = -[H_P][\phi_c]\{z_c\} - [H_V][\phi_c]\{\dot{z}_c\} \tag{93}$$

A simple procedure to eliminate spillover was suggested in Ref. 45. Premultiplying $[\phi_c]^T [M]$ to Eq. (88) and using the relations in Eqs. (89) results in

$$\{z_c\} = [\phi_c]^T [M] \{q\} \tag{94}$$

Substituting Eq. (94) into Eq. (93) gives

$$\{U^*\} = - [H_P] [\phi_c] [\phi_c]^T [M] \{q\} - [H_V] [\phi_c] [\phi_c]^T [M] \{\dot{q}\} \tag{95}$$

$$= - [H_P^*] \{q\} - [H_V^*] \{\dot{q}\}$$

where

$$[H_P^*] = [H_P] [\phi_c] [\phi_c]^T [M] \tag{96a}$$

and

$$[H_V^*] = [H_V] [\phi_c] [\phi_c]^T [M] \tag{96b}$$

In summary, destabilization of uncontrolled higher modes can be prevented by using the truncated feedback gain matrices shown above since these matrices are orthogonal to the uncontrolled higher modes $[\phi_u]$ and will eliminate observation spillover.

3.4 State Estimator

When using state space feedback, the state vector has to be estimated from the measured outputs, in order to implement the control law. Figure 3 shows the schematics of this procedure. The observer or state estimator is a dynamic system that reconstructs the complete state vector from those components measured by the sensors. This section gives the basic steps for the construction of such an observer.

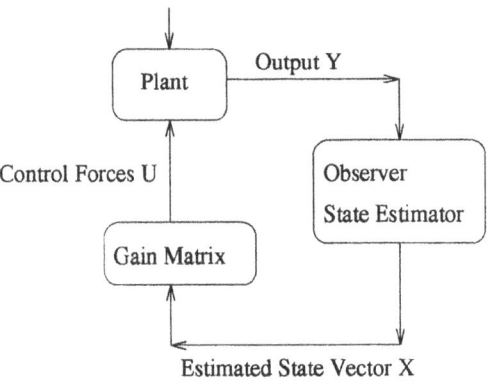

FIGURE 3
Schematic of State Feedback Control

Consider the dynamic system described by the equations (31), that is assumed to be completely observable. The observers most commonly used are of the non-statistical Luenberger type (Refs. 37, 39-41). The matrix $[C_C]$ (see Eq. (11)) is assumed to be a full rank matrix. Considering the transformation

$$\{X\} = [[T_1][T_2]] \begin{bmatrix} [I_1] & -[\Sigma] \\ [0] & [I_2] \end{bmatrix} \begin{Bmatrix} \{Z\} \\ \{Y\} \end{Bmatrix}$$
(97)

where $[I_1]$ and $[I_2]$ are identity matrices and $\{Z\}$ is a vector with $n_2 = 2n - n$, components. With this notation n_2 is the dimension of the vector of unknown or unmeasured components of the state vector $\{X\}$. Denoting

$$[T] = [[T_1][T_2]]$$
(98a)

and

$$[\bar{\Sigma}] = \begin{bmatrix} [I_1] & -[\Sigma] \\ [0] & [I_2] \end{bmatrix}$$
(98b)

if $[T]$ is non singular

$$\begin{Bmatrix} \{Z\} \\ \{Y\} \end{Bmatrix} = [\bar{\Sigma}]^{-1} [T]^{-1} \{X\}$$
(99)

but, since

$$[\bar{\Sigma}]^{-1} = \begin{bmatrix} [I_1] & [\Sigma] \\ [0] & [I_2] \end{bmatrix}$$
(100)

the following relation is obtained

$$\{Z\} = [[I_1][\Sigma]] [T]^{-1} \{X\}$$
(101)

From Eq. (97), if $\{Z\}$ is known or estimated, then the state vector $\{X\}$ can be obtained. The task is then to construct an estimate of $\{Z\}$, denoted by $\{\hat{Z}\}$.

In Eq. (101), let

$$[\tilde{T}_Z] = [[I_1][\Sigma]] [T]^{-1}$$
(102)

then

$$\{Z\} = [\tilde{T}_Z]\{X\}$$
(103)

On the other hand from Eq. (97) it follows that

$$\{X\} = [T_1]\{Z\} + [[T_2] - [T_1][\Sigma]]\{Y\}$$
(104)

and letting

$$[T_Z] = [T_1]$$
(105a)

and

$$[T_Y] = [T_2] - [T_1][\Sigma]\{Y\}$$
(105b)

gives

$$\{X\} = [T_Z]\{Z\} + [T_Y]\{Y\}$$
(106)

Differentiating Eq. (106) and eliminating $\{\dot{X}\}$ from Eqs. (31) gives

$$\{\dot{Z}\} = [\tilde{T}_Z][A][T_Z]\{Z\} + [\tilde{T}_Z][A][T_Y]\{Y\} + [\tilde{T}_Z][B_C]\{U\} + [\tilde{T}_Z]\{f\}. \tag{107}$$

Consider then, the estimate $\{\hat{Z}\}$ of $\{Z\}$ given by the equation

$$\frac{d}{dt}\{\hat{Z}\} = [A_e]\{\hat{Z}\} + [\tilde{T}_Z][A][T_Y][Y] + [\tilde{T}_Z][B_C]\{U\} + [\tilde{T}_Z]\{f\} \tag{108}$$

where $[A_e]$ is a matrix to be determined.

From Eq. (106) it is seen that the error in the estimation of $\{X\}$ is given by

$$\{e_X\} = \{X\} - \{\hat{X}\} = [T_Z](\{Z\} - \{\hat{Z}\}) \tag{109a}$$

or

$$\{e_X\} = [T_Z]\{e_Z\} \tag{109b}$$

where $\{e_Z\} = \{Z\} - \{\hat{Z}\}$ is the error in the estimate of $\{Z\}$. Then from Eqs. (107) and (108)

$$\{\dot{e}_Z\} = [\tilde{T}_Z][A][T_Z]\{e_Z\} + ([\tilde{T}_Z][A][T_Z] - [A_e])\{\hat{Z}\} \tag{110}$$

if the $[A_e]$ matrix is defined by

$$[A_e] = [\tilde{T}_Z][A][T_Z] \tag{111}$$

then

$$\{\dot{e}_Z\} = [A_e]\{e_Z\} \tag{112}$$

If at the initial time $\{e_Z\} = \{0\}$, then the error is zero, and the estimated state given by the observer is the real state. The usual case is that the initial state is not known, so it is required that $\{\hat{Z}\} \to \{Z\}$ when $t \to \infty$, according to Eqs. (109), this is achieved by choosing $[T_Z]$ and $[\tilde{T}_Z]$ which depend on $[T]$ and $[\Sigma]$ (see Eqs. (102) and (105)) such that $[A_e]$ is strongly stable. From the definitions of $[\tilde{T}_Z]$ and $[T_Z]$ the matrix $[A_e]$ is given by

$$[A_e] = [[I_1][\Sigma]][T]^{-1}[A][T] \begin{bmatrix} [I_1] \\ [0] \end{bmatrix} \tag{113}$$

Let

$$[\bar{A}] = [T]^{-1}[A][T] \equiv \begin{bmatrix} \bar{A}_{11} & \bar{A}_{12} \\ \bar{A}_{21} & \bar{A}_{22} \end{bmatrix} \tag{114}$$

then

$$[A_e] = [\bar{A}_{11}] + [\Sigma][\bar{A}_{21}] \tag{115}$$

So, $[\Sigma]$ must be chosen to stabilize $[\bar{A}_{11}]$. Then from Eq. (97)

$$\{Y\} = [C_C][T_1]\{Z\} + [C_C]([T_2] - [T_1][\Sigma])\{Y\} \tag{116}$$

Therefore, in order to satisfy Eq. (116), the following conditions must hold

$$[C_C][T_1] = [0] \tag{117a}$$

$$[C_C]([T_2] - [T_1][\Sigma]) = [C_C][T_2] = [I_1] \tag{117b}$$

and $[T]$ has to satisfy the condition

$$[C_C][T] = [[0][I_1]] \tag{118}$$

that is, in order to completely define the observer, a non-singular matrix $[T]$ has to be found satisfying Eq. (118) and a matrix $[\Sigma]$ has to be found to stabilize $[A_e]$. $[T]^{-1}$ can be written as

$$[T]^{-1} = \begin{bmatrix} S \\ C_C \end{bmatrix} \tag{119}$$

where $[S]$ is a $n_2 \times n_2$ matrix that has to be chosen such that $[T]^{-1}$ is made to be non-singular. Clearly the matrix $[S]$ is not unique. If $[C_C]$ consists only of zeroes and ones, a procedure to find $[S]$ is straight forward, since it can be constructed by placing zeroes and ones in the correct positions.

The remaining problem is to design the matrix $[\Sigma]$ so that it stabilizes the pair $(\overline{A}_{11}, \overline{A}_{21})$. The existence of such a matrix, to a given degree of stabilization, requires the pair $(\overline{A}_{21}, \overline{A}_{11})$ to be completely observable, i.e.

$$\text{Rank} \begin{bmatrix} [\overline{A}_{21}] \\ [\overline{A}_{21}][\overline{A}_{11}] \\ \cdot \\ \cdot \\ \cdot \\ [\overline{A}_{21}][\overline{A}_{11}]^{n_2-1} \end{bmatrix} = n_2 \tag{120}$$

Given the condition (120) $[\Sigma]$ can be chosen to be

$$[\Sigma] = -[P][\overline{A}_{21}]^T \tag{121}$$

where $[P]$ satisfies the Riccati equation

$$[P][\overline{A}_{11}]^T + [\overline{A}_{11}][P] - [P][\overline{A}_{21}]^T[\overline{A}_{21}][P] + [Q] = [0] \tag{122}$$

and $[Q] > 0$ is an arbitrary matrix, (typically a positive definite diagonal matrix). The degree of stability is increased with the norm of $[Q]$. If a preassigned degree of stability is required, Eq. (122) can be replaced by

$$[P]\left([\overline{A}_{11}]^T + \alpha[I_1]\right) + ([\overline{A}_{11}] + \alpha[I_1])[P] - [P][\overline{A}_{21}]^T[\overline{A}_{21}][P] + [Q] = [0] \tag{123}$$

where $\alpha > 0$. In this case the closed loop eigenvalues λ are stable and $R_e\lambda < -\alpha$.

According to the previous results, the equation for the observer is given by

$$\frac{d}{dt}\{\hat{Z}\} = [A_e]\{\hat{Z}\} + [H_e]\{\overline{Y}\} + [B_e]\{U\} + \{S_e\}\{f\} \tag{124}$$

where

$$[A_e] = [\overline{A}_{11}] - [P][\overline{A}_{21}]^T [\overline{A}_{21}] \tag{125a}$$

$$[H_e] = -[\overline{A}_{21}] + [\overline{A}_{12}] - [P][\overline{A}_{21}]^T [\overline{A}_{22}] \tag{125b}$$

$$[B_e] = [\overline{B}_1] - [P][\overline{A}_{21}]^T [B_2] \qquad ([\overline{B}] = [T]^{-1}[B]) \tag{125c}$$

$$[S_e] = [I_1] - [P][\overline{A}_{21}]^T \tag{125d}$$

Condition (120) is equivalent to the condition of complete observability of the system. Therefore, if observability constraints, given by the observability modal indices in Eq. (45) are imposed for the structural/control synthesis, the feasibility of solving Eq. (122) or Eq. (123) is insured.

4. DESIGN OPTIMIZATION PROCEDURE

The key to a tractable structural/control synthesis approach lies in the replacement of the original implicit nonlinear design problem with a sequence of explicit approximate problems of reduced dimensionality. The generation of these approximate problems is accomplished through the applications of techniques referred to as approximate concepts (Refs. 6, 41, 42). Primarily, these techniques serve to (1) reduce the number of design variables and constraints in the design problem and (2) reduce the required number of detailed (exact) constraint and objective function evaluations. There are various methods available for this purpose. The ones that are used most often include design variable linking, temporary constraint deletion and explicit first order approximations.

Figure 4 summarizes the optimization procedure. At the beginning of each design stage the structure/control system undergoes full static, dynamic response, natural frequency, and complex eigenvalue analyses, etc. in order to determine the values of the behavior constraints. In order to lower the cost of the sensitivity analysis, the constraints that are not active or potentially active are then deleted from the design problem for that stage. Then, first order analytical sensitivities of the retained response quantities with respect to structural properties and controller variables are calculated and used to construct an approximate optimization problem. The upper bound constraints are formulated as

$$g_k^U(y) = R(y) - 1 \le 0 \tag{126}$$

and the lower bound constraints are formulated as

$$g_k^L(y) = 1 - R(y) \le 0 \tag{127}$$

where $R(y)$ is the response ratio defined as

$$R(y) = \frac{f(y)}{f_a} \tag{128}$$

where $f(y)$ is the value of the constrained quantity and f_a is the allowable value.

Each approximate problem in the sequence is then created using the aforementioned sensitivities to construct explicit approximations of $f(y)$. Design variable move limits are used to insure the quality of the approximations. Although these problems are nonlinear, they are explicit and therefore are very inexpensive to solve compared to the solution of the actual problem. A complete description of the approximate problem is presented in the next sections. Each approximate optimization problem is solved using an optimization algorithm to generate a new design point which is then analyzed at the beginning of the next stage. This process is continued until the objective function changes by less than some prescribed value for three consecutive stages.

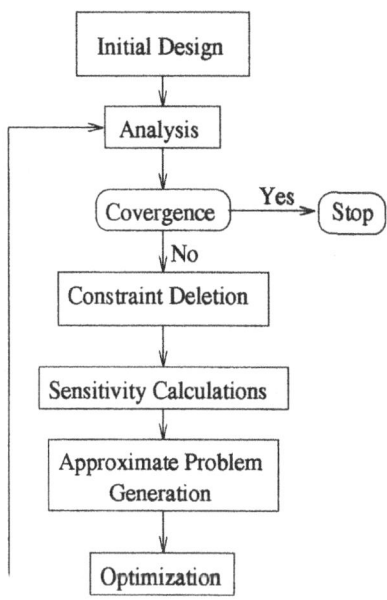

FIGURE 4
Optimization Procedure

4.1 Temporary Constraint Deletion

Since the driving constraints are not known at the outset of the design process, the synthesis problem statement may contain a large number of inequality constraints. In order to reduce the number of constraints, and the associated computational burden, it is possible to temporarily ignore certain constraints which are not expected to currently participate in the design. In effect, this procedure reduces the number of constraints by approximating the critical constraint set. The criteria by which particular constraints are judged to be participating (active) or non-participating (passive) form the basis of the constraint deletion technique. Various criteria are conceivable, however, an effective strategy consists of deleting all constraints with response ratios $(R(y_0))$ less than a specified constraint truncation parameter CTP (Ref. 6). The value of CTP may, in general, be chosen separately for each constraint type and may change during the design process.

4.2 Objective and Constraint Approximation

A key element in the efficient solution of structural/control optimization problems lies in the construction of accurate explicit objective and constraint function approximations. This is particularly true in the case of the behavior constraint functions because, in general, exact evaluation of these constraints requires the complete analysis of the system. Various methods are available for the construction of these approximations. The most commonly used techniques require only the function value and its first derivatives to construct the approximations.

A common approximate form is constructed by expanding the objective function and constraints in first order Taylor series about the current design point

$$f \approx f_o + \sum_{i=1}^{NDV} \frac{\partial f(y_o)}{\partial y_i} (y_i - y_{oi}) \tag{129}$$

where f is the constraint or objective function, f_o is the value of the function at the current design point or base design for the corresponding design cycle, y_i are the design variables, y_o is the vector of design variables at the current design point, and NDV is the number of design variables. An approximate problem generated in this form is known as a Linear Program and it can be solved by an efficient algorithm called the Simplex Method (Ref. 46).

It can be shown that a first order expansion in the reciprocals of the design variables such as

$$f \approx f_o + \sum_{i=1}^{NDV} (-y_{oi}^2) \frac{\partial f(y_o)}{\partial y_i} (1/y_i - 1/y_{oi}) \tag{130}$$

is more accurate for certain combinations of constraints and design variables. An approximate problem made up of functions of the form of Eq. (130) is explicit but nonlinear and it must be solved using a nonlinear optimization algorithm.

In many cases, however, these approximations are not sufficiently robust. It has been found that using a mix of direct and reciprocal approximations can lead to a more conservative approximate problem. The basic idea of these hybrid approximations, introduced in Ref. 47, is to select direct or reciprocal first order representations for each design variable, always using the more conservative of the two approximations. It is shown in Ref. 47 that this choice depends solely on the signs of the constraint function first derivatives and design variable values. The approximations in this case have the form

$$f \approx f_0 + \sum_{i=1}^{NDV} \frac{\partial f(y_o)}{\partial y_i} B_i \tag{131a}$$

where

$$B_i = \begin{cases} (y_i - y_{oi}) & \text{if } y_i \dfrac{\partial f(y_o)}{\partial y_i} \geq 0 \\[2ex] -y_{oi}^2 (1/y_i - 1/y_{oi}) & \text{if } y_i \dfrac{\partial f(y_o)}{\partial y_i} < 0 \end{cases} \tag{131b}$$

Numerical experience has shown this approximation to be quite robust, yielding good results for a significant class of structural synthesis problems (Ref. 48).

When choosing the form of the approximation, it is important to keep in mind the difference between accuracy and conservativeness. Picking the more accurate form, if it is known, will usually lead to faster overall convergence. However, picking an nonconservative form of the approximations can lead to very slow convergence. If the accuracy of the approximation is poor, move limits may have to be drastically reduced, thus increasing the number of analyses required to solve the optimization problem.

Highly accurate approximations can be constructed using the concepts of intermediate design variables and intermediate response quantities, which were introduced in Ref. 42. These approximations will be, in general, very nonlinear and therefore, the optimization procedure used to solve the approximate problems has to be looked at more carefully. The main concepts in these approaches are; (1) accurate approximations of response quantities with respect to structural design variables are generated by use of intermediate design variables, and (2) intermediate response quantities are approximated rather than the constraint functions themselves. The approximate constraint functions are then expressed explicitly using the approximated response quantities. This causes the explicit approximate constraint functions to be more accurate, but very nonlinear.

For example, response quantities must frequently be approximated with respect to deign variables that describe structural elements. Because of certain inherent nonlinearities the choice of intermediate design variables and the form of the approximation becomes important. Consider the case of beam elements. The designer would like to use variables that describe the actual beam element, such as web heights and flange thicknesses, called cross-sectional dimensions (CSD's). There are two drawbacks to this approach. The first drawback is that when the finite element method is used to generate analytical sensitivities of the desired response quantities, these sensitivities are with respect to the beam element section properties (SP's). This difficulty can be overcome by chain ruling the sensitivities using partial derivatives of the section properties with respect to the cross sectional dimensions. Defining the section properties as the x variables and the cross sectional dimensions as the y variables, the chain ruled sensitivity of the response quantity with respect to a cross sectional dimension y becomes:

$$\frac{\partial f(y_o)}{\partial y} = \sum_{i=1}^{4} \frac{\partial f(X_i(y_o))}{\partial x_i(y_o)} \frac{\partial x_i(y_o)}{\partial y} \tag{132}$$

where f is the response quantity, y_o is the vector of values of the beam's CSD's where the approximation is being made, and i sums over the four centroidal principal section properties (x_i) of the beam. The pertinent centroidal section properties are the area, the principal moments of inertia, and the polar moment of inertia (A, I_{yy}, I_{zz}, J). It must be noted that the section properties can be very nonlinear functions of the cross sectional dimensions. Therefore, the quality of first order approximations constructed using sensitivities that have been found via chain ruling, in the manner of Eq. (132), may be very poor.

A second example of a drawback that CSD's pose as design variables can be observed from the properties of the dynamic response and real and complex eigenvalue analyses. Examining the single degree of freedom equation for the real eigenvalue

$$\lambda = \omega^2 = \frac{k}{m} \tag{133}$$

it is seen that it is a function of both mass and stiffness. The mass and stiffness of a beam are both nonlinear functions of each cross sectional dimension of the beam. So as a beam CSD is increased, both the mass and stiffness increases. The effect of these increases may be to increase or decrease the eigenvalues of the system. In fact, the eigenvalue may first increase and then decrease as the CSD is increased due to the nonlinearities of the mass and stiffness with respect to the CSD. These nonlinearities can cause the approximations to be very poor.

The reason for the poor quality of an approximation with respect to a beam's CSD (height) is shown graphically in Fig. 5. This is a plot of the real part of the complex eigenvalue of the eighth mode of a ten element cantilever beam structure verses a 14% variation in the height of one of the beam elements. Note that, even though the range of the variation is small, the curve is nonlinear and non-monotonic. Therefore, a first order direct or reciprocal approximation of this response quantity with respect to the CSD variable (height) would be very poor. The reason that the curve is not monotonic is that as the area of the beam increases the real part of the complex eigenvalue decreases, but as the bending moment increases the real part increases.

It is concluded from the above discussion that it may be better to use the beam section properties as design variables whenever a response quantity is a function of both the structures mass and stiffness. The problem with this approach is that, although the approximations are good, the designer still needs to recover the beam CSD's in order to have the structure built. This can be done by making a linear approximation of the cross sectional dimensions with respect to the section properties as described in Ref. 48. But because the relation between the CSD's and the SP's is so nonlinear, these approximations are not very good and lead to large errors. The result is that the SP's calculated from the recovered CSD's may be significantly different from the original SP's. Therefore, using the section properties as alternative design variables does not solve the problem of poor approximations.

The solution to the problem of poor approximations resulting from the use of beam type elements is to use the cross sectional dimensions as design variables and make the approximations with respect to the section properties. Linear approximations of this type have the following form:

$$\tilde{f} = f_o + \sum_{j=1}^{NE} \sum_{i=1}^{4} \frac{\partial f(x_{ij}(y_{oj}))}{\partial x_{ij}(y_{oj})} [x_{ij}(y_j) - x_{ij}(y_{oj})] \tag{134}$$

where $\tilde{f}$ is the approximate response quantity, f_o is the initial value of the response quantity, and NE is the number of elements. Note that although the form of the approximation is linear, the approximation itself is nonlinear because of the explicit nonlinear dependence of the beam's four section properties (x_{ij}) on the beam's vector of cross sectional dimensions (y_j). This type of approximation is more accurate and more nonlinear than either approximations with respect to the SP's or CSD's. It should also be noted that this type of nonlinear approximation can lead to explicit approximate problems that are not convex and therefore may contain local optima. Experience shows that the benefit of the better approximation outweighs the problems that may arise from the nonconvexity. This is because the approximations capture the nonlinearity of the actual problem.

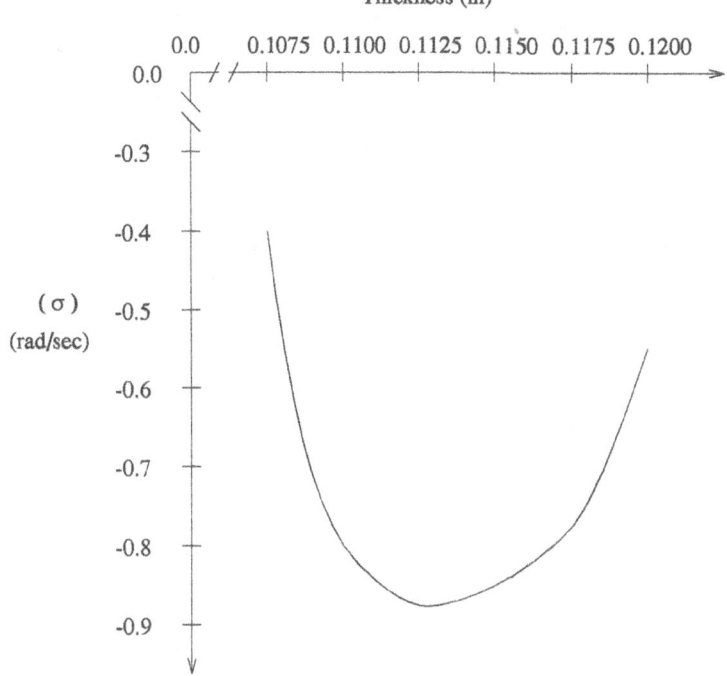

FIGURE 5
Real Part of Complex Eigenvalue Versus Beam Height

Gradients of approximations having the form of Eq. (134) with respect to CSD design variables are needed for efficient optimization. These can be found by taking the derivative of the approximate response quantity with respect to an individual cross sectional dimension (y). This results in:

$$\frac{\partial \tilde{f}}{\partial y} = \sum_{i=1}^{4} \frac{\partial f(x_i(y_o))}{\partial x_i(y_o)} \frac{\partial x_i(y)}{\partial y} \tag{135}$$

Note that the closed form expression for $\partial x_i(y)/\partial y$ can be easily obtained because the section properties (x_i) are explicit functions of the CSD (y) variables.

Note that in the case of axial element areas and control system gains, the intermediate design variables of choice (x) and the actual design variables (y) are the same.

The second concept is to approximate selected intermediate response qualities instead of the actual constraint functions. For example, it was shown in Ref. 49 that the quality of static stress constraint representations can be improved by approximating the element forces and then calculating the value of the stresses explicitly (i.e. the element forces are taken as intermediate response quantities). This idea of approximating intermediate response quantities also has merit in dynamic response analysis. This is because the constrained responses are often the amplitudes of the dynamic displacements and element forces. Calculating the amplitudes involves taking the square root of the sum of the squares of the real and imaginary components of the response. It is clear that this calculation is inherently nonlinear, and this nonlinearity is even more pronounced when the amplitude of the response is near zero.

Capturing the nonlinearity of actuator force amplitudes, particularly when they are near zero, it is important when total control force is taken as the objective function, because optimization tends to drive some of the control forces towards zero. The poor quality of the linear approximation can be seen in Fig. 6. If a linear approximation of the amplitude of the actuator force with respect to the velocity gain is generated at gain A, and a minimum move limit if 0.01 is used, then there is a large difference between the approximate and exact value of the force at gain B. However, if the sine and cosine components of the force are approximated by linear functions and the square root of the sum of the squares of their approximate values is calculated at gain B, this value will be very close to the actual value of the amplitude of the force at this point.

In general, the concepts of intermediate response quantities and intermediate design variables can be summarized as follows. Consider the notation (for constraint g_q)

$$R_k \quad : \quad \text{intermediate response quantities} \quad k \in K_q$$

$$x_i \quad : \quad \text{intermediate design variables} \quad i \in I_q$$

$$y_j \quad : \quad \text{actual design variables} \quad j \in J$$

Then the constraint is written as

$$g_q(R, x, y) \quad : \quad \text{explicit in} \quad R_k, x_i, y_j$$

$$x_i = x_i(y) \quad : \quad \text{explicit in} \quad y_j$$

$$R_k = R_k(x) \quad : \quad \text{implicit in} \quad x_i$$

The intermediate response quantities are approximated explicitly in terms of the intermediate design variables x_i's

$$R_k \approx \tilde{R}_k(x) \tag{136}$$

and then the approximate constraint is calculated as

$$\tilde{g}_q(y) = g_k(\tilde{R}, x, y) \tag{137}$$

with this approximation all essential nonlinearities have been kept, and only implicit quantities are approximated. The approximation of R_k can be linear, reciprocal, or hybrid.

The choice of intermediate response quantities and intermediate design variables depends on the type of response to be approximated. The reader can refer to Ref. 49 for static stress constraints, Ref. 50 for the frequency constraints, Ref. 8 for control force constraints and complex eigenvalues, Ref. 51 for steady state harmonic displacements and Ref. 52 for transient displacements.

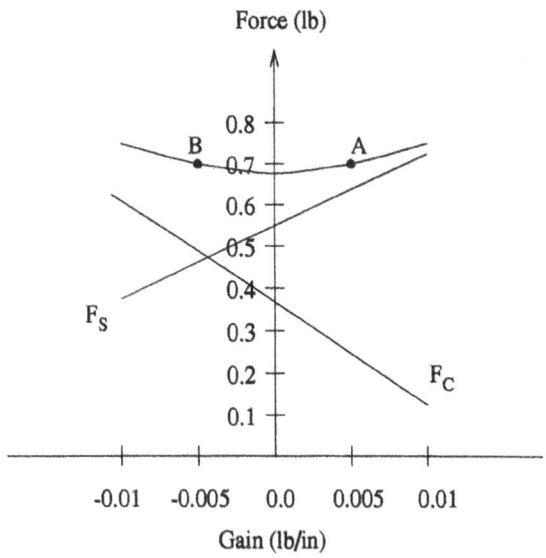

Force (lb)

FIGURE 6
Components of Controller Forces Versus Velocity Gain

5. NUMERICAL EXAMPLES

The following examples illustrate the application of the integrated approach to the optimization of a structural/control system. These examples involve direct output feedback and state feedback control laws, and are designed to show the flexibility offered by the integrated design procedure.

5.1 Cantilever Beam: Mass Minimization, Direct Output Feedback

The first problem, drawn from Ref. 8, is that of finding the minimum mass design of the 10-m cantilevered beam shown in Fig. 7. The beam is modeled with 10 equal length beam type finite elements. The beam is constrained so that only in-plane vertical motion is allowed. A concentrated mass of 200 kg is located at the midspan of the beam. The beam is loaded by a vertical harmonic load of 4000 N at 3.9 Hz applied at the tip; 2% structural damping is assumed. The design variables for this problem are the web and flange thicknesses t_h, t_b of the beam elements and the position and velocity gains of a single collocated control element located at the tip of the beam (see Fig. 7).

In this example problem, the web and flange thickness variables are linked along the entire length of the beam. The initial value of the beam element design variables are taken as 5.0 cm with side constraints imposed so that 0.5 cm $\leq t_h$, $t_b \leq 10.0$ cm. The magnitude of the tip displacement of the beam is constrained to be ≤ 10.0 cm, and the first frequency is constrained to be ≥ 4.0 Hz. Move limits of 60% are imposed on the design variables during each stage with a minimum move limit of 0.1.

Three runs were made for this example problem, all starting from the same initial structural design. In the first run, no control elements are used. In the second run, an axial controller is located at the tip of the beam in order to control the vertical displacement at the tip. The initial values of the position and velocity gain design variables in the second

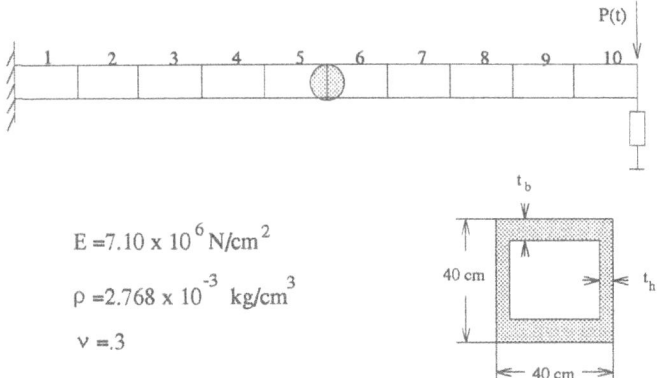

$E = 7.10 \times 10^6 \, \text{N/cm}^2$

$\rho = 2.768 \times 10^{-3} \, \text{kg/cm}^3$

$\nu = .3$

FIGURE 7
Cantilever Beam

and third runs are $h_p = 20.0$ N/cm and $h_v = 2.0$ N-s/cm. Although there are no constraints placed on the control gains, the magnitude of the control force is constrained to be ≤ 1000 N. The third run is the same as the second except that the real part of the complex eigenvalue (stability measure) of the first mode is constrained to be ≤ 1.0 rad/s.

The final designs and final design response ratios for all three runs are given in Tables 2 and 3. The iteration history plots are shown in Fig. 8. In all three cases, the tip displacement constraint is critical. This is to be expected because the minimization of the mass of the structure results in lower structural stiffness. In the second and third runs, where the controller is used, the control force constraint is also critical. This is because the control force increases as the structural stiffness decreases. Note that the addition of the critical actuator force constraints does not slow down the convergence of the synthesis process.

TABLE 2
Final Designs
Cantilevered Beam: Mass Minimization

Element type	Element number	Design variables	Final Design		
			Uncontrolled	Controlled	Controlled stability constraint
Frame	1-10	t_b, cm	1.995	1.588	1.617
		t_h, cm	0.500[a]	0.500[a]	0.500[a]
Control	1	h_p, N/cm		100.324	86.48
		h_v, N-s/cm		0.217	2.067
Mass, kg			541.4	453.5	459.8

[a]Lower bound value.

The position gain is much larger than the velocity gain in the final design of the second run. This is because the frequency of the harmonic load is well below the first mode structural frequency, which results in more stiffness augmentation, rather than damping augmentation. The mass of the final design of the second run is lower than that of the first run because of the stiffness augmentation provided by the controller.

TABLE 3
Final Design Response Ratios
Cantilevered Beam: Mass Minimization

Constraint	Response Ratio		
	Uncontrolled	Controlled	Controlled stability constraint
Tip displacement	1.000[a]	0.996[a]	0.998[a]
Frequency	0.653	0.705	0.700
Control force		1.001[a]	1.000[a]
Stability			0.999[a]

[a]Critical constraint.

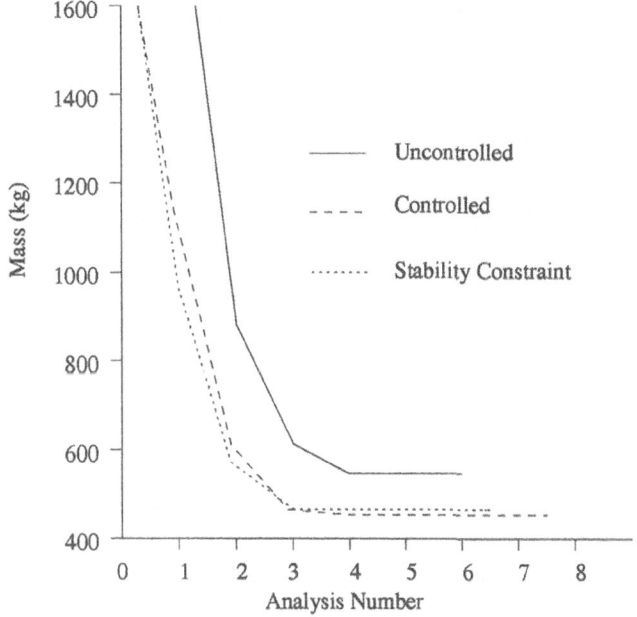

FIGURE 8
Iteration Histories, Cantilever Beam

In the final design of the second run, the value of the real part of the complex eigenvalue of the first mode is -0.35 rad/s. The stability constraint in the third run requires almost three times as much damping in this mode [i.e., $\sigma_1 \leq -1.0$ rad/s]. The controller velocity gain for the final design of the third run is much higher than that at the end of the second run due to this increased damping requirement. Because of the constraint on the control force, the position gain is forced to decrease from the value obtained in the second run. This results in an increase in the structural mass due to the loss of stiffness augmentation provided by the controllers.

In all three runs, the web thickness takes on its minimum gauge. This is because the flange's stiffness to mass ratio is much higher than that of the web.

5.2 Draper/RPL Structure: Control Force Minimization, Direct Output Feedback

The Draper/RPL structure (Ref. 8) consists of a massive central hub surrounded by four flexible appendages that have nonstructural mass attached to their free ends (see Fig. 9 and Ref. 11). The entire structure is free to rotate about the central axis of the hub. Table 4 lists the important parameters. Because of the symmetry of the structure, only half of it needs to be used for analysis purposes. The analysis model is shown in Fig. 9. There are four sensors on each arm at radial positions of $r = 24$ in. for sensors 2 and 6, $r = 36$ in. for sensors 3 and 7, $r = 43.6$ in. for sensors 4 and 8, and $r = 55.2$ in. for sensors 5 and 9. These sensors measure displacement and velocity in the circumferential direction. There is an additional rotational position and velocity sensor located at the axis of the hub. There are three actuators that apply torques to the structure. Actuator 1 is located at the axis of the hub and actuators 2 and 3 are located at $r = 36$ in. on arms 1 and 2, respectively. Twenty-one controllers, with sensor-actuator pairs defined in Table 5, will be used to control the structure. The controller and actuator configuration is the same on opposing arms. This actuator-sensor configuration is the same as that used in Ref. 11. Each arm of the structure is modeled with five beam type finite elements. The finite element nodes are located at the edge of the hub, at each sensor location, and at the tip mass.

TABLE 4
Draper/RPL Dimensions

Hub radius, R	12 in.
Rotary inertia of hub	1152 slug-in.^{2}
Mass density of beams	0.003021 slug/in.^{3}
Elastic modulus of arms	1.1×10^7 lb/in.^{3}
Arm thickness, t	0.125 in.
Arm height, h	6.0 in.
Arm length, l	48 in.
Tip mass	0.156941 slug
Rotary inertia of tip mass about its axis	0.2592 slug-in.^{2}

TABLE 5
Draper/RPL Structure Controller Sensor-Actuator
Definitions

	Sensor								
Actuator	1	2	3	4	5	6	7	8	9
1	1	4	–	5	–	2	–	3	–
2	–	18	19	20	21	14	15	16	17
3	–	10	11	13	13	6	7	8	9

The first nine open-loop frequencies of the structure, based on a finite element method analysis with 25 DOF, are shown in Table 6. The first frequency corresponds to a rigid-body mode of the structure rotating about its hub axis. Note that the next eight frequencies occur in four closely spaced pairs.

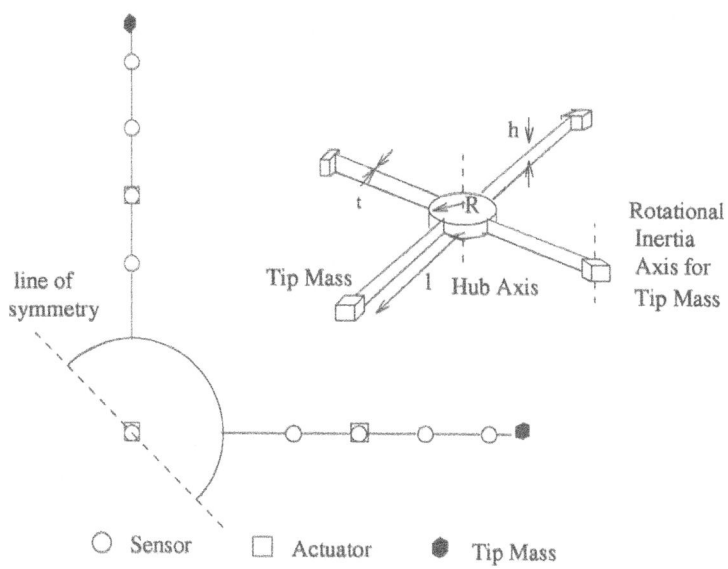

line of symmetry

Tip Mass

Hub Axis

h

R

t

l

Rotational
Inertia
Axis for
Tip Mass

○ Sensor   □ Actuator   ● Tip Mass

FIGURE 9
Draper/RPL Structure and Analysis Model
TABLE 6
Draper/RPL Structure Open-Loop Frequencies

Mode	Frequency (Hz)
1	0.00
2	0.70
3	1.26
4	8.18
5	8.40
6	24.87
7	25.00
8	47.65
9	47.74

The structure is loaded by a 720.0 in.-lb torque at 3.0 Hz applied at the hub axis of the structure. Structural damping of 0.1% is assumed. The total control force is the objective to be minimized. The design variables are the 42 position and velocity gains. The initial values for the gains are $h_P = 10.0$ lb-in./rad and $h_V = 1.0$ lb-in./rad for the controller 1 and $h_P = 1.0$ lb and h_V 1.0 lb-s for controllers 2-21. The initial gain values were chosen so that the system has no zero frequency closed-loop modes.

TABLE 7
Final Design Controller Forces
Draper/RPL Structure

Control Element	Controller Torque, in.-lb
1	0.005
2	0.005
3	0.005
4	0.006
5	0.003
6	2.555
7	0.082
8	4.300
9	0.399
10	0.052
11	0.006
12	0.007
13	0.001
14	0.004
15	0.005
16	0.002
17	0.014
18	0.004
19	0.012
20	0.739
21	0.003
Total	16.319

In this example, the total control force is minimized subject to constraints on the complex eigenvalues of the first nine modes and a tip displacement constraint. The damped frequency of mode 1 is constrained to be above 0.3 Hz and the damping ratios of the first nine modes are constrained as the following: above 0.03 for modes 1-3, above 0.01 for modes 4 and 5, above 0.002 for modes 6 and 7, and above 0.0015 for modes 8 and 9. These are the same damping ratio constraints as those used in Ref. 11. The tip displacement of arm 2 is constrained to be ≤ 0.1 in. The final design control element forces and final design response ratio values are shown in Tables 7 and 8. The iteration history plot is shown in Fig. 10. At the final design, five of the nine damping ratio constraints are critical (see Table 8). The control force of many of the control elements is near zero (see Table 7). Control elements 6, 8, 9, and 20 supply most of the control force in the final design. The tip displacement constraint is also critical.

In this problem, because the objective function is total control torque, the torque and, hence, the gains of the individual control elements are driven toward zero. The control torque is a very nonlinear function of the gains when they are near zero. Therefore, the high-quality nonlinear approximations, based on the concept of intermediate response quantities, must be used in order to achieve rapid convergence. The effect of this approximation on the convergence rate is shown graphically in Fig. 10. In this plot, the iteration

TABLE 8
Final Design Response Ratio Values
Draper/RPL Structure

Constraint	Response Ratio
Damped frequency (1)	0.898
Damping ratio (1)	0.769
Damping ratio (2)	0.997[a]
Damping ratio (3)	0.998[a]
Damping ratio (4)	0.997[a]
Damping ratio (5)	0.094
Damping ratio (6)	-4.990
Damping ratio (7)	-28.931
Damping ratio (8)	0.990[a]
Damping ratio (9)	1.000[a]
Tip displacement	0.999

[a]Constraint is critical.

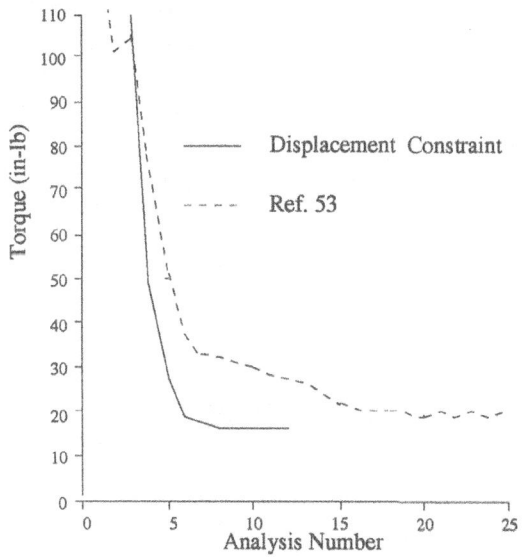

FIGURE 10
Iteration Histories, Draper/RPL Structure

history is shown, for the same problem, prior to the introduction of the refined approximations. The results reported in Ref. 53 were generated without the use of approximations based on intermediate response quantities. Note the oscillation near the optimum. This is caused by the poor quality of the approximations used in Ref. 53. The introduction of refined approximations not only reduces the number of analyses needed for convergence by a factor of 2, but it also leads to an objective function value that is 13.3% lower (16.32 in.-lb compared with 18.82 in.-lb).

5.3 Antenna Structure, State Feedback

An antenna structure (Refs. 30 and 31) is chosen as the example for state feedback control. It consists of eight aluminum beams ($E = 7.3 \times 10^6 \, N/cm^2$, $\rho = 2.77 \times 10^{-3} \, kg/cm^3$, $v = 0.325$) which have thin walled hollow box beam cross sections (see Figure 11). This structure is constrained to move vertically (**Y** - direction) only, so each nodal point has 3 degrees of freedom (translation in the **Y** direction and rotation about the **X** and **Z** reference axes shown in Figure 11) resulting in a total of eighteen degrees of freedom ($n = 18$). Four translational actuators ($n_a = 4$) weighting 4 kg each are attached to nodes 3, 5, 6 and 7. These actuators are oriented so that the force they generate acts in the vertical direction (degrees of freedom 4, 10, 13, and 16). Two ramp type transient loads are applied to node 3 at the same time. One is a vertical force ($f_1(t)$) and the other is a moment ($f_2(t)$) about a line parallel to the **X** reference axis but passing through node 3 which gives anti symmetric excitation. These loads are given by

$$f_1(t) = 333.3t \, N, \quad f_2(t) = 10.0 \times f_1(t) \, N \cdot cm$$

for $0 \le t \le 0.3$ seconds, and $f_1(t) = f_2(t) = 0$ for $t > 0.3$ seconds (see Figure 11). Transient response is considered for the time interval $0 \le t \le 2$ seconds and 20 out of 36 complex modes are used to calculate the peak response values. Passive damping is assumed to be zero. This assumption tends to be conservative for transient response constraints and it is further justified by observing that the magnitude of damping inherent to the structural system is likely to be small compared with that introduced by an active control system.

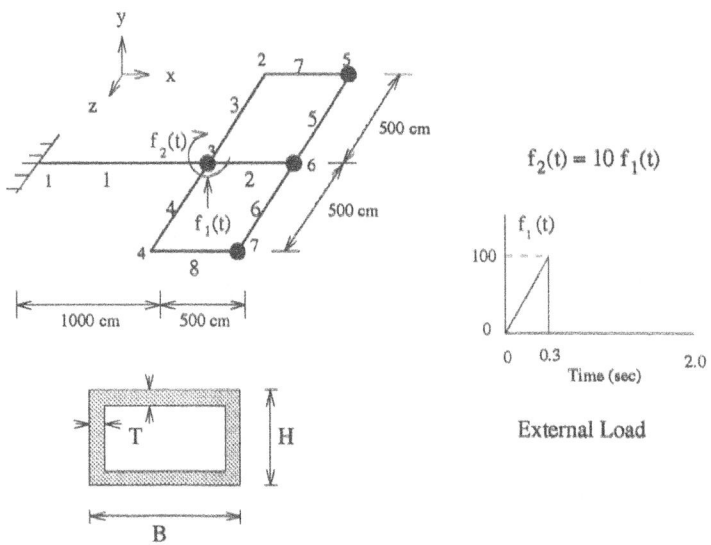

Box Beam Cross Section

FIGURE 11
Antenna Structure

Flange and web thicknesses are constrained to be the same, so there are three structural design variables for each finite element B, H and T). Structural linking is also used to make the structure remain symmetric with respect to the **XY** plane, which results in the total 15 independent structural design variables. The initial structure is uniform (B = H = 20.0 cm, T = 0.5 cm), and the side constraints are 10.0 cm $\leq$ **B**, **H** $\leq$ 25.0 cm, and 0.1 cm $\leq$ T $\leq$ 1.0 cm. The maximum number of control design variables is relatively large ($n_a \times 2n = 144$).

In all cases behavior constraints are imposed on: (1) the real part of all the retained complex modes ($\sigma_i \leq -0.5$); (2) the fourth and fifth damped frequencies ($\omega_{d_4} \geq 8.0$, $\omega_{d_5} \geq 9.25$ Hz); (3) the peak displacement of nodes 2, 4, 5 and 7 ($|q_i(t)| \leq 1.0$ cm, $i = 4$, 10, 13 and 16); and (4) the peak actuator force ($|U_j(t)| \leq 8.5$ N, $j = 1, 2, 3$ and 4).

Cases 1-10: Control Design Variable Linking

Cases 1-10 correspond to the 10 different control design variable linking schemes given in Table 1. Initial start-up gains are computed by solving 10 sets of 2×2 Riccati equations ($r = 10$). The 2×2 state weighting matrices are set to be $[Q_i] = \text{Diag}$ $[\omega_i^2, 1]$, $i = 1, 2, ..., r$ so that the first term of Eq. (77) represents a total (strain and kinetic) modal energy, and the control weighting coefficients (γ_i's) are chosen to be 1/400. In Figure 12, the initial closed-loop eigenvalues ($\lambda_i = \sigma_i + j\omega_{di}$) obtained by solving 10 sets of 2×2 Riccati equations are compared with those obtained from a full order Riccati equation solution, and as can be seen from the plot, these two solution methods give almost the same values for the lowest 10 modes.

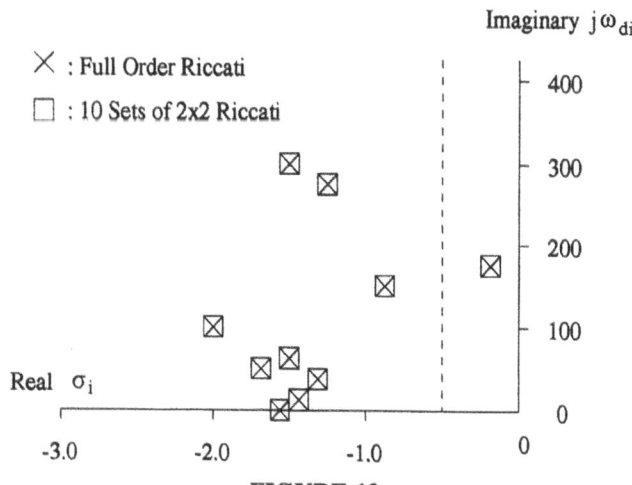

FIGURE 12
Comparison of Initial Closed-Loop Eigenvalues
Antenna Structure

At the initial design constraints on the real part of the 8-th closed loop eigenvalue and two of the damped frequencies are violated ($\sigma_g = -0.205 > -0.5$, $\omega_{d_4} = 6.74 < 8.0$ Hz and

$\omega_{d_5} = 7.61 < 9.25$ Hz). Furthermore, three of the peak control force constraints are initially violated ($|U_2(t)|_{max} = 8.76 > 8.5$ N, $|U_3(t)|_{max} = 9.26 > 8.5$ N and $|U_4(t)|_{max} = 9.33 > 8.5$ N), but all the peak displacement constraints are satisfied.

Case 1 does not use any control design variable linking scheme, therefore, 15 structural cross sectional dimensions and all 144 elements of the feedback gain matrix are treated as independent design variables. Since this case has full design freedom, the optimal solution for this case has the lowest value for all the cases. Iteration histories are shown in Table 9 and final structural designs in Table 10. Design masses include the fixed actuator masses as well as variable structural masses.

TABLE 9
Iteration Histories for Antenna Example: Cases 1-10

	Total Mass (kg)									
Analysis	Case 1	Case 2	Case 3	Case 4	Case 5	Case 6	Case 7	Case 8	Case 9	Case 10
1	502.14	502.14	502.14	502.14	502.14	502.14	205.14	502.14	502.14	502.14
2	462.20	462.39	466.52	462.25	455.96	469.64	465.14	485.24	474.45	484.61
3	345.34	343.47	349.71	351.13	359.45	376.65	361.18	383.18	374.55	387.64
4	254.17	259.92	261.19	263.19	283.10	293.90	280.72	302.66	291.97	297.69
5	213.91	218.31	235.33	230.47	231.57	237.33	231.87	241.31	234.26	240.14
6	184.11	191.87	210.26	213.59	212.83	218.09	212.22	216.56	214.15	215.05
7	174.34	176.98	199.19	200.50	204.21	205.11	204.82	208.12	207.01	208.60
8	168.93	170.94	192.61	195.82	200.82	201.28	201.05	206.07	207.00	206.33
9	169.35	167.01	193.88	194.60	198.91	200.51	200.69	205.78	204.60	206.07
10	165.61	166.99	190.74	192.95	198.71	200.35	200.63	204.22	204.47	206.06
11	164.42	165.43	188.11	191.28	198.45	200.27	200.59	204.19	204.46	206.06
12	164.08	165.03	185.03	189.94	198.01			204.16		
13	164.17	164.84	185.75	189.03	197.42					
14	163.99	164.64	184.26	187.51	196.93					
15	163.76	164.39	182.27	186.92	196.46					
16	163.20	164.19	180.79	186.66	196.32					
17	163.10	164.10	179.90	186.50	196.31					
18	163.11	164.05	179.79	186.34						
19			179.64							
Option*	(1)	(2)	(3)	(4)	(5)	(6)	(7)	(8)	(9)	(10)
CDV's**	(144)	(72)	(36)	(18)	(8)	(4)	(4)	(2)	(2)	(1)

* : Control design variable linking option number
** : Number of independent control design variables

As the freedom in the design space is reduced by imposing more restrictive control design variable linking schemes (from case 1 to case 10), it can be clearly seen from the results that: (1) the number of independent control variables in the optimization loop decreases (from 144 to 1); (2) the optimum mass increases (from 163.11 kg to 206.06 kg); and (3) total number of iterations decreases and the convergence becomes more robust. Even though there is more than 20 percent of difference in the optimum mass between case 1 and case 10, all cases show a similar trend in the final structural design. Namely, widths and depths of finite elements 1, 2, 5, and 6 take on their upper bound values and thicknesses of elements 3, 4, 7, and 8 move to their lower bound values (see Table 10).

TABLE 10
Final Designs for Antenna Example: Cases 1-10

		Cross Sectional Dimensions (cm)									
Case		Elem. 1	Elem. 2	Elem. 3,4	Elem. 5,6	Elem. 7,8	Case	Elem. 1	Elem. 2	Elem. 3,4	Elem. 5,6	Elem. 7,8
B		25.00[b]	25.00[b]	23.24	25.00[b]	19.31		25.00[b]	25.00[b]	22.04	25.00[b]	10.84
H	1	25.00[b]	25.00[b]	25.00[b]	25.00[b]	25.00[b]	6	25.00[b]	25.00[b]	25.00[b]	25.00[b]	24.83
T		0.100[a]	0.1108	0.100[a]	0.1935	0.100[a]		0.2166	0.1496	0.100[a]	0.2134	0.100[a]
B		25.00[b]	25.00[b]	24.11	25.00[b]	18.68		25.00[b]	25.00[b]	20.95	25.00[b]	12.08
H	2	25.00[b]	25.00[b]	25.00[b]	25.00[b]	24.99	7	25.00[b]	25.00[b]	25.00[b]	25.00[b]	24.86
T		0.1046	0.1129	0.100[a]	0.1908	0.100[a]		0.2054	0.1360	0.100[a]	0.2323	0.100[a]
B		25.00[b]	25.00[b]	17.68	25.00[b]	18.87		25.00[b]	25.00[b]	19.04	25.00[b]	14.43
H	3	25.00[b]	25.00[b]	24.41	25.00[b]	25.00[b]	8	25.00[b]	25.00[b]	25.00[b]	25.00[b]	25.00[b]
T		0.1393	0.1263	0.100[a]	0.2202	0.100[a]		0.1790	0.1115	0.100[a]	0.2832	0.100[a]
B		25.00[b]	25d.00	16.67	25.00[b]	18.56		25.00[b]	25.00[b]	18.47	25.00[b]	14.50
H	4	25.00[b]	[b]	21.92	25.00[b]	25.00[b]	9	25.00[b]	25.00[b]	25.00[b]	25.00[b]	25.00[b]
T		0.1454	25.00[b] 0.1213	0.100[a]	0.2490	0.100[a]		0.1752	0.1129	0.100[a]	0.2883	0.100[a]
B		25.00[b]	25.00[b]	21.55	25.00[b]	13.82		25.00[b]	25.00[b]	19.13	25.00[b]	15.80
H	5	25.00[b]	25.00[b]	25.00[b]	25.00[b]	25.00[b]	10	25.00[b]	25.00[b]	25.00[b]	25.00[b]	25.00[b]
T		0.1748	0.1152	0.1152	0.2526	0.100[a]		0.1904	0.100[a]	0.100[a]	0.2815	0.100[a]

[a] indicates lower bound value, [b] indicates upper bound value

It is interesting to compare the optimal weight using these 10 linking schemes with the case of using the Riccati equation to determine the control gains as a function of the structural parameters. This case was solved in Ref. 40 giving an optimal mass of 241.75 kg which is 15% higher than the one control variable case (case 10). This clearly shows that true integration of the control and structural system design can dramatically improve the overall characteristics of the final design.

Cases 11-12: Arbitrary Start Up Gains

This section examines the first method for initializing the control gain matrix before imposing control variable linking. In cases 11-12 two different sets of start up gains are chosen and four iterations are allowed without any control design variable linking (number of CDV = 144) and then the last linking option is imposed (number of CDV = 1). In case 11 the initial start up gains are the same as in cases 1-10 (obtained by solving 10 sets of decoupled Riccati equations). In case 12, the start up gain matrix is chosen in a manner similar to that which would be used if direct output feedback control was being employed ($H(1, 4) = H(2, 10) = H(3, 13) = H(4, 16) = 500$ kg/sec^2, $H(1, 22) = H(2,28) = H(2, 28) = H(3, 31) = H(4, 34) = 500$ kg/sec., $H(j,i) = 0$ elsewhere). In case 12 large move limits are used for small feedback gain elements (190 percent or absolute value of 100) in order to give more freedom at the beginning. The iteration history and the final structural designs are given in Tables 11 and 12.

TABLE 11
Iteration Histories for Antenna Example: Cases 11-13

	Total Mass (kg)		
Analysis	Case 11	Case 12	Case 13
1	502.14	502.14	502.14
2	462.20	527.16	484.61
3	345.34	544.29	385.71
4	254.17	461.82	287.94
5	213.91	399.83	239.17
6	184.11	346.49	211.97
7	174.34	309.54	203.64
8	172.94	276.72	201.39
9	174.24	245.00	201.29
10	174.18	225.77	201.03
11	174.18	207.78	201.00
12		202.86	200.99
13		201.34	
14		200.61	
15		199.42	
16		199.42	
17		199.42	

TABLE 12
Final Designs for Antenna Example: Cases 11-13

	Case	Cross Sectional Dimensions (cm)				
		Element 1	Element 2	Elements 3,4	Elements 5,6	Elements 7,8
B		25.00^b	25.00^b	20.21	25.00^b	17.10
H	11	25.00^b	24.90	25.00^b	25.00^b	23.99
T		0.1124	0.1572	0.100^a	0.2112	0.100^a
B		25.00^b	25.00^b	25.00^b	25.00^b	12.45
H	12	25.00^b	25.00^b	25.00^b	21.11	25.00^b
T		0.2032	0.1921	0.100^a	0.2092	0.100^a
B		25.00^b	25.00^b	20.36	25.00^b	15.71
H	13	25.00^b	25.00^b	25.00^b	25.00^b	25.00^b
T		0.1736	0.100^a	0.100^a	0.2774	0.100^a

[a] indicates lower bound value   [b] indicates upper bound value

TABLE 13
Iteration Histories for Antenna Example: Cases 14-23

	Total Mass (kg)									
Analysis	Case 14 (1*)	Case 15 (2*)	Case 16 (3*)	Case 17 (4*)	Case 18 (5*)	Case 19 (6*)	Case 20 (7*)	Case 21 (8*)	Case 22 (9*)	Case 23 (10*)
1	502.14	502.14	502.14	502.14	502.14	502.14	502.14	502.14	502.14	502.14
2	484.61	470.88	471.87	444.12	454.83	453.14	449.81	450.11	453.15	469.10
3	387.64	328.11	328.47	303.50	346.62	299.54	293.56	294.62	307.14	350.45
4	297.69	260.67	254.76	233.56	268.66	228.72	217.54	218.41	228.31	268.09
5	240.14	217.39	214.99	193.83	215.70	187.80	184.73	180.21	188.59	215.51
6	215.05	199.59	200.57	178.76	185.24	175.65	176.29	174.64	175.95	181.84
7	208.60	193.04	192.60	174.18	175.46	173.27	173.07	174.70	173.41	173.50
8	206.33	191.20	190.47	173.05	176.56	172.41	172.28	174.38	172.68	171.19
9	206.07	190.90	190.23	172.50	173.25	172.29	171.93	173.93	171.10	171.02
10	206.06	190.80	190.14	174.46	172.66	172.17	171.80	171.27	170.84	170.96
11	206.06	190.70	189.94	172.46	172.46		171.73	170.97	170.70	170.92
12			189.85		172.37			170.84	170.65	
13			189.81		172.32			170.81		

* : Number of independent control design variables

TABLE 14
Final Structural Designs, Cases 14-23

		Cross Sectional Dimensions (cm)									
	Case	Elem. 1	Elem. 2	Elem. 3,4	Elem. 5,6	Elem. 7,8	Case	Elem. 1	Elem. 2	Elem. 3,4	Elem. 5,6	Elem. 7,8
B		25.00[b]	25.00[b]	19.62	25.00[b]	15.51		25.00[b]	25.00[b]	18.72	25.00[b]	21.03
H	14	25.00[b]	25.00[b]	25.00[b]	25.00[b]	25.00[b]	19	25.00[b]	25.00[b]	25.00[b]	25.00[b]	25.00[b]
T		0.1899	0.100[a]	0.100[a]	0.2816	0.100[a]		0.100[a]	0.1107	0.100[a]	0.2330	0.100[a]
B		25.00[b]	25.00[b]	19.78	25.00[b]	24.94		25.00[b]	25.00[b]	17.65	25.00[b]	20.47
H	15	25.00[b]	25.00[b]	25.00[b]	23.05	22.23	20	25.00[b]	25.00[b]	24.89	25.00[b]	25.00[b]
T		0.1830	0.1326	0.100[a]	0.2103	0.100[a]		0.100[a]	0.1165	0.100[a]	0.2318	0.100[a]
B		25.00[b]	25.00[b]	20.03	25.00[b]	22.86		25.00[b]	25.00[b]	14.95	25.00[b]	23.42
H	16	25.00[b]	25.00[b]	25.00[b]	23.37	21.94	21	25.00[b]	25.00[b]	23.00	25.00[b]	25.00[b]
T		0.1869	0.1413	0.100[a]	0.2012	0.100[a]		0.1073	0.1244	0.100[a]	0.2203	0.100[a]
B		25.00[b]	25.00[b]	20.90	25.00[b]	20.51		25.00[b]	25.00[b]	14.62	25.00[b]	22.12
H	17	25.00[b]	25.00[b]	25.00[b]	25.00[b]	25.00[b]	22	25.00[b]	25.00[b]	22.36	25.00[b]	25.00[b]
T		0.100[a]	0.100[a]	0.100[a]	0.2356	0.100[a]		0.1013	0.1319	0.100[a]	0.2267	0.100[a]
B		25.00[b]	25.00[b]	18.85	25.00[b]	23.70		25.00[b]	25.00[b]	19.96	25.00[b]	20.29
H	18	25.00[b]	25.00[b]	25.00[b]	25.00[b]	25.00[b]	23	25.00[b]	25.00[b]	25.00[b]	25.00[b]	25.00[b]
T		0.1025	0.1036	0.100[a]	0.2287	0.100[a]		0.100[a]	0.1058	0.100[a]	0.2296	0.100[a]

[a] indicates lower bound value,   [b] indicates upper bound value

Case 13: Updating Fixed Ratios in Gain Matrices

Case 13 is identical to case 10 except that the feedback gain matrix is resolved at the beginning of each iteration to update the fixed ratios in the gain matrix. After 10 updating stages all the coefficients of the control force weighting matrix converge within 3 percent. The iteration history and the final structural designs are given in Tables 11 and 12. These last three cases show a similar pattern in the final designs with a difference in weight of 15% between case 13 and case 11.

TABLE 15
Iteration Histories for Antenna Example: Cases 24-33

	Total Mass (kg)									
Analysis	Case 24 (2*)	Case 25 (4*)	Case 26 (6*)	Case 27 (8*)	Case 28 (10*)	Case 29 (12*)	Case 30 (14*)	Case 31 (16*)	Case 32 (18*)	Case 33 (20*)
1	502.14	502.14	502.14	502.14	502.14	502.14	205.14	502.14	502.14	502.14
2	485.24	482.76	462.89	420.35	470.72	475.80	474.66	474.47	466.40	434.64
3	383.18	343.98	339.86	288.81	326.51	301.89	304.10	304.66	304.00	287.18
4	302.66	276.53	260.17	229.95	247.00	228.23	230.32	231.61	226.62	224.61
5	241.31	225.43	222.33	196.72	207.82	188.16	197.86	197.41	185.11	183.32
6	216.56	200.56	202.26	180.59	185.02	175.46	178.77	178.20	174.03	172.69
7	208.12	189.44	191.17	174.55	173.76	174.21	172.28	172.25	172.93	171.31
8	206.07	187.41	186.95	172.76	171.10	173.71	171.81	171.09	172.39	169.42
9	205.78	186.31	185.45	171.64	171.25	172.95	169.76	170.49	172.02	166.81
10	204.22	185.54	183.60	170.30	170.49	171.19	171.33	170.21	171.80	166.79
11	204.19	185.21	181.66	170.41	169.99	170.91	170.02	169.72	170.04	166.22
12	204.16	185.13	181.23	170.37	169.80	170.69	169.07	169.16	169.67	165.66
13		185.09	180.70		169.64	169.68	167.59	167.78	169.37	165.61
14			178.65		169.55	169.49	166.91	167.26	169.17	166.50
15			178.06			168.38	166.23	167.00	168.92	165.30
16			177.86			167.40	166.17	166.65	167.74	165.14
17			177.31			167.01	166.01	166.29	166.91	164.95
18			176.72			166.85		165.69	165.84	164.86
19			176.67			166.76		165.46	165.64	164.80
20			176.56					165.34	165.49	
21								165.29	165.38	

* : Number of independent control design variables

Cases 14-23, Linking on [H]

Cases 14-23 select participation coefficients α_i's of $[H^\circ]^{(i)}$ (see Eq. (86)) as design variables. These α_i's are further linked so that in each case the number of independent control design variables is different. The basic scheme employed here is to treat the first $(K-1)$ variables $\alpha_1, \ldots, \alpha_{K-1}$ as independent and then link all the remaining variables $\alpha_K = \alpha_{K+1} = \cdots = \alpha_{10}$ so that the total number of independent control design variables after linking is K. For example, when $K = 1$ there is only one independent design variable after linking, since $\alpha_1 = \alpha_2 = \cdots = \alpha_{10}$. When $K = 2$ there are two independent design variables after linking, namely α_1 and $\alpha_2 = \alpha_3 = \cdots = \alpha_{10}$. Finally, when $K = 10$ there will be ten independent design variables after linking, namely the participation coefficients of the ten basis matrices in Eq. (86). Iteration histories and final structural designs for cases 14-23 are given in Tables 13 and 14.

Cases 24-33, Linking on $[H_P]$ and $[H_v]$

Cases 24-33 are the same as cases 14-23 except that position and velocity parts of the gain matrix are separated. Namely, the α_i's of both $[H_P^o]^{(i)}$ and $[H_V^o]^{(i)}$ (see Eq. (87)) are candidates for design variables, so that the maximum number of independent control design variableqqs after linking is doubled from 10 to 20. Iteration histories and final structural designs are given in Tables 15 and 16.

Comparing Tables 9, 13 and 15 shows that using the 2×2 Riccati basis matrices (cases 14-33) gives better results than the row and column linking options given in Table 1. That is, the basis matrix linking (see Eqs. 86 and 87) gives better results, in the sense that with the same or fewer independent control design variables, significantly lower final design mass can be achieved.

TABLE 16
Final Structural Designs, Cases 24-33

	Case	Elem. 1	Elem. 2	Elem. 3,4	Elem. 5,6	Elem. 7,8	Case	Elem. 1	Elem. 2	Elem. 3,4	Elem. 5,6	Elem. 7,8
				Cross Sectional Dimensions (cm)								
B		25.00^b	25.00^b	18.99	25.00^b	14.13		25.00^b	25.00^b	13.02	25.00^b	24.32
H	24	25.00^b	25.00^b	25.00^b	25.00^b	25.00^b	29	25.00^b	25.00^b	24.12	24.93	25.00^b
T		0.1777	0.1149	0.100^a	0.2835	0.100^a		0.100^a	0.1341	0.100^a	0.2082	0.100^a
B		25.00^b	25.00^b	19.57	25.00^b	25.00^b		25.00^b	25.00^b	13.64	25.00^b	24.70
H	25	25.00^b	25.00^b	25.00^b	22.84	22.14	30	25.00^b	25.00^b	24.06	25.00^b	25.00^b
T		0.1687	0.1361	0.100^a	0.2032	0.100^a		0.100^a	0.1289	0.100^a	0.2061	0.100^a
B		25.00^b	25.00^b	13.22	25.00^b	25.00^b		25.00^b	25.00^b	15.33	25.00^b	25.00^b
H	26	25.00^b	25.00^b	25.00^b	23.70	24.65	31	25.00^b	25.00^b	23.14	25.00^b	25.00^b
T		0.1239	0.1259	0.100^a	0.2271	0.100^a		0.100^a	0.1213	0.100^a	0.2047	0.100^a
B		25.00^b	25.00^b	22.11	25.00^b	20.00		25.00^b	25.00^b	14.99	25.00^b	24.29
H	27	25.00^b	25.00^b	25.00^b	25.00^b	25.00^b	32	25.00^b	25.00^b	23.31	25.00^b	25.00^b
T		0.100^a	0.1021	0.100^a	0.2256	0.100^a		0.100^a	0.1186	0.100^a	0.2084	0.100^a
B		25.00^b	25.00^b	19.45	25.00^b	21.32		25.00^b	25.00^b	14.91	25.00^b	24.62
H	28	25.00^b	25.00^b	24.70	25.00^b	25.00^b	33	25.00^b	25.00^b	22.93	25.00^b	25.00^b
T		0.100^a	0.1051	0.100^a	0.2246	0.100^a		0.100^a	0.1205	0.100^a	0.2054	0.100^a

[a] indicates lower bound value [b] indicates upper bound value

6. SUMMARY

The present chapter examines the integrated approach to the simultaneous optimization of structural and control systems. The design problem is posed as a general non-linear mathematical programming problem in which structural variables and control variables are treated simultaneously as design variables. Using this approach, complete integration of the design space is achieved. Constraints on pure structural responses, as well as constraints associated to the control system behavior and control-structure interaction can be imposed simultaneously into the design problem.

The two most usual ways of implementing a feedback control law (direct output feedback, state feedback) are discussed in detail. A complete analysis is given on how to choose design variables and associated constraints. In particular, the concepts of stability, observability, and controllability are presented in a way that can be incorporated to the integrated design process. For the case of state feedback, in which the gains cannot be used directly as design variables, linking schemes are presented which allow the integration with a reasonable amount of computational effort.

The general design problem is solved through the iterative construction and solution of a sequence of explicit approximate problems (approximation concepts approach). Each approximate problem is generated through the application of the concepts of intermediate design variables and intermediate response quantities.

The methodology presented is applied in the last section to the solution of a number of example problems that show the flexibility offered by the integrated design. Different kinds of constraints are considered for both direct output feedback and state feedback control laws. The use of the approximation concepts approach is reflected in the small number of system analyses required to obtained a near optimal solution.

In summary, the emerging field of actively controlled structures or intelligent structures is very promising and a wide variety of applications are at hand. There are still several areas of research such as considering reliability of the system, integrating to the design the actuator material and configuration, extending the design space to include observer variables, considering time delay for transient constraints, etc. The further development of these areas in addition to the improvement in the design techniques will stimulate the use of intelligent structures in all future engineering applications. Some of the applications currently being considered include: control of shape and vibration of large space optics support structures, automotive active suspensions, active attenuation of sound and transmission from and through panels, and active vibration isolation of engines.

7. REFERENCES

1. Haftka, R.T., Martinovic, Z.N. and Hallauer, W.L., "Enhanced Vibration Controllability by Minor Structural Modification," *AIAA Journal*, Vol. 23, No. 8, Aug. 1985, pp. 1260-1266.

2. Junkins, J.L., Bodden, D.S., and Turner, J.D., "A Unified Approach to Structure and Control System Design Iteration," *Proceedings of the 4th International Conference on Applied Numerical Modeling,* Tainan, Taiwan, Dec. 1984, pp. 483-490.

3. Junkins, J.L. and Rew, D.W., "Unified Optimization of Structures and Controllers," *Large Space Structures: Dynamics and Control*, edited by Atluri, S.N. and Amos, A.K., Springer-Verlag, Berlin, 1988, pp. 323-353.

4. Lim, K.B. and Junkins, J.L., "Robustness Optimization of Structural and Controller Parameters," *Journal of Guidance and Control*, Vol. 12, No. 1, Jan.-Feb., 1989, pp. 89-96.

5. Lust, R.V. and Schmit, L.A., "Control-Augmented Structural Synthesis," *AIAA Journal*, Vol. 26, No. 1, Jan. 1988, pp. 86-94.

6. Lust, R.V. and Schmit, L.A., "Control-Augmented Structural Synthesis," NASA CR 4132, April 1988.

7. Thomas, H.L., "Improved Approximations for Simultaneous Structural and Control System Synthesis," PhD Dissertation, University of California, Los Angeles, 1990.

8. Thomas, H.L., Sepulveda, A.E., and Schmit, L.A., "Improved Approximations for Control Augmented Structural Optimization," *AIAA Journal*, Vol. 30, No. 1, Jan. 1992, pp. 171-179.

9. McLaren, M.D. and Slater, G.L., "A Covariance Approach to Integrated Control/Structure Optimization," *AIAA Dynamics Specialist Conference*, AIAA-90-1211-CP, Long Beach, CA, April 5-6, 1990, pp. 189-205.

10. Hale, A.L., Lisowski, R.J., and Dahl, W.E., "Optimal Simultaneous Structural and Control Design of Maneuvering Flexible Spacecraft," *Journal of Guidance, Control and Dynamics*, Vol. 8, No. 1, Jan.-Feb. 1985, pp. 86-93.

11. Bodden, D.S. and Junkins, J.L., "Eigenvalue Optimization Algorithms for Structure/Controller Design Iterations," *Journal of Guidance, Control, and Dynamics*, Vol. 8, No. 6, Nov.-Dec. 1985, pp. 697-706.

12. Venkayya, V.B. and Tischler, V.A., "Frequency Control and its Effect on the Dynamic Response of Flexible Structures," *AIAA Journal*, Vol. 23, No. 11, Nov. 1985, pp. 1768-1774.

13. Khot, N.S., Venkayya, V.B., and Eastep, F.E., "Optimal Structural Modifications to Enhance the Active Vibration Control of Flexible Structures," *AIAA Journal*, Vol. 24, No. 8, Aug. 1986, pp. 1368-1374.

14. Khot, N.S. and Grandhi, R.V., "An Integrated Approach to Structure and Control Design of Space Structures," *Recent Developments in Structural Optimization*, edited by Cheng, F.Y., ASCE Publication, 1986, pp. 26-39.

15. Khot, N.S., Grandhi, R.V., and Venkayya, V.B., "Structures and Control Optimization of Space Structures," *Proceedings of the AIAA/ASME/ASCE/AHS 28th Structures, Structural Dynamics and Materials Conference*, Monterey, CA, April 1987, pp. 850-860.

16. Grandhi, R.V., Structural and Control Optimization of Space Structures," *Computers and Structures*, Vol. 31, No. 2, 1989, pp. 139-150.

17. Khot, N.S., "Minimum Weight and Optimal Control Design of Space Structures," *Computer Aided Optimal Design: Structural and Mechanical Systems*, NATO ASI Series, Vol. F27, edited by Mota Soares, CA, Springer-Verlag, Berlin, 1987, pp. 389-403.

18. Khot, N.S., Öz, H., Grandhi, R.V., Eastep, F.E., and Venkayya, V.B., "Optimal Structural Design with Control Gain Norm Constraints," *AIAA Journal*, Vol. 26, No. 5, May 1988, pp. 604-611.

19. Belvin, W.K. and Park, K.C., "Structural Tailoring and Feedback Control Synthesis: An Interdisciplinary Approach," *Journal of Guidance, Control, and Dynamics*, Vol. 13, No. 3, May-June, 1990, pp. 424-429.

20. Miller, D.F. and Shim, J., "Gradient-Based Combined Structural and Control Optimization," *Journal of Guidance*, Vol. 10, No. 3, May-June, 1987, pp. 291-298.

21. Salama, M., Garba, J., Demsetz, L., and Udwadia, F., "Simultaneous Optimization of Controlled Structures," *Computational Mechanics*, Vol. 3, 1988, pp. 275-282.

22. Becus, G.A., Lui, C.Y., Venkayya, V.B., and Tischler, V.A., "Simultaneous Structural and Control Optimization via Linear Quadratic Regulator Eigenstructure Assignment," *Proceedings of the 58th Shock and Vibration Symposium*, Huntsville, AL, Oct. 1987, pp. 225-232.

314

23. Onoda, J. and Haftka, R.T., "An Approach to Structure/Control Simultaneous Optimization for Large Flexible Spacecraft," *AIAA Journal*, Vol. 25, No. 8, Aug. 1987, pp. 1133-1138.

24. Rao, S.S., "Combined Structural and Control Optimization for Flexible Structures," *Engineering Optimization*, Vol. 13, 1988, pp. 1-16.

25. Rao, S.S., Venkayya, V.B., and Khot, N.S., "Game Theory Approach for the Integrated Design of Structures and Controls," *AIAA Journal*, Vol. 26, No. 4, 1988, pp. 463-469.

26. Rao, S.S., Pan, T.S., and Venkayya, V.B., "Robustness Improvement of Actively Controlled Structures Through Structural Modifications," *AIAA Journal*, Vol. 28, No. 2, Feb. 1990, pp. 353-361.

27. Eastep, F., Khot, N.S., and Grandhi, R., "Improving the Active Vibrational Control of Large Space Structures Through Structural Modifications," *Acta Astronautica*, Vol. 15, No. 6/7, 1987, pp. 383-389.

28. Manning, R.A. and Schmit, L.A., "Control Augmented Structural/Synthesis with Transient Response Constraints," *AIAA Journal*, Vol. 28, No. 5, May 1990, pp. 883-891.

29. Sepulveda, A.E. and Schmit, L.A., "Optimal Placement of Actuators and Sensors in Control Augmented Structural Optimization," *International Journal for Numerical Methods in Engineering*, Vol. 32, No. 6, October 1991, pp. 1165-1187.

30. Jin, I.M. and Schmit, L.A., "Control Design Variable Linking for Optimization of Structural/Control Systems," *AIAA Journal*, Vol. 30, No. 7, July 1992, pp. 1892-1900.

31. Jin, I.M. and Schmit, L.A., "Improved Control Design Variable Linking for Optimization of Structural/Control Systems," *AIAA Journal*, Vol. 31, No. 11, Nov. 1993, pp. 2111-2120.

32. Jin, I.M. and Sepulveda, A.E., "Structural/Control System Optimization with Variable Actuator Masses," *Proc. of the 34th AIAA/ASME/ASCE/AHS/ASC Structures, Structural Dynamics and Materials Conference*, La Jolla, CA, April 19-22, 1993, pp. 1443-1451.

33. Sepulveda, A.E., Jin, I.K., and Schmit, L.A., "Optimal Placement of Active Elements in Control Augmented Structural Synthesis," *AIAA Journal*, Vol. 31, No. 10, October, 1993, pp. 1906-1915.

34. Sepulveda, A.E. and Jin, I.K., "Design of Structure/Control with Transient Response Constraints Exhibiting Relative Minima," *Proc. of the 4th AIAA/USAF/NASA/OAI Symposium on Multidisciplinary Analysis and Optimization*, Cleveland, OH, September 21-23, 1992.

35. Basile, G. and Marro, G., "Self-Bounded Controlled Invariant Subspaces: A Straight-Forward Approach to Constrained Controllability," *J. of Optimization Theory and Applications*, Vol. 38, No. 1, September 1982, pp. 71-83.

36. Barmish, B.R. and Schmitendorf, W.E., "The Associated Disturbance-Free System: A Means for Investigating the Controllability of a Disturbed System," *J. Of Optimization Theory and Applications*, Vol. 38, No. 4, December 1982, pp. 525-540.

37. Luenberger, D.G., "Introduction to Dynamic Systems: Theory, Models and Applications," Wiley, 1979.

38. Müller, P.C. and Weber, H.I., "Analysis and Optimization of Certain Qualities of Controllability and Observability for Linear Dynamical Systems," *Automotive*, Vol. 8, 1972, pp. 237-246.

39. Johnson, C.D., "Theory of Disturbance-Accomodating Controllers," Control and Dynamic Systems, Advances in Theory and Applications, Edited by C.T. Leonds, Academic Press, Vol. 12, 1976.

40. Sepulveda, A.E., "Optimal Placement of Actuators and Sensors in Control Augmented Structural Optimization," PhD Dissertation, University of California, Los Angeles, 1990.

41. Schmit, L.A. and Farshi, B., "Some Approximation Concepts for Efficient Structural Synthesis," *AIAA Journal*, Vol. 12, May 1974, pp. 692-699.

42. Schmit, L.A. and Miura, H., "Approximation Concepts for Efficient Structural Synthesis," NASA CR 2552, March 1976.

43. Khot, N.S. and Abkyankar, N.S., "Integrated Optimum Structural and Control Design," *AIAA Progress in Astronautics and Aeronautics*, Edited by M.P. Kamat, Vol. 150, 1993, pp. 743-767.

44. Balas, M.J., "Active Control of Flexible Systems," *Journal of Optimization Theory and Applications*, Vol. 25, No. 3, July 1978, pp. 415-436.

45. Jin, I.M. and Sepulveda, A.E., "Structural/Control System Optimization with Variable Actuator Masses," *Proc. of the 34th AIAA/ASME/ASCE/AHS/ASC Structures, Structural Dynamics and Materials Conference*, La Jolla, CA, April 19-22, 1993, pp. 1443-1451.

46. Dantzig, G.B., Linear Programming and Extensions, Princeton University Press, Princeton, NJ, 1963.

47. Starnes, J.R. Jr. and Haftka, R.T., "Preliminary Design of Composite Wings for Buckling, Stress, and Displacement Constraints," *Journal of Aircraft*, Vol. 16, Aug. 1979, pp. 564-570.

48. Lust, R.V. and Schmit, L.A., "Alternative Approximation Concepts for Space Frame Synthesis," NASA CR 172526, March, 1985.

49. Vanderplaats, G.N. and Salajegheh, E., "A New Approximation Method for Stress Constraints in Structural Synthesis," *AIAA Journal*, Vol. 27, March 1989, pp. 352-358.

50. Canfield, R.A., "An Approximation Function for Frequency Constrained Structural Optimization," *AIAA Journal*, Vol. 28, No. 6, June 1990, pp. 1116-1122.

51. Thomas, H.L., Sepulveda, A.E., and Schmit, L.A., "Improved Approximations for Dynamic Displacements Using Intermediate Response Quantities," *Proceedings of the Third NASA/Air Force Symposium on Recent Advances in Multidisciplinary Analysis and Optimization*, San Francisco, CA, Sept. 24-26, 1990, pp. 95-104.

52. Sepulveda, A.E., Thomas, H.L., and Schmit, L.A., "Improved Transient Response Approximations for Control Augmented Structural Optimization," *Proc. of the Second Pan American Congress of Applied Mechanics*, Jan. 2-5, Valparaiso, Chile, 1991, pp. 611-614.

53. Thomas, H.L. and Schmit, L.A., "Control Augmented Structural Synthesis with Dynamic Stability Constraints," *Proceedings of the AIAA/ASME/ASCE/AHS/ASE 30th Structures, Structural Dynamics and Materials Conference*, Washington, DC, April 1989, pp. 521-531.

CONTACT SHAPE OPTIMIZATION

JAROSLAV HASLINGER

Faculty of Mathematics and Physics
Charles University, Prague
Ke Karlovu 5, 12000 Praha 2, Czech Republic
and
University of Jyväskylä
P.O. Box 35
FIN–40351 Jyväskylä, Finland

INTRODUCTION

Shape optimization is a branch of the optimal control theory in which the control variable is connected with the geometry of the problem. The aim is to find a shape from an a priori defined class of domains, for wich the corresponding cost functional attains its minimum. Shape optimization of mechanical systems, behaviour of which is described by equations, has been very well analyzed from the mathematical, as well as from the mechanical point of view, see [1], [2], [3] and references therein. The aim of this contribution is to extend results to the case, in which the system is described by the so called *variational inequalities*. There are two reasons for doing that: 1) The behavior of many mechanical models can be described just using the framework of variational inequalities, see [4], [5]. 2) The optimal control of systems governed by variational inequalities has one specific feature, compared with problems governed by equations, namely the whole problem is *nonsmooth*. By nonsmoothness we mean the fact that the mapping: *Control variable → state → cost functional* is *not continuously differentiable*. Therefore we have to be very careful with the choice of numerical minimization methods. In classical optimization problems we also often meet the situation when the resulting function which has to be minimized is nonsmooth in the above mentioned sense. But this is due to the nonsmoothness of the cost functional itself, or by the presence of constraints, etc. Such kind of explicit nonsmoothness can be usually overcome by introducing dummy variables, e.g. The source of the nonsmoothness in our kind of problems is different, namely it is due to the nonsmoothness of the inner mapping: control variable → state, i.e. this phenomena is more implicit.

The aim of this contribution is to present mathematical theory for such type of problems. Here we restricted ourselves to one of the most typical applications of the theory of variational inequalities in mechanics of solids, namely to the so called

J. Herskovits (ed.), Advances in Structural Optimization, 317–347.

contact problems. We were inspired by [6], where the problem of designing a contact part of two deformable bodies, separated by an initial gap, was solved. The aim was to find a shape of boundaries in the vicinity of the contact area in such a way that the reaction forces are evenly distributed along it.

The contribution is organized as follows: in Section 1, we briefly repeat the classical and variational definition of contact problems and their approximation by finite elements. The material for this part can be found in [7], [8]. In Section 2, the abstract setting of optimal shape design problems is presented and sufficient conditions, guaranteeing the existence of its solution are presented. The results of this section are then used in Section 3, where the shape optimization of a deformable body, supported by a perfectly rigid half plane is defined. Section 4 deals with finite element approximation of the problem in question. Next two sections are the most important part of the contribution. In Section 5 we analyze in more details how the solution of the discrete problem depends on design variables. We explain why the solution is not continuously differentiable, when this situation occurs and how the directional derivatives can be computed. In Section 6 we present two examples of the cost functionals, the choice of which seems to be very important in practical applications, namely *the total potential* and the *reciprocal energy* of the system. Finally, results of a model example are presented in Section 7.

The theoretical background for the material presented in Sections 2–6 can be found in [9].

Very brief survey of results from the theory of elliptic variational inequalities and notations is available at the end of the contribution.

1. VARIATIONAL FORMULATION OF CONTACT PROBLEMS

Let us assume two deformable bodies in mutual contact, subjected to body force F and surface tractions P on a part of their boundaries. The aim is to determine the equilibrium state of this system. It is clear that the deformation depends, besides other, on the reaction forces acting on the contact part. Unfortunately, these forces are not known a priori. More precisely, at points which are not in contact, reactions are equal to zero, while at points being in contact, only the pressure may occur, but its magnitude is also not known a priori. All these facts make the whole problem more involved. Next, we present the classical formulation of the problem, assuming the plane case only with *linearly elastic bodies* in contact.

Let both bodies be represented by domains $\Omega', \Omega'' \subset \mathbf{R}^2$, with boundaries $\partial\Omega'$, $\partial\Omega''$, respectively. To simplify the formulation of contact conditions we shall consider the case, when there is no gap between Ω' and Ω''. Denote by $\Gamma_K = \partial\Omega' \cap \partial\Omega''$ the contact part of both bodies and let us consider the case when Γ_K cannot expend beyond a certain domain, which is determined by the geometric situation in the vicinity of Γ_K (see Fig. 1).

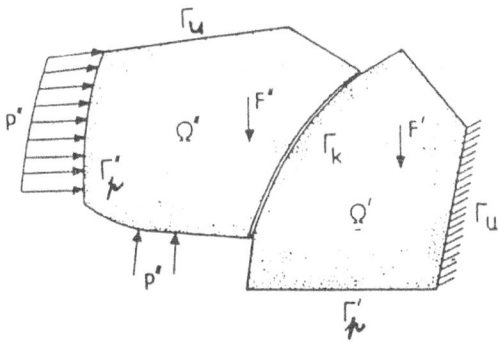

Figure 1.

We split the boundaries $\partial\Omega'$ and $\partial\Omega''$ as follows:

$$\partial\Omega' = \Gamma'_u \cup \Gamma'_p \cup \Gamma_K$$
$$\partial\Omega'' = \Gamma''_u \cup \Gamma''_p \cup \Gamma_K$$

where Γ'_u, Γ'_p are *open, non-empty* parts of $\partial\Omega'$ (and analogously for Γ''_u, Γ''_p).
We say that *the unilateral contact* occurs on Γ_K, iff

(1.1) $$u'_n + u''_n \leq 0 \quad \text{on } \Gamma_K,$$

where $u'_n - u' \cdot n'$, $u''_n = u'' \cdot n''$ are the normal components of the displacement fields $u' = [u'_1, u'_2]$, $u'' = [u''_1, u''_2]$ related to Ω', Ω'', respectively (note that $n' = -n''$).

REMARK 1.1: The condition (1.1) is the *linearization* of the real non-penetration condition, valid for "flat" arc Γ_K. For the detailed derivation and justification see [7], [8]. $\square$

The kinematical condition (1.1) has to be completed, namely:

(1.2) $$T'_n = T''_n, \; T'_t = T''_t \quad \text{on } \Gamma_K \quad (\textit{action and reaction law}),$$

where T'_n and T'_t are the normal, the tangential component of the stress vector T', respectively (analogously T''_n, T''_t). Their common value will be denoted by T_n and T_t, respectively.

Assuming that surfaces of Ω' and Ω'' are perfectly smooth, i.e. no friction occurs, the tangential component T_t vanishes on Γ_K, while the normal component T_n cannot be tension:

(1.3) $$T_t = 0, \; T_n \leq 0 \quad \text{on } \Gamma_K.$$

Finally, the following *complementarity condition* holds:

$$(1.4) \qquad\qquad (u'_n + u''_n)T_n = 0 \quad \text{on } \Gamma_K.$$

The condition (1.4) results from the following argument: at the points where no contact occurs, i.e. $u'_n + u''_n < 0$, no contact forces can arise, i.e. $T_n = 0$.

On the parts Γ'_u and Γ''_u zero displacements are prescribed:

$$(1.5) \qquad\qquad u'_i = 0 \text{ on } \Gamma'_u, \quad u''_i = 0 \text{ on } \Gamma''_u; \quad i = 1, 2,$$

while on the parts Γ'_p and Γ''_p surface tractions are acting:

$$(1.6) \qquad\qquad \tau'_{ij} n'_j = P'_i \text{ on } \Gamma'_p, \quad \tau''_{ij} n''_j = P''_i \text{ on } \Gamma''_p; \quad i = 1, 2.$$

Here the symbols τ'_{ij} and τ''_{ij} stand for the ij-th component of the stress tensor, related to the strain tensor $\varepsilon' = \{\varepsilon'_{ij}(u')\}_{i,j=1}^2$, $\varepsilon'' = \{\varepsilon''_{ij}(u'')\}_{i,j=1}^2$ by means of the linear Hooke's law

$$(1.7) \qquad\qquad \tau^M_{ij} = c^M_{ijk\ell}\varepsilon^M_{k\ell}(u^M), \quad M =',\,'',$$

where $c^M_{ijk\ell}$ are elasticity coefficients of the material in Ω^M, $M =',\,''$ and

$$\varepsilon^M_{k\ell}(u^M) = \frac{1}{2}\left(\frac{\partial u^M_k}{\partial x_\ell} + \frac{\partial u^M_\ell}{\partial x_k}\right), \quad M = ',\,''.$$

We suppose that the elasticity coefficients are bounded measurable functions, satisfying the usual symmetry and ellipticity conditions:

$$(1.8) \qquad c^M_{ijk\ell} = c^M_{jikm} = c^M_{kmij}, \quad M = ',\,''$$

$$(1.9) \qquad \exists c_0 = \text{const} > 0: c^M_{ijk\ell}(x)e_{ij}e_{k\ell} \geq c_0 e_{ij}e_{ij}, \quad M = ',\,''$$

holds for all symmetric matrices $e = (e_{ij})_{i,j=1}^2$ and a.c. in Ω^M, $M = ',\,''$.

By the *classical solution* of the contact problem of two deformable bodies we mean the problem of finding a displacement field $u = [u', u'']$ satisfying all the boundary conditions (1.1)-(1.6), the constitutive law (1.7) and the system of the equations of equilibrium

$$(1.10) \qquad\qquad \frac{\partial \tau^M_{ij}}{\partial x_j} + F^M_i = 0 \quad \text{in } \Omega^M, \; M = ',\,''.$$

REMARK 1.2: We shall study also the case when the influence of friction is taken into account. We shall consider the model with "given friction" when the unknown normal component of the stress vector T_n is replaced by a given slip stress $g \geq 0$.

By means of this simple model, the classical Coulomb's law of friction can be numerically realized (see [7]). Consequently, the boundary condition (1.3) has to be replaced by

$$(1.3)' \quad \begin{cases} T_n \le 0, \ |T_t| \le \mathcal{F}g \quad \text{on } \Gamma_K; \\ \text{if } |T_t(x)| < \mathcal{F}g \Rightarrow (u_t' + u_t'')(x) = 0; \\ \text{if } |T_t(x)| = \mathcal{F}g \Rightarrow \exists \lambda(x) \ge 0 : (u_t' + u_t'')(x) = -\lambda(x)T_t(x), \end{cases}$$

where $\mathcal{F}$ is the coefficient of friction, u_t^M is the tangential component of u^M, $M = ', ''$ and $g \ge 0$ is a non-negative function on Γ_K. Let us notice that the frictionless case can be treated as the special case of the model with given friction by setting $g \equiv 0$ and consequently, both cases will be formulated simultaneously. □

REMARK 1.3: If one of the bodies becomes perfectly rigid, say Ω'', then setting $u'' \equiv 0$ in (1.1)–(1.10), the resulting system of boundary conditions and equations defines the so called *Signorini problem*. □

REMARK 1.4: If both bodies in contact are separated by an initial gap, when the range of the contact zone may expand during the deformation process, the kinematical condition (1.1) has to be changed taking into account the distance between contact zones of the both bodies (for details see [7], [8]). □

In order to give the variational formulation of our problem, we introduce new notations. Let

$$(1.11) \quad \begin{aligned} V = \{ v \equiv [v', v''] \in (H^1(\Omega'))^2 \times (H^1(\Omega''))^2 \mid \\ v^M = 0 \text{ on } \Gamma_u^M, \ M = ', '' \} \end{aligned}$$

be the set of *virtual displacements* and

$$(1.12) \quad K = \{ v \in V \mid v_n' + v_n'' \le 0 \text{ on } \Gamma_K \}$$

its *convex* subset of kinematically admissible displacements.
Let us define the functional of *total potential energy*

$$(1.13) \quad \mathcal{J}(v) = \frac{1}{2}a(v, v) + j(v) - \langle f, v \rangle,$$

where

$$(1.14) \quad a(u, v) = \int_{\Omega' \cup \Omega''} c_{ijk\ell}\varepsilon_{ij}(u)\varepsilon_{ij}(v) \, dx$$

$$(1.15) \quad \langle f, v \rangle = \int_{\Omega' \cup \Omega''} F_i v_i \, dx + \int_{\Gamma_p' \cup \Gamma_p''} P_i v_i \, ds$$

and

$$(1.16) \qquad j(v) = \int_{\Gamma_K} g|v'_t + v''_t|\, ds.$$

The symbol $c_{ijk\ell}$ is defined as follows: $c_{ijk\ell}|_{\Omega^M} = c^M_{ijk\ell}$, $M = '$, $''$ (and analogously for other quantities).

REMARK 1.5: For the sake of simplicity of notations we assume the coefficient of friction $\mathcal{F}$ to be identically equal to 1 in (1.16). □

The notation in (1.11)–(1.16) corresponds exactly to this one from Appendix A.

The *variational solution* of our problem is defined as an element $u \in K$ minimizing $\mathcal{J}$ over K, that is

$$(1.17) \qquad \mathcal{J}(u) \leq \mathcal{J}(v) \quad \forall v \in K,$$

or equivalently

$$(1.18) \qquad u \in K : a(u, v - u) + j(v) - j(u) \geq \langle f, v - u \rangle \quad \forall v \in K$$

(see Appendix A).

REMARK 1.6: Using formally the integration by parts in (1.18) together with a suitable choice of test functions v, we immediately obtain (1.1)–(1.7) and (1.3)' (for details see [7], [8]). □

Due to the Korn's inequality (see [10]), the bilinear form a is V-elliptic. Moreover, the functional j is convex and continuous in V. On the basis of results presented in Appendix A we may conclude that the problem (1.17) has a *unique solution*.

REMARK 1.7: Here we assumed *the coercive case*, only, when the bilinear form a is coercive on V, i.e.

$$(1.19) \qquad \exists \alpha = \mathrm{const} > 0 : a(v, v) \geq \alpha \|v\|^2 \quad \forall v \in V,$$

where $\|v\|$ is the norm in V, defined as follows:

$$\|v\| = \left\{ \|v'\|^2_{1,\Omega'} + \|v''\|^2_{1,\Omega''} \right\}^{1/2}, \quad v = [v', v''] \in V$$

and $\| \ \|_{1,\Omega^M}$ is the usual norm in $(H^1(\Omega^M))^2$, $M = '$, $''$ (see Appendix B).

Sometimes, because of boundary conditions, we have to treat the so called *semi-coercive* cases, when the bilinear form a satisfies a weaker assumption then (1.19), namely

$$(1.20) \qquad \exists \alpha = \mathrm{const} > 0 : a(v, v) \geq \alpha |v|^2 \quad \forall v \in V,$$

where $|\ |$ is a seminorm in V (see Appendix B). To guarantee the existence and the uniqueness of the solution, some additional restrictions on the linear form f have to be added (for details see [7]). □

Now we shall briefly describe the finite element approximation of (1.17). To this end we assume that Ω' and Ω'' are bounded domains with *polygonal* boundaries, so that

$$\overline{\Gamma}_K = \bigcup_{i=1}^{m} \overline{\Gamma}_{K,i},$$

where $\overline{\Gamma}_{K,i}$ is a closed line segment with an initial point A_i and an endpoint A_{i+1}. By T_h', T_h'' we denote triangulations of Ω' and Ω'', respectively, keeping the usual rules for their construction. Moreover, we shall suppose that nodes lying on Γ_K belong to both the triangulations. With a pair $\{T_h', T_h''\}$ the space of piecewise linear vector functions, vanishing on Γ_u' and Γ_u'' will be associated:

$$V_h = \left\{ v_h \in (C(\overline{\Omega}'))^2 \times (C(\overline{\Omega}''))^2 \cap V \mid v_h|_T \in (P_1(T))^2 \ \forall T \in T_h \right\}.$$

Let $a_j^i, j = 1, ..., m_i$ be the vertices of $\{T_h', T_h''\}$ lying on $\overline{\Gamma}_{K,i}$ ($a_1^i \equiv A_i$, $a_{m_i}^i \equiv A_{i+1}$), $i = 1, ..., m$ and let n^i be the unit vector of the outer normal of the part $\Gamma_{K,i}$ oriented with respect to Ω'. We define

$$K_h = \left\{ v_h \in V_h \mid n^i \cdot (v_h' - v_h'')(a_j^i) \leq 0, \ i = 1, ..., m, \ j = 1, ..., m_i \right\}.$$

It is easy to see that K_h is an *internal approximation* of K, i.e. $K_h \subset K \ \forall h > 0$. The set K_h will be used for the approximation of our problem. More precisely, we look for $u_h \in K_h$ such that

(1.21) $$\mathcal{J}(u_h) \leq \mathcal{J}(v_h) \quad \forall v_h \in K_h$$

or equivalently

(1.22) $$u_h \in K_h : a(u_h, v_h - u_h) + j(v_h) - j(u_h) \geq \langle f, v_h - u_h \rangle$$
$$\forall v_h \in K_h.$$

Using results of Appendix A it is possible to prove the convergence of u_h to u in the norm of V. Detailed analysis, together with the error estimates can be found in [7] and [8].

In inclusion, we say several words about the numerical realization of (1.21). For the sake of simplicity we assume the frictionless case only, i.e. $g \equiv 0$. It is readily seen that the problem in a finite dimension leads to the following quadratic programming problem: to find the minimum of the quadratic function

$$\mathcal{L}(\vec{x}) = \frac{1}{2}(\vec{x}, A\vec{x})_{\mathbf{R}^n} - (\vec{F}, \vec{x})_{\mathbf{R}^n}$$

on the closed convex subset

$$\mathcal{K} = \left\{ \vec{x} \in \mathbf{R}^n \mid B\vec{x} \leq \vec{d} \right\},$$

where A is an $n \times n$ stiffness matrix, $\vec{F} \in \mathbb{R}^n$ is the vector arising by the integration of the body and surface forces, B is generally a rectangular $m \times n$ matrix and $\vec{d} \in \mathbb{R}^m$ is a given vector. In problems where there is no initial gap, the vector $\vec{d} \equiv 0$, otherwise the components of $\vec{d}$ are determined by distances of the contact surfaces of the both bodies. Matrix B, defining the convex set $\mathcal{K}$, is sparse: each row contains at most four nonzero entries (components of the vector n), while in each column there are at most two such entries. Methods for the numerical realization of this problem are presented in **[7]**, **[8]**, **[11]**, **[12]**, e.g.

2. ABSTRACT SETTING OF OPTIMAL SHAPE DESIGN PROBLEMS

Let $\mathcal{O}$ be a family of *admissible domains*, that is, $\mathcal{O}$ contains all possible candidates in the optimization process. In $\mathcal{O}$ we introduce a notion of the convergence of domains by specifying the meaning of such convergence.

REMARK 2.1: Convergence of domains may be defined in various ways, for example by the convergence of their characteristic functions in appropriate functional spaces. One of the main difficulties is the fact that $\mathcal{O}$ is not closed with respect to such convergence, in general, that is, the limit set need not be an element of $\mathcal{O}$ (even not a domain). The concrete choice of $\mathcal{O}$ should reflect the physical reality, as well. Mathematical models, describing the behaviour of a mechanical system, are usually valid under some restrictions. The restrictions, concerning the geometry of the problem, should be taken into account when $\mathcal{O}$ is defined. □

With any $\Omega \in \mathcal{O}$ we associate a Hilbert space $V(\Omega)$ – space of functions with finite energy defined on Ω. Besides of the convergence of domains in $\mathcal{O}$, we have to define convergence of functions belonging to spaces $V(\Omega)$ for different $\Omega \in \mathcal{O}$. The main difficulty here is the fact that the domain of definition of such functions varies. One of possible ways, how to overcome this difficulty is to extend functions from the domain of their definition on a larger domain $\widehat{\Omega}$, containing all $\Omega \in \mathcal{O}$ and to define:

$$y_n \to y, \quad y_n \in V(\Omega_n),\ y \in V(\Omega) \Longleftrightarrow \tilde{y}_n \to \tilde{y} \text{ in } V(\widehat{\Omega}),$$

where the symbol "~" means an appropriate extension of y_n from Ω_n on $\widehat{\Omega}$ and the convergence in $\widehat{\Omega}$ can be already introduced in a standard way since $\widehat{\Omega}$ is fixed (for details see **[9]**).

With any $\Omega \in \mathcal{O}$ we associate the unique solution $u(\Omega)$ of *a state problem*, given by PDE, variational inequality,.... In other words, we define the mapping $u \colon \mathcal{O} \to \{V(\Omega)\}$ as

$$u \colon \Omega \mapsto u(\Omega) \in V(\Omega).$$

Let $G = \{(\Omega, u(\Omega)),\ \Omega \in \mathcal{O}\}$ be the graph of such a mapping. Finally, let $I \colon (\Omega, y) \mapsto \mathbb{R}^1$, $\Omega \in \mathcal{O}$, $y \in V(\Omega)$ be *a cost functional* and write $\mathcal{J}(\Omega) = I(\Omega, u(\Omega))$ for $\Omega \in \mathcal{O}$.

The *abstract optimal shape* design problem now reads as follows:

$$(\text{P}) \quad \begin{cases} \text{find } \Omega^* \in \mathcal{O} \text{ such that} \\ \mathcal{J}(\Omega^*) \leq \mathcal{J}(\Omega) \quad \forall \Omega \in \mathcal{O}. \end{cases}$$

A domain Ω^* (if it exists) will be called *an optimal one* with respect to the choice of I and $\mathcal{O}$.

REMARK 2.2: Let us stress the fact that the notion "to be optimal" is *relative*, that is, without specifying $\mathcal{O}$ (if I is given) the problem (P) is not well posed. □

Imposing additional compactness property on G and on lower continuity of I, one can prove the existence of at least one optimal domain Ω^* (for more details see [9]). Usually there is no uniqueness of the solution.

3. CONTACT SHAPE OPTIMIZATION

This part deals with the shape optimization of systems, the state of which is described by the mathematical model, presented in Section 1. For the sake of simplicity we shall assume Signorini problems only, i.e. the second body Ω'' will be replaced by a perfectly rigid foundation, given by the *half space*. This simple shape of the foundation makes possible to formulate the non-penetration condition *exactly*. To fix the ideas, let us consider the situation, shown in Fig. 2.

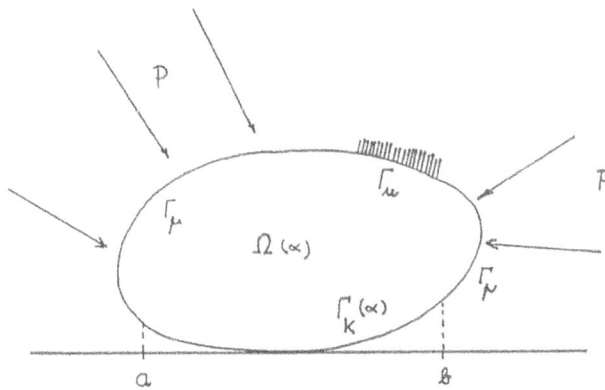

Figure 2.

The boundary $\partial \Omega$ will be decomposed as follows: $\partial \Omega = \overline{\Gamma}_u \cup \overline{\Gamma}_p \cup \overline{\Gamma}_K(\alpha)$. On each part different boundary conditions will be prescribed. The contact part $\Gamma_K(\alpha)$, the object of the optimization, will be given by the graph of a non-negative function α defined over a fixed interval (a, b), the domain itself will be denoted by $\Omega(\alpha)$:

$$\Gamma_K(\alpha) = \big\{ [x_1, x_2] \mid x_2 = \alpha(x_1) \; x_1 \in (a, b) \big\}.$$

In order to specify the set $\mathcal{O}$ of all admissible domains $\Omega(\alpha)$, we shall specify the class of admissible functions α, as follows:

$$\text{(3.1)} \qquad U_{\text{ad}} = \left\{ \alpha \in C^{0,1}([a,b]) \mid 0 \le \alpha \le \gamma, \right.$$
$$\left. |\alpha'(x_1)| \le C_0, \ \text{meas}\, \Omega(\alpha) = C_1 \right\}$$

in frictionless case (i.e. $g \equiv 0$) and

$$\text{(3.2)} \qquad U_{\text{ad}} = \left\{ \alpha \in C^{1,1}([a,b]) \mid 0 \le \alpha \le \gamma, \ |\alpha'(x_1)| \le C_0, \right.$$
$$\left. |\alpha''(x_1)| \le C_1, \ \text{meas}\, \Omega(\alpha) = C_2 \right\}$$

in the case of Signorini problem with given function.

THE INTERPRETATION OF U_{ad}:
Let U_{ad} be given by (3.1). Then U_{ad} contains all Lipschitz continuous functions α, satisfying the box constraints $\alpha \in [0, \gamma]$, where $\gamma > 0$ is a given number, further constant volume constraint defined by a given number $C_1 > 0$. To avoid undesired oscilations of the contact part $\Gamma_K(\alpha)$, the condition restricting the derivative of α has been included. This restriction is important not only from the mathematical point of view, but also from the reason of validity of the physical model. As we have already mentioned in Section 1, the real non-penetration condition is replaced by the linearized one. Such approximation is good if the number C_0 is sufficiently small.

The interpretation of the set U_{ad} given by (3.2) is almost the same. We only require higher smoothness of the functions α, describing the contact part $\Gamma_K(\alpha)$. $\square$

Referring to the notations of Section 2, we set

$$\text{(3.3)} \qquad \mathcal{O} = \left\{ \Omega(\alpha), \ \alpha \in U_{\text{ad}} \right\},$$

where U_{ad} is given by (3.1) or (3.2).

If $\{\Omega_n\}_{n=1}^{\infty}$ is a sequence of domains belonging to $\mathcal{O}$, $\Omega_n = \Omega(\alpha_n)$, we say that $\Omega_n \to \Omega$ if and only if

$$\alpha_n \rightrightarrows \alpha \quad \text{(uniformly) in } [a,b]$$

when U_{ad} is given by (3.1) or

$$\alpha_n \rightrightarrows \alpha, \quad \alpha_n' \rightrightarrows \alpha', \quad \text{(uniformly) in } [a,b]$$

when U_{ad} is given by (3.2).

With any $\Omega(\alpha) \in \mathcal{O}$ the space of virtual displacements $V(\alpha)$ will be associated:

$$V(\alpha) = \left\{ v \in (H^1(\Omega(\alpha)))^2 \mid v_i = 0 \text{ on } \Gamma_u, \ i = 1, 2 \right\}.$$

Let $\{y_n\}_{n=1}^\infty$ be a sequence of functions belonging to $V(\Omega_n)$. We say that the sequence $\{y_n\}$ converges to a function $y \in V(\alpha)$ if and only if there exist extensions of y_n and y (denoted by $\tilde{y}_n, \tilde{y}$) from Ω_n and Ω, respectively on a domain $\hat{\Omega}$, containing $\Omega(\alpha)$ for all $\alpha \in U_{ad}$ and such that

$$(3.4) \qquad \tilde{y}_n, \tilde{y} \in (H^1(\hat{\Omega}))^2 \quad \forall n,$$

$$(3.5) \qquad \tilde{y}_n \rightharpoonup \tilde{y} \quad \text{(weakly) in } (H^1(\hat{\Omega}))^2.$$

Now, let $\alpha \in U_{ad}$ be fixed. On any $\Omega(\alpha)$ we define the following state problem, the variational formulation of which reads as follows:

$$(\mathcal{P}(\alpha)) \qquad \begin{cases} \text{find } u(\alpha) \in K(\alpha) \text{ such that} \\ \mathcal{J}_\alpha(u(\alpha)) \le \mathcal{J}_\alpha(v) \quad \forall v \in K(\alpha), \end{cases}$$

or equivalently

$$(\mathcal{P}(\alpha))' \qquad u(\alpha) \in K(\alpha) : a_\alpha(u(\alpha), v - u(\alpha)) + j_\alpha(v) - j_\alpha(u(\alpha))$$
$$\ge \langle f, v - u(\alpha) \rangle_\alpha \qquad \forall v \in K(\alpha).$$

Here

$$K(\alpha) = \big\{ v \in (H^1(\Omega(\alpha)))^2 \mid v_i = 0 \text{ on } \Gamma_u, \ i = 1, 2,$$
$$v_2(x_1, \alpha(x_1)) \ge -\alpha(x_1) \ \forall x_1 \in (a, b) \big\},$$

$$\mathcal{J}_\alpha(v) = \frac{1}{2} a_\alpha(v, v) + j_\alpha(v) - \langle f, v \rangle_\alpha,$$

$$a_\alpha(u, v) = \int_{\Omega(\alpha)} c_{ijk\ell} \varepsilon_{ij}(u) \varepsilon_{k\ell}(v) \, dx,$$

$$j_\alpha(v) = \int_{\Gamma_K(\alpha)} g|v_t| \, ds, \quad \langle f, v \rangle_\alpha = \int_{\Omega(\alpha)} F_i v_i \, dx + \int_{\Gamma_p} P_i v_i \, ds.$$

We suppose that the elasticity coefficients $c_{ijk\ell}$ and the body forces F are defined in $\hat{\Omega}$.

Finally, let $I : (\alpha, y) \mapsto \mathbf{R}^1$, $\alpha \in U_{ad}$, $y \in V(\alpha)$ be a *cost functional* and write $\mathcal{C}(\alpha) = I(\alpha, u(\alpha))$ with $u(\alpha) \in K(\alpha)$ being the solution of $(\mathcal{P}(\alpha))$. In accordance with Section 2 we define the problem

$$(\mathbf{P}) \qquad \begin{cases} \text{find } \alpha^* \in U_{ad} \text{ such that} \\ \mathcal{C}(\alpha^*) \le \mathcal{C}(\alpha) \quad \forall \alpha \in U_{ad}. \end{cases}$$

The domain $\Omega^* = \Omega(\alpha^*)$ will be called *optimal* with respect to U_{ad} and I. To guarantee the existence of at least one solution of $(\mathbf{P})$ we need some additional assumptions, concerning I, namely

$$(3.6) \qquad \left. \begin{cases} \text{if } \alpha_n \rightrightarrows \alpha \text{ (uniformly) in } [a, b], \ \alpha_n, \alpha \in U_{ad} \\ \hat{y}_n \rightharpoonup \hat{y} \text{ (weakly) in } (H^1(\hat{\Omega}))^2, \ \hat{y}_n, \hat{y} \in (H^1(\hat{\Omega}))^2 \end{cases} \right\} \Longrightarrow$$

$$\Longrightarrow \liminf_{n \to \infty} I(\alpha_n, \hat{y}_n|_{\Omega_n}) \ge I(\alpha, \hat{y}|_{\Omega(\alpha)}),$$

where $\Omega_n = \Omega(\alpha_n)$. Then one can prove

THEOREM 3.1. *Let (3.6) be satisfied. Then the problem (P) has at least one solution.*

Sketch of the proof (for the detailed proof see [9]). There are 4 important steps:
(i) First we have to show that the system $\mathcal{O}$ is *compact* with respect to the corresponding convergences of domains. This follows immediately from the definition of U_{ad} and Arzela–Ascoli theorem.
(ii) We have to show that if the sequence $\{\alpha_n\}$, $\alpha_n \in U_{\mathrm{ad}}$, is bounded, then the sequence $\{u(\alpha_n)\}_{n=1}^\infty$ of the corresponding solutions of $(\mathcal{P}(\alpha_n))$ is bounded as well, that is

$$\exists c = \mathrm{const} > 0 : \|u(\alpha_n)\|_{(H^1(\Omega_n))^2} \le c \quad \forall n.$$

To prove this we have to use the nontrivial fact that the constant in Korn's inequality can be estimated uniformly with respect to $\Omega \in \mathcal{O}$.
(iii) Next step is to show that solutions of the Signorini problem depend *continuously on changes of domains*. More precisely:

let $u(\alpha_n)$ be the solution of $(\mathcal{P}(\alpha_n))$ on $\Omega(\alpha_n)$.

If $\Omega(\alpha_n) \to \Omega(\alpha)$ and $\tilde{u}(\alpha_n) \rightharpoonup \hat{u}$ (weakly) in $(H^1(\widehat{\Omega}))^2$,

where the symbol " ˜ " means an appropriate extension of $u(\alpha_n)$ on $\widehat{\Omega}$, then the function $u \equiv \hat{u}|_{\Omega(\alpha)}$ solves $(\mathcal{P}(\alpha))$ on $\Omega(\alpha)$. To establish such kind of a result in the case when friction is taken into account, higher smoothness of admissible shapes is needed. That is why we introduced an alternative definition of U_{ad} by (3.2).
(iv) The last step is to use the lower continuity of the cost functional I. □

REMARK 3.1: As we have already mentioned, there is no uniqueness of the solution, in general. □

In Section 5 we shall discuss two important choices of the cost functionals.

REMARK 3.2: Here, we discussed the case, in which the shape of the deformable body was designed. It would be possible to fix the geometry of the body and to try to optimize the shape of a rigid foundation. This case is much simpler from the theoretical as well as the practical point of view. □

4. APPROXIMATION OF THE PROBLEM (P)

The aim of this section is to define the finite element approximation of the problem (P) and to mention results, concerning the mutual relation between (P) and (P)$_h$ for $h \to 0_+$. For the sake of simplicity we restrict our discussions to frictionless case only, when U_{ad} is given by (3.1).

Let $D_h : a = x_0 < x_1 < \cdots < x_{D(h)} = b$ be a partition of the interval $[a, b]$ and define

$$U_{\mathrm{ad}}^h = \left\{ \alpha_h \in C([a, b]) \mid \alpha_h|_{\overline{x_{i-1} x_i}} \in P_1, \ i = 1, ..., D(h) \right\} \cap U_{\mathrm{ad}},$$

i.e. U^h_{ad} contains all piecewise linear functions over D_h belonging to U_{ad}.

REMARK 4.1: The realization of U^h_{ad} is simple. The piecewise linear function α_h belongs to U^h_{ad} iff

$$0 \le \alpha_h(x_i) \le \gamma \quad \forall i = 0, ..., D(h),$$
$$\left|\alpha_h(x_i) - \alpha_h(x_{i-1})\right| \le C_0(x_i - x_{i-1}) \quad \forall i = 1, ..., D(h),$$
$$\int_a^b \alpha_h(x_1)\,dx_1 = \text{const.}$$

The last integral can be evaluated exactly by means of the trapezoid formula. □

Instead of $\mathcal{O}$ we shall assume its approximation $\mathcal{O}_h$, defined as follows:

$$\mathcal{O}_h = \left\{ \Omega(\alpha_h), \; \alpha_h \in U^h_{\text{ad}} \right\}.$$

To simplify the presentation of results, we shall assume bodies, the shape of which is shown in Fig. 3, i.e. $\Omega(\alpha)$ is a punch for any $\alpha \in U_{\text{ad}}$ with one curved side only, namely the contact zone $\Gamma_K(\alpha)$. For such a shape of $\Omega(\alpha)$, the corresponding domains $\Omega(\alpha_h)$, $\alpha_h \in U^h_{\text{ad}}$ are *polygonal*.

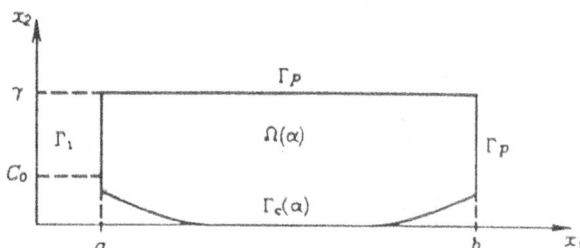

Figure 3.

Let $T(h, \alpha_h)$ be a triangulation of $\overline{\Omega(\alpha_h)}$. Besides of the usual requirements on the mutual position of triangles, we shall need additional requirements, which are necessary for convergence analysis:

(4.1) for any $h > 0$ fixed, the triangulations $T(h, \alpha_h)$ are *topologically equivalent* for any $\alpha_h \in U^h_{\text{ad}}$, i.e. $T(h, \alpha_h)$ have the same number of nodes and the nodes have the same neighbours for any $\alpha_h \in U^h_{\text{ad}}$;

(4.2) For any $h > 0$ fixed, the triangulations $T(h, \alpha_h)$ depend continuously on $\alpha_h \in U^h_{\text{ad}}$;

(4.3) The system of triangulations $\{T(h, \alpha_h)\}$ is *uniformly regular* with respect to $h \to 0_+$ and $\alpha_h \in U^h_{\text{ad}}$, i.e. the minimum angle condition is satisfied uniformly with respect to $h > 0$ and $\alpha_h \in U^h_{\text{ad}}$.

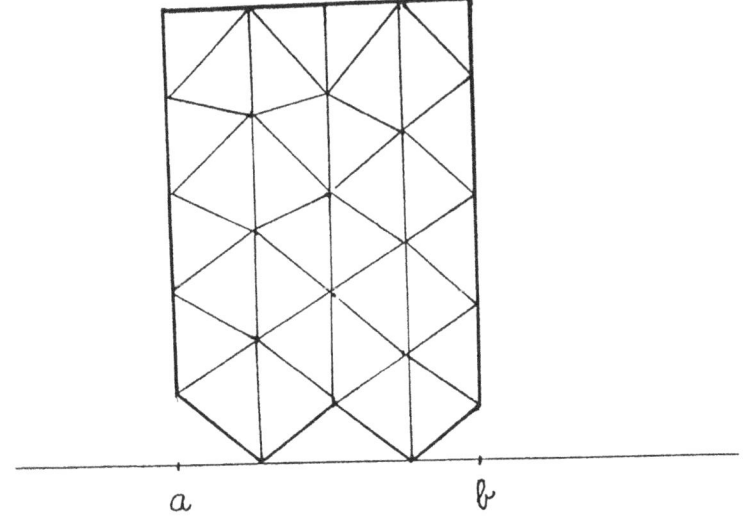

a   b

Figure 4.

In our special case, shown in Fig. 3, a family $\{T(h, \alpha_h)\}$, satisfying all the above mentioned assumptions, can be constructed (see Fig. 4).

The triangulations for such type of domains can be easily parametrized by means of the so called *principle moving nodes* N_i^D, $i = 0, ..., D(h)$, where $N_i^D = [x_i, \alpha_h(x_i)]$, which are allowed to move in x_2-direction only and the position of other nodes of $T(h, \alpha_h)$ is uniquelly determined by $\{N_i^D\}$. The domain $\Omega(\alpha_h)$ with the triangulation $T(h, \alpha_h)$ will be denoted by Ω_h, in what follows.

Finite elements for the approximation of $\mathcal{P}(\alpha)$, introduced in Section 3 will be used. Let $\alpha_h \in U_{ad}^h$ be given and construct Ω_h. We define

$$V_h(\alpha_h) = \left\{ v_h \in (C(\overline{\Omega}_h))^2 \mid v_h \in (P_1(T))^2 \; \forall T \in T(h, \alpha_h), \; v_h = 0 \text{ on } \overline{\Gamma}_u \right\}$$

and

$$K_h(\alpha_h) = \left\{ v_h \in V_h(\alpha_h) \mid v_{h2}(x_i, \alpha_h(x_i)) \geq -\alpha_h(x_i) \; \forall x_i \in \overline{\Gamma}_K(\alpha_h) \setminus \overline{\Gamma}_u \right\}.$$

It is readily seen that $K_h(\alpha) \subset K(\alpha_h)$. Instead of $(\mathcal{P}(\alpha))$ we assume the problem

$(\mathcal{P}(\alpha_h))_h$
$$\begin{cases} \text{find } u_h(\alpha_h) \in K_h(\alpha_h) \text{ such that} \\ \mathcal{J}_{\alpha_h}(u_h(\alpha_h)) \leq \mathcal{J}_{\alpha_h}(v_h) \quad \forall v_h \in K_h(\alpha_h), \end{cases}$$

or equivalently

$(\mathcal{P}(\alpha_h))_h$
$$u_h \in K_h(\alpha_h) : a_{\alpha_h}(u_h(\alpha_h), v_h - u_h(\alpha_h))$$
$$\geq \langle f, v_h - u_h(\alpha_h) \rangle_{\alpha_h} \quad \forall v_h \in K_h(\alpha_h).$$

The meaning of the symbols is exactly the same as in Section 3.

Now, let $I_h \colon (\alpha_h, y_h) \to \mathbf{R}^1$, $\alpha_h \in U_{\mathrm{ad}}^h$, $y_h \in V_h(\alpha_h)$ be an approximation of the cost functional, which is lower continuous in the sense of (3.6).

The approximation of (P) reads as follows:

$$
(\mathbf{P})_h \qquad
\begin{cases}
\text{find } \alpha_h^* \in U_{\mathrm{ad}}^h \text{ such that} \\
\mathcal{C}_h(\alpha_h^*) \le \mathcal{C}_h(\alpha_h) \quad \forall \alpha_h \in U_{\mathrm{ad}}^h,
\end{cases}
$$

where $\mathcal{C}_h(\alpha_h) \equiv I_h(\alpha_h, u_h(\alpha_h))$ with $u_h(\alpha_h)$ being the solution of $(\mathcal{P}(\alpha_h))_h$.

On the basis of the condition (4.2) one can show that $(\mathbf{P})_h$ has at least one solution α_h^*. The domain $\Omega_h^* \equiv \Omega(\alpha_h^*)$ will be called the *approximated optimal shape*.

In [9] the mutual relation between $(\mathbf{P})$ and $(\mathbf{P})_h$ for $h \to 0_+$ is studied. It is shown that these problems are closed each other on subsequences, provided that (4.1)–(4.3) hold.

5. SENSITIVITY ANALYSIS

The aim of this part is to present results, concerning the sensitivity of the solution of contact problems with respect to changes of the shape of Ω. Just at this point the contact shape optimization differs considerably from shape optimization with classical state problems, described by equations. To see which kind of the problem is numerically realized, we rewrite the problem $(\mathbf{P})_h$ in an algebraic form.

Let us assume that $h > 0$ is fixed, i.e. the dimension of $V_h(\alpha_h)$ equals $n(h)$ for any $\alpha_h \in U_{\mathrm{ad}}^h$. For the sake of simplicity we shall write n instead of $n(h)$.

Let $\vec{\alpha} = (\alpha_0, ..., \alpha_D)$ be a vector with components $\alpha_i = \alpha(x_i)$, $i = 0, ..., D$, called *discrete design* variables. Then the shape of $\Omega(\alpha_h)$ can be identified with $\vec{\alpha} \in \mathcal{U}$, where

$$
(5.1) \qquad \mathcal{U} = \Big\{ \vec{\alpha} \in \mathbf{R}^{D+1} \mid 0 \le \alpha_i \le \gamma, \ i = 0, ..., D;
$$

$$
|\alpha_i - \alpha_{i-1}| \le C_0(x_i - x_{i-1}), \ i = 1, ..., D; \ \sum_{i=0}^{D+1} \omega_i \alpha_i = \mathrm{const} \Big\}
$$

when U_{ad} is given by (3.1) or

$$
(5.2) \qquad \mathcal{U} = \Big\{ \vec{\alpha} \in \mathbf{R}^{D+1} \mid 0 \le \alpha_i \le \gamma, \ i = 0, ..., D;
$$

$$
|\alpha_i - \alpha_{i-1}| \le C_0(x_i - x_{i-1}), \ i = 1, ..., D;
$$

$$
|\alpha_{i+1} - 2\alpha_i + \alpha_{i-1}| \le C_1(x_{i+1} - x_i)(x_i - x_{i-1}),
$$

$$
i = 1, ..., D-1; \ \sum_{i=0}^{D} \omega_i \alpha_i = \mathrm{const} \Big\}
$$

for U_{ad} given by (3.2).

REMARK 5.1: The last condition appearing in the definition of $\mathcal{U}$ comes from the trapezoid formula, used for the evaluation of $\int_a^b \alpha_h \, dx_1$.

To approximate the second derivative of α in (3.2), the second central difference has been used in (5.2). $\qquad\square$

Let $\vec{\alpha} \in \mathcal{U}$ be given. Then the finite element approximation of the contact problem expressed in algebraic form reads as follows:

$$(\mathcal{P}(\vec{\alpha})) \qquad \begin{cases} \text{find } \vec{x}(\vec{\alpha}) \in \mathcal{K}(\vec{\alpha}) \text{ such that} \\ \mathcal{L}_{\vec{\alpha}}(\vec{x}(\vec{\alpha})) \leq \mathcal{L}_{\vec{\alpha}}(\vec{y}) \quad \forall \vec{y} \in \mathcal{K}(\vec{\alpha}), \end{cases}$$

where

$$\mathcal{L}_{\vec{\alpha}}(\vec{x}) \equiv \frac{1}{2}(\vec{x}, A(\vec{\alpha})\vec{x})_{\mathbf{R}^n} + \sum_{j_i \in I_1} \omega_{j_i}(\vec{\alpha})|x_{j_i}| - (\vec{F}(\vec{\alpha}), \vec{x})_{\mathbf{R}^n}$$

and

$$\mathcal{K}(\vec{\alpha}) = \left\{ \vec{x} \in \mathbf{R}^n \mid x_{j_i} \geq -\alpha_i, \ j_i \in I_2 \right\}.$$

Here $A(\vec{\alpha})$ and $\vec{F}(\vec{\alpha})$ means the stiffness matrix and the force vector, respectively, both depending on the discrete design variable $\vec{\alpha}$. I_1 and I_2 are disjoint sets, containing indices of x_1 components and x_2 components, respectively, of the nodal displacement field $\vec{x}$ at nodes of Γ_K, where the unilateral boundary conditions are prescribed. The second term in the definition of $\mathcal{L}_{\vec{\alpha}}$ comes from the approximation of j_{α_h} by using a suitable quadrature formula. To simplify the presentation we set $g \equiv 1$ along $\Gamma_K(\alpha) \ \forall \alpha \in U_{\mathrm{ad}}$. If frictionless case is assumed, this term is omitted.

Let $C : [\vec{\alpha}, \vec{x}] \mapsto \mathbf{R}^1$, $\vec{\alpha} \in \mathcal{U}$, $\vec{x} \in \mathbf{R}^n$ be the cost functional $\mathcal{C}_h$ expressed in terms of discrete design variables and nodal displacements.

Thus, we are led to the following nonlinear mathematical programming problem:

$$(\mathsf{P}) \qquad \begin{cases} \text{find } \vec{\alpha}^* \in \mathcal{U} \text{ such that} \\ C(\vec{\alpha}^*, \vec{x}(\vec{\alpha}^*)) \leq C(\vec{\alpha}, \vec{x}(\vec{\alpha})) \quad \forall \vec{\alpha} \in \mathcal{U}, \end{cases}$$

where $\vec{x}(\vec{\alpha}) \in \mathcal{K}(\vec{\alpha})$ solves $(\mathcal{P}(\vec{\alpha}))$.

To discover a solution $\vec{\alpha}^*$ of (P), a suitable mathematical programming method will be used. For the most of such methods, gradient informations are necessary. This is the main difficulty of our problem, since as we shall see the mapping $\vec{\alpha} \mapsto \vec{x}(\vec{\alpha})$ *is not differentiable*, in general and consequently the function $\vec{\alpha} \mapsto C(\vec{\alpha}, \vec{x}(\vec{\alpha}))$ is not, as well. To understand this phenomena, let us consider the classical state problem, given by the equation

$$(5.3) \qquad\qquad A(\vec{\alpha})\vec{x}(\vec{\alpha}) = \vec{F}(\vec{\alpha}).$$

The differentiability of $\vec{x}(\vec{\alpha})$ is a consequence of the implicit function theorem and the derivative of $\vec{x}(\vec{\alpha})$ can be easily found by the differentiation of (5.3). When $\vec{x}(\vec{\alpha})$

is the solution of the variational inequality $(\mathcal{P}(\vec{\alpha}))$, the situation is more involved. According to Appendix A (A.14), the solution $\vec{x}(\vec{\alpha})$ is given by the projection $P_{\mathcal{K}(\vec{\alpha})}$ of a certain element onto the convex set $\mathcal{K}(\vec{\alpha})$. The presence of the mapping $P_{\mathcal{K}(\vec{\alpha})}$ makes things more complicated. This mapping is only Lipschitz continuous as follows from (A.8) and consequently *differentiable only* almost everywhere. Summarizing, there are points in which $\vec{x}(\vec{\alpha})$ has no gradient. Therefore, the application of classical gradient type methods may fail.

On the other hand, the situation in our case is not so bad. It is possible to prove (see [13], [14]) that the state $\vec{x}(\vec{\alpha})$ is at least *directionally differentiable* at any point $\vec{\alpha} \in \mathcal{U}$ and in any direction $\vec{\beta}$, i.e.

$$\lim_{t \downarrow 0_+} \frac{\vec{x}(\vec{\alpha} + t\vec{\beta}) - \vec{x}(\vec{\alpha})}{t} \equiv \vec{x}'(\vec{\alpha}, \vec{\beta}) \equiv \vec{x}'$$

exists and is finite.

In what follows we explain, how to recognize the case when $\vec{x}(\vec{\alpha})$ is differentiable in classical sense and how to compute the directional derivative $\vec{x}'(\vec{\alpha}, \vec{\beta})$. We use a new formulation of the state problem $(\mathcal{P}(\vec{\alpha}))$.

To this end we introduce 2 sets of Lagrangian multipliers, one to release the kinematical constraint $\vec{x} \in \mathcal{K}(\vec{\alpha})$ and the second one to regularize the terms, containing the absolute value.

Denote by

$$\Lambda_1 = \left\{ \vec{\mu} \in \mathbf{R}^p \mid |\mu_i| \leq 1 \ \forall i \right\}$$
$$\Lambda_2 = \left\{ \vec{\mu} \in \mathbf{R}^p \mid \mu_i \geq 0 \ \forall i \right\},$$

where $p = \mathrm{card}(I_1) = \mathrm{card}(I_2)$. Let $\mathcal{H} \colon \mathbf{R}^n \times \Lambda_1 \times \Lambda_2$ be the Lagrangian function defined by

$$\mathcal{H}(\vec{\alpha}, \vec{x}, \vec{\mu}_1, \vec{\mu}_2) = \frac{1}{2}(\vec{x}, A(\vec{\alpha})\vec{x})_{\mathbf{R}^n} - (\vec{F}(\vec{\alpha}), \vec{x})_{\mathbf{R}^n}$$
$$+ \sum_{j_i \in I_1} \omega_i(\vec{\alpha})\mu_i^1 x_{j_i} - \sum_{j_i \in I_2} \mu_i^2 (x_{j_i} + \alpha_i),$$

where $\vec{\mu}_1 = (\mu_1^1, ..., \mu_p^1) \in \Lambda_1$, $\vec{\mu}_2 = (\mu_1^2, ..., \mu_p^2) \in \Lambda_2$ with $\vec{\alpha} \in \mathcal{U}$ fixed.

Let $(\vec{x}^*(\vec{\alpha}), \vec{\lambda}_1(\vec{\alpha}), \vec{\lambda}_2(\vec{\alpha})) \equiv (\vec{x}^*, \vec{\lambda}_1, \vec{\lambda}_2)$ be a saddle point of $\mathcal{H}$ on $\mathbf{R}^n \times \Lambda_1 \times \Lambda_2$, i.e.

(5.4) $\qquad \mathcal{H}(\vec{\alpha}, \vec{x}^*, \vec{\mu}_1, \vec{\mu}_2) \leq \mathcal{H}(\vec{\alpha}, \vec{x}^*, \vec{\lambda}_1, \vec{\lambda}_2) \leq \mathcal{H}(\vec{\alpha}, \vec{y}, \vec{\lambda}_1, \vec{\lambda}_2)$

$$\forall \vec{y} \in \mathbf{R}^n, \ \forall(\vec{\mu}_1, \vec{\mu}_2) \in \Lambda_1 \times \Lambda_2.$$

Then one can easily check that

$$
(5.5) \qquad
\begin{cases}
\vec{x}^* = \vec{x}(\vec{\alpha}), \\
-\omega_i(\vec{\alpha})\lambda_i^1 = r_{j_i}(\vec{\alpha}) & j_i \in I_1, \\
\lambda_i^2 = r_{j_i}(\vec{\alpha}) & j_i \in I_2,
\end{cases}
$$

where $\vec{x}(\vec{\alpha}) \in \mathcal{K}(\vec{\alpha})$ solves $(\mathcal{P}(\vec{\alpha}))$ and $r_i(\vec{\alpha})$ is the i-th component of the residual vector $\vec{r}(\vec{\alpha}) \equiv A(\vec{\alpha})\vec{x}(\vec{\alpha}) - \vec{F}(\vec{\alpha})$.

An equivalent expression of (5.4) is given by the following system of Karush–Kuhn–Tucker conditions

$$
(5.6) \qquad
\begin{cases}
a_{ij}(\vec{\alpha})x_j(\vec{\alpha}) = F_i(\vec{\alpha}) & i \notin I_1 \cup I_2 \\
a_{j_i k}x_k(\vec{\alpha}) = F_{j_i}(\vec{\alpha}) - q_i(\vec{\alpha}) & j_i \in I_1 \\
a_{j_i k}x_k(\vec{\alpha}) = F_{j_i}(\vec{\alpha}) + \lambda_i^2 & j_i \in I_2 \\
\sum_{j_i \in I_1} \omega_i(\vec{\alpha})(\mu_i^1 - \lambda_i^1)x_{j_i}(\vec{\alpha}) \leq 0 & \forall \vec{\mu}_1 \in \Lambda_1 \\
\sum_{j_i \in I_2} (\mu_i^2 - \lambda_i^2)(x_{j_i}(\vec{\alpha}) + \alpha_i) \geq 0 & \forall \vec{\mu}_2 \in \Lambda_2
\end{cases}
$$

where $q_i(\vec{\alpha})$ on the right hand side of $(5.6)_2$ equals $\omega_i(\vec{\alpha})\lambda_i(\vec{\alpha})$ (no sum).

REMARK 5.2: From (5.2) and (5.3), the physical interpretation of Lagrange multipliers is readily seen: $\vec{\lambda}^1$ and $\vec{\lambda}^2$ are related to the tangential and normal component of the stress vector, respectively. $\qquad\square$

Let the solution $\vec{x}(\vec{\alpha})$ of $(\mathcal{P}(\vec{\alpha}))$ be known. Then the sets I_1 and I_2 can be decomposed as follows:

$$
I_1 = I_1^+(\vec{\alpha}) \cup I_1^0(\vec{\alpha}) \cup I_1^-(\vec{\alpha})
$$
$$
I_2 = I_2^+(\vec{\alpha}) \cup I_2^0(\vec{\alpha}) \cup I_2^-(\vec{\alpha}),
$$

where

$$
I_1^+(\vec{\alpha}) = \big\{ j_i \in I_1 \mid x_{j_i}(\vec{\alpha}) \neq 0 \big\}
$$
$$
I_1^0(\vec{\alpha}) = \big\{ j_i \in I_1 \mid x_{j_i}(\vec{\alpha}) = 0, \ |\lambda_i^1(\vec{\alpha})| = 1 \big\}
$$
$$
I_1^-(\vec{\alpha}) = \big\{ j_i \in I_1 \mid x_{j_i}(\vec{\alpha}) = 0, \ |\lambda_i^1(\vec{\alpha})| < 1 \big\}
$$

and analogously

$$
I_2^+(\vec{\alpha}) = \big\{ j_i \in I_2 \mid x_{j_i}(\vec{\alpha}) > -\alpha_i \big\}
$$
$$
I_2^0(\vec{\alpha}) = \big\{ j_i \in I_2 \mid x_{j_i}(\vec{\alpha}) = -\alpha_i, \ \lambda_i^2(\vec{\alpha}) = 0 \big\}
$$
$$
I_2^-(\vec{\alpha}) = \big\{ j_i \in I_2 \mid x_{j_i}(\vec{\alpha}) = -\alpha_i, \ \lambda_i^2(\vec{\alpha}) > 0 \big\}.
$$

REMARK 5.3: The mechanical interpretation of the previous decomposition is the following:

$I_1^+(\vec{\alpha})$ is the set of nodes at which *slip occurs*;

$I_1^-(\vec{\alpha})$ is the set of nodes with *"strong" stick*, when the tangential forces are below a priori given bounds;

$I_1^0(\vec{\alpha})$ is the set of node with *"tight" stick*. The bounds for the tangential forces are achieved but the node still remains in stick.

Similar interpretation holds for the subsets of I_2:

$I_2^+(\vec{\alpha})$ is the set of nodes with *non-active constraints*, i.e. no contact is realized there;

$I_2^0(\vec{\alpha})$ corresponds to the so called *semiactive constraints* (nodes with tight contact);

$I_2^+(\vec{\alpha})$ is the set of nodes with *strong contact*, or *strongly active constraints*. □

From (5.6) it is easy to see that $\vec{x}(\vec{\alpha})$, $\vec{\lambda}_1(\vec{\alpha})$ and $\vec{\lambda}_2(\vec{\alpha})$ are *continuous* functions of the discrete design variable $\vec{\alpha}$ and consequently if $j_i \in I_k^+(\vec{\alpha})$ or $I_k^-(\vec{\alpha})$, $k = 1, 2$, then j_i belongs to $I_k^+(\vec{\alpha} + t\vec{\beta})$ or $I_k^-(\vec{\alpha} + t\vec{\beta})$ as well, where $\vec{\beta} \in \mathbf{R}^{D+1}$ is an arbitrary vector and $t > 0$ is sufficiently small. But this is no longer true if $j_i \in I_k^0(\vec{\alpha})$, $k = 1, 2$. After small perturbation of $\vec{\alpha}$, the corresponding index j_i may belong to any of the three subsets $I_k^+(\vec{\alpha})$, $I_k^-(\vec{\alpha})$, $I_k^0(\vec{\alpha})$, $k = 1, 2$. Summarizing: if subsets $I_1^0(\vec{\alpha})$ and $I_2^0(\vec{\alpha})$ are empty, i.e. there are no points with tight stick or tight contact, the decomposition of I_1 and I_2 is *stable* with respect to small perturbations of $\vec{\alpha}$ and the corresponding derivatives of $\vec{x}(\vec{\alpha})$ can be computed exactly in the same way as in the classical linear elasticity problem with bilateral conditions. At such a point $\vec{\alpha}$, the solution $\vec{x}(\vec{\alpha})$ is continuously differentiable.

Thus the non-differentiability may occur only if one of the subsets $I_1^0(\vec{\alpha})$, $I_2^0(\vec{\alpha})$ is *non-empty*. Let us recall that in such a case, the corresponding solution $\vec{x}(\vec{\alpha})$ is still directionally differentiable. The same can be proven for the corresponding Lagrange multipliers $\vec{\lambda}_1(\vec{\alpha})$ and $\vec{\lambda}_2(\vec{\alpha})$. Denote the corresponding directional derivatives by $\vec{x}'$, $\vec{\lambda}_1'$ and $\vec{\lambda}_2'$, respectively. The formal differentiation in (5.6) yields

$$(5.7) \quad \begin{cases} a_{ij}(\vec{\alpha})x_j'(\vec{\alpha}) = F_i'(\vec{\alpha}) - a_{ij}'(\vec{\alpha})x_j(\vec{\alpha}) & i \notin I_1 \cup I_2, \\ a_{j_i k}(\vec{\alpha})x_k'(\vec{\alpha}) = F_{j_i}'(\vec{\alpha}) - a_{j_i k}'(\vec{\alpha})x_k(\vec{\alpha}) \\ \qquad\qquad -\omega_i'(\vec{\alpha})\lambda_i^1 - \omega_i(\vec{\alpha})(\lambda_i^1)' & j_i \in I_1, \\ a_{j_i k}(\vec{\alpha})x_k'(\vec{\alpha}) = F_{j_i}'(\vec{\alpha}) - a_{j_i k}'x_k(\vec{\alpha}) + (\lambda_i^2)' & j_i \in I_2. \end{cases}$$

Using the continuity of $\vec{x}(\vec{\alpha})$, $\vec{\lambda}_1(\vec{\alpha})$ and $\vec{\lambda}_2(\vec{\alpha})$ with respect to $\vec{\alpha}$, we easily see that

$$(5.8) \qquad\qquad x_{j_i}'(\vec{\alpha}) = 0 \quad \text{if } j_i \in I_1^-(\vec{\alpha}),$$

$$(5.9) \qquad\qquad (\lambda_i^1)'(\vec{\alpha}) = 0 \quad \text{if } j_i \in I_1^+(\vec{\alpha}).$$

Let $j_i \in I_1^0(\vec{\alpha})$ and $\lambda_i^1(\vec{\alpha}) = 1$. Then we immediately obtain

$$(5.10) \qquad\qquad (\lambda_i^1)'(\vec{\alpha}) \leq 0, \quad x_{j_i}'(\vec{\alpha}) \leq 0, \quad (\lambda_i^1)'x_{j_i}'(\vec{\alpha}) = 0.$$

If $j_i \in I_1^0(\vec{\alpha})$ and $\lambda_i^1(\vec{\alpha}) = -1$, the sign $\leq$ in (5.10) has to be replaced by $\geq$. Similar analysis can be made for the index set I_2.

Indeed:

$$
(5.11) \qquad \begin{cases} (\lambda_i^2)'(\vec{\alpha}) = 0 & \text{if } j_i \in I_2^+(\vec{\alpha}), \\ x_{j_i}'(\vec{\alpha}) = -\beta_i & \text{if } j_i \in I_2^-(\vec{\alpha}), \end{cases}
$$

assuming that the derivative in the direction $\vec{\beta}$ is evaluated, $\vec{\beta} = (\beta_0, ..., \beta_D)$. Finally, if $j_i \in I_2^0(\vec{\alpha})$ then it is easy to show that

$$
(5.12) \qquad (\lambda_i^2)'(\vec{\alpha}) \geq 0, \quad x_{j_i}'(\vec{\alpha}) \geq -\beta_i, \quad (\lambda_i^2)'(x_{j_i}'(\vec{\alpha}) + \beta_i) = 0.
$$

From (5.7)–(5.12) it follows that the directional derivative $\vec{x}' \equiv \vec{x}'(\vec{\alpha}, \vec{\beta})$ is the element of the closed convex set $\mathcal{K}(\vec{\alpha}, \vec{\beta})$, the definition of which depends on the vectors $\vec{\alpha}, \vec{\beta}$, i.e. at which point and in which direction the derivative is computed:

$$
\begin{aligned}
\mathcal{K}(\vec{\alpha}, \vec{\beta}) = \big\{ \vec{z} \in \mathbf{R}^n \,|\, & z_i = 0 \quad \forall i \in I_1^-(\vec{\alpha}), \\
& z_{j_i} \leq 0 \quad \forall j_i \in I_1^0(\vec{\alpha}), \ \lambda_i^1(\vec{\alpha}) = 1, \\
& z_{j_i} \geq 0 \quad \forall j_i \in I_1^0(\vec{\alpha}), \ \lambda_i^1(\vec{\alpha}) = -1, \\
& z_{j_i} = -\beta_i \quad \forall j_i \in I_2^-(\vec{\alpha}), \\
& z_{j_i} \geq -\beta_i \quad \forall j_i \in I_2^0(\vec{\alpha}) \big\}.
\end{aligned}
$$

Moreover, (5.10), (5.12) together with (5.7) show us that $\vec{x}'$ is a unique minimizer of the quadratic functional

$$
\begin{aligned}
Q(\vec{z}) = \frac{1}{2} (\vec{z}, A(\vec{\alpha})\vec{z})_{\mathbf{R}^n} &- (F'(\vec{\alpha}) - A'(\vec{\alpha})\vec{x}(\vec{\alpha}), \vec{z}(\vec{\alpha}))_{\mathbf{R}^n} \\
&+ \sum_{j_i \in I_1} r_{j_i}(\vec{\alpha}) \frac{\omega_i'(\vec{\alpha})}{\omega_i(\vec{\alpha})} z_{j_i}
\end{aligned}
$$

over the set $\mathcal{K}(\vec{\alpha}, \vec{\beta})$. The symbols $F'(\vec{\alpha})$, $A'(\vec{\alpha})$ and $\omega_i'(\vec{\alpha})$ stand for the directional derivatives of $\vec{F}(\vec{\alpha})$, $A(\vec{\alpha})$ and $\omega_i(\vec{\alpha})$ at the point $\vec{\alpha}$ and the direction $\vec{\beta}$. Assuming that the mapping $\vec{\alpha} \mapsto \vec{F}(\vec{\alpha})$ is continuously differentiable, it holds

$$
F'(\vec{\alpha}) = (\nabla_{\vec{\alpha}} \vec{F}(\vec{\alpha}), \vec{\beta})_{\mathbf{R}^{D+1}}
$$

and similarly for $A'(\vec{\alpha})$ and $\omega_i'(\vec{\alpha})$.

On the basis of results presented here, we conclude that the mapping $\vec{\alpha} \mapsto \mathbb{C}(\vec{\alpha}, \vec{x}(\vec{\alpha}))$ is not continuously differentiable at each $\vec{\alpha} \in \mathcal{U}$, in general. This explains why computed results usually are not satisfactory, when the classical gradient type methods are used. There are two ways how to overcome this difficulty. The first one is to *regularize* our state problem by using the penalty approach, e.g. The unilateral condition on $\Gamma_K(\alpha)$ can be penalized so that the constrained optimization problem is replaced by a sequence of unconstrained ones, solutions of which are differentiable functions of α. For details see [9]. The advantage of this approach is the fact that the problem is transformed into smooth one. Therefore classical gradient type method can be applied. This approach however has several drawbacks. First of all, the unconstrained problem which has to be solved is no longer linear. Secondly, in order to obtain a good approximation of the original constrained problem, quite small penalty parameter has to be used. Then, theoretically smooth problem, behaves practically as a nonsmooth one, again. The second way is to use *nonsmooth optimization methods*. These methods are very well adapted for the minimization of functions which are not continuously differentiable. Their theoretical background can be found in [15], [16]. The use of such kind of methods in structural optimization is just at the beginning, but the results are very promising [21]. There are many situations, when classical smooth methods fail, the resulting designs are very far from to be optimal, while nonsmooth methods discover clearly better design. On the other hand it is difficult to apply such methods as a black box, without deeper knowledge of their theoretical background. But this is a usual price which has to be paid when more sophisticated approaches are used.

For some special choices of the cost functional, however the function $\vec{x} \mapsto C(\vec{\alpha}, \vec{x}(\vec{\alpha}))$ can be *continuously differentiable*, regardless the fact that the inner mapping $\vec{x} \mapsto \vec{x}(\vec{\alpha})$ is not. This is, for example, the case when the cost functional is equal to the value of total potential energy, evaluated at the equilibrium state, i.e.

$$(5.13) \qquad C(\vec{\alpha}, \vec{x}(\vec{\alpha})) \equiv \mathcal{L}_{\vec{\alpha}}(\vec{x}(\vec{\alpha})) = \frac{1}{2}(\vec{x}(\vec{\alpha}), A(\vec{\alpha})\vec{x}(\vec{\alpha}))_{\mathbf{R}^n}$$
$$+ \sum_{j_i \in I_1} \omega_{j_i}(\vec{\alpha})|x_{j_i}(\vec{\alpha})| - (\vec{F}(\vec{\alpha}), \vec{x}(\vec{\alpha}))_{\mathbf{R}^n}.$$

We shall study the properties of this cost functional in more details in the next section.

Indeed, let $\vec{\alpha} \in \mathcal{U}$ and $\vec{\beta} \in \mathbf{R}^{D+1}$ be given and let us evaluate $C'(\vec{\alpha}, \vec{\beta})$. Differentiating (5.13), we obtain (see [9]):

$$(5.14) \qquad C'(\vec{\alpha}, \vec{\beta}) = \frac{1}{2}(\vec{x}(\vec{\alpha}), A'(\vec{\alpha}, \vec{\beta})\vec{x}(\vec{\alpha}))_{\mathbf{R}^n} - (\vec{F}'(\vec{\alpha}, \vec{\beta}), \vec{x}(\vec{\alpha}))_{\mathbf{R}^n}$$
$$- \sum_{j_i \in I_2} \beta_i r_{j_i}(\vec{\alpha}) + \sum_{j_i \in I_1} \omega_i'(\vec{\alpha}, \vec{\beta})|x_{j_i}(\vec{\alpha})|,$$

where $A'(\vec{\alpha}, \vec{\beta})$ is the directional derivative of A at $\vec{\alpha}$ and in the direction $\vec{\beta}$ (the same for $\vec{F}$ and ω_i) and $r(\vec{\alpha}) = A(\vec{\alpha})\vec{x}(\vec{\alpha}) - \vec{F}(\vec{\alpha})$ is the residual vector. The expression (5.14) does not contain the directional derivative of $\vec{x}(\vec{\alpha})$. Therefore the cost functional (5.13) is *once continuously differentiable* and consequently, classical smooth methods for its minimization can be used.

6. TOTAL POTENTIAL AND RECIPROCAL ENERGY AS OBJECTIVE
FUNCTIONALS IN CONTACT SHAPE OPTIMIZATION

From previous section we already know that the total potential energy evaluated in the equilibrium state is once continuously differentiable function of the design variable. This functional has been proposed by authors in [6] in order to obtain constant stress distribution along $\Gamma_K(\alpha)$. Let us analyse the problem in more details. We shall consider the optimal shape design problem for frictionless Signorini problem, introduced in Section 3.

Let $\alpha \in U_{\mathrm{ad}}$, where U_{ad} is given by (3.1), let $u(\alpha) \in K(\alpha)$ be a solution of $(\mathcal{P}(\alpha))$ and denote

$$(6.1) \qquad \mathcal{C}(\alpha) \equiv \mathcal{J}_\alpha(u(\alpha)) = \frac{1}{2} a_\alpha(u(\alpha), u(\alpha)) - \langle f, u(\alpha) \rangle_\alpha.$$

Our aim will be to compute the directional derivative of $\mathcal{C}$ and to analyse the corresponding optimality conditions, valid at an optimal point. Let $\mathcal{V} = (0, \mathcal{V}_2)$, $\mathcal{V}_2 \in H^1(\Omega(\alpha))$, $\mathcal{V}_2|_{\partial\Omega(\alpha)\backslash\Gamma_K(\alpha)} \equiv 0$ be a vector field and assume the mapping $F_t \colon \mathbf{R}^2 \to \mathbf{R}^2$ given by $F_t = id + t\mathcal{V}$, $t > 0$. The deformation of $\Omega(\alpha)$ by means of F_t will be denoted by Ω_t, i.e. $\Omega_t = F_t(\Omega(\alpha))$. Define also the set $V_t(\alpha)$ similarly to $V(\alpha)$ but with $\Omega(\alpha)$ changed by $\Omega_t(\alpha)$ and set

$$K_t(\alpha) = \big\{ v \in V_t(\alpha) \mid v_2(x_1, \alpha(x_1)) + t\mathcal{V}_2(x_1, \alpha(x_1)) \geq$$
$$\geq -\alpha(x_1) - t\mathcal{V}_2(x_1, \alpha(x_1)) \ x_1 \in (a, b) \big\}.$$

The convex set $K_t(\alpha)$ contains all kinematically admissible displacements on the deformed $\Gamma_K(\alpha_t) \equiv F_t(\Gamma_K(\alpha))$. Let $u_t(\alpha) \in K_t(\alpha)$ be the solution of the Signorini problem without friction on Ω_t and set $\mathcal{C}_t(\alpha) \equiv \mathcal{J}_t(u_t(\alpha))$, where $\mathcal{J}_t$ is the total potential energy functional on Ω_t. Our goal is to compute $\dot{\mathcal{C}}(\alpha) = \frac{d}{dt}\mathcal{C}_t(\alpha)|_{t=0}$.

Before doing that we reformulate the state problem by introducing Lagrange multipliers in order to release the kinematical constraint on $\Gamma_K(\alpha_t)$. Define the Lagrangian L_t on $V_t(\alpha) \times L^2_+(a, b)$ as follows:

$$L_t(v, \mu) = \mathcal{J}_t(v) - \int_a^b \mu(v_2 + \alpha + t\mathcal{V}_2)\, dx_1^t,$$

where $L^2_+(a, b)$ denotes the set of all non-negative square integrable functions in (a, b). A direct calculation yields the following result (for details see [17]):

$$(6.2) \qquad \dot{\mathcal{C}}(\alpha) = \frac{1}{2} \int_{\Gamma_K(\alpha)} \tau_{ij}(u(\alpha))\varepsilon_{ij}(u(\alpha))\mathcal{V}_k n_k\, ds - \int_{\Gamma_K(\alpha)} F_i u_i(\alpha)\mathcal{V}_k n_k\, ds$$

$$- \int_{\Gamma_K(\alpha)} \tau_{2i}(u(\alpha))n_i \mathcal{V}_2\, ds - \int_{\Gamma_K(\alpha)} \tau_{2i}(u(\alpha))n_i \frac{\partial u_2(\alpha)}{\partial x_2}\mathcal{V}_2\, ds.$$

Such expression of $\dot{\mathcal{C}}(\alpha)$ in terms of boundary integrals will be useful in what follows. From here we see again that the cost functional (6.1) is once continuously differentiable, the fact already observed in finite dimension. Let us assume that *the displacement gradient and the displacement itself are small compared to unity.* The second assumption is a consequence of the first one, assuming $\Gamma_u \neq \emptyset$. Thus (6.2) consists of two parts: one dominating (the last term on the right hand side of (6.2)), coming from the explicit presence of the design variable in the definition of $K(\alpha)$ and one lower order part which comes from the moving boundary (the first three terms). Consequently

$$(6.3) \qquad \dot{\mathcal{C}}(\alpha) \approx - \int_{\Gamma_K(\alpha)} \tau_{2i}(u(\alpha)) n_i V_2 \, ds.$$

Assuming that the unit normal $n \approx (0, -1)$, (6.3) can be expressed as

$$(6.4) \qquad \dot{\mathcal{C}}(\alpha) \approx \int_{\Gamma_K(\alpha)} \tau_{22}(u(\alpha)) V_2 \, ds \equiv \int_{\Gamma_K(\alpha)} T_2(u) V_2 \, ds.$$

In order to remove the constant volume constraint, appearing in the definition of U_{ad}, another Lagrange function will be introduced:

$$\mathcal{L}_t(\alpha) = \mathcal{C}_t(\alpha) + \lambda(\mathrm{meas}\,\Omega_t(\alpha) - C_1), \quad t > 0, \ \lambda \in \mathbf{R}^1.$$

Let $\alpha^* \in U_{\mathrm{ad}}$ be an optimal solution and introduce the set

$$\overline{I}(\alpha^*) = \left\{ x_1 \in (a, b) \mid 0 < \alpha^*(x_1) < C_0, \ |\alpha^{*\prime}(x_1)| < C_1 \right\}.$$

The set $\overline{I}(\alpha^*)$ contains points of (a, b), where the inequality constraints are non-active. Assume that $\overline{I}(\alpha^*)$ is a non-empty interval. Let $\overline{V} = (0, \overline{V}_2)$ be any vector field such that $\mathrm{supp}\,\overline{V}_2(\cdot, \alpha^*(\cdot)) \subset \overline{I}(\alpha^*)$, i.e. $\overline{V}$ deforms $\Gamma_K(\alpha^*)$ at points from $\Xi(\alpha^*) = \{(x_1, \alpha^*(x_1)) \mid x_1 \in \overline{I}(\alpha^*)\}$, only. Then the necessary optimality condition for α^* to be an optimal solution is the existence of $\lambda^* \in \mathbf{R}^1$ such that

$$(6.5) \qquad \dot{\mathcal{L}}(\alpha^*) \equiv \frac{d}{dt} \mathcal{L}_t(\alpha^*)\Big|_{t=0} = \dot{\mathcal{C}}(\alpha) + \lambda^* \frac{d}{dt} \mathrm{meas}\,\Omega_t(\alpha^*)\Big|_{t=0} = 0$$

for any vector field $\overline{V}$ with the above mentioned property. It is easy to show that

$$\frac{d}{dt} \mathrm{meas}\,\Omega_t(\alpha^*)\Big|_{t=0} = \int_{\Xi(\alpha^*)} \overline{V}_2 n_2 \, ds \approx - \int_{\Xi(\alpha^*)} \overline{V}_2 \, ds$$

if $n \approx (0, -1)$. From this, (6.4) and (6.5) we arrive at

$$(6.6) \qquad T_2(u) \approx \lambda^* = \mathrm{const} \quad \text{on } \Xi(\alpha^*),$$

i.e. contact forces are constant along $\Xi(\alpha^*)$. As we shall see in next section, numerical results are in very good agreement with (6.6).

The fact that the cost functional (6.1) has something to do with stress distribution along $\Gamma_K(\alpha)$ is not clear from the beginning at all and a long computation was necessary to see that. That is why we introduce a new functional, where (at least at the beginning, again) this link will be more explicit.

The natural idea is to define the new cost functional as follows:

$$(6.7) \qquad \mathcal{C}(\alpha, u(\alpha)) = \int_{\Gamma_K(\alpha)} \|T(u)\|^2 \, ds,$$

where $\| \ \|$ stands for the norm of the stress vector $T = (T_1, T_2)$. The choice of (6.7) is nothing else but the least square method for the minimization of the stress vector on $\Gamma_K(\alpha)$. Unfortunately, the problem from the mathematical point of view is not well defined, due to the fact that stress vector T is not square integrable function, in general. Some other norm has to be introduced. We sketch the main idea of the approach. For more details we refer to [18].

Let $M(\alpha) \subset \partial\Omega(\alpha)$ be an open part of the boundary and such that $M(\alpha) \cap \Gamma_u = \emptyset$ while $M(\alpha) \cap \Gamma_K(\alpha) \neq \emptyset$ and denote

$$V_M(\alpha) = \{ v \in (H^1(\Omega(\alpha))^2 \mid v_i = 0 \ i = 1, 2 \text{ on } \partial\Omega(\alpha) \setminus \overline{M}(\alpha) \}.$$

Let $W_M(\alpha)$ be the space of all traces of functions belonging to $V_M(\alpha)$:

$$W_M(\alpha) = \{ w \text{ defined on } \partial\Omega(\alpha) \mid \exists v \in V_M(\alpha) : w = v \text{ on } \partial\Omega(\alpha) \}$$

and let $W'_M(\alpha)$ be the dual space on $W_M(\alpha)$. Then it is possible to show that $T(u) \in W'_M(\alpha)$. A natural idea arises, namely to replace the L^2-norm in (6.7) by the corresponding dual norm in $W'_M(\alpha)$:

$$(6.8) \qquad \mathcal{C}(\alpha, u(\alpha)) \equiv \|T(u)\|_{*,M(\alpha)}$$

or more generally

$$(6.9) \qquad \mathcal{C}(\alpha, u(\alpha)) \equiv \|T(u) - z_d\|_{*,M(\alpha)},$$

where z_d is a given element and $\| \ \|_{*,M(\alpha)}$ denotes the dual norm in $W'_M(\alpha)$ defined as follows

$$\|f\|_{*,M(\alpha)} = \sup_{w \neq 0} \frac{\langle f, w \rangle}{\|w\|_{W_M(\alpha)}}.$$

The problem ($\mathbb{P}$) is now well defined from the mathematical point of view, unfortunately the dual norm, introduced above does not suit very well for practical calculations. It can be shown however that the dual norm can be expressed in an equivalent way as follows:

$$\left\|T(u) - z_d\right\|_{*,M(\alpha)}^2 = \left|z(u)\right|_{1,\Omega(\alpha)}^2,$$

where $z(u) \in V_M(\alpha)$ is the unique solution of an auxiliary elasticity problem

(6.10)
$$\int_{\Omega(\alpha)} \tau_{ij}(z)\varepsilon_{ij}(v)\, dx = \langle T(u) - z_d, v\rangle \quad \forall v \in V_M(\alpha)$$

and $|\cdot|_{1,\Omega(\alpha)} \equiv (\tau_{ij}(\cdot), \varepsilon_{ij}(\cdot))_{0,\Omega}^{1/2}$. Thus, we are led to the following expression of the cost functional:

(6.11)
$$\mathcal{C}(\alpha, u(\alpha)) = \left|z(u)\right|_{1,\Omega(\alpha)}^2,$$

where $z(u) \in V_M(\alpha)$ solves (6.10).

In what follows we make the physical interpretation of the cost functional (6.11). Using Lagrangian multiplier technique, we can write

$$\inf_{v\in K(\alpha)} \mathcal{J}_\alpha(v) = \inf_{v\in V(\alpha)} \sup_{\lambda\in\Lambda} \left\{ \mathcal{J}_\alpha(v) - \int_{\Gamma_K(\alpha)} \lambda(v_2 + \alpha)\, ds\right\},$$

where

$$\Lambda = \left\{\lambda \in L^2(\Gamma_K(\alpha)) \mid \lambda \geq 0 \text{ a.e. on } \Gamma_K(\alpha)\right\}.$$

Interchanging inf sup formulation by sup inf we obtain the so called *reciprocal variational formulation* (for details see [19], e.g.):

(6.12)
$$\sup_{\lambda\in\Lambda} \inf_{v\in V(\alpha)} \left\{ \mathcal{J}_\alpha(v) - \int_{\Gamma_K(\alpha)} \lambda(v_2 + \alpha)\, ds\right\} \equiv \sup_{\lambda\in\Lambda} \Pi_R(\lambda),$$

where $\Pi_R(\lambda)$ is the so called *reciprocal energy functional*. A direct calculation shows that Π_R is a quadratic functional again and

$$\Pi_R(\lambda) = \frac{1}{2}\beta_\alpha(\lambda, \lambda) - \xi_\alpha(\lambda),$$

where

$$\beta_\alpha(\mu, \lambda) = (\mu, G(\lambda))_{0,\Gamma_K(\alpha)}$$
$$\xi_\alpha(\mu) = -(\mu, G(F, P))_{0,\Gamma_K(\alpha)} - (\mu_2, \alpha)_{0,\Gamma_K(\alpha)},$$

where $(\mu, \lambda) \in \Lambda \times \Lambda$. Here $G: V'(\alpha) \to V(\alpha)$ is the Green's operator, which with any $\lambda \in \Lambda$ and given forces F, P associates the unique solution of the linear elasticity problem

$$z \in V_M(\alpha) : a_\alpha(z, v) = (\lambda, v)_{0, \Gamma_K(\alpha)} + \int_{\Omega(\alpha)} F_i v_i \, dx + \int_{\Gamma_p} P_i v_i \, ds$$

$$\forall v \in V_M(\alpha).$$

Using the linearity of this problem, one can write

$$z = G(\lambda) + G(F, P),$$

which explains the meaning of symbols introduced before. It is well-known that (6.12) has a unique solution $\lambda^* = T_2(u)$. From this and Betti's reciprocal theorem, we finally obtain

$$\mathcal{J}_\alpha(u(\alpha)) = \frac{1}{2}(T(u), G(T(u)))_{0, \Gamma_K(\alpha)} - \frac{1}{2}\langle f, G(F, P)\rangle_\alpha$$

and

(6.13) $$\Pi_R(u) = -\frac{1}{2}(T(u), G(T(u)))_{0, \Gamma_K(\alpha)}.$$

The right hand side of (6.13) can be interpreted as the energy stored in the body due to contact forces, had no other forces been presented.

If $M(\alpha) \equiv \partial\Omega(\alpha) \setminus \overline{\Gamma}_u$ (i.e. $V_M(\alpha) = V(\alpha)$), $z_d \equiv P$ on Γ_p and $z_d \equiv 0$ on $\Gamma_K(\alpha)$ it is readily seen that

(6.14) $$\|T(u) - z_d\|_{*, M(\alpha)} \equiv (T(u), G(T(u)))_{0, \Gamma_K(\alpha)},$$

i.e. the cost functional (6.11) has precisely meaning stated above. Also in this case, when (6.14) holds, the cost functional (6.11) is once continuously differentiable (see [18], [20]).

REMARK 6.1: All our considerations here are formal from the mathematical point of view. They are valid, if the solution of our problem is sufficiently smooth. If not, the L^2 scalar product has to be replaced by the corresponding duality in the appropriate space of traces. □

7. NUMERICAL EXAMPLE

Here we present numerical results of one very simple model example of contact shape optimization, using two different cost functionals, studied in the previous section, namely *total potential energy* and *reciprocal energy* functionals. Augmented-Lagrangian techniques (to solve contact problem) and an interior point programming algorithm (for shape optimization) have been used to obtain numerical results. For more details we refer to [20].

A body Ω in undeformed state is given by

$$\Omega = \{[x, y] \in \mathbf{R}^2 \mid 0 < x < 4,\ 0 < y < 1\}$$

and the rigid foundation S by

$$S = \{[x, y] \in \mathbf{R}^2 \mid (x + r)^2 + (x - 2)^2 < r^2\}, \quad r = 40.025.$$

Material constants are $E = 2150$, $\nu = 0.29$ and surface tractions $P = (-5.78, -5.78)$ are acting on $\Gamma_p = \{[x, y] \in \mathbf{R}^2 \mid 2 < x < 4,\ y = 1\}$ (see Fig. 5).

The triangulation of the initial configuration is seen from Fig. 6. The final (optimal) configuration with respect to potential energy and reciprocal energy, respectively, is presented in Fig. 7 and 8, respectively. Finally, the distribution of normal forces along a contact surface is shown in Fig. 9. From here we clearly see that the total potential energy functional yields a uniform contact stress distribution as claimed before.

The problem (A.18) can be solved numerically. Indeed, as V_h is finite-dimensional, there exists its basis, the elements of which are denoted by $\varphi_1, \varphi_2, ..., \varphi_{n(h)}$. Therefore the isomorfism $T : V_h \to \mathbf{R}^{n(h)}$ can be defined in a usual way:

$$T v_h = (\alpha_1, ..., \alpha_{n(h)}) \in \mathbf{R}^{n(h)}, \quad v_h \in V_h,$$

where $\alpha_1, ..., \alpha_{n(h)}$ are coordinates of v_h with respect to the basis $\varphi_1, ..., \varphi_{n(h)}$, i.e.

$$v_h = \sum_{j-1}^{n(h)} \alpha_j \varphi_j.$$

Denote by

$$\mathcal{K} = \{\vec{x} \in \mathbf{R}^{n(h)} \mid T^{-1}\vec{x} \in K_h\},$$

where T^{-1} stands for the inverse of T. It is easy to see that $\mathcal{K}$ is the closed convex subset of $\mathbf{R}^{n(h)}$. Finally, let

$$\mathcal{L}(\vec{x}) \equiv \mathcal{J}(T^{-1}\vec{x}) = \frac{1}{2}(\vec{x}, A\vec{x})_{\mathbf{R}^{n(h)}} - (\vec{F}, \vec{x})_{\mathbf{R}^{n(h)}}$$

where A is the stiffness matrix and $\vec{F}$ the force vector. Then an equivalent expression of (A.18) reads as follows:

(A.22) \qquad find $\vec{x}^* \in \mathcal{K}$ such that $\mathcal{L}(\vec{x}^*) \leq \mathcal{L}(\vec{x}) \quad \forall \vec{x} \in \mathcal{K}$.

This is a non-linear programming problem, which can be numerically realized by various minimization methods. Their concrete choice depends on the character of the convex set $\mathcal{K}$, among others.

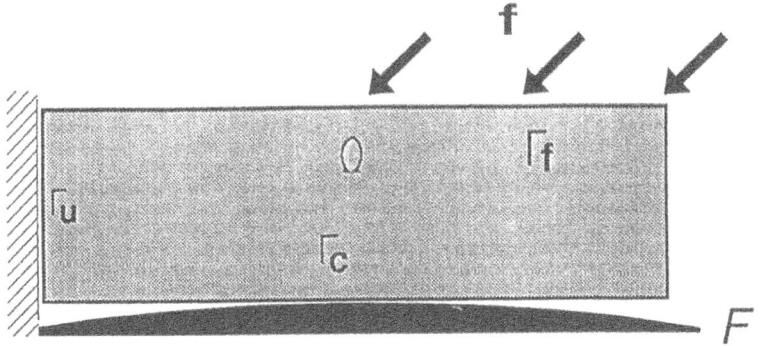

Figure 5.

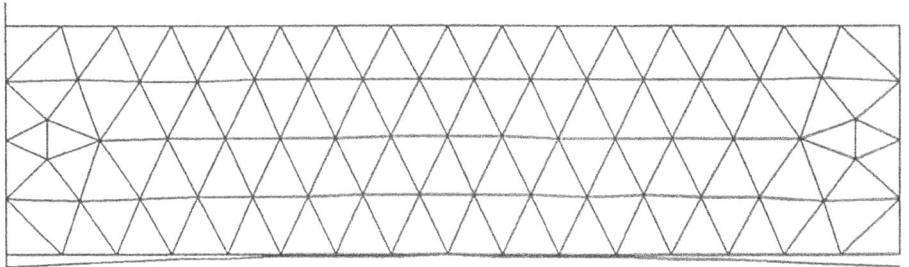

Figure 6.

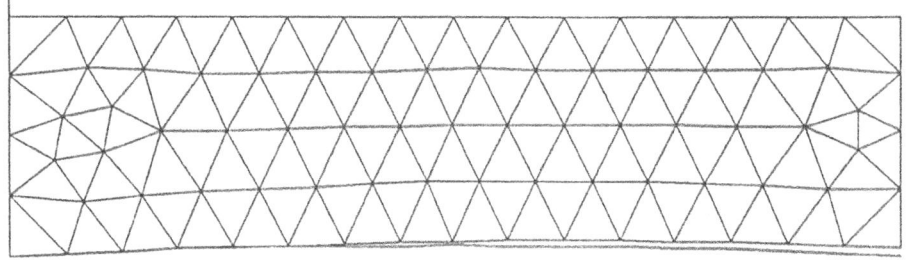

Figure 7.

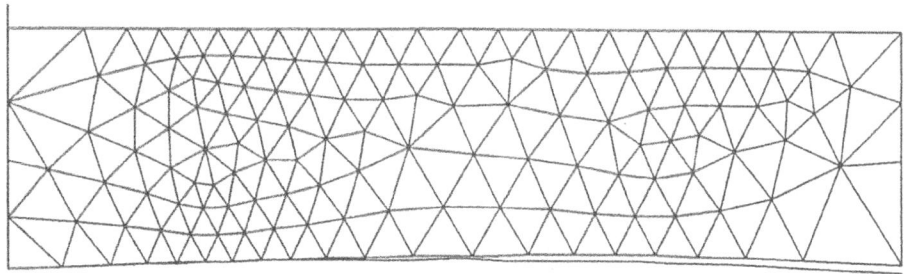

Figure 8.

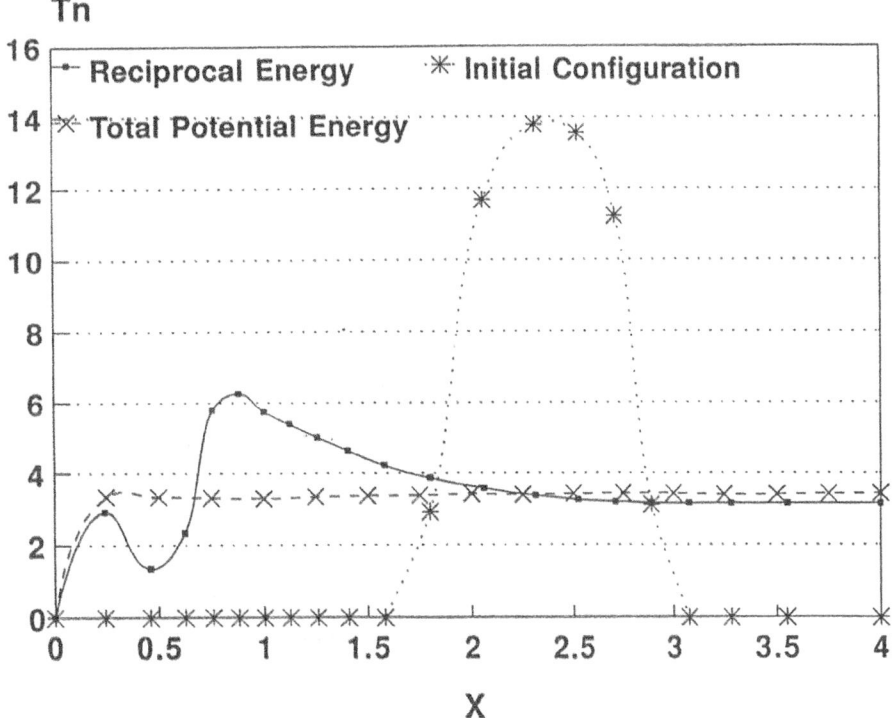

Figure 9.

The theory of variational inequalities together with their approximation and the numerical realization can be found in [7], [11], [12] and references therein.

APPENDIX B

Let $\Omega \subset \mathbf{R}^n$ be a bounded domain. The set of all square integrable functions in Ω will be denoted by $L^2(\Omega)$. It is well known that $L^2(\Omega)$ is a Hilbert space with the scalar product

$$(f,g)_0 \equiv \int_\Omega fg\,dx.$$

Let k be an integer. By $H^k(\Omega)$ we denote the space of functions, the all derivatives of which up to the order k are square integrable in Ω. It is well known that $H^k(\Omega)$ is a Hilbert space with the scalar product

$$(f,g)_k = \sum_{|\alpha|\le k} \int_\Omega D^\alpha f D^\alpha g\,dx,$$

where $\alpha = (\alpha_1,...,\alpha_n)$ is a multiindex, $|\alpha| = \sum_{i=1}^n \alpha_i$ and $D^\alpha f = \dfrac{\partial^{|\alpha|} f}{\partial x_1^{\alpha_1}...\partial x_n^{\alpha_n}}$.

The corresponding norm will be denoted by $\|\cdot\|_k$, i.e.

$$\|v\|_k = \left\{ \sum_{|\alpha|\le k} \int_\Omega |D^\alpha v|^2\,dx \right\}^{1/2}$$

The expression

$$|v|_j \equiv \left(\sum_{|\alpha|=j} \int_\Omega |D^\alpha v|^2\,dx \right)^{1/2}$$

will be called the j-th seminorm of v.

If $k = 0$ then $H^0(\Omega) \equiv L^2(\Omega)$. The scalar product and the norm in cartesian product $(H^k(\Omega))^m$ are defined as follows:

$$\|f\|_k = \left(\sum_{i=1}^m \|f_i\|_k^2 \right)^{1/2}$$

$$(f,g) = \sum_{i=1}^m (f_i, g_i)_k, \qquad \begin{array}{l} f = (f_1,...,f_m) \in (H^k(\Omega))^m \\ g = (g_1,...,g_m) \in (H^k(\Omega))^m. \end{array}$$

REFERENCES

1. Pironneau, O. (1984), "Optimal shape design for elliptic systems," Springer series in Computational Physics, Springer-Verlag, New York.

2. Banichuk, N. V. (1983), "Problems and methods of optimal structural design," Plenum Press, New York.

3. Haug, E. J., Choi, K. K., Komkov (1986), "Design sensitivity analysis of structural systems," Academic Press, Orlando.

4. Duvaut, G., Lions, J. L. (1976), "Inequalities in mechanics and physics," Grundlehren der mathematischen Wissenschaften 219, Springer-Verlag, Berlin.

5. Panagiotopoulos, P. D. (1985), "Inequality problems in mechanics and applications," Birkhäuser, Boston.

6. Benedict, R. L., Taylor, J. E. (1981), *Optimal design for elastic bodies in contact*, in "Optimization of distributed parameter structures, Part II," Haug, E. J. and Céa, J. (eds.), Nato Advances Study Institute Series, Series E, Sijthoff & Noordhoff, Alphen aan der Rijn, pp. 1553–1599.

7. Hlavaček, I., Haslinger, J., Nečas, J., Lovišek, J. (1988), "Numerical solution of variational inequalities," Springer Series in Applied Mathematical Sciences 66, Springer-Verlag, New York.

8. Kikuchi, N., Oden, J. T. (1988), "Study of variational inequalities and finite element methods," Mathematics and Computer Sciences for Engineers, SIAM.

9. Haslinger, J., Neittaanmäki, P. (1988), "Finite element approximation for optimal shape design: Theory and applications," J. Wiley & Sons.

10. Nečas, J. (1967), "Les méthodes directes en théorie des equations elliptiques," Masson, Paris.

11. Céa, J. (1978), "Lectures on optimization theory and algorithms," Springer-Verlag, Berlin.

12. Glowinski, R., Lions, J. L., Trémolières, R. (1981), "Numerical analysis of variational inequalities, studies in mathematics and its applications," North Holland Publishing Company, Amsterdam.

13. Mignot, F. (1976), *Contrôle dans les inéquations variationnelles elliptiques*, J. Funct. Anal. **22**, 25–39.

14. Sokolowski, J. (1981), *Sensitivity analysis for a class of variational inequalities*, in (Haug, Céa, 1981, Part II, see [6]).

15. Kiwiel, K. C. (1985), "Methods of descent for non-differentiable optimization," Lecture Notes in Mathematics 1133, Springer-Verlag, New York.

16. Lemarechal, Cl., Miffeir, R. (1978), "Nonsmooth optimization," Pergamon Press, Oxford.

17. Klarbring, A., Haslinger, J. (1993), *On almost constant contact stress distribution by shape optimization*, Struct. Optim. **3**, 213–216.

18. Haslinger, J., Klarbring, A., *Shape optimization in unilateral contact problems using generalized reciprocal energy as objective functional*, J. Nonlinear Anal. Theory Meth. Appl. (to appear).

19. Haslinger, J., Panagiotopoulos, P. D. (1984), *Approximation of contact problems with friction by reciprocal variational formulations*, Prod. Roy. Soc. Edinburgh **98A**, 365–383.

20. Feijoo, R. A., Fancello, E. A. (1992), *A finite element approach for an optimal shape design in contact problems*, in "Contact Mechanics International Symposium," ed. A. Curnier, EPFL, Lausanne, Switzerland, pp. 287–306.

21. Mäkelä, M. M., Neittaanmäki, P. (1992), "Nonsmooth optimization: Analysis and algorithms with applications to optimal control," World Scientific Publishing Co, Singapore.

FIRST AND SECOND ORDER DESIGN SENSITIVITY AT A BIFURCATION POINT

Zenon MRÓZ and Jarosław PIEKARSKI

Institute of Fundamental Technological Research,
Polish Academy of Sciences, Warsaw, Poland

ABSTRACT

The variation of critical load factor corresponding to bifurcation point is considered assuming variation of a structure parameter s. Two cases are considered: regular sensitivity case when the bifurcation point is preserved for varying parameter s, and the singular sensitivity case when the limit points evolve from the bifurcation point. First and second order sensitivity expressions are derived analytically. A simple example is provided in order to illustrate the general theory. Next, the sensitivity of vibration frequency is studied and the bifurcation load sensitivity is generated in the limiting case.

1 INTRODUCTION

A non-linear elastic structure subjected to increasing load usually deforms initially in a regular deformation regime for which uniqueness and stability is preserved. However, for a sufficiently high load value a critical state is reached, such as a single or multi-modal bifurcation point and a limit point. The fundamental problem is to determine critical load value and post-critical response when the deformation process continues beyond the critical state. An equally important problem is associated with the analysis of sensitivity of the critical load and the associated failure mode due to variation of structural parameters. In the optimization procedure the sensitivity expressions are fundamental both for generating effective redesign procedures and for study of structure response when the design parameters vary near the optimal design.

J. Herskovits (ed.), Advances in Structural Optimization, 349–380.
© *1995 Kluwer Academic Publishers.*

The present paper follows the previous study of Mróz and Haftka [1993] concerned with the sensitivity analysis of non–linear structures in regular and critical states. It is assumed that an elastic discrete structure response is governed by the potential energy which is a function of n generalized coordinates q_i, load factor λ, and the structure parameter s. The incremental equilibrium equations of first, second and third order can then be generated by the potential energy and both regular and critical states can be identified, cf. general methodology developed by Thompson and Hunt [1973]. The first order sensitivity expressions were derived by Mróz and Haftka [1993] in a general case following earlier results obtained by Dems and Mróz [1989], Mróz and Haftka [1988], and Haftka, Cohen and Mróz [1990]. A recent study by Godoy, Tarocco and Feijoo [1993] is concerned with the same problem using operator form of equilibrium equations. The present paper supplements the previous treatments and provides the formulae for second order sensitivity derivatives for regular and singular sensitivity cases. The regular sensitivity case occurs when the bifurcation point is preserved and evolves with the variation of structure parameter s. On the other hand, the singular sensitivity case occurs when the bifurcation point vanishes and limit points evolve from the critical state. We shall discuss both cases and derive the sensitivity derivatives with respect to the design parameter s.

2 CRITICAL LOAD SENSITIVITY

2.1 FUNDAMENTAL RELATIONS

In order to make this presentation self-contained, let us briefly derive the first order sensitivity expressions following the analysis of Mróz and Haftka [1993].

Consider a nonlinear elastic structure for which the potential energy $V(\boldsymbol{q}, \lambda, s)$ depends on the generalized coordinate vector $\boldsymbol{q}$, load factor λ, and the structure parameter s. The equilibrium conditions for such structure are:

$$V_i(\boldsymbol{q}, \lambda, s) = 0 , \quad i = 1, 2, \ldots, n \tag{1}$$

The first, second and third order equilibrium equations are

$$V_{ij}\dot{q}_j + V_{i\lambda}\dot{\lambda} + V_{is}\dot{s} = 0 \tag{2}$$

$$V_{ij}\ddot{q}_j + V_{ijk}\dot{q}_j\dot{q}_k + 2V_{ij\lambda}\dot{q}_j\dot{\lambda} + 2V_{ijs}\dot{q}_j\dot{s} + \\ + 2V_{is\lambda}\dot{s}\dot{\lambda} + V_{i\lambda\lambda}\dot{\lambda}^2 + V_{iss}\dot{s}^2 + V_{i\lambda}\ddot{\lambda} + V_{is}\ddot{s} = 0 \tag{3}$$

$$V_{ij}\ddot{q}_j + V_{ijkl}\dot{q}_j\dot{q}_k\dot{q}_l + 3V_{ijk}\dot{q}_j\ddot{q}_k + 3V_{ijk\lambda}\dot{q}_j\dot{q}_k\dot{\lambda} + 3V_{ijks}\dot{q}_j\dot{q}_k\dot{s} +$$
$$+ 3V_{ij\lambda}\ddot{q}_j\dot{\lambda} + 3V_{ijs}\ddot{q}_j\dot{s} + 3V_{ij\lambda}\dot{q}_j\ddot{\lambda} + 3V_{ijs}\dot{q}_j\ddot{s} + V_{i\lambda}\dddot{\lambda} +$$
$$+ V_{is}\dddot{s} + 3V_{ij\lambda\lambda}\dot{q}_j\dot{\lambda}^2 + 3V_{ijss}\dot{q}_j\dot{s}^2 + 6V_{ij\lambda s}\dot{q}_j\dot{\lambda}\dot{s} +$$
$$+ 3V_{i\lambda\lambda}\dot{\lambda}\ddot{\lambda} + 3V_{iss}\dot{s}\ddot{s} + 3V_{i\lambda s}\dot{\lambda}\ddot{s} + 3V_{i\lambda s}\ddot{\lambda}\dot{s} +$$
$$+ 3V_{i\lambda\lambda s}\dot{\lambda}^2\dot{s} + 3V_{i\lambda ss}\dot{\lambda}\dot{s}^2 + V_{i\lambda\lambda\lambda}\dot{\lambda}^3 + V_{isss}\dot{s}^3 = 0 \tag{4}$$

where dot denotes the derivative with respect to the path parameter η at the equilibrium state, so that $q = q(\eta)$, $\lambda = \lambda(\eta)$, $s = s(\eta)$. The subscripts indicate partial derivatives with respect to generalized coordinates, load factor λ, and structure parameter s.

The bifurcation point conditions are:

$$V_{ij}^c q_{1j} = 0, \quad V_{i\lambda}^c q_{1i} = 0 \tag{5}$$

where q_{1j} is the eigenvector. The superscript c denotes evaluation of function at the critical state.

Further, if a *regular sensitivity bifurcation point* occurs, then

$$V_{is}^c q_{1i} = 0 \tag{6}$$

and $\dot{\lambda} = \dot{\lambda}^c$ or $\dot{s} = \dot{s}^c$ at the bifurcation point need not vanish. On the other hand at the *singular sensitivity bifurcation point* there is

$$V_{is}^c q_{1i} \neq 0, \quad \dot{s} = \dot{s}^c = 0 \tag{7}$$

and the bifurcation point disappears for varying s, with limit points evolving from the critical states, Fig.1.

The length of eigenvector q_1 can be arbitrarily chosen, with specified uniqueness condition. We use here the normalization condition

$$T_{ij}q_{1i}\dot{q}_{1j} = 0 \tag{8}$$

where T_{ij} is a positive definite symmetric matrix.

Consider the critical state path following critical points with varying s, so that

$$\dot{q}_j^c = q_j^{c\prime}\dot{\lambda}^c + q_{s_j}^c\dot{s}^c \tag{9}$$

where $q_j^c(\lambda, s)$ denotes the critical state path, $q_j^{c\prime}$ denotes the derivative of critical state path with respect to load parameter λ, $q_{s_j}^c$ is the derivative with respect to s.
The derivatives of coordinate vector, occurring in (9) can be calculated specializing respectively equation (2).

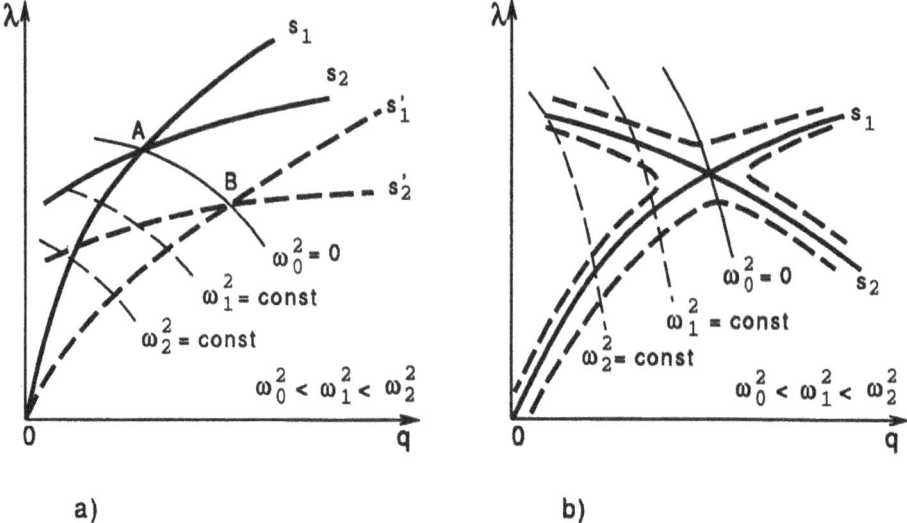

Fig.1 Load – deformation paths passing through bifurcation point,
a) regular sensitivity bifurcation case
b) singular sensitivity bifurcation case

Setting $\dot{s} = 0$, we have

$$V_{ij}^c q_j^{c\prime} + V_{i\lambda}^c = 0 \tag{10}$$

so $q_j^{c\prime}$ is identical to the prebuckling path derivative with respect to the load parameter $q_{0j}^\prime$.

Since V_{ij}^c is a singular matrix at the critical state, we may resort to Eq.(3) with $\dot{s} = 0$, multiplied by q_{1i} so that

$$V_{ijk}^c q_{1i} q_j^{c\prime} q_k^{c\prime} + 2V_{ij\lambda}^c q_{1i} q_j^{c\prime} + V_{i\lambda\lambda}^c q_{1i} = 0 \tag{11}$$

Equation (11) provides the complementary condition to (10). However, it is non-linear with respect to $q_j^{c\prime}$.

Setting now in (2) $\dot{\lambda} = 0$, we obtain the formula allowing for calculation of the derivative q_s^c

$$V_{ij}^c q_{sj}^c + V_{is}^c = 0 \tag{12}$$

Similarly as previously, we provide from (3) the complementary condition to the set of equations (12)

$$V_{ijk}^c q_{1i} q_{sj}^c q_{sk}^c + 2V_{ijs}^c q_{1i} q_{sj}^c + V_{iss}^c q_{1i} = 0 \tag{13}$$

2.1.1 DIRECT APPROACH

The fundamental relations used in direct calculation of first and second order critical load sensitivity are obtained by differentiating the critical state equation (5a) with respect to the path parameter, so that

$$V_{ij}^c \dot{q}_{1j} + V_{ijk}^c q_{1j} \dot{q}_k^c + V_{ij\lambda}^c q_{1j} \dot{\lambda}^c + V_{ijs}^c q_{1j} \dot{s}^c = 0 \tag{14}$$

Differentiating (14) once again, we obtain

$$
\begin{aligned}
V_{ijkl}^c q_{1j} \dot{q}_k^c \dot{q}_l^c + V_{ij\lambda\lambda}^c q_{1j} \left(\dot{\lambda}^c \right)^2 + V_{ijss}^c q_{1j} \left(\dot{s}^c \right)^2 + \\
+ 2V_{ijk\lambda}^c q_{1j} \dot{q}_k^c \dot{\lambda}^c + 2V_{ijks}^c q_{1j} \dot{q}_k^c \dot{s}^c + 2V_{ij\lambda s}^c q_{1j} \dot{\lambda}^c \dot{s}^c + \\
+ V_{ijk}^c q_{1j} \ddot{q}_k^c + V_{ij\lambda}^c q_{1j} \ddot{\lambda}^c + V_{ijs}^c q_{1j} \ddot{s}^c + V_{ij}^c \ddot{q}_{1j} + \\
+ 2V_{ijk}^c \dot{q}_{1j} \dot{q}_k^c + 2V_{ij\lambda}^c \dot{q}_{1j} \dot{\lambda}^c + 2V_{ijs}^c \dot{q}_{1j} \dot{s}^c = 0
\end{aligned}
\tag{15}
$$

2.1.2 ADJOINT APPROACH

The critical state and equilibrium conditions can be expressed in terms of the Lagrangian form

$$E = V_{ij}^c q_{1i} q_{1j} - \mu_i V_i^c = 0 \tag{16}$$

where μ is the Lagrange multiplier vector. The length of eigenvector q_1 can be arbitrarily chosen, with specified uniqueness condition. The normalization condition (8) can be used.

After differentiation with respect to the critical state path parameter, equation (16) takes the form

$$
\begin{aligned}
\dot{E} = V_{ijk}^c q_{1i} q_{1j} \dot{q}_k^c + V_{ij\lambda}^c q_{1i} q_{1j} \dot{\lambda}^c + V_{ijs}^c q_{1i} q_{1j} \dot{s}^c + 2V_{ij}^c q_{1i} \dot{q}_{1j} + \\
- \mu_i \left(V_{ij}^c \dot{q}_j^c + V_{i\lambda}^c \dot{\lambda}^c + V_{is}^c \dot{s}^c \right) = 0
\end{aligned}
\tag{17}
$$

Differentiating (17) once again, we obtain

$$\ddot{E} = V_{ijkl}^c q_{1i} q_{1j} \dot{q}_k^c \dot{q}_l^c + V_{ij\lambda\lambda}^c q_{1i} q_{1j} \left(\dot{\lambda}^c \right)^2 + V_{ijss}^c q_{1i} q_{1j} \left(\dot{s}^c \right)^2 +$$

$$+2V^c_{ijk\lambda}q_{1i}q_{1j}\dot{q}^c_k\dot{\lambda}^c + 2V^c_{ijks}q_{1i}q_{1j}\dot{q}^c_k\dot{s}^c + 2V^c_{ij\lambda s}q_{1i}q_{1j}\dot{\lambda}^c\dot{s}^c +$$
$$+4V^c_{ijk}q_{1i}\dot{q}_{1j}\dot{q}^c_k + 4V^c_{ij\lambda}q_{1i}\dot{q}_{1j}\dot{\lambda}^c + 4V^c_{ijs}q_{1i}\dot{q}_{1j}\dot{s}^c + 2V^c_{ij}\dot{q}_{1i}\dot{q}_{1j} +$$
$$+ V^c_{ijk}q_{1i}q_{1j}\ddot{q}^c_k + V^c_{ij\lambda}q_{1i}q_{1j}\ddot{\lambda}^c + V^c_{ijs}q_{1i}q_{1j}\ddot{s}^c + 2V^c_{ij}q_{1i}\ddot{q}_{1j} + \qquad (18)$$
$$-\mu_i\left(V^c_{ijk}\dot{q}^c_j\dot{q}^c_k + V^c_{i\lambda\lambda}\left(\dot{\lambda}^c\right)^2 + V^c_{iss}\left(\dot{s}^c\right)^2 + 2V^c_{ij\lambda}\dot{q}^c_j\dot{\lambda}^c + 2V^c_{ijs}\dot{q}^c_j\dot{s}^c +\right.$$
$$\left. +2V^c_{i\lambda s}\dot{\lambda}^c\dot{s}^c + V^c_{ij}\ddot{q}^c_j + V^c_{i\lambda}\ddot{\lambda}^c + V^c_{is}\ddot{s}^c\right) = 0$$

Differentiating the critical state condition (5a), multiplying it by the eigenvector q_{1i} and substituting into (18), provides

$$V^c_{ijkl}q_{1i}q_{1j}\dot{q}^c_k\dot{q}^c_l + V^c_{ij\lambda\lambda}q_{1i}q_{1j}\left(\dot{\lambda}^c\right)^2 + V^c_{ijss}q_{1i}q_{1j}\left(\dot{s}^c\right)^2 + 2V^c_{ijk\lambda}q_{1i}q_{1j}\dot{q}^c_k\dot{\lambda}^c +$$
$$+2V^c_{ijks}q_{1i}q_{1j}\dot{q}^c_k\dot{s}^c + 2V^c_{ij\lambda s}q_{1i}q_{1j}\dot{\lambda}^c\dot{s}^c - 2V^c_{ij}\dot{q}_{1i}\dot{q}_{1j} +$$
$$+ V^c_{ijk}q_{1i}q_{1j}\ddot{q}^c_k + V^c_{ij\lambda}q_{1i}q_{1j}\ddot{\lambda}^c + V^c_{ijs}q_{1i}q_{1j}\ddot{s}^c + 2V^c_{ij}q_{1i}\ddot{q}_{1j} + \qquad (19)$$
$$-\mu_i\left(V^c_{ijk}\dot{q}^c_j\dot{q}^c_k + V^c_{i\lambda\lambda}\left(\dot{\lambda}^c\right)^2 + V^c_{iss}\left(\dot{s}^c\right)^2 + 2V^c_{ij\lambda}\dot{q}^c_j\dot{\lambda}^c + 2V^c_{ijs}\dot{q}^c_j\dot{s}^c +\right.$$
$$\left. +2V^c_{i\lambda s}\dot{\lambda}^c\dot{s}^c + V^c_{ij}\ddot{q}^c_j + V^c_{i\lambda}\ddot{\lambda}^c + V^c_{is}\ddot{s}^c\right) = 0$$

As it will be shown in following sections, the adjoint field needs to be defined by the equation

$$V^c_{ijk}q_{1i}q_{1j} - \mu_i V^c_{ik} = 0 \qquad (20)$$

However, this equation is singular and has the solution only if

$$V^c_{ijk}q_{1i}q_{1j}q_{1k} = 0 \qquad (21)$$

that is for the symmetric bifurcation case. In view of (21), we can take the orthogonality condition as the complementary condition to the system (20), thus

$$\mu_i q_{1i} = 0 \qquad (22)$$

2.2   REGULAR SENSITIVITY CASE

2.2.1   DIRECT APPROACH

First order sensitivity

Multiplying equation (14) by the eigenvector q_{1i}, introducing the decomposition (9), and setting the structure parameter s as the path parameter, so $\dot{s}^c = 1$, $\dot{\lambda}^c = \lambda^c_s$, we obtain

$$\lambda_s^c = -\frac{V_{ijk}^c q_{1i} q_{1j} q_{sk}^c + V_{ijs}^c q_{1i} q_{1j}}{V_{ijk}^c q_{1i} q_{1j} q_k^{c\prime} + V_{ij\lambda}^c q_{1i} q_{1j}} \tag{23}$$

The derivative $q_j^{c\prime}$ is to be determined from (10) and (11), and q_{sk}^c from (12) and (13).

Second order sensitivity

The second order derivative of the coordinate vector with respect to the evolution parameter along the critical state path C_s can be written as follows:

$$\ddot{q}_j^c = q_j^{c\prime} \ddot{\lambda}^c + q_{sj}^c \ddot{s}^c + q_j^{c\prime\prime} \left(\dot{\lambda}^c \right)^2 + q_{ssj} \left(\dot{s}^c \right)^2 + 2 q_{sj}^{c\prime} \dot{\lambda}^c \dot{s}^c \tag{24}$$

The eigenvector derivative along the critical state path can be decomposed into two terms, so that

$$\dot{q}_{1j} = q_{1j}^\prime \dot{\lambda}^c + q_{1sj} \dot{s}^c \tag{25}$$

Evaluating (15) at the regular bifurcation point, subtracting (14) multiplied by $2\dot{q}_{1i}$, substituting decompositions (9), (24) and (25), and taking s as the critical state path parameter (so $\dot{s}^c = 1$, $\ddot{s}^c = 0$), we obtain

$$
\begin{aligned}
V_{ijkl}^c q_{1i} q_{1j} \left(q_k^{c\prime} \lambda_s^c + q_{sk}^c \right) \left(q_l^{c\prime} \lambda_s^c + q_{sl}^c \right) + V_{ij\lambda\lambda}^c q_{1i} q_{1j} \left(\lambda_s^c \right)^2 + \\
+ V_{ijss}^c q_{1i} q_{1j} + 2 V_{ijk\lambda}^c q_{1i} q_{1j} \left(q_k^{c\prime} \lambda_s^c + q_{sk}^c \right) \lambda_s^c + \\
+ 2 V_{ijks}^c q_{1i} q_{1j} \left(q_k^{c\prime} \lambda_s^c + q_{sk}^c \right) + 2 V_{ij\lambda s}^c q_{1i} q_{1j} \lambda_s^c + \\
- 2 V_{ij}^c \left(q_{1i}^\prime \lambda_s^c + q_{1si} \right) \left(q_{1j}^\prime \lambda_s^c + q_{1sj} \right) + V_{ij\lambda}^c q_{1i} q_{1j} \lambda_{ss}^c + \\
+ V_{ijk}^c q_{1i} q_{1j} \left(q_k^{c\prime} \lambda_{ss}^c + q_{0k}^{\prime\prime} \left(\lambda_s^c \right)^2 + q_{ssk}^c + 2 q_{0sk}^{c\prime} \lambda_s^c \right) = 0
\end{aligned} \tag{26}
$$

so there is

$$\lambda_{ss}^c = -\frac{A \left(\lambda_s^c \right)^2 + B \lambda_s^c + C}{V_{ijk}^c q_{1i} q_{1j} q_k^{c\prime} + V_{ij\lambda}^c q_{1i} q_{1j}} \tag{27}$$

where

$$
\begin{aligned}
A &= V_{ijkl}^c q_{1i} q_{1j} q_k^{c\prime} q_l^{c\prime} + V_{ij\lambda\lambda}^c q_{1i} q_{1j} + \\
&\quad + 2 V_{ijk\lambda}^c q_{1i} q_{1j} q_k^{c\prime} + V_{ijk}^c q_{1i} q_{1j} q_k^{c\prime\prime} - 2 V_{ij}^c q_{1i}^\prime q_{1j}^\prime \\
B &= 2 V_{ijkl}^c q_{1i} q_{1j} q_k^{c\prime} q_{sl}^c + 2 V_{ijk\lambda}^c q_{1i} q_{1j} q_{sk}^c + 2 V_{ijks}^c q_{1i} q_{1j} q_k^{c\prime} + \\
&\quad + 2 V_{ij\lambda s}^c q_{1i} q_{1j} + 2 V_{ijk}^c q_{1i} q_{1j} q_{sk}^{c\prime} - 4 V_{ij}^c q_{1i}^\prime q_{1sj} \\
C &= V_{ijkl}^c q_{1i} q_{1j} q_{sk}^c q_{sl}^c + V_{ijss}^c q_{1i} q_{1j} + 2 V_{ijks}^c q_{1i} q_{1j} q_{sk}^c + \\
&\quad - 2 V_{ij}^c q_{1si} q_{1sj} + V_{ijk}^c q_{1i} q_{1j} q_{ssk}^c
\end{aligned} \tag{28}
$$

The derivatives $q_j^{c\prime}$ and q_{sj}^c are given by (10), (11), (12), and (13).

To obtain $q_j^{c''}$, equation (3) can be specialized by evaluating it along the fundamental loading path C_l, so that $\dot{\lambda}^c = 1$, $\ddot{\lambda} = \dddot{\lambda} = \ldots = 0$, and $\dot{s} = \ddot{s} = \dddot{s} = \ldots = 0$. Then we have

$$V_{ij}^c q_j^{c''} + V_{ijk}^c q_j^{c'} q_k^{c'} + 2V_{ij\lambda}^c q_j^{c'} + V_{i\lambda\lambda}^c = 0 \tag{29}$$

At the bifurcation point this system of equations is singular. As the complementary condition, we can take equation (4), specialized in the same way as (29) and multiplied by the eigenvector q_{1i}, so that

$$3V_{ijk}^c q_{1i} q_j^{c'} q_j^{c''} + 3V_{ij\lambda}^c q_{1i} q_j^{c''} + V_{ijkl}^c q_{1i} q_j^{c'} q_k^{c'} q_l^{c'} + \\ 3V_{ijk\lambda}^c q_{1i} q_j^{c'} q_k^{c'} + 3V_{ij\lambda\lambda}^c q_{1i} q_j^{c'} + V_{i\lambda\lambda\lambda}^c q_{1i} = 0 \tag{30}$$

Next, specializing (3) and (4) along the transformation path C_t with $\lambda = const$, $\dot{\lambda} = \ddot{\lambda} = \ldots = 0$ and s taken as the path parameter, we can generate the system of equations allowing to obtain the derivative q_{ssj}^c, namely

$$V_{ij}^c q_{ssj}^c + V_{ijk}^c q_{sj}^c q_{sk}^c + 2V_{ijs}^c q_{sj}^c + V_{iss}^c = 0 \tag{31}$$

and

$$3V_{ijk}^c q_{1i} q_{sj}^c q_{ssk}^c + 3V_{ijs}^c q_{1i} q_{ssj}^c + V_{ijkl}^c q_{1i} q_{sj}^c q_{sk}^c q_{sl}^c + \\ + 3V_{ijks}^c q_{1i} q_{sj}^c q_{sk}^c + 3V_{ijss}^c q_{1i} q_{sj}^c + V_{isss}^c q_{1i} = 0 \tag{32}$$

Using the decompositions (9) and (24), and taking into account equations (10), (12), (29), (31) and (3), we obtain

$$V_{ij}^c q_{sj}^{c'} + V_{ijk}^c q_j^{c'} q_{sk}^c + V_{ij\lambda}^c q_{sj}^c + V_{ijs}^c q_j^{c'} + V_{is\lambda}^c = 0 \tag{33}$$

This equation can be used to determine the mixed derivative $q_{sj}^{c'}$.

The complementary condition is provided by taking (4) evaluated along the critical state path C_s, setting there s as the path parameter and using the decompositions (9) and (24). Then, in view of (30) and (32) we can write

$$V_{ijkl}^c q_{1i} q_j^{c'} q_{sk}^c (q_k^{c'} \lambda_s^c + q_{sk}^c) + 2V_{ijk}^c q_{1i} q_{sj}^{c'} (q_l^{c'} \lambda_s^c + q_{sl}^c) + \\ + V_{ijk}^c q_{1i} \left(q_{sj}^c q_{0k}'' \lambda_s^c + q_j^{c'} q_{ssk}^c \right) + V_{ijk\lambda}^c q_{1i} q_{sj}^c (2q_k^{c'} \lambda_s^c + q_{sk}^c) + \\ + V_{ijks}^c q_{1i} q_j^{c'} (q_k^{c'} \lambda_s^c + 2q_{sk}^c) + V_{ij\lambda}^c q_{1i} (2q_{sk}^{c'} \lambda_s^c + q_{ssk}^c) + \\ + V_{ijs}^c q_{1i} (q_{0k}'' \lambda_s^c + 2q_{sk}^{c'}) + 2V_{ij\lambda s}^c q_{1i} \left(q_j^{c'} \lambda_s^c + q_{sj}^c \right) + \\ + V_{ij\lambda\lambda}^c q_{1i} q_{sj}^c \lambda_s^c + V_{ijss}^c q_{1i} q_j^{c'} + V_{i\lambda\lambda s}^c q_{1i} \lambda_s^c + V_{i\lambda ss}^c q_{1i} = 0 \tag{34}$$

The partial derivatives of the eigenvector $\dot{q}_{1j}$ can be determined by differentiating the critical state condition (5) and specializing it subsequently along the fundamental loading path C_l and along the transformation path C_t at constant load factor.

The complementary condition to such systems is provided by differentiating the eigenvector normalization condition (8).

2.2.2 ADJOINT APPROACH

First order sensitivity

Let us consider now equation (17). Assuming the structure parameter s as the path parameter ($\dot{s}^c = 1$, $\ddot{s}^c = 0$, $\dddot{s}^c = 0$, ...) and using the decomposition (8), one obtains

$$\left(V^c_{ijk}q_{1i}q_{1j} - \mu_i V^c_{ik}\right) q^c_{sk} + \lambda^c_s \left[\left(V^c_{ijk}q^{c\prime}_k + V^c_{ij\lambda}\right) q_{1i}q_{1j} + \right.$$
$$\left. - \mu_i \left(V^c_{ij}q^{c\prime}_j + V^c_{i\lambda}\right)\right] + V^c_{ijs}q_{1i}q_{1j} - \mu_i V^c_{is} = 0 \tag{35}$$

If the adjoint field μ is defined by (20), the coefficient of q^c_{sk} in (35) vanishes and we now obtain an alternative expression for λ^c_s, namely

$$\lambda^c_s = -\frac{V^c_{ijs}q_{1i}q_{1j} - \mu_i V^c_{is}}{V^c_{ijk}q_{1i}q_{1j}q^{c\prime}_k + V^c_{ij\lambda}q_{1i}q_{1j}} \tag{36}$$

Thus instead of solution of (12) for q^c_s, we may solve (20) and determine the adjoint field μ. As it was shown by Mróz and Haftka [1994], this field is identical to post-buckling field associated with the symmetric bifurcation point.

Another sensitivity formula can be derived from (17) by setting s as the path parameter, so we obtain

$$\left(V^c_{ijk}q_{1i}q_{1j} - \mu_i V^c_{ik}\right) \dot{q}^c_k + \lambda^c_s \left(V^c_{ij\lambda}q_{1i}q_{1j} - \mu_i V^c_{i\lambda}\right) +$$
$$+ V^c_{ijs}q_{1i}q_{1j} - \mu_i V^c_{is} = 0 \tag{37}$$

The first term of this equation is identical to the left hand side of equation (20), so it vanishes and we obtain

$$\lambda^c_s = -\frac{V^c_{ijs}q_{1i}q_{1j} - \mu_i V^c_{is}}{V^c_{ij\lambda}q_{1i}q_{1j} - \mu_i V^c_{i\lambda}} \tag{38}$$

The formula (38) may prove simpler in calculation as it requires only the determination of the adjoint state μ and the eigenvector q_1.

Second order sensitivity

Let us rewrite (19) using the decomposition (9) and setting s as the critical state path parameter. Then, we have

$$
\begin{aligned}
& V_{ijkl}^c q_{1i} q_{1j} \left(q_k^{c\prime} \lambda_s^c + q_{sk}^c \right) \left(q_l^{c\prime} \lambda_s^c + q_{sl}^c \right) + V_{ij\lambda\lambda}^c q_{1i} q_{1j} \left(\lambda_s^c \right)^2 + V_{ijss}^c q_{1i} q_{1j} + \\
& \quad 2 V_{ijk\lambda}^c q_{1i} q_{1j} \left(q_k^{c\prime} \lambda_s^c + q_{sk}^c \right) \lambda_s^c + 2 V_{ijks}^c q_{1i} q_{1j} \left(q_k^{c\prime} \lambda_s^c + q_{sk}^c \right) + \\
& \quad\quad + 2 V_{ij\lambda s}^c q_{1i} q_{1j} \lambda_s^c - 2 V_{ij}^c \left(q_{1i}^\prime \lambda_s^c + q_{1si} \right) \left(q_{1j}^\prime \lambda_s^c + q_{1sj} \right) + \\
& \quad\quad\quad + V_{ijk}^c q_{1i} q_{1j} \ddot{q}_k^c + V_{ij\lambda}^c q_{1i} q_{1j} \lambda_{ss}^c + \\
& \quad - \mu_i \left[V_{ijk}^c \left(q_j^{c\prime} \lambda_s^c + q_{sj}^c \right) \left(q_k^{c\prime} \lambda_s^c + q_{sk}^c \right) + V_{i\lambda\lambda}^c \left(\lambda_s^c \right)^2 + V_{iss}^c + \right. \\
& \quad\quad 2 V_{ij\lambda}^c \left(q_j^{c\prime} \lambda_s^c + q_{sj}^c \right) \lambda_s^c + 2 V_{ijs}^c \left(q_j^{c\prime} \lambda_s^c + q_{sj}^c \right) + \\
& \quad\quad\quad \left. + 2 V_{i\lambda s}^c \lambda_s^c + V_{ij}^c \ddot{q}_j^c + V_{i\lambda}^c \lambda_{ss}^c \right] = 0
\end{aligned}
\tag{39}
$$

Introducing the adjoint system (20), identical to that used in the first order sensitivity analysis, the term with $\ddot{q}_j^c$ can be eliminated and the second order sensitivity of λ^c is

$$
\lambda_{ss}^c = - \frac{\left(\lambda_s^c \right)^2 \left(A_1 - \mu_i A_{2i} \right) + \lambda_s^c \left(B_1 - \mu_i B_{2i} \right) + C_1 - \mu_i C_{2i}}{V_{ij\lambda}^c q_{1i} q_{1j} - \mu_i V_{i\lambda}^c}
\tag{40}
$$

where

$$
\begin{aligned}
A_1 &= V_{ijkl}^c q_{1i} q_{1j} q_{0k}^\prime q_{0l}^\prime + V_{ij\lambda\lambda}^c q_{1i} q_{1j} + 2 V_{ijk\lambda}^c q_{1i} q_{1j} q_{0k}^\prime - 2 V_{ij}^c q_{1i}^\prime q_{1j}^\prime \\
A_{2i} &= V_{ijk}^c q_{0j}^\prime q_{0l}^\prime + V_{i\lambda\lambda}^c + V_{ij\lambda}^c q_{0j}^\prime \\
B_1 &= 2 V_{ijkl}^c q_{1i} q_{1j} q_{0k}^\prime q_{sl}^c + 2 V_{ijk\lambda}^c q_{1i} q_{1j} q_{sk}^c + \\
& \quad + 2 V_{ijks}^c q_{1i} q_{1j} q_{0k}^\prime + 2 V_{ij\lambda s}^c q_{1i} q_{1j} - 4 V_{ij}^c q_{1i}^\prime q_{1sj} \\
B_{2i} &= 2 V_{ijk}^c q_{0j}^\prime q_{sk}^c + 2 V_{ij\lambda}^c q_{sj}^c + 2 V_{ijs}^c q_{0j}^\prime + 2 V_{i\lambda s}^c \\
C_1 &= V_{ijkl}^c q_{1i} q_{1j} q_{sk}^c q_{sl}^c + V_{ijss}^c q_{1i} q_{1j} + 2 V_{ijks}^c q_{1i} q_{1j} q_{sk}^c - 2 V_{ij}^c q_{1si} q_{1sj} \\
C_{2i} &= V_{ijk}^c q_{sj}^c q_{sk}^c + 2 V_{ijs}^c q_{sj}^c + V_{iss}^c
\end{aligned}
\tag{41}
$$

This method can be used in order to avoid the determination of $\ddot{q}^c$.

2.3   SINGULAR SENSITIVITY AT THE BIFURCATION POINT

2.3.1   DIRECT APPROACH

When the conditions (7) occur, then $\dot{s} = \dot{s}^m = 0$ and the parameter s cannot be the path parameter. Assume as the evolution parameter η one of path coordinates or path length, so that derivatives with respect to η exist at the critical state. Equations (2) and (9) now provide

$$V_{ij}^c q_{sj}^c \dot{s}^m = 0 \tag{42}$$

so the vector q_{sj}^c is the eigenvector of the tangent matrix V_{ij}^c. We can write

$$q_{sj}^c \dot{s}^m = \alpha^m q_{1j} = q_{1j}^m \tag{43}$$

and

$$\dot{q}_j^c = \dot{\lambda}^m q_j^{c\prime} + \alpha^m q_{1j} = \dot{\lambda}^m q_j^{c\prime} + q_{1j}^m \tag{44}$$

where q_{1j} is the normalized eigenvector and α^m is the scaling factor depending on the critical state path parameterization.

For instance, let η be the length of the critical state path. Then there is

$$\dot{q}_j^c \dot{q}_j^c = 1 \tag{45}$$

Applying to (45) the decomposition (44), we find that α^m is the root of the second order polynomial

$$q_{1j}q_{1j} \left(\alpha^m\right)^2 + 2q_j^{c\prime} q_{1j} \dot{\lambda}^m \alpha^m + \left[q_j^{c\prime} q_j^{c\prime} \left(\dot{\lambda}^m\right)^2 - 1\right] = 0 \tag{46}$$

First order sensitivity

Substituting (44) into (14) multiplied by the eigenvector q_{1i}, we have

$$\dot{\lambda}^m = -\frac{V_{ijk}^c q_{1i} q_{1j} q_{1k}^m}{V_{ijk}^c q_{1i} q_{1j} q_k^{c\prime} + V_{ij\lambda}^c q_{1i} q_{1j}} \tag{47}$$

$\dot{\lambda}^m$ and $\dot{s}^m$ denote the values of $\dot{\lambda}$ and $\dot{s}$ on the critical state path C_s. From the equilibrium equation (3) along this path we obtain

$$\dot{s}^m = -\frac{V_{ijk}^c q_{1i} \dot{q}_j^c \dot{q}_k^c + 2V_{ij\lambda}^c q_{1j} \dot{q}_j^c \dot{\lambda}^m + V_{i\lambda\lambda}^c q_{1i} (\dot{\lambda}^m)^2}{V_{is}^c q_{1i}} \tag{48}$$

Second order sensitivity

Let us extract the derivatives along the fundamental path from the second order coordinate derivative at the singular bifurcation point

$$\ddot{q}_j^c = q_j^{c\prime} \ddot{\lambda}^m + q_j^{c\prime\prime} \left(\dot{\lambda}^m\right)^2 + q_{2j}^m \tag{49}$$

The derivative $q_j^{c\prime}$ can be determined from (10) and (11), and the second order derivative $q_j^{c\prime\prime}$ from (29) and (30).

Specializing (3) to the singular sensitivity case, next introducing the decompositions (44) and (49), and taking into account (10) and (29), we obtain the system of equations permitting to calculate the vector $\boldsymbol{q}_2^m$

$$V_{ij}^c q_{2j}^m + 2\left(V_{ijk}^c q^{c\prime} q_{1j}^m + V_{ij\lambda} q_{1j}^m\right)\dot{\lambda}^m + V_{ijk}^c q_{1j}^m q_{1k}^m + V_{is}^c \dot{s}^m = 0 \tag{50}$$

The system of equations (50) is singular, so the complementary equation is required. This equation can be generated from the parametrisation of the critical state path.

If, for instance, η is the length of the critical state path, equation (45) occurs. Differentiating (45), we obtain the required condition, that has the form

$$\dot{q}_j^c \ddot{q}_j^c = 0 \tag{51}$$

Introducing to (51) the decompositions (44) and (49) we obtain

$$q_{2j}^m \left(q_j^{c\prime}\dot{\lambda}^m + q_{1j}^m\right) + q_j^{c\prime} q_j^{c\prime}\dot{\lambda}^m\ddot{\lambda}^m + q_j^{c\prime} q_j^{c\prime\prime}\left(\dot{\lambda}^m\right)^3 + \\ + q_j^{c\prime} q_{1j}^m\ddot{\lambda}^m + q_j^{c\prime\prime} q_{1j}^m\left(\dot{\lambda}^m\right)^2 = 0 \tag{52}$$

Now, specializing (4) to the singular sensitivity case, next multiplying it by the eigenvector q_{1i}, we obtain the formula permitting to calculate the third order derivative $\ddot{s}^m$

$$
\ddot{s}^m = \frac{1}{V_{is}^c q_{1j}}\Big[V_{ijkl}^c q_{1i}\dot{q}_j\dot{q}_k\dot{q}_l + 3V_{ijk}^c q_{1i}\dot{q}_j\ddot{q}_k + 3V_{ijk\lambda}^c q_{1i}\dot{q}_j\dot{q}_k\dot{\lambda}^m + \\
+ 3V_{ij\lambda}^c q_{1i}\ddot{q}_j\dot{\lambda}^m + 3V_{ij\lambda}^c q_{1i}\dot{q}_j\ddot{\lambda} + 3V_{ijs}^c q_{1i}\dot{q}_j\ddot{s}^m + V_{is}^c q_{1i}\dddot{s} + \\
+ 3V_{ij\lambda\lambda}^c \dot{q}_j\dot{\lambda}^2 + 3V_{i\lambda\lambda}^c q_{1i}\dot{\lambda}^m\ddot{\lambda}^m + 3V_{i\lambda s}^c q_{1i}\dot{\lambda}^m\ddot{s}^m + V_{i\lambda\lambda\lambda}^c q_{1i}\dot{\lambda}^3 \Big] \tag{53}
$$

2.3.2 ADJOINT APPROACH

First order sensitivity

Applying (44) to (17) and introducing the adjoint field (20), it is seen that only symmetric case can be treated, so that (21) is satisfied. Then, we have

$$\dot{\lambda}^m = 0 \tag{54}$$

$$\ddot{s}^m = 0 \tag{55}$$

and

$$\ddot{q}_j^c = q_{1j}^m \qquad (56)$$

To determine $\ddot{\lambda}^m$, the second order sensitivity should be determined.

Second order sensitivity

Let us write the critical state condition for the singular sensitivity case satisfying (7), (54), (55) and (56). We have

$$V_{ijkl}^c q_{1i} q_{1j} q_{1k}^m q_{1l}^m - 2V_{ij}^c \dot{q}_{1i} \dot{q}_{1j} + V_{ijk}^c q_{1i} q_{1j} \ddot{q}_k^c + V_{ij\lambda}^c q_{1i} q_{1j} \ddot{\lambda}^m +$$
$$- \mu_i \left[V_{ijk}^c q_{1j}^m q_{1k}^m + V_{ij}^c \ddot{q}_j^c + V_{i\lambda}^c \ddot{\lambda}^m \right] = 0 \qquad (57)$$

This equation can also be presented in the form

$$\ddot{\lambda}^m \left(V_{ij\lambda}^c q_{1i} q_{1j} - \mu_i V_{i\lambda}^c \right) + V_{ijkl}^c q_{1i} q_{1j} q_{1k}^m q_{1l}^m - 2^c_{ij} \dot{q}_{1i} \dot{q}_{1j} +$$
$$- \mu_i V_{ijk}^c q_{1j}^m q_{1k}^m + \ddot{q}_i^c \left(V_{ijk}^c q_{1j} q_{1k} - \mu_i V_{ij}^c \right) = 0 \qquad (58)$$

In order to avoid determination of $\ddot{q}_i^c$, let us introduce the adjoint system specified by (20), so the following expression for $\ddot{\lambda}^m$ is obtained

$$\ddot{\lambda}^m = - \frac{D_1 - \mu_i D_{2i}}{V_{ij\lambda}^c q_{1i} q_{1j} - \mu_i V_{i\lambda}^c} \qquad (59)$$

where

$$\begin{aligned} D_1 &= V_{ijkl}^c q_{1i} q_{1j} q_{1k}^m q_{1l}^m - 2V_{ij}^c \dot{q}_{1i} \dot{q}_{1j} \\ D_{2i} &= V_{ijk}^c q_{1j}^m q_{1k}^m \end{aligned} \qquad (60)$$

The derivative $\dot{q}_{1j}$ of the eigenvector can be specified from the equation

$$V_{ij}^c \dot{q}_{1j} + V_{ijk}^c q_{1j} q_{1k}^m = 0 \qquad (61)$$

with the differentiated normalization condition taken as the complementary condition

$$T_{ij} q_{1i} \dot{q}_{1j} = 0 \qquad (62)$$

Multiplying (4) by q_{1i} and taking into account (7) and the symmetric bifurcation conditions, we find

$$\ddot{s}^m = - \frac{1}{V_{is}^c q_{1i}} \left(V_{ijkl}^c q_{1i} q_{1j}^m q_{1k}^m q_{1l}^m + 3V_{ijk}^c q_{1i} q_{1j}^m \ddot{q}_k^c + 3V_{ij\lambda}^c q_{1i} q_{1j}^m \ddot{\lambda}^m \right) \qquad (63)$$

Let us decompose the derivative $\ddot{q}_j^c$ into two parts, orthogonal and collinear to the eigenvector q_{1j}, so that

$$\ddot{q}_j^c = \beta q_{1j} + q_{2j} \qquad (64)$$

where

$$q_{1j}q_{2j} = 0 \qquad (65)$$

In view of (21), equation (63) can be rewritten as

$$\ddot{s}^m = -\frac{1}{V_{is}^c q_{1i}}\left(V_{ijkl}^c q_{1i}q_{1j}^m q_{1k}^m q_{1l}^m + 3V_{ijk}^c q_{1i}q_{1j}^m q_{2k} + 3V_{ij\lambda}^c q_{1i}q_{1j}^m \ddot{\lambda}\right) \qquad (66)$$

To calculate q_{2j} we can specialize (3) to the symmetric bifurcation case, applying the decomposition (64), so that

$$V_{ij}^c q_{2j} + V_{ijk}^c q_{1j}^m q_{1k}^m + V_{i\lambda}^c \ddot{\lambda}^m = 0 \qquad (67)$$

and we take (65) as the complementary condition.

2.3.3   SINGULAR SENSITIVITY REPRESENTATION

First order sensitivity representation

i) *Non–symmetric bifurcation case*

At a non-symmetric bifurcation point there is $V_{ijk}^c q_{1i}q_{1j}q_{1k} \neq 0$, and $\dot{\lambda}^m \neq 0$, $\ddot{s}^m \neq 0$. Then the power expansion of the critical load increment $\Delta\lambda = \lambda^m - \lambda^c$ and of the structural parameter increment $\Delta s = s^m - s^c$ are written as

$$\Delta\lambda = \lambda^m - \lambda^c = \dot{\lambda}^m \Delta\eta + \frac{1}{2}\ddot{\lambda}^m \Delta\eta^2 + \frac{1}{6}\dddot{\lambda}^m \Delta\eta^3 + \ldots \qquad (68)$$

$$\Delta s = s^m - s^c = \frac{1}{2}\ddot{s}^m \Delta\eta^2 + \frac{1}{6}\dddot{s}^m \Delta\eta^3 + \ldots \qquad (69)$$

Considering only first terms in the expansions (68) and (69) we obtain the well-known square root sensitivity typical for geometric imperfection, cf. Thompson and Hunt [1973]

$$\Delta\lambda = \pm\dot{\lambda}^m \sqrt{\frac{2\Delta s}{\ddot{s}^m}} \qquad (70)$$

ii) *Symmetric bifurcation case*

In the symmetric bifurcation case there is $V_{ijk}^c q_{1i}q_{1j}q_{1k} = 0$, and also $\dot{\lambda}^m = 0$, $\ddot{s}^m = 0$. Now the first order term in (68) and the second order term in (69) vanishes. Taking into account the subsequent terms (second order term in (68) and third order term in (69)) we obtain the sensitivity formula for symmetric case

$$\Delta\lambda = \frac{\ddot{\lambda}^m}{2}\left(\frac{6\Delta s}{\ddot{s}^m}\right)^{\frac{2}{3}} \tag{71}$$

Second order sensitivity representation

i) *Non–symmetric bifurcation case*

Let us consider again the power expansions (68) and (69). Taking into account only the terms up to the second order and eliminating $\Delta\eta$, we find the first estimation of the second order critical load sensitivity

$$\Delta\lambda = \pm\dot{\lambda}^m\sqrt{\frac{2\Delta s}{\ddot{s}^m}} + \frac{\ddot{\lambda}^m}{\ddot{s}^m}\Delta s \tag{72}$$

A more precise estimation of the second order critical load sensitivity in the singular case can be obtained by introducing a new structural parameter t that is regular with respect to the path parameter η.

The power expansion of $\Delta\lambda = \lambda^m - \lambda^c$ along the critical path with respect to t is written as

$$\Delta\lambda = \lambda_t^m\Delta t + \frac{1}{2}\lambda_{tt}^m\Delta t^2 + \frac{1}{6}\lambda_{ttt}^m\Delta\eta^3 + \dots \tag{73}$$

where $(\ )_t$ denotes derivatives with respect to t.

Applying the chain rule in derivatives of (73), we can write

$$\lambda_t^m = \dot{\lambda}^m\frac{d\eta}{dt} \tag{74}$$

$$\lambda_{tt}^m = \ddot{\lambda}^m\left(\frac{d\eta}{dt}\right)^2 + \dot{\lambda}^m\frac{d^2\eta}{dt^2} \tag{75}$$

The power expansions of the structural parameter increment Δt with respect to path parameter η is written as follows

$$\Delta t = \dot{t}^m\Delta\eta + \frac{1}{2}\ddot{t}^m\Delta\eta^2 + \frac{1}{6}\dddot{t}^m\Delta\eta^3 + \dots \tag{76}$$

We can assume t such, that

$$(\Delta t)^2 = (t^m - t^c)^2 = \Delta s \tag{77}$$

From (76), it follows that

$$(\Delta t)^2 = \left(\dot{t}^m\right)^2 \Delta\eta^2 + \dot{t}^m \ddot{t}^m \Delta\eta^3 + \left(\frac{1}{4}\ddot{t}^m + \frac{1}{3}\dot{t}^m \dddot{t}^m\right)\Delta\eta^4 + \ldots \tag{78}$$

In view of (77), after comparing the terms in (69) and in (78), we find

$$\left(\dot{t}^m\right)^2 = \frac{\ddot{s}^m}{2} \tag{79}$$

$$\ddot{t}^m = \frac{1}{6}\frac{\dddot{s}^m}{\dot{t}^m} \tag{80}$$

We can demonstrate, that

$$\frac{d\eta}{dt} = \frac{1}{\dot{t}^m} \quad, \qquad \frac{d^2\eta}{dt^2} = -\frac{\ddot{t}^m}{\left(\dot{t}^m\right)^3} \tag{81}$$

so, there is

$$\frac{d\eta}{dt} = \pm\sqrt{\frac{2}{\ddot{s}^m}} \quad, \qquad \frac{d^2\eta}{dt^2} = -\frac{2}{3}\frac{\dddot{s}^m}{\left(\ddot{s}^m\right)^3} \tag{82}$$

Introducing (82) to (74) and (75) we obtain the first and second order t–sensitivity

$$\lambda_t^m = \pm\dot{\lambda}^m\sqrt{\frac{2}{\ddot{s}^m}} \tag{83}$$

$$\lambda_{tt}^m = \frac{2\ddot{\lambda}^m}{\ddot{s}^m} - \frac{2}{3}\frac{\dot{\lambda}^m \dddot{s}^m}{\left(\ddot{s}^m\right)^2} \tag{84}$$

In view of (83) and (84), neglecting terms of orders higher than 2, from (68) the sensitivity expression for non–symmetric bifurcation is obtained

$$\Delta\lambda = \pm\dot{\lambda}^m\sqrt{\frac{2\Delta s}{\ddot{s}^m}} + \left[\frac{\ddot{\lambda}^m}{\ddot{s}^m} - \frac{1}{3}\frac{\dot{\lambda}^m \dddot{s}^m}{\left(\ddot{s}^m\right)^2}\right]\Delta s \tag{85}$$

Formula (85) coincides with (70) when the terms with Δs are neglected, and with (72) when $\dddot{s}^m = 0$ is assumed.

ii) *Symmetric bifurcation case*

Let us recall, that for a symmetric bifurcation case there is $\dot{\lambda}^m = 0$ and $\ddot{s}^m = 0$. Now, instead of (77), let us assume

$$(\Delta t)^3 = (t^m - t_c)^2 = \Delta s \tag{86}$$

so there is

$$(\Delta t)^3 = \left(\dot{t}^m\right)^3 \Delta\eta^3 + \frac{3}{2}\left(\dot{t}^m\right)^2 \ddot{t}^m \Delta\eta^4 + \frac{1}{2}\left(\dot{t}^m\right)^2 \dddot{t}^m \Delta\eta^5 + \dots \tag{87}$$

In view of (86), comparing (87) with (69), we obtain

$$\left(\dot{t}^m\right)^3 = \frac{\dddot{s}^m}{6} \tag{88}$$

$$\ddot{t}^m = \frac{1}{36} \frac{s^{(4)m}}{\left(\dot{t}^m\right)^2} \tag{89}$$

Using (74), (75), and (81), with $\lambda^m = 0$, we have

$$\lambda_t^m = 0 \ , \quad \lambda_{tt}^m = -\frac{\ddot{\lambda}^m}{\left(\dot{t}^m\right)^2} = -\ddot{\lambda}^m \left(\frac{6}{\dddot{s}^m}\right)^{\frac{2}{3}} \tag{90}$$

Applying the chain rule to (75) we obtain

$$\lambda_{ttt}^m = \dddot{\lambda}^m \left(\frac{d\eta}{dt}\right)^3 + 3\frac{d\eta}{dt}\frac{d^2\eta}{dt^2}\ddot{\lambda}^m + \frac{d^3\eta}{dt^3}\dot{\lambda}^m \tag{91}$$

In view of (81) and (90), (91) provides

$$\lambda_{ttt}^m = \frac{6\,\dddot{\lambda}^m}{\dddot{s}^m} - 3\ddot{\lambda}^m \frac{s^{(4)m}}{\left(\dddot{s}^m\right)^2} \tag{92}$$

Now, (73) can can be presented in the form

$$\Delta\lambda = \frac{\ddot{\lambda}^m}{2}\left(\frac{6\Delta s}{\dddot{s}^m}\right)^{\frac{2}{3}} + \left[\frac{\dddot{\lambda}^m}{\dddot{s}^m} - \frac{\ddot{\lambda}^m}{2}\frac{s^{(4)m}}{\left(\dddot{s}^m\right)^2}\right]\Delta s + \dots \tag{93}$$

To calculate the second term of (93), the derivatives $\dddot{\lambda}^m$ and $s^{(4)m}$ should be specified. The first singular term is identical to (71).

3   SENSITIVITY OF EIGENFREQUENCIES

Let us now consider small harmonic oscillations

$$q_i = q_i^0 + u_i \cos\omega t \tag{94}$$

superposed on the initial equilibrium state specified by q_i^0, λ^0, and the equilibrium condition

$$V_i \left(q_i^0, \lambda^0, s \right) = 0 \qquad (95)$$

The equations of motion

$$V_{ij}u_j - \omega^2 M_{ij}u_j = 0 \qquad (96)$$

specify the eigenvector $u_i = u_i(\lambda, s)$ and the fundamental eigenfrequency $\omega = \omega(\lambda, s)$. Here $M_{ij} = M_{ij}(s)$ denotes the mass matrix of a discrete system.

Multiplying (96) by u_i one obtains the Rayleigh quotient

$$\omega^2 = \frac{V_{ij}u_iu_j}{M_{ij}u_iu_j} \qquad (97)$$

3.1 DIRECT APPROACH

3.1.1 FIRST ORDER EIGENFREQUENCY SENSITIVITY

Differentiating (96) with respect to the evolution parameter η, we obtain

$$V_{ij}\dot{u}_j - \omega^2 M_{ij}\dot{u}_j + V_{ijk}u_j\dot{q}_k^0 + V_{ij\lambda}u_j\dot{\lambda} + V_{ijs}u_j\dot{s} +$$
$$-\omega_\lambda^2 M_{ij}u_j\dot{\lambda} - \omega_s^2 M_{ij}u_j\dot{s} - \omega^2 M_{ijs}u_j\dot{s} = 0 \qquad (98)$$

As the coordinate vector is a function of the load parameter λ and the structural parameter s, we can write

$$\dot{q}_j^0 = q_j^{0\prime}\dot{\lambda} + q_{sj}^0\dot{s} \qquad (99)$$

As it was shown by Mróz [1991], multiplying now the equation (98) by the eigenvector u_i and specializing this equation along the fundamental loading path C_l ($\dot{s} = 0$, $\dot{\lambda} \neq 0$), we obtain the eigenfrequency derivative with respect to the load parameter

$$\omega_\lambda^2 = \frac{V_{ijk}u_iu_jq_k^{0\prime} + V_{ij\lambda}u_iu_j}{M_{ij}u_iu_j} \qquad (100)$$

Applying now (98) for the transformation path C_t under constant load ($\dot{\lambda} = 0$, $\dot{s} \neq 0$) we obtain the eigenfrequency sensitivity expression with respect to the structure parameter s

$$\omega_s^2 = \frac{V_{ijk}u_iu_jq_{sk}^0 + V_{ijs}u_iu_j - \omega^2 M_{ijs}u_iu_j}{M_{ij}u_iu_j} \qquad (101)$$

Let us consider the constant frequency path C_ω, ($\omega^2 = const$). Along this path, we have the relation

$$\omega_\lambda^2\dot{\lambda} + \omega_s^2\dot{s} = 0 \qquad (102)$$

Setting s as the path parameter and using the coordinate vector derivative decomposition (99), we obtain the formula specifying the variation of load parameter along the path C_w, namely

$$\lambda_s = -\frac{\omega_s^2}{\omega_\lambda^2} = -\frac{V_{ijk}u_i u_j q_{sk}^0 + V_{ijs}u_i u_j - \omega^2 M_{ijs}u_i u_j}{V_{ijk}u_i u_j q_k^{0\prime} + V_{ij\lambda}u_i u_j} \tag{103}$$

The sensitivity derivative of the state solution $q_j^{0\prime}$ and q_{sj}^0 are obtained from the equation (3) evaluated along the C_l and C_t respectively, thus

$$\begin{aligned}
V_{ij}q_j^{0\prime} + V_{i\lambda} &= 0 \\
V_{ij}q_{sj}^0 + V_{is} &= 0
\end{aligned} \tag{104}$$

At the critical state we need to provide the complementary conditions, for instance, (11) and (13).

The critical load sensitivity is obtained from (103) by setting there $\omega^2 = 0$. The expression (103) then coincides with (23).

3.1.2 SECOND ORDER EIGENFREQUENCY SENSITIVITY

Differentiating (96) second time, one obtains

$$\begin{aligned}
&V_{ij}\ddot{u}_j - \omega^2 M_{ij}\ddot{u}_j + 2\Big[V_{ijk}\dot{u}_j\dot{q}_k^0 + V_{ij\lambda}\dot{u}_j\dot{\lambda} + V_{ijs}\dot{u}_j\dot{s} - \omega_\lambda^2 M_{ij}\dot{u}_j\dot{\lambda}+ \\
&-\omega_s^2 M_{ij}\dot{u}_j\dot{s} - \omega^2 M_{ijs}\dot{u}_j\dot{s} \Big] + V_{ijk}u_j\ddot{q}_k^0 + V_{ij\lambda}u_j\ddot{\lambda} + V_{ijs}u_j\ddot{s}+ \\
&-\omega_\lambda^2 M_{ij}u_j\ddot{\lambda} - \omega_s^2 M_{ij}u_j\ddot{s} - \omega^2 M_{ijs}u_j\ddot{s} + V_{ijkl}u_j\dot{q}_k^0\dot{q}_l^0 + V_{ij\lambda\lambda}u_j\dot{\lambda}^2+ \\
&+V_{ijss}u_j\dot{s}^2 + 2V_{ijk\lambda}u_j\dot{q}_k^0\dot{\lambda} + 2V_{ijks}u_j\dot{q}_k^0\dot{s} + 2V_{ij\lambda s}u_j\dot{\lambda}\dot{s}+ \\
&-\omega_{\lambda\lambda}^2 M_{ij}u_j\dot{\lambda}^2 - \omega_{ss}^2 M_{ij}u_j\dot{s}^2 - 2\omega_{\lambda s}^2 M_{ij}u_j\dot{\lambda}\dot{s}+ \\
&-2\omega_\lambda^2 M_{ijs}u_j\dot{\lambda}\dot{s} - 2\omega_s^2 M_{ijs}u_j\dot{s}^2 - \omega^2 M_{ijss}u_j\dot{s}^2 = 0
\end{aligned} \tag{105}$$

Multiplying this equation by the eigenvector u_i and subtracting (98) multiplied by the doubled eigenvector derivative $2\dot{u}_i$, we obtain

$$\begin{aligned}
&-2V_{ij}\dot{u}_i\dot{u}_j + 2\omega^2 M_{ij}\dot{u}_i\dot{u}_j + V_{ijk}u_i u_j\ddot{q}_k^0 + V_{ij\lambda}u_i u_j\ddot{\lambda}+ \\
&+V_{ijs}u_i u_j\ddot{s} - \omega_\lambda^2 M_{ij}u_i u_j\ddot{\lambda} - \omega_s^2 M_{ij}u_i u_j\ddot{s} - \omega^2 M_{ijs}u_i u_j\ddot{s}+ \\
&+V_{ijkl}u_i u_j\dot{q}_k^0\dot{q}_l^0 + V_{ij\lambda\lambda}u_i u_j\dot{\lambda}^2 + V_{ijss}u_i u_j\dot{s}^2+ \\
&+2V_{ijk\lambda}u_i u_j\dot{q}_k^0\dot{\lambda} + 2V_{ijks}u_i u_j\dot{q}_k^0\dot{s} + 2V_{ij\lambda s}u_i u_j\dot{\lambda}\dot{s}+ \\
&-\omega_{\lambda\lambda}^2 M_{ij}u_i u_j\dot{\lambda}^2 - \omega_{ss}^2 M_{ij}u_i u_j\dot{s}^2 - 2\omega_{\lambda s}^2 M_{ij}u_i u_j\dot{\lambda}\dot{s}+ \\
&-2\omega_\lambda^2 M_{ijs}u_i u_j\dot{\lambda}\dot{s} - 2\omega_s^2 M_{ijs}u_i u_j\dot{s}^2 - \omega^2 M_{ijss}u_i u_j\dot{s}^2 = 0
\end{aligned} \tag{106}$$

Differentiating (99), we can write

$$\ddot{q}_j^0 = q_j^{0\prime}\ddot{\lambda} + q_{sj}^0\ddot{s} + q_j^{0\prime\prime}\dot{\lambda}^2 + q_{ssj}^0\dot{s}^2 + 2q_{sj}^{0\prime}\dot{\lambda}\dot{s} \tag{107}$$

The eigenvector derivative can be decomposed as follows

$$\dot{u}_i = u_i'\dot{\lambda} + u_{si}\dot{s} \tag{108}$$

Introducing (99), (107) and (108) into (105), we obtain

$$
\begin{aligned}
-2V_{ij}\left(u_i'\dot{\lambda} + u_{si}\dot{s}\right)\left(u_j'\dot{\lambda} + u_{sj}\dot{s}\right) + \\
+2\omega^2 M_{ij}\left(u_i'\dot{\lambda} + u_{si}\dot{s}\right)\left(u_j'\dot{\lambda} + u_{sj}\dot{s}\right) + \\
+V_{ijk}u_iu_j\left(q_k^{0\prime}\ddot{\lambda} + q_{sk}^0\ddot{s} + q_k^{0\prime\prime}\dot{\lambda}^2 + q_{ssk}^0\dot{s}^2 + 2q_{sk}^{0\prime}\dot{\lambda}\dot{s}\right) + \\
+V_{ij\lambda}u_iu_j\ddot{\lambda} + V_{ijs}u_iu_j\ddot{s} + V_{ijkl}u_iu_j\left(q_k^{0\prime}\dot{\lambda} + q_{sk}^0\dot{s}\right)\left(q_l^{0\prime}\dot{\lambda} + q_{sl}^0\dot{s}\right) + \\
-\omega_\lambda^2 M_{ij}u_iu_j\dot{\lambda} - \omega_s^2 M_{ij}u_iu_j\ddot{s} - \omega^2 M_{ijs}u_iu_j\ddot{s} + \\
+V_{ij\lambda\lambda}u_iu_j\dot{\lambda}^2 + V_{ijss}u_iu_j\dot{s}^2 + 2V_{ijk\lambda}u_iu_j\left(q_k^{0\prime}\dot{\lambda} + q_{sk}^0\dot{s}\right)\dot{\lambda} + \\
+2V_{ijks}u_iu_j\left(q_k^{0\prime}\dot{\lambda} + q_{sk}^0\dot{s}\right)\dot{s} + 2V_{ij\lambda s}u_iu_j\dot{\lambda}\dot{s} + \\
-\omega_{\lambda\lambda}^2 M_{ij}u_iu_j\dot{\lambda}^2 - \omega_{ss}^2 M_{ij}u_iu_j\dot{s}^2 - 2\omega_{\lambda s}^2 M_{ij}u_iu_j\dot{\lambda}\dot{s} + \\
-2\omega_\lambda^2 M_{ijs}u_iu_j\dot{\lambda}\dot{s} - 2\omega_s^2 M_{ijs}u_iu_j\dot{s}^2 - \omega^2 M_{ijss}u_iu_j\dot{s}^2 = 0
\end{aligned} \tag{109}
$$

Let us specialize the equation (109) along the fundamental loading path C_l, setting λ as the evolution parameter and $s = const.$ Then (109) is reduced to

$$
\begin{aligned}
-2V_{ij}u_i'u_j' + 2\omega^2 M_{ij}u_i'u_j' + V_{ijk}u_iu_jq_k^{0\prime\prime} + V_{ijkl}u_iu_jq_k^{0\prime}q_l^{0\prime} + \\
+V_{ij\lambda\lambda}u_iu_j + 2V_{ijk\lambda}u_iu_jq_k^{0\prime} - \omega_{\lambda\lambda}^2 M_{ij}u_iu_j = 0
\end{aligned} \tag{110}
$$

so the second order derivative with respect to the load parameter is written as

$$
\begin{aligned}
\omega_{\lambda\lambda}^2 = \frac{1}{M_{ij}u_iu_j}\Big[-2V_{ij}u_i'u_j' + 2\omega^2 M_{ij}u_i'u_j' + V_{ijk}u_iu_jq_k^{0\prime\prime} + \\
+V_{ijkl}u_iu_jq_k^{0\prime}q_l^{0\prime} + V_{ij\lambda\lambda}u_iu_j + 2V_{ijk\lambda}u_iu_jq_k^{0\prime}\Big]
\end{aligned} \tag{111}
$$

Let us now evaluate (109) along the transformation path C_t, with s taken as the path parameter and $\lambda = const.$ Then we obtain

$$
\begin{aligned}
-2V_{ij}u_{si}u_{sj} + 2\omega^2 M_{ij}u_{si}u_{sj} + V_{ijk}u_iu_jq_{ssk}^0 + \\
+V_{ijkl}u_iu_jq_{sk}^0q_{sl}^0 + V_{ijss}u_iu_j + 2V_{ijks}u_iu_jq_{sk}^0 \\
-\omega_{ss}^2 M_{ij}u_iu_j - 2\omega_s^2 M_{ijs}u_iu_j - \omega^2 M_{ijss}u_iu_j = 0
\end{aligned} \tag{112}
$$

and from (112) it follows, that

$$\omega_{ss}^2 = \frac{1}{M_{ij}u_iu_j}\Big[-2V_{ij}u_{si}u_{sj} + 2\omega^2 M_{ij}u_{si}u_{sj}+$$
$$V_{ijk}u_iu_jq_{ssk}^0 + + V_{ijkl}u_iu_jq_{sk}^0q_{sl}^0 + V_{ijss}u_iu_j+ \qquad (113)$$
$$+2V_{ijks}u_iu_jq_{sk}^0 - 2\omega_s^2 M_{ijs}u_iu_j - \omega^2 M_{ijss}u_iu_j\Big]$$

By virtue of (98), (99), (110) and (112), we find, that in (109) the terms with $\ddot{\lambda}$, $\ddot{s}$, $\dot{\lambda}^2$ and $\dot{s}^2$ vanish, so we have

$$-2V_{ij}u_i'u_{sj} + 2\omega^2 M_{ij}u_i'u_{sj} + V_{ijk}u_iu_jq_{sk}^{0\prime}+$$
$$+V_{ijkl}u_iu_jq_k^{0\prime}q_{sl}^0 + V_{ijk\lambda}u_iu_jq_{sk}^0 + V_{ijks}u_iu_jq_k^{0\prime}+ \qquad (114)$$
$$+V_{ij\lambda s}u_iu_j - \omega_{\lambda s}^2 M_{ij}u_iu_j - \omega_\lambda^2 M_{ijs}u_iu_j = 0$$

and now the mixed second order eigenfrequency derivative is written as follows

$$\omega_{\lambda s}^2 = \frac{1}{M_{ij}u_iu_j}\Big[-2V_{ij}u_i'u_{sj} + 2\omega^2 M_{ij}u_i'u_{sj}+$$
$$+V_{ijk}u_iu_jq_{sk}^{0\prime} + V_{ijkl}u_iu_jq_k^{0\prime}q_{sl}^0 + V_{ijk\lambda}u_iu_jq_{sk}^0+ \qquad (115)$$
$$+V_{ijks}u_iu_jq_k^{0\prime} + V_{ij\lambda s}u_iu_j - \omega_\lambda^2 M_{ijs}u_iu_j\Big]$$

Let us return to equation (102) occurring along the path C_ω. Differentiating the constant eigenfrequency constraint (102), we obtain

$$\omega_\lambda^2\ddot{\lambda} + \omega_s^2\ddot{s} + \omega_{\lambda\lambda}^2\dot{\lambda}^2 + \omega_{ss}^2\dot{s}^2 + 2\omega_{\lambda s}^2\dot{\lambda}\dot{s} = 0 \qquad (116)$$

Setting s as the path parameter, we can obtain the second order load sensitivity along the constant eigenfrequency path C_ω

$$\lambda_{ss} = \frac{\omega_{\lambda\lambda}^2\lambda_s^2 + 2\omega_{\lambda s}^2\lambda_s + \omega_{ss}^2}{\omega_\lambda^2} \qquad (117)$$

Let us note, that for $\omega^2 = 0$ equation (117) provides the second order critical load sensitivity derivative with respect to the structure parameter s and coincides with the formula (27) provided the mass matrix does not depend on the structure parameter s.

3.2 ADJOINT APPROACH

3.2.1 FIRST ORDER EIGENFREQUENCY SENSITIVITY

In order to eliminate the evaluation of the sensitivity derivative of state solution $\dot{q}_j^0$ let us regard the equilibrium equation (95) as constraint condition imposed on (96).

Using the Lagrange multiplier vector μ_i, we can write

$$V_{ij}u_iu_j - \omega^2 M_{ij}u_iu_j - \mu_i V_i = 0 \tag{118}$$

Differentiating this equation with respect to the path parameter one obtains

$$(V_{ij}u_i - \omega^2 M_{ij}u_i)\,\dot{u}_j + (V_{ijk}u_iu_j - \mu_i V_{ik})\,\dot{q}_k^0 +$$
$$+ (V_{ij\lambda}u_iu_j - \omega_\lambda^2 M_{ij}u_iu_j - \mu_i V_{i\lambda})\,\dot{\lambda} + \tag{119}$$
$$+ (V_{ijs}u_iu_j\dot{s} - \omega_s^2 M_{ij}u_iu_j - \omega^2 M_{ijs}u_j\dot{s} - \mu_i V_{is})\,\dot{s} = 0$$

In view of (96) the first term of (119) vanishes.

Let us define the adjoint field μ_j such, that the term with $\dot{q}_j^0$ vanishes, so that

$$V_{ijk}u_ju_k - V_{ij}\mu_j = 0 \tag{120}$$

Now, evaluating Eq.(119) along the fundamental loading path C_l (so $\dot{s} = 0$) we obtain the derivative ω_λ^2

$$\omega_\lambda^2 = \frac{V_{ij\lambda}u_iu_j - \mu_i V_{i\lambda}}{M_{ij}u_iu_j} \tag{121}$$

Specializing (119) to the transformation path C_t with load factor constant, we obtain

$$\omega_s^2 = \frac{V_{ijs}u_iu_j - \omega^2 M_{ijs}u_iu_j - \mu_i V_{is}}{M_{ij}u_iu_j} \tag{122}$$

Setting s as the parameter of constant eigenfrequency path C_ω, we obtain the load factor variation along C_ω

$$\lambda_s = -\frac{V_{ijs}u_iu_j - \omega^2 M_{ijs}u_iu_j - \mu_i V_{is}}{V_{ij\lambda}u_iu_j - \mu_i V_{i\lambda}} \tag{123}$$

3.2.2 SECOND ORDER EIGENFREQUENCY SENSITIVITY

Differentiating (119), one obtains

$$2V_{ij}u_i\ddot{u}_j - 2\omega^2 M_{ij}u_i\ddot{u}_j + 2V_{ij}\dot{u}_i\dot{u}_j - 2\omega^2 M_{ij}\dot{u}_i\dot{u}_j +$$
$$+4\Big[V_{ijk}u_i\dot{u}_j\dot{q}_k^0 + V_{ij\lambda}u_i\dot{u}_j\dot{\lambda} + V_{ijs}u_i\dot{u}_j\dot{s} - \omega_\lambda^2 M_{ij}u_i\dot{u}_j\dot{\lambda} +$$
$$-\omega_s^2 M_{ij}u_i\dot{u}_j\dot{s} - \omega^2 M_{ijs}u_i\dot{u}_j\dot{s}\Big] + V_{ijk}u_iu_j\ddot{q}_k^0 +$$
$$+V_{ij\lambda}u_iu_j\ddot{\lambda} + V_{ijs}u_iu_j\ddot{s} - \omega_\lambda^2 M_{ij}u_iu_j\ddot{\lambda} - \omega_s^2 M_{ij}u_iu_j\ddot{s} +$$

$$-\omega^2 M_{ijs} u_i u_j \ddot{s} + V_{ijkl} u_i u_j \dot{q}_k^0 \dot{q}_l^0 + V_{ij\lambda\lambda} u_i u_j \dot{\lambda}^2 + V_{ijss} u_i u_j \dot{s}^2 +$$
$$+2V_{ijk\lambda} u_i u_j \dot{q}_k^0 \dot{\lambda} + V_{ij\lambda s} u_i u_j + V_{ijk\lambda} u_i u_j q_{sk}^0 + 2V_{ijks} u_i u_j \dot{q}_k^0 \dot{s} +$$
$$+2V_{ij\lambda s} u_i u_j \dot{\lambda} \dot{s} - \omega_{\lambda\lambda}^2 M_{ij} u_i u_j \dot{\lambda}^2 - 2\omega_{\lambda s}^2 M_{ij} u_i u_j \dot{\lambda} \dot{s} +$$
$$-2\omega_\lambda^2 M_{ijs} u_i u_j \dot{\lambda} \dot{s} - \omega_{ss}^2 M_{ij} u_i u_j \dot{s}^2 - \omega^2 M_{ijss} u_i u_j \dot{s}^2 +$$
$$-2\omega_s^2 M_{ijs} u_i u_j \dot{s}^2 - \mu_i \left(V_{ij} \ddot{q}_j^0 + V_{i\lambda} \ddot{\lambda} + V_{is} \ddot{s} + V_{ijk} \dot{q}_j^0 \dot{q}_k^0 + \right.$$
$$\left. +V_{i\lambda\lambda} \dot{\lambda}^2 + V_{iss} \dot{s}^2 2 V_{ij\lambda} \dot{q}_j^0 \dot{\lambda} + 2 V_{ijs} \dot{q}_j^0 \dot{s} + 2 V_{i\lambda s} \dot{\lambda} \dot{s} \right) = 0 \quad (124)$$

In view of (96), the term with $\ddot{u}_j$ vanishes. Next, we subtract from (124) equation (100) multiplied by $4\dot{u}_i$, use the adjoint field defined by Eq.(120), and equations (121), (122), so the terms containing $\ddot{q}_j^0$, $\ddot{\lambda}$, and $\ddot{s}$ are eliminated. Then (124) takes the form

$$-2V_{ij} \dot{u}_i \dot{u}_j + 2\omega^2 M_{ij} \dot{u}_i \ddot{u}_j + +V_{ijkl} u_i u_j \dot{q}_k^0 \dot{q}_l^0 + V_{ij\lambda\lambda} u_i u_j \dot{\lambda}^2 +$$
$$+V_{ijss} u_i u_j \dot{s}^2 + 2V_{ijk\lambda} u_i u_j \dot{q}_k^0 \dot{\lambda} + 2V_{ijks} u_i u_j \dot{q}_k^0 \dot{s} + 2V_{ij\lambda s} u_i u_j \dot{\lambda} \dot{s} +$$
$$-\omega_{\lambda\lambda}^2 M_{ij} u_i u_j \dot{\lambda}^2 - 2\omega_{\lambda s}^2 M_{ij} u_i u_j \dot{\lambda} \dot{s} - 2\omega_\lambda^2 M_{ijs} u_i u_j \dot{\lambda} \dot{s} +$$
$$-\omega_{ss}^2 M_{ij} u_i u_j \dot{s}^2 - \omega^2 M_{ijss} u_i u_j \dot{s}^2 - 2\omega_s^2 M_{ijs} u_i u_j \dot{s}^2 +$$
$$-\mu_i \left(V_{ijk} \dot{q}_j^0 \dot{q}_k^0 + V_{i\lambda\lambda} \dot{\lambda}^2 + V_{iss} \dot{s}^2 + 2V_{ij\lambda} \dot{q}_j^0 \dot{\lambda} + \right.$$
$$\left. +2V_{ijs} \dot{q}_j^0 \dot{s} + 2V_{i\lambda s} \dot{\lambda} \dot{s} \right) = 0 \quad (125)$$

Specializing (125) to the fundamental loading path C_l, we obtain the second order eigenfrequency derivative with respect to the load factor

$$\omega_{\lambda\lambda}^2 = \frac{1}{M_{ij} u_i u_j} \left[-2V_{ij} u_i' u_j' + 2\omega^2 M_{ij} u_i' u_j' + V_{ijkl} u_i u_j q_k^{0\prime} q_l^{0\prime} + \right.$$
$$\left. +V_{ij\lambda\lambda} u_i u_j + 2V_{ijk\lambda} u_i u_j q_k^{0\prime} - \mu_i \left(V_{i\lambda\lambda} + V_{ijk} q_j^0 q_k^{0\prime} + 2V_{ij\lambda} q_j^{0\prime} \right) \right] \quad (126)$$

Evaluating now (126) along the transformation path C_t with s taken as the path parameter and the load factor constant, we obtain

$$\omega_{ss}^2 = \frac{1}{M_{ij} u_i u_j} \left[-2V_{ij} u_{si} u_{sj} + 2\omega^2 M_{ij} u_{si} u_{sj} + V_{ijkl} u_i u_j q_{sk}^0 q_{sl}^0 + \right.$$
$$+V_{ijss} u_i u_j + 2V_{ijks} u_i u_j q_{sk}^0 - \omega^2 M_{ijss} u_i u_j - 2\omega_s^2 M_{ijs} u_i u_j +$$
$$\left. -\mu_i \left(V_{iss} + V_{ijk} q_{sj}^0 q_{sk}^0 + 2V_{ijs} q_{sj}^0 \right) \right] \quad (127)$$

In view of (126) and (127), evaluating (125) along the constant eigenfrequency path C_ω and setting s as the path parameter, the mixed derivative $\omega_{\lambda s}^2$ is obtained

$$\omega_{\lambda s}^2 = \frac{2}{M_{ij} u_i u_j} \left[-2V_{ij} u_i' u_{sj} + 2\omega^2 M_{ij} u_i' u_{sj} + V_{ijkl} u_i u_j q_k^{0\prime} q_{sl}^0 + \right.$$
$$+ V_{ij\lambda s} u_i u_j + V_{ijk\lambda} u_i u_j q_{sk}^0 + V_{ijks} u_i u_j q_k^{0\prime} - \omega_\lambda^2 M_{ijs} u_i u_j +$$
$$\left. -\mu_i \left(V_{i\lambda s} + V_{ijk} q_{sj}^{0\prime} q_{sk}^0 + V_{ij\lambda} q_{sj}^0 + V_{ijs} q_j^{0\prime} \right) \right] \quad (128)$$

Now, using (117) the second order load sensitivity with respect to the structure parameter s can be specified.

The formulae (126), (127) and (128) do not require computations of the second order state equilibrium state sensitivity $\ddot{q}_j^0$. The first order derivatives $q_j^{0\prime}$ and $q_{s_j}^0$ can be obtained from (3), specialized for C_l and C_t, respectively.

The eigenmode derivatives u_j' and u_{s_j} can be specified from (98). But such systems are singular, so differentiated normalization condition can be used as the complementary equation.

The sensitivity derivatives of load factor λ_s and λ_{ss} coincide with those, obtained by the static critical state analysis (cf. (38) and (40)) by assuming $\omega^2 = 0$ at the critical state. Let us note, that for this case the motion equation (96) is reduced to the critical state equation (5a), so the eigenvector q_{1j} and eigenmode u_j are identical. Also, at the critical state, the adjoint field μ_i is the same for both, critical state and eigenfrequency analysis.

4 EXAMPLE

In order to illustrate the general derivations, let us consider a simple two degree of freedom system composed of an extensible bar OA with a horizontal spring attached to its tip A and to a movable support B, Fig.2. Denote by L the initial bar length, by K_1 its axial stiffness, by K_2 the spring stiffness and by $\bar{b}$ the initial horizontal position of the tip A. As the generalized coordinates we assume the axial displacement $\bar{q}_1$ and horizontal position $\bar{q}_2$ of the bar tip A.

The bifurcation appears in the system when $\bar{b} = 0$, and such structure will be called perfect system. It is seen in view of (6) and (7) that the regular sensitivity case occurs when $\alpha = K_2/K_1$ is the structural parameter and $\bar{b} = 0$. On the other hand, the singular sensitivity case occurs when $b = \bar{b}/L$ is a structural parameter undergoing variation. The first and second order sensitivity of the critical load of the system in the bifurcation point with respect to α and b will be specified.

In Fig.3 the behaviour of the system for the stiffness parameter $\alpha = 1.0$ and several values of the geometric imperfection parameter b are presented.

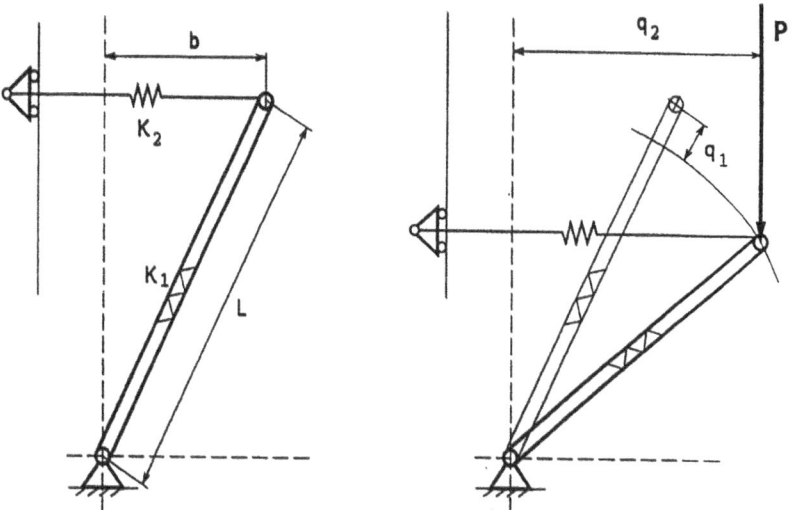

Fig.2 Unloaded and loaded structure

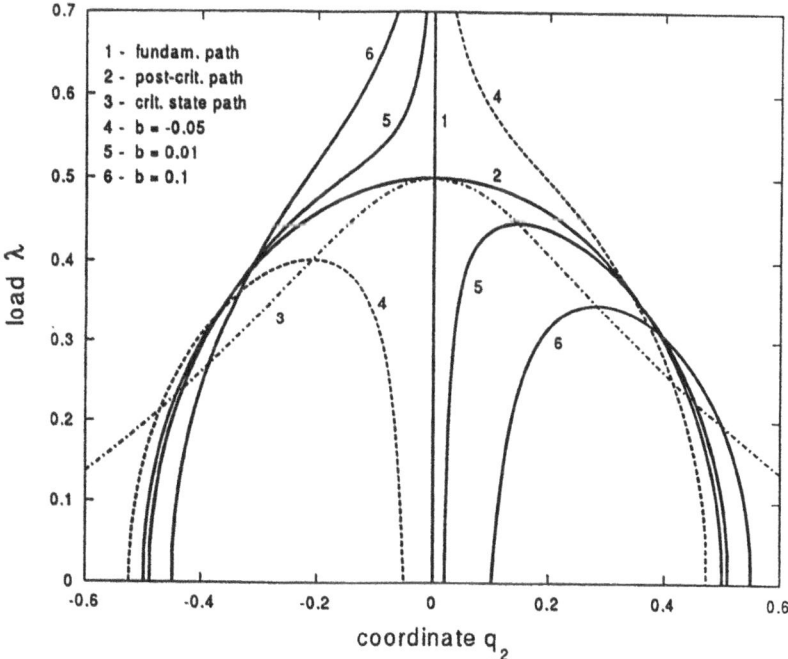

Fig.3 Equilibrium paths in $(q_2 - \lambda)$ plane

The potential energy of the system equals

$$\bar{V} = \frac{1}{2}K_1\bar{q}_1^2 + \frac{1}{2}K_2\left(\bar{q}_2 - \bar{b}\right)^2 - P\bar{w}$$

or, written in non-dimensional variables, has the form

$$V = \frac{\bar{V}}{K_1 L^2} = \frac{1}{2}q_1^2 + \frac{1}{2}\alpha\left(q_2 - b\right)^2 - \lambda w \qquad (129)$$

where P denotes the vertical loading, independent of bar configuration, λ is the non-dimensional load parameter and q_1 and q_2 are non-dimensional coordinates,

$$\lambda = \frac{P}{K_1 L}, \qquad q_1 = \frac{\bar{q}_1}{L}, \qquad q_2 = \frac{\bar{q}_2}{L} \qquad (130)$$

and

$$w = \frac{\bar{w}}{L} = \sqrt{1 - b^2} - a$$
$$a = \sqrt{\left(1 - q_1^2\right)^2 - q_2^2} \qquad (131)$$

The equilibrium equations are

$$V_1 = \frac{\partial V}{\partial q_1} = q_1 - \lambda\frac{1 - q_1}{a} = 0$$
$$V_2 = \frac{\partial V}{\partial q_2} = \alpha\left(q_2 - b\right) - \lambda\frac{q_2}{a} = 0 \qquad (132)$$

and the matrices appearing in the first order incremental equations (2) have the form

$$V_{ij} = \begin{bmatrix} 1 - \lambda\dfrac{q_2^2}{a^3} & -\lambda\dfrac{q_2\left(1 - q_1\right)}{a^3} \\ -\lambda\dfrac{q_2\left(1 - q_1\right)}{a^3} & \alpha - \lambda\dfrac{\left(1 - q_1\right)^2}{a^3} \end{bmatrix}, \qquad V_{i\lambda} = \begin{bmatrix} -\dfrac{1 - q_1}{a} \\ -\dfrac{q_2}{a} \end{bmatrix} \qquad (133)$$

The derivatives of equilibrium equation with respect to α and b are

$$V_{i\alpha} = \begin{bmatrix} 0 \\ q_2 - b \end{bmatrix}, \qquad V_{ib} = \begin{bmatrix} 0 \\ -\alpha \end{bmatrix} \qquad (134)$$

The initial state, defined as a load-free state is described by

$$\lambda = 0, \quad q_1 = 0, \quad q_2 - b = 0 \qquad (135)$$

The critical state condition is

$$det\, V_{ij}^c = \alpha - \lambda\frac{\left(1 - q_1\right)^2}{a^3} - \alpha\frac{q_2^2}{a^3} = 0 \qquad (136)$$

The fundamental path occurs, when the system is perfect ($b = 0$), and it is described by the set of equilibrium equations

$$q_1 - \lambda = 0 , \qquad q_2 = 0 \tag{137}$$

For the perfect system the critical state occurs, when the variables take the values

$$q_1^c = \frac{\alpha}{1 + \alpha} , \quad q_2^c = 0 , \quad \lambda^c = \frac{\alpha}{1 + \alpha} \tag{138}$$

Solving (5a) at the bifurcation point, we find the eigenvector of the matrix V_{ij}^c, which, after normalization is as follows

$$\boldsymbol{x}_1 = [0, \ 1]^T \tag{139}$$

Using (5b), it can be shown, that (138) describes the bifurcation point.

The adjoint system (20) for this structure is written as follows

$$\begin{bmatrix} 1 & 0 \\ 0 & 0 \end{bmatrix} \begin{bmatrix} \mu_1 \\ \mu_2 \end{bmatrix} - \begin{bmatrix} -\alpha(\alpha + 1) \\ 0 \end{bmatrix} = 0 \tag{140}$$

The solution of (140), orthogonal to q_1 is

$$\mu_1 = [-\alpha(\alpha + 1) , \ 0]^T \tag{141}$$

Using (6) and (7), we can see that for varying α, equation (138) specifies the regular sensitivity point, and for varying b it corresponds to the singular sensitivity point.

Let us calculate the sensitivity of the critical load at the regular bifurcation point, setting α as a structural parameter. For this case the matrix V_{ij}^c is constant, so the derivative of its eigenvector $\dot{\boldsymbol{x}}_1$ vanishes. Using (38) we can compute now the first order sensitivity of the critical load in the form

$$\lambda_{c\alpha} = \frac{1}{(1 + \alpha)^2} \tag{142}$$

The sensitivity of the coordinate vector $\dot{\boldsymbol{q}}$ is obtained by solving (2) with $\eta = \alpha$ and using (3) multiplied by $\boldsymbol{q}_1$ as the additional condition. Then

$$\dot{q}_1 = \frac{1}{(1 + \alpha)^2} , \qquad \dot{q}_2 = 0 \tag{143}$$

Now, from (40) we obtain the second order critical load sensitivity

$$\lambda^c_{\alpha\alpha} = -\frac{2}{(1+\alpha)^2} \tag{144}$$

To analyze the singular sensitivity case, we set $s = b$. As this is the symmetric bifurcation case, there is $\dot{\lambda}^m = 0$, $\ddot{b}^m = 0$. The derivative of the eigenvector is

$$\dot{x}_1 = [\alpha(1+\alpha) , \ 0]^T \tag{145}$$

Taking q_2 coordinate as the critical state path parameter, we find the scalar factor occurring in (43) $\alpha^m = 1$, and

$$\dot{q} = x_1 \tag{146}$$

Now $\ddot{\lambda}^m$ can be determined from (58), thus

$$\ddot{\lambda}^m = -3\alpha(1+\alpha) \tag{147}$$

Next, the vector q_2 can be calculated from (67) and (65) resulting in the formula

$$q_2 = [-2\alpha(\alpha+1) , \ 0]^T \tag{148}$$

From (63) the derivative $\ddot{b}^m$ is obtained, so that

$$\ddot{b}^m = 6(\alpha+1)^3 \tag{149}$$

Using (147),(148) and (149) in (72), the critical load sensitivity is now specified , namely

$$\Delta\lambda = -\frac{3}{2}\frac{\alpha}{2\alpha+1}b^{\frac{2}{3}} \tag{150}$$

Let us now specify the eigenfrequency sensitivity, assuming the unitary mass matrix

$$M_{ij} = \delta_{ij} \tag{151}$$

We perform the calculations for $b = 0$.

Solving the characteristic equation (96), we find the first normalized eigenmode

$$u = [0 , \ 1]^T \tag{152}$$

and the coordinate variables take the values

$$q_1 = \lambda = \frac{\alpha - \omega^2}{1 + \alpha - \omega^2} \tag{153}$$

$$q_2 = 0$$

Their partial derivatives are

$$q' = [1, \, 0]^T$$

$$q_\alpha = 0 \tag{154}$$

$$q_b = \left[0, \, \frac{\alpha}{\omega^2}\right]^T$$

Now, using formula (100) we find the eigenfrequency partial derivative with respect to the load parameter

$$\omega_\lambda^2 = -\left(1 + \alpha - \omega^2\right)^2 \tag{155}$$

With α taken as the path parameter (101), we have

$$\omega_\alpha^2 = 1 \tag{156}$$

and with $b = s$, there is

$$\omega_b^2 = 0 \tag{157}$$

In view of (103), we can write the critical load sensitivity with respect to the structure parameters, namely

$$\lambda_\alpha = -\left(1 + \alpha - \omega^2\right)^{-2} \tag{158}$$

$$\lambda_b = 0 \tag{159}$$

In order to determine second order sensitivities we start with the eigenvector derivatives. Solving differentiated equation (96), properly specialized, we obtain

$$u' = 0 \tag{160}$$

$$u_\alpha = 0 \tag{161}$$

$$u_b = \left[\frac{\alpha\left(\alpha - \omega^2\right)\left(1 + \alpha - \omega^2\right)}{\left(\omega^2\right)^2}, \, 0\right]^T \tag{162}$$

Solving now Eq. (3), specialized for particular cases, we obtain

$$q^{0\prime\prime} = 0 \tag{163}$$

$$q_{\alpha\alpha}^0 = 0 \tag{164}$$

$$q_\alpha^{0\prime} = 0 \tag{165}$$

$$q_{bb}^0 = \left[\frac{\alpha^2\left(\alpha - \omega^2\right)\left(1 + \alpha - \omega^2\right)}{\left(\omega^2\right)^2}, \, 0\right]^T \tag{166}$$

$$q_b^{0\prime} = \left[0, \, \left(\frac{1 + \alpha - \omega^2}{\omega^2}\right)^2\right]^T \tag{167}$$

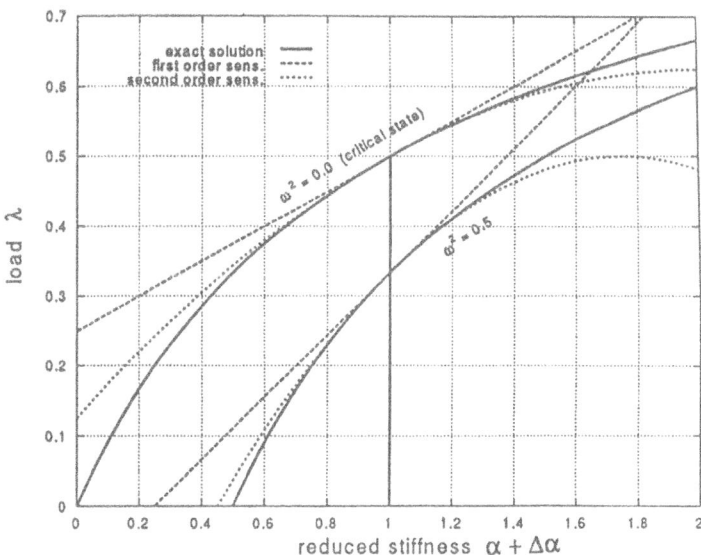

Fig.4 First and second order sensitivity diagrams for $\alpha = 1.0$; (regular case).

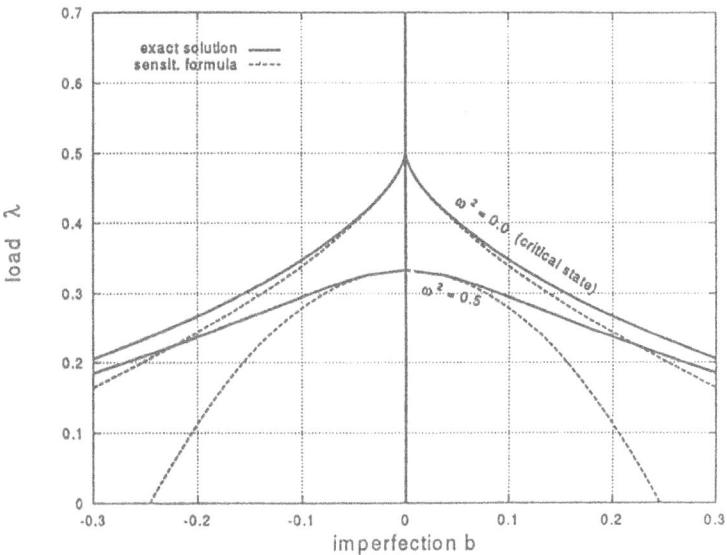

Fig.5 Sensitivity diagram for $\alpha = 1.0$; (singular case).

Now, in view of (111), (113) and (115) the second order eigenfrequency sensitivity is written as follows

$$\omega_{\lambda\lambda}^2 = -2\left(1 + \alpha - \omega^2\right)^3 \tag{168}$$

$$\omega_{\alpha\alpha}^2 = 0 \tag{169}$$

$$\omega_{\lambda\alpha}^2 = 0 \tag{170}$$

$$\omega_{bb}^2 = -\alpha^2 \left(\alpha - \omega^2\right)\left(\frac{1 + \alpha - \omega^2}{\omega^2}\right)^2 \left[\frac{2\left(\alpha - \omega^2\right)}{1 - \omega^2} + \alpha - \omega^2 + 3\right] \tag{171}$$

$$\omega_{\lambda b}^2 = 0 \tag{172}$$

Using (117), the second order load sensitivity can be determined.

For $s = \alpha$, there is

$$\lambda_{\alpha\alpha} = -\frac{\omega_{\lambda\lambda}^2}{\omega_\lambda^2}\lambda_\alpha^2 = -\frac{2}{\left(1 + \alpha - \omega^2\right)^3} \tag{173}$$

and if $s = b$ we have

$$\lambda_{bb} = -\frac{\omega_{bb}^2}{\omega_\lambda^2} = -\frac{\alpha^2 \left(\alpha - \omega^2\right)}{\left(\omega^2\right)^2}\left[\frac{2\left(\alpha - \omega^2\right)}{1 - \omega^2} + \alpha - \omega^2 + 3\right] \tag{174}$$

The load parameter - structure parameter diagrams are presented in Fig.4 (the structural parameter is α) and in Fig.5 (the structural parameter is b), for two values of eigenfrequency ω^2. The value $\omega^2 = 0$ represents the critical state.

The exact solutions are obtained by direct differentiation of equilibrium equations. The estimation formulae are written as follows:

$$\Delta\lambda = \lambda_\alpha\Delta\alpha + \frac{1}{2}\lambda_{\alpha\alpha}\left(\Delta\alpha\right)^2$$

when α is the structural parameter, and

$$\Delta\lambda = \frac{1}{2}\lambda_{bb}b^2$$

when b is the structural parameter and $\omega^2 \neq 0$. The estimation formula, when singular sensitivity case occurs, provides (150).

5 CONCLUDING REMARKS

The procedure of generating first and second order sensitivity of critical load at the bifurcation point and of free frequency, presented in this paper, could be implemented in any numerical code concerned with the non-linear analysis of structures.

Both regular and singular sensitivity cases can be treated in uniform way. The adjoint method is vary useful, especially for the symmetric bifurcation case. The adjoint field is then identical to the second order post-buckling field, cf. Mróz and Haftka [1994]. The variation of load along constant frequency paths provides assessment the critical buckling load by assuming the state of vanishing frequency.

AKNOWLEGMENT

This research presented in this paper was supported by Science Research Council (KBN) Grant number 30939-91-01.

REFERENCES

[1973]   J.H.T.THOMPSON, G.W.HUNT, A general theory of elastic stability, J.Wiley Sons,

[1988]   Z.MRÓZ, R.T.HAFTKA: Sensitivity of buckling loads and vibration frequencies of plates, J.Singer Ann.Vol., Elsevier Sci. Publ.,

[1989]   K.DEMS, Z.MRÓZ: Sensitivity of buckling load and vibration frequency with respect to shape of stiffened and unstiffened plate, J. Mech.Struct.Machines, **17**, 4, pp.431-457,

[1991]   Z.MRÓZ: Sensitivity analysis for vibration and stability of structures, in "Optimization of Large Systems", G.Rozvany (Ed.), NATO, DF6 ASI, Lect.Notes,, **2**, pp.139–157,

[1993]   L.A.GODOY, E.O.TARROCO, R.A.FEIJOO: Second order sensitivity analysis in vibration and buckling problems, Int.J.Num.Meth.Eng., (in print)

[1994]   Z.MRÓZ, R.T.HAFTKA: Design sensitivity analysis of non-linear structures in regular and critical states, Int.J.Solids Struct., (in print)

SHELL OPTIMIZATION

Some Remarks

B. ROUSSELET AND S. MEHREZ
Laboratoire J. Dieudonné U.R.A. C.N.R.S. n⁰ 168
Université de Nice BP 71 F-06108 NICE CEDEX 2

A. MYSLINSKI
Systems Research Institute
ul. Newelska 6, 01-447 Varsovie Pologne

AND

J. PIEKARSKI
IPPT PAN, ul. Swietokrzyska 21 00-049 Varsovie Pologne

1. Introduction

This is not a report of the subject of shape optimization of shells; it gives simply an idea of what has been done in our laboratory or in connection with it. It should be clear that this work stems from previous works on shape optimization developed previously in "Laboratoire J.A. Dieudonné" and in other laboratories. We quote some landmarks [7], [8], [9], [13], [20],[21], [27], [38], [39], [40], [41], [42], [43]. A survey of optimal structural design of shells may be found in ([55]).

There are only a few papers dealing with the shape optimization of nonlinear elastic systems (see [15, 33, 45]); here we present some results for optimizing shells in linear elasticity and new results to optimize geometrically non linear shells.

We present in the next section equations of geometrically non linear shells; in section 3 some linear models; in section 4 sensitivity computations: after recalling some basic general results, we apply them to the linear cases of section 3. A finite strain beam model is described in section 5; its design sensitivity is then described in section 6. In section 7 we turn back to the model of section 2 and we give a preliminary result for shape sensitivity of geometrically non linear shells.

J. Herskovits (ed.), Advances in Structural Optimization, 381–412.
© *1995 Kluwer Academic Publishers.*

2. Equations of geometrically non linear shells

It is well known that there are several mechanical models with many options. We present the analysis of geometrically non linear shells following R. Valid [53], [54]; this presentation uses intrinsic notations (coordinate free); sometimes we use local coordinates as in [4] ; this notation comes from Koiter [22] .

We consider only the case of Kirchoff-Love hypothesis; they can be briefly stated as:

– the normals to the middle-surface are rigid bodies;
– the normals remain normal to the deformed middle-surface (i.e. no transverse shearing)

With these hypothesis, it is known that the virtual strain energy may be expressed with symmetric **surface Cauchy stress**:

– membrane stress resultant n
– bending stress couple m

$$\delta w = \int_\Sigma Tr(n\delta\gamma + m\delta\chi)d\Sigma \qquad (1)$$

where Σ is the *deformed* middle surface and

– $\delta\gamma$ is the first variation of the first fundamental form (arc lengths)
– $\delta\chi$ is the first variation of the second fundamental form (curvatures)

If we denote by $v = \delta m$ the virtual displacement of a point m of the mid-surface, we have

$$\delta\gamma = \frac{1}{2}\left[\Pi\frac{\partial v}{\partial m} + \overline{\Pi\frac{\partial v}{\partial m}}\right], \quad \delta\chi = \overline{\frac{\partial v}{\partial m}}\frac{\partial N}{\partial m} + \Pi\frac{\partial\delta N}{\partial m} \qquad (2)$$

with Π the normal projection on the tangent plane at m , N is the unit normal at m and $\overline{A}$ means *transposition of A*; note that $\overline{N}N = 1$ but $N\overline{N}$ is the orthogonal projection on N so that:

$$v = N\overline{N}v + \Pi v \quad \text{for any vector v} \qquad (3)$$

The hypothesis that normals remain normal gives

$$\delta N = -\overline{\frac{\partial v}{\partial m}}N \qquad (4)$$

Expression in local basis

Denote by a_1, a_2 a basis of the tangent plane and by a_3 the unit normal vector; in the following we use mainly Koiter's [22] notation in particular the rule of repeated indexes to mean summation; greek indexes are assumed to run from 1 to 2. a^α denotes the dual basis: $a^\alpha.a_\beta = \delta^\alpha_\beta$ so that

$$v = v_\alpha a^\alpha + v_3 a^3 \quad \delta\gamma = \delta\gamma_{\alpha\beta}\, a^\alpha \otimes a^\beta \tag{5}$$

with

$$\delta\gamma_{\alpha\beta} = \frac{1}{2}(v_{\alpha|\beta} + v_{\beta|\alpha}) - b_{\alpha\beta}v_3 \tag{6}$$

and similarly

$$\delta\chi_{\alpha\beta} = \frac{1}{2}\left[-2v_{3|\alpha\beta} - (b^\lambda_\alpha v_\lambda)_{|\beta} - (b^\lambda_\beta v_\lambda)_{|\alpha} + b^\lambda_\alpha v_{\lambda|\beta} + b^\lambda_\beta v_{\lambda|\alpha} - b^\lambda_\alpha b_{\lambda\beta}v_3\right] \tag{7}$$

as in Koiter [23].

For simplicity we assume linear elasticity (hyperelasticity would work similarly):

$$n = hC\gamma \quad m = \frac{h^3}{3}C\rho \tag{8}$$

with γ and ρ the *linear* strain tensors (identical to $\delta\gamma$ and $\delta\chi$).

In case of *large* displacements (and large strains), the equations are practically written in the reference state Σ_0; as usual this is written with *Piola-Kirchoff* surface stresses

$$\mathcal{N} = det(\frac{\partial m}{\partial m_0})\frac{\partial m_0}{\partial m}\, n\, \frac{\overline{\partial m_0}}{\partial m} \tag{9}$$

$$\mathcal{M} = det(\frac{\partial m}{\partial m_0})\frac{\partial m_0}{\partial m}\, m\, \frac{\overline{\partial m_0}}{\partial m} \tag{10}$$

$$\Gamma = \frac{1}{2}[\frac{\overline{\partial m}}{\partial m_0}\frac{\partial m}{\partial m_0} - id_{m_0}] \tag{11}$$

$$\mathcal{K} = \frac{\overline{\partial m}}{\partial m_0}\frac{\partial N}{\partial m}\frac{\partial m}{\partial m_0} - \frac{\partial N_0}{\partial m_0} \tag{12}$$

with N_0 unit normal at m_0 to Σ_0 . For linearization, it is clever to notice that:

$$\overline{dm_0}\Gamma dm_0 = \frac{1}{2}(\overline{dm}dm - \overline{dm_0}dm_0) \tag{13}$$

$$\overline{dm_0}\mathcal{K}dm_0 = \overline{dm}\frac{\partial N}{\partial m}dm - \overline{dm_0}\frac{\partial N_0}{\partial m_0}dm_0 \tag{14}$$

Principle of virtual work

Set $v(m)$, a virtual displacement at point m; we denote

$$\delta\Gamma = \frac{\partial v}{\partial m}\, v(m) \tag{15}$$

and we obtain

$$\delta\Gamma = \frac{1}{2}\left(\overline{\frac{\partial m}{\partial m_0}\frac{\partial v}{\partial m_0}} + \overline{\frac{\partial v}{\partial m_0}\frac{\partial m}{\partial m_0}}\right) \tag{16}$$

$$\delta K = \overline{\frac{\partial v}{\partial m_0}\frac{\partial N}{\partial m_0}} + \overline{\frac{\partial m}{\partial m_0}\frac{\partial \delta N}{\partial m_0}} \tag{17}$$

and

$$\delta N = -\overline{\frac{\partial v}{\partial m}}N \tag{18}$$

if we assume the material is hyperelastic with strain energy density $\alpha(\Gamma, K)$, the stress-strain law may be written:

$$\mathcal{N} = \frac{\partial \alpha}{\partial \Gamma}, \quad \mathcal{M} = \frac{\partial \alpha}{\partial K} \tag{19}$$

Finally the principle of virtual work may be written: $G(\phi_0, \phi)\, v = 0$ for any v kinematically admissible, where

$$G(\phi_0, \phi)\, v = \int_{\Sigma_0} Tr(\mathcal{N}\delta\Gamma + \mathcal{M}\delta K)d\Sigma_0 - \int_{\Sigma_0} fvd\Sigma_0 - \int_{\partial\Sigma_0 F} FvdS_0 \tag{20}$$

3. Shells in linear elasticity

3.1. GENERAL SHELLS

Koiter's model of thin elastic shells uses intrinsic geometrical properties of the middle surface of the undeformed shell; a presentation is to be found in [4] with a detailed study of finite element approximations of general shells. Geometrical properties may be explicitly given for a surface parametrized in $\mathbf{R}^3$ with a mapping Φ from Ω a plane domain.

3.2. SHELLS OF REVOLUTION

Geometrical properties may be explicitly given for a surface parametrized in $\mathbf{R}^3$ with a mapping Φ from Ω a plane domain. Here we assume that the midsurface is of revolution; by using a Fourier series expansion for applied loads and displacements, we substitute a two dimensions problem by a sequence of one dimension equations: one per harmonic. We assume that the middle surface is parametrized in the following way

$$\phi(\xi^1, \xi^2) = r(\xi^1)cos(\xi^2)\vec{e_1} + r(\xi^1)sin(\xi^2)\vec{e_2} + \psi(\xi^1)\vec{e_3}.$$

Then a basis of the tangent space is:

$$\vec{a_1} = \vec{\phi}_{,1} = \begin{pmatrix} r'cos\xi^2 \\ r'sin\xi^2 \\ \psi' \end{pmatrix}, \ \vec{a_2} = \vec{\phi}_{,2} = \begin{pmatrix} -rsin\xi^2 \\ rcos\xi^2 \\ 0 \end{pmatrix}$$

The first fundamental form is:

$$a_{11} = r'^2 + \psi'^2, \ a_{12} = a_{21} = 0, \ a_{22} = r^2$$

and then

$$\vec{a^1} = \frac{\vec{a_1}}{a_{11}}, \ \vec{a^2} = \frac{\vec{a_2}}{a_{22}}$$

$$a^{11} = \frac{1}{a_{11}}, \ a^{22} = \frac{1}{a_{22}}, \ a^{12} = a^{21} = 0$$

so that

$$a = |\vec{a_1} \ X \ \vec{a_2}|^2 = r^2 a_{11} \ \text{and} \ dS = r\sqrt{r'^2 + \Psi'^2} \ d\xi^1 d\xi^2$$

the normal vector is

$$\vec{a_3} = \frac{1}{\sqrt{a_{11}}} \ a \begin{pmatrix} -\psi'cos\xi^2 \\ -\psi'sin\xi^2 \\ r' \end{pmatrix}$$

The curvature tensor may be written here

$$b_{11} = \frac{1}{\sqrt{a_{11}}} \ (- \psi' \ r" + \psi" \ r') = r" \ cos \ \eta + \psi" \ sin \ \eta$$

$$b_{22} = \frac{r\psi'}{\sqrt{a_{11}}} = - r \ cos \ \eta$$

$$b_{12} = b_{21} = 0$$

with η angle of - e^3 and a_1 ; expressions of b^α_λ and $b^{\alpha\beta}$ may then be derived easily.

Similarly *Christoffel symbols* are $\Gamma^\alpha_{\beta\lambda} = \vec{a^\alpha}.\vec{a_{\beta,\lambda}}$ and we get

$$\Gamma^1_{11} = \frac{1}{a_{11}}(r"r' + \psi'\psi") \qquad (21)$$

$$\Gamma^1_{22} = -\frac{r'r}{a_{11}} = -\frac{rsin\eta}{\sqrt{a_{11}}}, \ \Gamma^2_{21} = \frac{r'}{r} = \frac{\sqrt{a_{11}}sin\eta}{r} \qquad (22)$$

the others being zero. In this presentation, to *simplify* we assume that the *load* is *axisymmetric* and that there is no torsion. We have

$$p = p^1 a_1 + p^3 a_3 \text{ and } u = u_1 a^1 + u_3 a_3,$$

where these functions are *independent* of ξ^2 . We then obtain simplified

expressions of *covariant derivatives*. Hence *strain tensor* is

$$\gamma_{11}(\vec{u}) = u_{1,1} - \Gamma_{11}^1 u_1 - b_{11} u_3 \quad \gamma_{12}(\vec{u}) = 0$$

$$\gamma_{22}(\vec{u}) = -\Gamma_{22}^1 u_1 - b_{22} u_3$$

and *change of curvature tensor* is

$$\bar{\rho}_{11}(\vec{u}) = u_{3,11} - \Gamma_{11}^1 u_{3,1} + (b_{1,1}^1 - 2\Gamma_{11}^1 b_1^1) u_1 + 2 b_1^1 u_{1,1} - b_1^1 b_{11} u_3$$

$$\bar{\rho}_{22}(\vec{u}) = -\Gamma_{22}^1 u_{3,1} - \Gamma_{22}^1 (b_1^1 + b_2^2) u_1 - b_2^2 b_{22} u_3$$

$$\bar{\rho}_{12}(\vec{u}) = 0$$

Stress-strain relation is then

$$n^{11} = \frac{eE}{1 - \nu^2} (a^{11} a^{11} \gamma_{11} + \nu a^{11} a^{22} \gamma_{22}) \tag{23}$$

$$n^{12} = \frac{eE}{1 + \nu} a^{11} a^{22} \gamma_{12} = 0 \tag{24}$$

$$n^{22} = \frac{eE}{1 - \nu^2} (\nu a^{11} a^{22} \gamma_{11} + a^{22} a^{22} \gamma_{22}) \tag{25}$$

and

$$m^{11} = \frac{e^3 E}{12(1-\nu^2)} (a^{11} a^{11} \bar{\rho}_{11} + \nu a^{11} a^{22} \bar{\rho}_{22})$$

$$m^{12} = \frac{e^3 E}{12(1+\nu)} a^{11} a^{22} \bar{\rho}_{12} = 0$$

$$m^{22} = \frac{e^3 E}{12(1-\nu^2)} (\nu a^{11} a^{22} \bar{\rho}_{11} + a^{22} a^{22} \bar{\rho}_{22})$$

Thus we have *strain energy:*

$$a(\vec{u}, \vec{v}) = a_m^0 (\vec{u}, \vec{v}) + a_f^0 (\vec{u}, \vec{v})$$

with $a_m^0(\vec{u}, \vec{v})$ *membrane energy* and $a_f^0(\vec{u}, \vec{v})$ *flexural energy* where

$$a_m(v, v) = \int_s (n^{11}\gamma_{11} + n^{22}\gamma_{22})\, ds \qquad (26)$$

and a similar expression for flexural energy; the arc length along the meridian line is $ds = r\sqrt{a_{11}}\, d\xi^1$

Virtual work of pressure load may be written

$$l(v) = -2\pi \int_{s_0}^{s_1} pv_3\, ds \qquad (27)$$

In case the load and hence the displacement is not axisymmetric, they are expanded in *Fourier series* and we obtain a one dimensional problem to solve per each harmonic; this is detailed in [28]

4. Sensitivity Computation

4.1. BACKGROUND

We outline here how to compute the design sensitivity; these general formulas can then be used in several examples presented later; we consider the nonlinear and the linear case. We denote the *functional* $j(\phi_0) = J(\phi_0; \phi)$ where ϕ_0 is the design variable and ϕ the state of the structure; note that in *shape* design ϕ_0 is the reference state.
    We denote the **state equation** (the principle of virtual work)

$$G(\phi_0; \phi)v = 0 \quad \text{for any v kinematically admissible.} \qquad (28)$$

and in the linear case:

$$a(\phi_0; \phi, v) - l(\phi_0; v) = 0 \qquad (29)$$

The design sensitivity of the state equation:

$$\frac{\partial G(\phi_0; \phi)}{\partial \phi_0}[v, \Delta\phi_0] + \frac{\partial G(\phi_0; \phi)}{\partial \phi}[v, \Delta\phi] = 0 \qquad (30)$$

and in the linear case:

$$\frac{\partial a(\phi_0; \phi, v)}{\partial \phi_0}\Delta\phi_0 - \frac{\partial l(\phi_0; v)}{\partial \phi_0}\Delta\phi_0 + a(\phi_0; \Delta\phi, v) = 0 \qquad (31)$$

We note that the last term of (30) is a bilinear form with respect to v and $\Delta\phi$. Moreover in case of hyperelasticity, in general, this form is symmetric being the second derivative of the energy; it is not always symmetric if ϕ lies on a manifold as in the beam model described later; in the case of linear elasticity this form is just the virtual variation of strain energy.

Design sensitivity of the functional:

$$\frac{\partial j}{\partial \phi_0} \Delta \phi_0 = \frac{\partial J}{\partial \phi_0} \Delta \phi_0 + \frac{\partial J}{\partial \phi} \Delta \phi \tag{32}$$

We note that in the last equation $\Delta \phi$ depends implicitly on $\Delta \phi_0$, this dependence can be made explicit using (30): either by inverting it (this is performed usually at the finite element level) or by introducing an adjoint state : ψ solution of the **adjoint equation:**

$$\frac{\partial G}{\partial \phi}[v, \psi] = -\frac{\partial J}{\partial \phi} v \quad \text{for any v kinematicaly admissible} \tag{33}$$

and in the linear case:

$$a(\phi_0; v, \psi) = -\frac{\partial J}{\partial \phi} v \tag{34}$$

so that *the design sensitivity of the functional* becomes completely explicit:

$$\frac{\partial j}{\partial \phi_0} \Delta \phi_0 = \frac{\partial J}{\partial \phi_0} \Delta \phi_0 + \frac{\partial G}{\partial \phi_0}[\psi, \Delta \phi_0] \tag{35}$$

and in the linear case:

$$\frac{\partial j}{\partial \phi_0} \Delta \phi_0 = \frac{\partial J}{\partial \phi_0} \Delta \phi_0 + \frac{\partial a(\phi_0; \phi, v)}{\partial \phi_0} \Delta \phi_0 \tag{36}$$

This formula may be used as soon as one has computed
 − the state ϕ of the structure
 − its adjoint state ψ : it is solution of a linear equation, but the left hand side *has not necessarily an inverse!* This will be the case for example at a turning or buckling state.

In the next subsection we consider the use of these formulas in several cases; for the linear case, implementation issues are discussed for example in [44].

4.2. GENERAL SHELLS IN LINEAR ELASTICITY

One of the first attempt to optimize the shape of the midsurface of a general shell in linear elasticity seems to be [14]; the shell model considered is Budiansky-Sanders. In [5], a methodology is proposed to optimize the shape of the midsurface; the model considered is the one of Koiter. Computational experiments are in preparation [6].

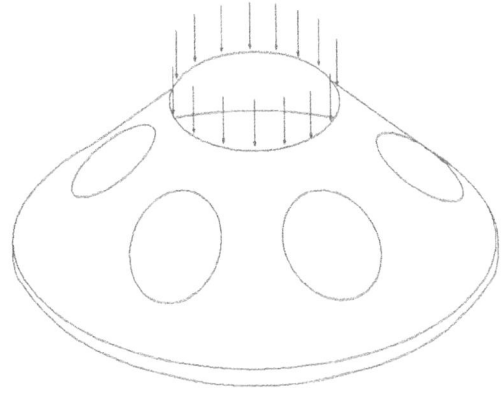

Figure 1. Conical shell with holes

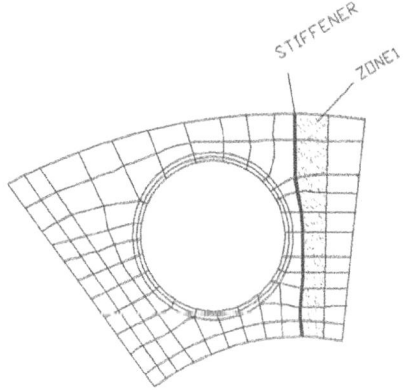

Figure 2. 60° sector of the shell

Example: thickness optimization of a conical shell.

We describe now an example of sizing optimization of a general shell. This example has been studied by Gauthier-Giuliano from Aerospatiale (Cannes la Bocca, France) in collaboration with B. Rousselet and supported by DRET. The analysis and design sensitivity have been performed using Nastran and the optimization has been conducted with a home made projected-gradient algorithm interfaced with Nastran.

The objective of this example is to minimize the mass of a conical shell with holes subjected to a uniform axial load applied on the top; the bottom is simply supported (see Fig. 1). The thickness of the skin is 3.2 mm and around the holes it is of 12 mm ; Young modulus is $E = 7. \times 10^{10} N/m^2$

For symmetry reasons a 60° sector is studied (see Fig. 2); we notice

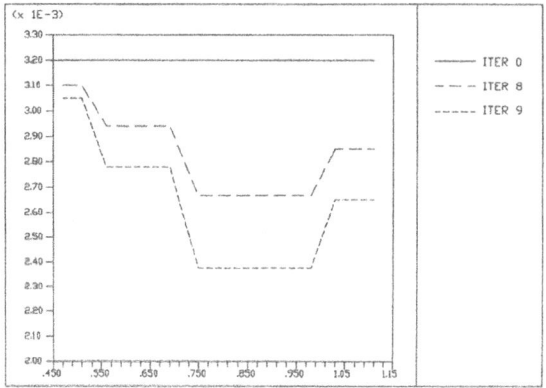

Figure 3. Thickness behaviour

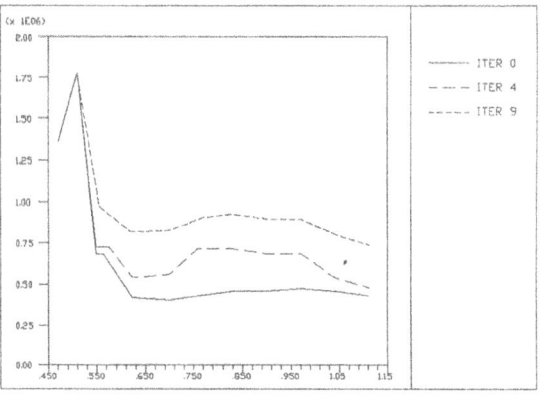

Figure 4. Von Mises stress behaviour

stiffeners: they are of square section of 40mm with thickness of 4mm; the top frame is identical but open on a side; $E = 1.5 \times 10^{11} N/m^2$; yield stress is $2. \times 10^6 N/m^2$; the uniform axial load is 6200 N.

The optimization is subjected to stress constraints: Von Mises in the skin and axial stress in stiffeners should be less than the yield stress. Ten optimization steps have been performed; Figs. 3 and 4 show the behaviour of thickness and Von Mises stress in zone 1 defined in Fig. 2; from these figures and others related to other zones, it is clear that the optimization process modifies the distribution of stress to minimize the mass; the mass decrease is of 31%.

4.3. OPTIMIZATION OF ARCHES IN LINEAR ELASTICITY

Due to its practical importance and its simplicity it is the first shell model we have tried to optimize with respect to the shape of its middle curve ([16]); the optimization of this model with nondifferentiable constraints is addressed in ([19]).

4.4. OPTIMIZATION OF SHELLS OF REVOLUTION IN LINEAR ELASTICITY

4.4.1. *Midsurface sensitivity.*

Computations to obtain midsurface sensitivity are quite intricate; here we just give an idea of the methodology. From now on we will denote derivatives with respect to Φ in a simplified way used in classical calculus of variations : Δ a(u , u) instead of the precise notation ∂_Φ a(Φ ; u $_\Phi$, λ_Φ) .ψ ;

We use freely midsurface sensitivity of general shells obtained in [5], and we particularise for axisymmetric shells; we concentrate on the membrane energy ; midsurface sensitivity of flexural energy may be computed in a similar way but computations are really more intricate. Also we shall restrict to axisymmetric load; extension to non axisymmetric loads can be dealt with the same methodology (see [29] and [28] for more details).

A key to get derivatives is to use intensively chain rule as noticed in [16] and [14] . Proof of differentiability has been studied in these papers. We can first notice that from (26), (23), (25), we obtain

$$\Delta a_m(v, v) =$$

$$\int_{\xi_0}^{\xi_1} e(2\frac{E}{1-\nu^2}a^{11}a^{11}\Delta\gamma_{11}\gamma_{11}\sqrt{a} + \gamma_{11}\gamma_{11}\Delta(\frac{E}{1-\nu^2}a^{11}a^{11}\sqrt{a}))\,d\xi$$

+ similar terms

We go on with the first integral, the other ones may be treated similarly; we have to compute two terms $\Delta\gamma_{11}$ and

$$\Delta(\frac{E}{1-\nu^2}a^{11}a^{11}\sqrt{a})$$

From $\gamma_{11}(\vec{u}) = u_{1,1}$ - $\Gamma_{11}^1 u_1$ - $b_{11}u_3$; we obtain $\Delta\gamma_{11}(\vec{u}) = $ - $\Delta\Gamma_{11}^1 u_1$ - $\Delta b_{11}u_3$
To compute $\Delta\Gamma_{11}^1$ we could use (21) to relate it to Δr and $\Delta\psi$ but it is more systematic to differentiate directly the general formula: $\Gamma_{\beta\lambda}^\alpha = \vec{a^\alpha}.\vec{a_{\beta,\lambda}}$. Noting that $\vec{a_\alpha} = \Phi_{,\alpha}$ yields $\Delta\vec{a_\alpha} = \Delta\Phi_{,\alpha}$, for general surfaces we then obtain easily the following

Lemma 4.1 $\Delta\vec{a_\alpha} = \Delta \Phi_{,\alpha}$ $\Delta a_{\alpha\beta} = \vec{a_\beta}.\Delta\vec{a_\alpha} + \vec{a_\alpha}.\Delta\vec{a_\beta}$

Other formulas for $\Delta\Gamma^\alpha_{\beta\lambda}$ and other tensors are to be found in [5]. For a shell of revolution we obtain the following formulas.

Proposition 4.1 $\Delta\sqrt{a} = a\vec{a}^\alpha.\Delta\vec{a}_\alpha$

$\Delta\Gamma^1_{11} = (-a^1\Gamma^1_{11} + a^{11} b_{11}\vec{a}_3).\Delta\Phi_{,1} + a^1.\Delta\Phi_{,11}$

$\Delta b_{11} = \vec{a}_3.(\Delta\Phi_{,11} - \Gamma^1_{11}\Delta\Phi_{,1})$

$\Delta(\frac{E}{1-\nu^2}a^{11}a^{11}\sqrt{a}) = (\frac{E}{1-\nu^2} a^{11}a^{11}\sqrt{a}) (3a^1.\Delta\Phi_{,1} - \Delta\Phi_{,2}.a^2)$

REMARK. We notice that these formulas are quite easy to compute as they involve geometric quantities already determined for the strain energy plus, naturally, derivatives of the variation of the mapping $\Delta\Phi$

Sensitivity of the pressure load.

From (27) we get $\Delta\,l(v) = -2\pi \int_{\xi_0}^{\xi_1} pv_3\Delta\sqrt{a}\ d\xi$ where $\Delta\sqrt{a}$ is given in previous proposition.

4.4.2. *Implementation.*

A conformal finite element has been implemented in Modulef library; in case the load and hence the displacement is not axisymmetric they are expanded in Fourier series; then for each harmonic , tangent displacement is approximated by a Lagrange element of degree one and normal displacement by a cubic Hermite element; an error estimate has been obtained in [28].

The geometry of the meridian line is defined by the explicit use of a mapping Φ as described in §2.2; this mapping is provided by the user of the library as a FORTRAN subprogram. This has been used previously in Modulef library for several general shell elements.

In practice the user does not always know an explicit mapping of the midline (or surface) of the shell but he knows some measurements points. It is then natural to use spline functions to get an approximate smooth mapping of the midline (or surface) and of the thickness ; a module has been implemented in Modulef see [36]. Moreover spline functions provide a natural structure of finite dimensional vector space in which to implement line or surface sensitivity; they provide easily a class of surfaces with the required smoothness property for shell analysis and midsurface sensitivity. This is also implemented in Modulef library. We describe some numerical experiments.

Half a sphere with internal pressure: it is clamped on its base circle; the meridian line is a quarter of a circle; the normal displacement is nearly constant far from the clamped point; in the neighbourhood of this clamped point it changes abruptly; the finite element computations have used the following data

— radius 1m; length 8m; thickness 0.01m

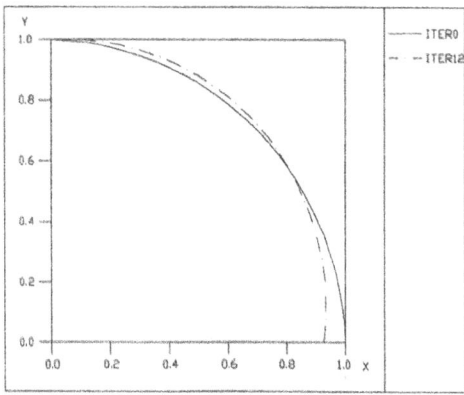

Figure 5. Half a sphere with internal pressure

- Young modulus $0.1\,10^7 Pa$; Poisson coefficient 0.3; internal pressure 1.Pa .
- simply supported in 0; symmetry conditions in 4 .
- 25 elements; and 20 spline nodes for optimization

we have minimized the work of the applied pressure with a constant volume of material; the design variables are the 20 spline nodes used to describe the shape of the meridian line; we show the evolution of this meridian line; we notice it is changing mainly near the clamped point; this is in agreement with the fact that the solution is nearly constant far from this point (Fig. 5).

Cylindrical vessel filled with liquid: this example is described in [52] p 514; it is an horizontal cylindrical vessel of length l simply supported at both end; the only applied load is the pressure of the liquid in a gravitational field; an explicit solution with Fourier series is provided in [52]; in case the vessel is full of water only two terms of these Fourier series are needed. We have checked the accuracy of our finite element program with the formulas of [52]; we have found very good agreement for the displacements and reasonable agreement for the stresses; the finite element computations have been performed with half a cylinder with symmetry conditions; with following data

- radius 1m; length 8m; thickness 0.001m
- Young modulus $0.1\,10^8 Pa$; Poisson coefficient 0.3; specific weight of liquid 1. .
- simply supported in 0; symmetry conditions in 4 .
- 29 elements; 116 degrees of freedom; two harmonics only are needed(see above)

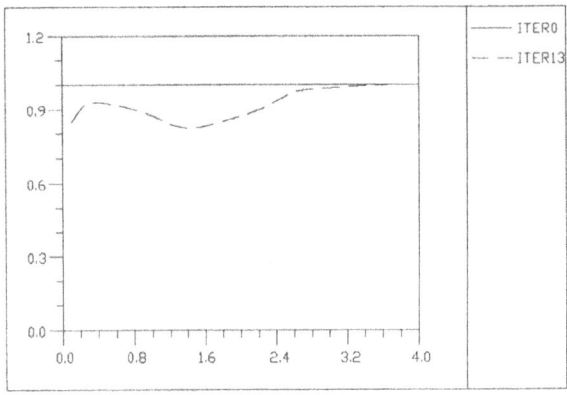

Figure 6. Cylindrical vessel filled with liquid

Optimization of the mid line of this vessel: we have minimized the work of the applied loads with a constraint on the volume of material; the design variables are 20 spline nodes to describe the shape of the mid line; we show the evolution of the mid line; we notice no modification around the symmetry point but an important change where the cylinder is supported; this is quite natural if we recall that the solution is nearly constant near the symmetry point and changes much where the vessel is supported (Fig. 6).

4.5. OPTIMIZATION OF SHELLS WITH INSTABILITIES

One chapter of the book [17] is devoted to the optimization of shells under stability constraints.

In [15] we consider the dependence of the buckling load of a nonshallow arch with respect to the shape of its midcurve; here the case of linear buckling is addressed; as is well known in this case the buckling load is the smallest eigenvalue of a boundary value problem; we give a mathematical proof of existence of this smallest eigenvalue and we prove its differentiability with respect to the shape of its midcurve; for a simple eigenvalue we get Frechet differentiability but directional derivative in case of repeated eigenvalue; in both case we can derive necessary optimality conditions.

In [31] a general design sensitivity formalism is introduced for non linear elastic structures reaching critical equilibrium states. A mathematical proof of differentiability of the load associated to a turning point has been obtained by Rousselet-Vlasak (in preparation); numerical results for nonlinear beams are also obtained.

In the next paragraph some numerical results will be presented for a general strain beam model up to a turning point. Computation of design sensitivity through buckling load is in progress using ideas of arclength method of Keller.

5. A finite strain beam model including shear

5.1. INTRODUCTION

This section[1] and the following deals with the formulation of an optimal design problem of a hyperelastic rod. They should be considered as an illustration of the methodology of section 4; in this example a mathematical treatment is given; details will be found in [35]. The rod is subjected to large displacement involving flexion, shear, torsion and longitudinal extension [1, 2, 24, 45, 48, 49, 50]. The equilibrium state of the rod is described by a system of two nonlinear coupled ordinary differential equations of the first order [1, 2, 50]. In this model the function describing the position of the point at the line of centroids of cross-sections of the rod as well as the orthonormal directors [1] are used to characterize the motion and deformation of the rod in an objective and intrinsic framework. Moreover the nonlinear geometry of the rod is taken into account exactly [50]. The existence of global solutions to the rod problem, i.e. the existence of global minimizers to the rod energy functional was studied in [1, 11, 24, 25, 50]. In [1] it was proved, that the energy functional corresponding to the system describing the equilibrium state of the rod has at least one global minimizer, in general, nonunique. Moreover under additional assumptions this global minimizer satisfies the Euler–Lagrange equations for the rod [1, 25]. The existence of the local solutions to the hyperelastic rod problem is based on Stopelli theorem [25]. If all data in the problem are smooth enough and the linearized state system of rod equations has a unique solution then the nonlinear system has a unique, enough regular solution in the neighbourhood of the linearization point (details will be found in [35]). The finite–dimensional approximation of the rod model and Newton type algorithm for solving this model numerically are proposed in [24, 48, 49].

5.2. KINEMATICS DESCRIPTION

Consider a rod in a reference configuration. This configuration is specified by enough regular mapping $\varphi_0(\xi)$ and an orthonormal pair of vector functions $E_1(\xi)$ and $E_2(\xi)$ of the real variable $\xi \in [0, L]$. The mapping :

$$\varphi_0 : R \ni \xi \to \varphi_0(\xi) \in R^3 \tag{37}$$

[1]this section and the following are the result of collaboration with Myslinski and Piekarski; the content of these 2 sections is presented in French in [34]

defines the line of centroids of cross–sections of a slender three dimensional body called a rod [1, 2, 50]. The line of centroids is assumed to be an arbitrary curve. This is different of the assumption of some authors [24, 48, 49]; **this curve is our design variable.** The cross–section $A(\xi)$ of the rod is defined as a plane passing through a point $\varphi_0(\xi) \in R^3$, $\xi \in [0, L]$ and normal to the vector $E_3(\xi) = E_1(\xi) \times E_2(\xi)$. $\varphi_0(\xi)$ is interpreted as the position in a reference configuration of the material point at the centroid line of the section $A(\xi)$ [1, 2, 48, 49]. $E_1(\xi)$ and $E_2(\xi)$ characterize the reference configuration of the section $A(\xi)$.

The rod can suffer flexure, torsion, shear and longitudinal extension [1, 2, 48]. After deformation the current configuration of the rod is characterized by a position vector function $\varphi(\xi)$ of the line of centroids and an orthonormal pair of vectors $t_1(\xi)$ and $t_2(\xi)$ defining the cross–section of the deformed rod. The functions φ, t_1, t_2 depending on $\xi \in [0, L]$ have similar interpretation as φ_0, E_1, E_2 [1, 2]. The cross–section $A(\xi)$ of the deformed rod is defined as a plane passing through a point of the deformed rod is defined as a plane passing through a point $\varphi(\xi) \in R^3$, $\xi \in [0, L]$ and normal to the vector $t_3(\xi) = t_1(\xi) \times t_2(\xi)$.

Moreover the current configuration $x(X)$ is assumed to satisfy the kinematics boundary conditions imposed to the rod. Assume these conditions are :

$$\varphi(0) = \varphi_0 \quad \varphi(L) = \varphi_0(L) \tag{38}$$

$$t_i(0) = t_i^0 \text{ for } i \text{ given in } \{1, 2, 3\} \tag{39}$$

where φ_0, φ_L and t_i^0, $i = 1, 2, 3$ are given vectors. The condition (38) imply the rod is fixed at both ends, the condition (39) imply the rod can rotate around t_i^0.

Note, that the orthonormal frames $E_i(\xi)$ and $t_i(\xi)$, $i = 1, 2, 3$, $\xi \in [0, L]$ are related by an orthogonal mapping $\Lambda(\xi)$, i.e.,

$$t_i(\xi) = \Lambda(\xi) E_i(\xi) \tag{40}$$

For details concerning the mapping $\Lambda(\xi)$ see [25, 48, 49, 50].

5.3. NONLINEAR ROD MODEL

5.3.1. *Resultant force and moment*
Let $T = T(X_1, X_2, \xi)$, $(X_1, X_2) \in A(\xi)$, $\xi \in [0, L]$ denote the first Piola–Kirchhoff stress tensor [1, 3, 11, 25, 50]. The resultant contact force n per unite of the reference archlength over the cross–section $A(\xi)$ in the current configuration is given by [2],

$$n(\xi) = \int_{A(\xi)} T(X_1, X_2, \xi) E_3(\xi) dX_1 dX_2 \tag{41}$$

The moment m per unit of the reference archlength over the cross–section $A(\xi)$ in the current configuration is given by [2]:

$$m(\xi) = \int_{A(\xi)} [x - \varphi(\xi)] \times T(X_1, X_2, \xi) E_3(\xi) dX_1 dX_2 \qquad (42)$$

The vector fields n and m take values in the current configuration, i.e. [48, 49],

$$n = N_i t_i(\xi) \quad m = M_i t_i(\xi) \qquad (43)$$

where N_i and M_i, $i = 1, 2, 3$ are given numbers. Using the orthogonal mapping (40) we can pull back [25, 48, 49, 50] the vector fields n and m to the reference configuration. We denote by N and M the vector fields in the reference configuration corresponding to the vector fields n and m respectively in the current configuration. The vector fields N and M are defined by :

$$n = \Lambda N \quad m = \Lambda M \qquad (44)$$

From the orthogonality of the mapping (40) as well as from (43), (44) it follows,

$$N = N_i E_i(\xi) \quad M = M_i E_i(\xi) \quad i = 1, 2, 3 \qquad (45)$$

i.e. the components of the force and moment vectors n as well as m associated with the moving frame $\{t_i(\xi)\}$, $i = 1, 2, 3$ equal to those of vector fields N and M, respectively, related to the reference frame $\{E_i(\xi)\}$, $i = 1, 2, 3$.

5.3.2. Strain measures

Let us introduce vector strain measures. By γ we denote the vector measuring the elongation and shear of the rod in the current configuration [45, 48, 49] :

$$\gamma = \frac{d\varphi}{ds} - t_3 \qquad (46)$$

By ω_0 we denote a vector field relating E_i' and E_i :

$$\frac{dE_i}{ds} = \omega_0 \times E_i \quad i = 1, 2, 3 \qquad (47)$$

We shall assume $\{E_i\}$, $i = 1, 2, 3$ is the Frenet[2] frame [10] of the undeformed line of centroids. Hence, assuming that the rod cross–section is regular enough, it follows [10, 26, 45] :

$$\omega_0 = \omega_0^i E_i \quad i = 1, 2, 3 \qquad (48)$$

---

[2]this will imply in practice that the cross section is circular in order to write the stored energy function as in (55)

$$\omega_0^1 = 0 \quad \omega_0^2 = \frac{1}{\varrho_0} \quad \omega_0^3 = \frac{1}{\tau_0}$$

where $\frac{1}{\varrho_0}$ is a curvature and $\frac{1}{\tau_0}$ is a torsion of the line of centroids in the reference configuration. $\frac{1}{\varrho_0}$ and $\frac{1}{\tau_0}$ are given by [26] :

$$\frac{1}{\varrho_0} = \| \frac{d^2 \varphi_0}{ds^2} \| \tag{49}$$

$$\frac{1}{\tau_0} = \| \{[\frac{d\varphi_0}{ds} \times \frac{d^2 \varphi_0}{ds^2}] \frac{d^3 \varphi_0}{ds^3}\} / \| \frac{d^2 \varphi_0}{ds^2} \| \tag{50}$$

The change of curvature and torsion are measured by the vector field ω such that [1, 2, 45, 48, 49, 50] :

$$\frac{dt_i}{ds} = (\omega + \omega_0) \times t_i \quad i = 1, 2, 3 \tag{51}$$

The vector fields γ and ω take values in the current configuration [45, 48, 49, 50]:

$$\gamma = \Gamma^i t_i(\xi) \quad \omega = \chi^i t_i(\xi) \quad i = 1, 2, 3 \tag{52}$$

Using the orthogonal mapping $\Lambda(\xi)$ defined by (40) we can pull back [26, 48, 49, 50] the vector fields γ and ω to the reference configuration. By $\Gamma(\xi)$ and $\chi(\xi)$ we denote vector fields corresponding to γ and ω respectively taking values in reference configuration. $\Gamma(\xi)$ and $\chi(\xi)$ are defined by :

$$\Gamma(\xi) = \Lambda^T \gamma(\xi) \quad \chi(\xi) = \Lambda^T \omega(\xi) \tag{53}$$
$$\Gamma(\xi) = \Gamma^i E_i(\xi) \quad \chi(\xi) = \chi^i E_i(\xi) \quad i = 1, 2, 3 \tag{54}$$

5.3.3. Stored energy function

We shall consider the rod made of hyperelastic material. To simplify we assume, the stored energy function [3, 11, 25] has the following form [24](Saint Venant and Kirchoff material):

$$w(\xi, \chi^i, \Gamma^j) = \frac{1}{2}\{GA_1(\Gamma^1)^2 + GA_2(\Gamma^2)^2 + EA(\Gamma^3 - 1)^2 + \tag{55}$$

$$EI_1(\chi^1)^2 + EI_2(\xi^2)^2 + GJ(\xi^3)^2\}$$

where $GA_i = GA_i(\xi)$, $EI_i = EI_i(\xi)$, $i = 1, 2$, $EA = EA(\xi)$, $GJ = GJ(\xi)$, are given coefficients.

5.3.4. Stress–strain relations

Using (41)–(55) we can formulate stress–strain relations in the current and reference configurations. These relations are given by [1, 2, 24, 45, 48, 49,

50],

$$N = \frac{\partial w(\xi, \chi^i, \Gamma^j)}{\partial \Gamma^i} E_i(\xi), \quad M = \frac{\partial w(\xi, \chi^i, \Gamma^j)}{\partial \chi^j} E_j(\xi \tag{56}$$

for $i, j = 1, 2, 3$.

5.3.5. *The equilibrium equations*

Let the rod be loaded by the resultant external force $\bar{n}$ as well as by the resultant external moment $\bar{m} = t_i \times f_i$, $f_i, i = 1, 2, 3$ given. The equilibrium equations take the form [1, 2, 24, 45, 48, 49, 50] :

Find vectors φ and T_i, $i = 1, 2, 3$ satisfying

$$n' + \bar{n} = 0 \text{ in } (0, L) \tag{57}$$

$$m' + \varphi' \times n + \bar{m} = 0 \text{ in } (0, L) \tag{58}$$

Moreover φ and t_i, $i = 1, 2, 3$ satisfy the boundary conditions (38), (39).

5.3.6. *Variational formulation*

We shall consider system (57), (58) with the boundary conditions (38), (39) in variational form. Let $E = R^3$. By K we denote the set of kinematically admissible deformation fields :

$$K = \{\{\varphi, t_1, t_2, t_3\} \in W^{1,p}(0, L; E^4) : \varphi(0) = \varphi_0, \varphi(L) = \varphi_L,$$
$$\forall i = 1, 2, 3 \ t_i(0) = l_i^0; \ \forall i, j = 1, 2, 3; \ \forall \xi \in [0, L]; \ t_i(\xi)t_j(\xi) = \delta_{ij}$$
$$\forall \xi \in [0, L] \ det(t_1(\xi), t_2(\xi), t_3(\xi)) > 0 \ \} \tag{59}$$

By $E(.,.) : K \to R$ we denote potential energy functional given by [2, 24, 45, 48] :

$$E(\varphi, t_i) = \int_0^L w(\xi, \chi^j(t_i), \Gamma^k(\varphi, t_i))d\xi - \int_0^L (\bar{n}\varphi + ft_i)d\xi \tag{60}$$

where $\bar{n} \in L^{p^*}(0, L; E)$ and $f \in L^{p^*}(0, L; E^3)$ are given, $p^* = \frac{p}{p-1}$.

The problem (57), (58) with boundary conditions (38), (39) is equivalent to the following optimization problem [2, 11, 12, 24, 25, 45, 48, 49, 50] :

Find $\{\varphi, t_i\} \in K$ such that :

$$E(\varphi, t_i) \leq E(z, d_i) \ \forall \{z, d_i\} \in K \tag{61}$$

Let us denote by $dK(\varphi, t_i)$ the space of kinematically admissible variations [24, 45, 48],

$$dK(\varphi, t_i) = \{\{\delta\varphi, \delta t_i\} \in W^{1,p}(0, L; E^4)$$
$$\delta\varphi(0) = \delta\varphi(L) = 0, \ \forall i = 1, 2, 3 \ \ \delta t_i(0) = 0 \tag{62}$$
$$\exists \theta \in W^{1,p}(0, L; E), \ \forall i = 1, 2, 3 \ \ \delta t_i = \theta \times t_i\}$$

$dK(\varphi, t_i)$ is the space of all tangent vectors to the set K at a point $\{\varphi, t_i\}$. $\delta(.)$ may be interpreted as a differentiation operator on K into a tangent direction to K [45, 48].

Lemma 5.1 *Problem (61) is equivalent to the following variational problem :*

$$Find \ \ \{\varphi, t_i\} \in K \ \ satisfying$$

$$\forall \{\delta\varphi, \delta t_i\} \in dK(\varphi, t_i) \ \ b(\varphi, t_i, \delta\varphi, \delta t_i) = l(\delta\varphi, \delta t_i) \tag{63}$$

where the form $b(., ., ., .) : [W^{1,p}(0, L; E^4)]^2 \to R$ *is given by :*

$$b(\varphi, t_i, \delta\varphi, \delta t_i) = \int_0^L (N\delta\Gamma + M\delta\chi)ds \tag{64}$$

where N and M are given by (45), (56). $\delta\Gamma$ and $\delta\chi$ are virtual variations of strains.

The existence of global solutions to the problem (63) is investigated in [24].

Since the problem (63) is one–dimensional from Antmans' results [1, 25] it follows that under the assumptions of Lemma 3.2 and if the stored energy function (55) is strictly convex with respect to φ' then the minimizer of the functional (60) satisfies also the system (57), (58) and is C^1 regular.

5.3.7. *The existence of local solutions*
Let us denote by $\tilde{\delta} : L^{p^*}(0, L) \to L^p(0, L)$ a differentiation operator at a point $\{\varphi, t_i\} \in K$ in a direction $\{\tilde{\delta}\varphi, \tilde{\delta}t_i\} \in dK(\varphi, t_i)$ defined as follows :

$$\tilde{\delta}[z(\varphi, t_i)] \stackrel{df}{=} \frac{\partial z}{\partial \varphi}\tilde{\delta}\varphi + \frac{\partial z}{\partial t_i}\tilde{\delta}t_i \tag{65}$$

By $a(., ., ., ., ., .) : [W^{1,p}(0, L; E^4)]^3 \to R$ we denote the linearized form of (64). The form a is given by :

$$a(\varphi, t_i, \tilde{\delta}\varphi, \tilde{\delta}t_i, \delta\varphi, \delta t_i) = \tilde{\delta}[b(\varphi, t_i, \delta\varphi, \delta t_i)](\tilde{\delta}\varphi, \tilde{\delta}t_i) \tag{66}$$

Let us calculate the form a given by (66), at a point $\varphi = \varphi_0$, $t_i = E_i$. First note, that [1, 2, 24, 45, 48] :

$$E_3 = \frac{d\varphi_0}{ds} \tag{67}$$

The linearized system (63) takes the form :

$$\text{Find } \{\tilde{\delta\varphi}, \tilde{\delta t_i}\} \in dK(\varphi_0, E_i) \text{ satisfying}$$

$$\forall\{\delta\varphi, \delta t_i\} \in dK(\varphi_0, E_i) \ a(\varphi_0, E_i, \tilde{\delta\varphi}, \tilde{\delta t_i}, \delta\varphi, \delta t_i) = l(\delta\varphi, \delta t_i) \qquad (68)$$

To investigate the existence of a solution to the system (68) we shall need :

Lemma 5.2 *There exists constant* $\alpha > 0$ *such that for all* $\{\delta\varphi, \delta t_i\} \in dK(\varphi_0, E_i)$ *the following condition holds :*

$$a(\varphi_0, E_i, \delta\varphi, \delta t_i, \delta\varphi, \delta t_i) \geq \alpha(\| \delta\varphi \|^2_{W^{1,2}(0,L)} + \| \delta t_i \|^2_{W^{1,2}(0,L)}) \qquad (69)$$

Proof : to be found in [35]

Lemma 5.3 *Problem (68) has a unique solution*
$\{\tilde{\delta\varphi}, \tilde{\delta t_i}\} \in dK(\varphi_0, E_i).$

Proof : follows from Lemma 3.3 and [11].

Proposition 5.1 *In a neighbourhood of* $\{\varphi_0, E_i\} \in K$, *the nonlinear system (57), (58) has a unique solution in* $\{\varphi, t_i\} \in K \cap W^{2,p}(0, L)$.

Proof: it is a consequence of the previous lemma and of the implicit function theorem as in three dimensional elasticity (see e.g. [11, 25]).

Details will be found in [35].

6. Shape optimization of the finite strain beam

An optimal design problem for hyperelastic rod consists in finding such reference configuration ϕ_0 of the rod to minimize -for example- its compliance [21, 20] measured as a distance between current and reference configurations occupied by the rod loaded by a given body force. The function ϕ_0 describing the position of a point on the line of centroids of cross–sections of the rod in its reference configuration is variable subject to optimization. In order to get easily existence of an optimal solution, it is assumed that this function is bounded and has bounded derivative. In literature most authors considered the shape optimization problems for linear elastic systems [21, 20, 51]. The aim of this section is to formulate the shape optimization problem for the highly nonlinear rod model. We have shown the existence of locally unique regular solution to the rod model as well as Lipschitz continuity of this solution with respect to the optimized variable. The design sensitivity analysis of the solution to the state system is performed and the directional derivative of the cost functional is calculated; a detailed proof will be found in [35]. The calculated derivative may be employed as an element of descent direction finding procedure for solving numerically this optimization problem; a detailed numerical study will be found in [37]; this derivative may also be used to derive necessary optimality conditions, see [35].

6.1. FORMULATION OF THE OPTIMIZATION PROBLEM

Let $\varphi_0 = \varphi_0(\xi)$ defined by (37) be the variable subject to optimization. φ_0 is assumed to satisfy :

$$c_1 \le \varphi_0(\xi) \le c_2, \quad | \frac{d\varphi_0}{d\xi} | \le c_3, \quad | \frac{d^2\varphi_0}{d\xi^2} | \le c_4 \tag{70}$$

and the constant volume condition [21, 20]:

$$\int_0^l A(\xi)\varphi_0{}'(\xi)d\xi = c_5 \tag{71}$$

where c_1, c_2, c_3, c_4, c_5 are given positive constants. By U_{ad} we denote the set of admissible designs :

$$U_{ad} = \{\varphi_0 \in C^{2,1}(0, L; E) : \varphi_0 \text{ satisfies (70) and (71)}\} \tag{72}$$

$C^{2,1}$ denotes a space of Lipschitz continuous functions having Lipschitz continuous first and second order derivatives [20]. U_{ad} is assumed to be nonempty. In order to underline the dependence of the solution $\{\varphi, t_i\} \in K \cap W^{2,p}(0, L)$ to the system (63) on $\varphi_0 \in U_{ad}$ we shall write :

$$b(\varphi_0, \varphi, t_i, \delta\varphi, \delta t_i) \text{ for } b(\varphi, t_i, \delta\varphi, \delta t_i) \tag{73}$$

$$l(\varphi_0, \delta\varphi, \delta t_i) \text{ for } l(\delta\varphi, \delta t_i) \tag{74}$$

The problem (63) takes the form :

For given $\varphi_0 \in U_{ad}$, find $\{\varphi, t_i\} \in K$ satisfying

$$b(\varphi_0, \varphi, t_i, \delta\varphi, \delta t_i) = l(\varphi_0, \delta\varphi, \delta t_i) \ \forall\{\delta\varphi, \delta t_i\} \in dK(\varphi, t_i) \tag{75}$$

From proposition 5.1 it follows that for a given point $\varphi_0 \in U_{ad}$, problem (75) has a unique solution $\{\varphi, t_i\} \in K \cap W^{2,p}(0, L)$, $p \ge 2$ in a neighbourhood of the linearization point $\{\varphi_0, E_i\} \in K$. We shall consider the following optimization problem :

Find $\varphi_0 \in U_{ad}$ minimizing the cost functional :

$$J(\varphi_0; \varphi, t_i) = \int_0^L \{(\varphi - \varphi_0)^2 + (t_i - E_i)^2\}ds \qquad (76)$$

on the set U_{ad}.where$\{\varphi, t_i\} \in K \cap W^{2,p}(0, L)$ is a

solution to the system (75).

The cost functional (76) is given as an example, the compliance of the rod would be an other simple one (see [21, 20]).

Lemma 6.1 $\{E_i\}$, $i = 1, 2, 3$, are Lipschitz continuous with respect to $\varphi_0 \in U_{ad}$.

Proof : Since $\varphi_0 \in C^{2,1}(0, L; E)$ and $\{E_i\}, i = 1, 2, 3$, are orthonormal vectors from (67) it follows that E_i, $i = 1, 2, 3$ are Lipschitz continuous functions with respect to $\varphi_0 \in U_{ad}$.

Let us assume :

$$\varphi, t_i, i = 1, 2, 3 \text{ are Lipschitz continuous}$$
$$\text{functions with respect to } \varphi_0 \in U_{ad} \qquad (77)$$

Lemma 6.2 If (77) is satisfied then there exists an optimal solution $\hat{\varphi}_0 \in U_{ad}$ to the problem (76).

Proof : From (77) follows the continuity of the cost functional (76) with respect to $\varphi_0 \in U_{ad}$. The set U_{ad} given by (72) is compact in C^2 topology. Hence by Weierstrass Theorem follows the existence of an optimal solution $\hat{\varphi}_0 \in U_{ad}$ to the problem (76).

We shall calculate the directional derivative of the cost functional (76). For the sake of simplicity we shall write for the cost functional (76) :

$$j(\varphi_0) = J(\varphi_0, \varphi, t_i) \qquad (78)$$

Lemma 6.3 If the Assumption (77) is satisfied then the directional derivative $dj(\varphi_0, \delta\varphi_0)$ of the cost functional (76) at a point $\varphi_0 \in U_{ad}$ in the direction $\delta\varphi_0 \in U_{ad}$ is given by :

$$dj(\varphi_0, \delta\varphi_0) = 2\int_0^L \{(\varphi - \varphi_0)(\frac{\partial\varphi}{\partial\varphi_0} - 1)\delta\varphi_0 + \qquad (79)$$

$$(t_i - E_i)(\frac{\partial t_i}{\partial\varphi_0} - \frac{\partial E_i}{\partial\varphi_0})\delta\varphi_0\}ds$$

$$+ \int_0^L [(\varphi - \varphi_0)^2 + (t_i - E_i)^2]E_3\frac{d\delta\varphi_0}{ds}ds \qquad (80)$$

Proof : Note [10, 26, 45] that,

$$ds = \| \frac{d\varphi_0}{d\xi} \| d\xi \qquad (81)$$

Hence

$$\frac{\partial}{\partial \varphi_0}(ds)\delta\varphi_0 = (E_3 \cdot \frac{d\delta\varphi_0}{ds})ds \qquad (82)$$

and from the use of the chain rule, follows (80).

In order to calculate the derivative (80) we have to verify the condition (77) as well as to calculate the derivatives $\frac{\partial \varphi}{\partial \varphi_0}$ and $\frac{\partial t_i}{\partial \varphi_0}$ of the solution $\{\varphi, t_i\} \in K \cap W^{2,p}(0, L)$ to the state system (75) and the derivatives $\frac{\partial E_i}{\partial \varphi_0}$ with respect to $\varphi_0 \in U_{ad}$. We shall do it in the next subsection.

6.2. DESIGN SENSITIVITY ANALYSIS OF SOLUTIONS TO THE STATE SYSTEM

Recall from subsection 5.3.7 that $\{\varphi, t_i\} \in K \cap W^{2,p}(0, L)$ is a unique local solution to the nonlinear system (75) in a neighbourhood of the reference configuration $\{\varphi_0, E_i\} \in K$. In this linearization point (69) holds.

Assuming that the solution $\{\varphi, t_i\} \in K \cap W^{2,p}(0, L)$ is sufficiently close to the linearization point $\{\varphi_0, E_i\} \in K$, from Stopelli Theorem [25] it follows that the condition (69) still holds at a point $\{\varphi, t_i\} \in K \cap W^{2,p}(0, L)$, i.e., we have :

$$a(\varphi_0, \varphi, t_i, \delta\varphi, \delta t_i, \delta\varphi, \delta t_i) > 0 \qquad (83)$$

for all $\{\delta\varphi, \delta t_i\} \in dK(\varphi, t_i)$ where a is given by (66). Using (83) we are able to prove :

Lemma 6.4 If $\{\varphi, t_i\} \in K \cap W^{2,p}(0, L)$ is a unique local solution to the system (75) then there exists the Frechet derivative of the mapping

$$U_{ad} \ni \varphi_0 \to (\varphi, t_i) \in K \qquad (84)$$

at a point $\{\varphi, t_i\} \in K \cap W^{2,p}(0, L)$ in a direction $\delta\varphi_0 \in U_{ad}$.

Proof : From Proposition 5.1 and [25] it follows that a point $\{\varphi, t_i\} \in K \cap W^{2,p}(0, L)$ is a unique solution to the system (75). Let us calculate the derivatives of the form (73) with respect to φ and t_i, respectively, at a point $\{\varphi, t_i\} \in K \cap W^{2,p}(0, L)$ in a direction $\{\tilde{\delta}\varphi, \tilde{\delta}t_i\} \in dK(\varphi, t_i)$; it is the bilinear form a, see (66).

From Lemma 5.2, it follows positive definiteness of this form. Hence by implicit function theorem [11, 25, 51] follows the Frechet differentiability of the mapping (84).

Note that from Lemma 6.4, it follows the Assumption (77) is satisfied.

6.3. NUMERICAL RESULTS

The numerical methodology for the analysis of the beam deformation uses previous results ([24],[49]); we have used an incremental method of predictor corrector type with use of archlength method to pass the turning points;

these algorithms are used with an approximation of the principle of virtual work with finite elements of degree one. The sensitivity computation is implemented using detailed formulas (see [37])based on results of this section; these formulas are numerically computed with the finite element approximation used in the analysis of the beam; we follow the lines of [44]. The optimization algorithm used is steepest descent method; it is well known for its slow convergence and ease of implementation; other experiments are in progress: here we present only plane problems. Next we describe the figures 7 and 8.

- a beam clamped at one end and with a point load at the other end; in this example the design variables are the vertical components of the nodes of the middle line; the initial shape is a straight segment.
- an arch clamped at both ends with snow load; this example is analysed with a linear model in [19]; we have considered it with our geometrically non linear model for increasing values of the load: $10., 10^3, 10^4, 2 \times 10^4$; the initial shape is a parabola $y = (x - 0.5)(x + 0.5)$; in the figure 7 the following curves have been drawn:

  - the parabola before and after loading with 2 of the load cases;
  - the parabola and the optimized shape with 2 of the load cases;
  - the parabola and two of the loaded optimized shapes;

  the first load case is identical to the one used in [19]; we notice that the optimized shapes are very close; despite the fact that the design variable are different: here position of finite element nodes and spline nodes in [19]; also the mechanical model includes here transverse shearing and not in [19].

- an arch clamped at both ends and loaded in the middle by an upward point load:

  - to validate the software we have first considered the case where the arch is approximated with only 2 elements; in this case we can obtain an explicit formula for the elastic energy and its optimal value with respect to the second coordinate of the mid-point; our numerical computation gives this value up to 10^{-6}! this is not presented in a figure;
  - a more precise computation has been performed with 26 elements for increasing values of the load: $10., 10^3, 10^4, 2.10^4$; in figure 8 we have drawn:

    * the parabola before and after loading with 2 of the load case;
    * the parabola and the optimized shape for 2 of the load cases;
    * two initial shapes loaded with 2.10^4 and the shapes obtained by optimization ; in this case we obtain a shape of arch which

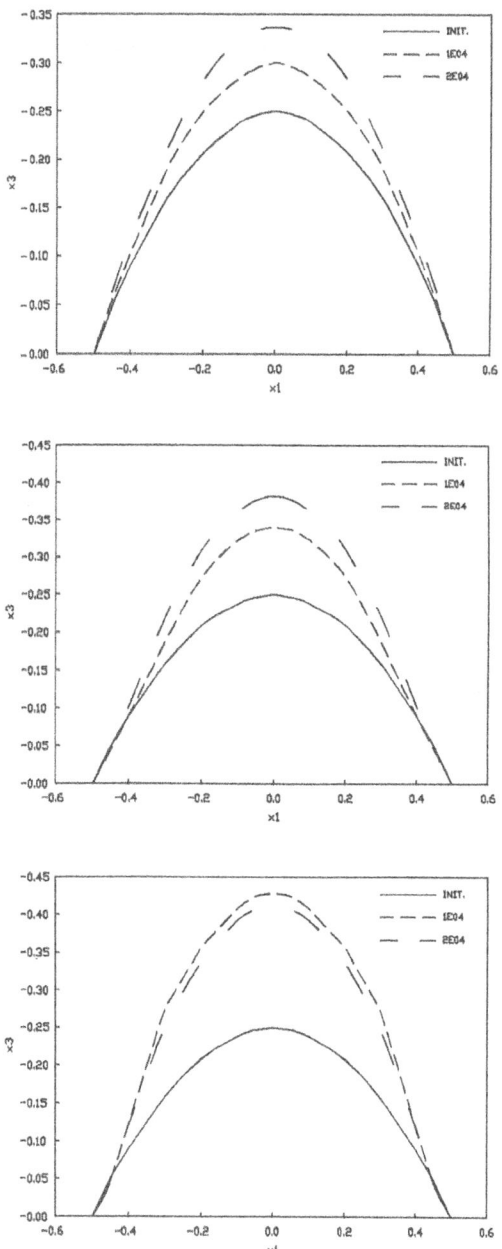

Figure 7. Arch clamped at both ends with snow load

is not stable; the adjoint state is not solvable; we have reached the limit of the use of our theoretical approach; improvements are in progress as explained previously.

7. Shape sensitivity for shells in large displacement

We sketch here how to perform this shape sensitivity for the model presented at the beginning. This case is obviously more difficult than the finite strain beam of section 5; in its full generality it seems considered here for the first time; the detailed mathematical proof is not yet completed as for the beam; this is just a first step to optimize the shape of geometrically nonlinear shells. The basic idea is to use intrinsic notation as presented in second section; in numerical implementation this will amount to use a fixed frame instead of a moving frame as in [5].

An important task as explained in section 4 is to compute $\frac{\partial G}{\partial \phi}$ and $\frac{\partial G}{\partial \phi_0}$. the first quantity is computed when one solves the equation with a Newton method; so we consider only the second quantity.

To compute $\frac{\partial G}{\partial \phi_0}$ we need to compute $\Delta \Gamma$ and $\Delta \mathcal{K}$ where Δ means sensitivity with respect to the reference midsurface:

$$\Delta \Gamma = \frac{\partial \Gamma}{\partial m_0} \Delta m_0, \quad \Delta \mathcal{K} = \frac{\partial \mathcal{K}}{\partial m_0} \Delta m_0 \tag{85}$$

Lemma 7.1 let Γ and $\mathcal{K}$ be defined in (11),(12) ; let m be fixed and assume that m_0 is perturbed with a vector field ψ: $M_0 = m_0 + \psi(m_0)$, then we have:

$$\Delta \Gamma = -(\overline{\Pi \frac{\partial \psi}{\partial m_0} \Gamma + \Gamma \Pi \frac{\partial \psi}{\partial m_0}}) - \frac{1}{2}(\overline{\Pi \frac{\partial \psi}{\partial m_0} + \Pi \frac{\partial \psi}{\partial m_0}}) \tag{86}$$

$$\delta \mathcal{K} = -(\overline{\Pi \frac{\partial \psi}{\partial m_0} \mathcal{K} + \mathcal{K} \Pi \frac{\partial \psi}{\partial m_0}}) - (\overline{\Pi \frac{\partial \psi}{\partial m_0} \frac{\partial N}{\partial m_0} + \Pi \frac{\partial \delta N_0}{\partial m_0}}) \tag{87}$$

with

$$\Delta N_0 = -\overline{\Pi \frac{\partial \psi}{\partial m_0}} N_0 \tag{88}$$

Comments and proof.

One should notice the similarity of structure of the second term of the right hand side of (86,87) with the expression of $\delta \gamma$ and $\delta \chi$ of (2); the first term of the right hand side comes from the fact that $\Gamma(M_0)$ and $\Gamma(m_0)$ are quadratic forms on different tangent spaces. We sketch the proof only for Γ; as noted in (13, 14), we have:

$$\overline{dM_0} \Gamma(M_0) dM_0 = \frac{1}{2}(\overline{dm\,dm} - \overline{dM_0\,dM_0}) \tag{89}$$

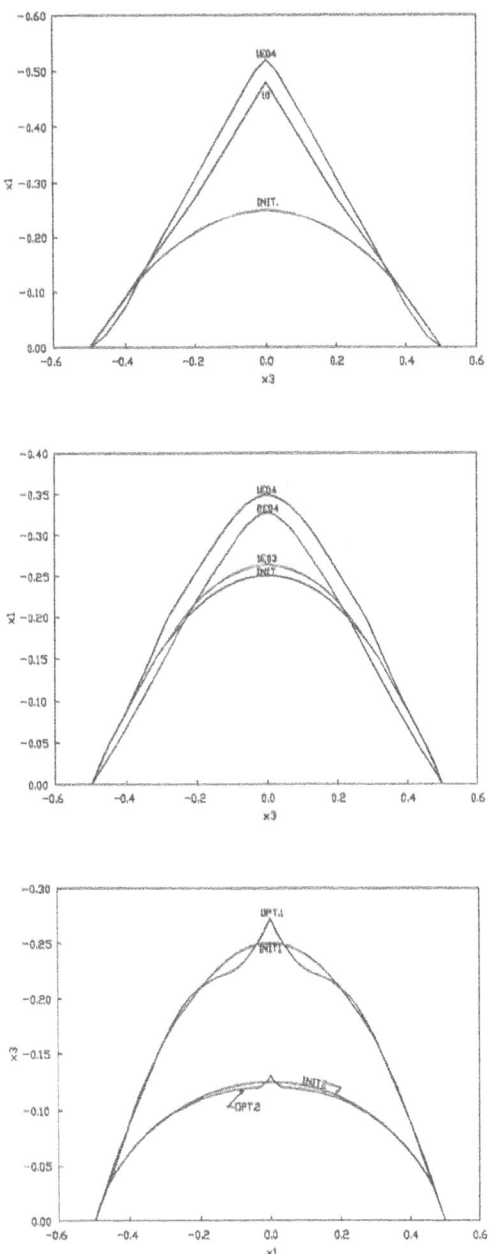

Figure 8. Arch clamped at both ends and loaded in the middle

and

$$\overline{dm_0\Gamma(M_0)dm_0} = \frac{1}{2}(\overline{dmdm} - \overline{dm_0dm_0}) \tag{90}$$

as

$$M_0 = m_0 + \psi(m_0) \quad dM_0 = dm_0 + \frac{\partial\psi(m_0)}{\partial m_0}dm_0 \tag{91}$$

we get:

$$\overline{dM_0dM_0} = \overline{dm_0dm_0} + \overline{dm_0\Pi\frac{\partial\psi}{\partial m_0}dm_0} + \overline{dm_0\Pi\frac{\partial\psi}{\partial m_0}dm_0} + o(\frac{\partial\psi}{\partial m_0}) \tag{92}$$

and

$$\overline{dM_0\Gamma(M_0)dM_0} = \overline{dm_0\Gamma(m_0)dm_0} +$$

$$\overline{dm_0\Gamma\Pi\frac{\partial\psi}{\partial m_0}dm_0} + \overline{dm_0\Pi\frac{\partial\psi}{\partial m_0}\Gamma dm_0} + \overline{dm_0\Delta\Gamma dm_0} + o(\frac{\partial\psi}{\partial m_0}) \tag{93}$$

We obtain (86), by combining (93), (92) and (89), (90). The computation of $\Delta\delta\Gamma$ and $\Delta\delta K$ is similar to expressions used in Newton's method, up to similar modifications.

These formulas should be used with general formulas for shape sensitivity of surface systems (see for example [46]):

$$\Delta(d\Sigma_0) = div_{\Sigma_0}\Delta m_0 d\Sigma_0 \tag{94}$$

where div_{Σ_0} is the tangential divergence of the reference surface. The details of mathematical proof and its numerical implementation are under study.

It is worth emphasizing the simplicity of formulas (86, 87); one should compare with formulas obtained in [5] and in subsection 4.4 for shells of revolution.

8. Conclusion

We have presented a general approach to design midsurface of shells and emphasize the case of designing meridian line of shells of revolution. For shells in linear elasticity, this approach is completely justified by mathematical arguments and yields formulas which may be systematically implemented in connection with finite elements approximation. The implementation for general shells is also in progress [6]. Midsurface has been also considered for eigenfrequencies ([28]). For shells in linear elasticity we have only presented the first steps; but it is clear that the use of intrinsic notations provides much clear and simple notations; the mathematical proof however is still in preparation.

As a first step, the shape optimization problem for nonlinear rod was formulated; here the situation is more complete; the existence of solutions to the state system was investigated. The differentiability of solutions to the state system with respect to the function describing the position of the line of centroids in the reference configuration of the rod was shown. The directional derivative of the cost functional was calculated. For details concerning the finite element approximation of the systems (75) and the application of Augmented Lagrangian approach or Newton algorithm for numerical solving of these systems see [24, 49]. Note that the calculated directional derivative has been used in numerical optimization algorithm as an element of a descent direction finding procedure. The optimization of a geometrically nonlinear beam with inextensibility constraints will be considered in [18].

References

1. S.S. Antman, Ordinary Differential Equations of One Dimensional Elasticity, Archive for Rational Mechanics and Analysis, vol. 61, pp. 307 – 393, 1976.
2. S.S. Antman, C.S. Kenney, Large Buckled States of Nonlineary Elastic Rods under Torsion, Thrust and Gravity, Archive for Rational Mechanics and Analysis, vol.76, pp. 289 – 338, 1981.
3. J.M. Ball, Convexity Conditions and Existence Theorems in Nonlinear Elasticity, Archive for Rational Mechanics and Analysis, vol. 63, pp. 337 – 403, 1977.
4. M. Bernadou, M. Boisserie, The finite element method in thin shell theory: application to arch dam simulation, Birkhauser, Boston, 1982.
5. M. Bernadou, F.J. Palma, B. Rousselet, Shape optimization of an elastic thin shell under various criteria, J. Structural optimization, vol 3, pp 7-21 (1991).
6. M. Bernadou, F.J. Palma, B. Rousselet, Shape optimization of an elastic thin shell under various criteria,numerical computations, in preparation.
7. Braibant-Fleury (1984). Shape optimal design using B-splines, Comp. Meth. in appl. mech. and eng., vol 44, pp 247-267.
8. Céa J, Gioan A., Michel J., Quelques résultats sur l'identification de domaines, Calcolo, III-IV, 1973.
9. Céa J. , Conception optimale ou identification de domaines: calcul rapide de la derivée directionnelle de la fonction cout, Math. modeling & numerical analysis, vol. 20, 1986, pp 371-402 .
10. H. Cartan, Formes Differentielles, Hermann, Paris, 1967.
11. Ph. Ciarlet, Mathematical Elasticity. Vol 1: Three Dimensional Elasticity, North–Holland, Amsterdam, 1988.
12. J. Cea, Optimisation.Theorie et Algorithmes, Dunod, Paris, 1971.
13. Chenais D., Sur une famille de variétés à bord lipschtziennes: une application à un problème d'identification de domaines, Annales de l'Institut fourier (1977), 27, pp 201-231.
14. D. Chenais, optimal design of mid surface of shells: differentiability proof and sensitivity computation, J. APPL. Math. Opt. vol 16, pp 93-133, 1987.
15. D.Chenais, B. Rousselet, Dependance of the Buckling Load of a Nonshallow Arch with Respect to the Shape of its Midcurve, Mathematical Modelling and Numerical Analysis, vol.24, pp. 307 – 341, 1990.
16. Chenais-Rousselet-Benedict (1988). Design sensitivity for arch structures with respect to midsurface shape under static loading, J.O.T.A., vol 58, pp 225-239.

17. A. Gajewski, M. Zyczkowski, Optimal structural design under stability constraints, Kluwer Ac. Publ., 1988.

18. P. Gosling, B.Rousselet, Thickness optimization of a geometrically non linear beam with inextensibility constraint, in preparation.

19. A. Habbal, Theoretical and numerical study of nonsmooth shape optimization applied to the arch problem, Mech. struct.& mach., 20(1), 93-117 (1992).

20. J. Haslinger, P. Neittaanmaki, Finite Element Approximation for Optimal Shape Design. Theory and Applications., John Wiley and Sons, Chichester, 1988.

21. E.J. Haug, K.K. Choi, V. Komkov, Design Sensitivity Analysis of Structural Systems, Academic Press, New York, 1986.

22. W.T. Koiter. On the nonlinear theory of thin elastic shells, Proc. Kon. Ned. Akad. Wetensch. B69, pp1-54, 1966.

23. W.T. Koiter. On the foundations of the linear theory of thin elastic shells, Proc. Kon. Ned. Akad. Wetensch. B73, pp169-195, 1970.

24. P. LeTallec, S. Mani, F.A. Rochinha, Finite Element Computation of Hyperbolic Rods in Large Displacement, Mathematical Modelling and Numerical Analysis, vol.26, pp. 595 – 625, 1992.

25. J.E. Marsden, T.J.R. Hughes, Mathematical Foundations of Elasticity, Prentice - Hall, New Yersey, 1983.

26. J.E. Marsden, A.J. Tromba, Vector Calculus, W.H. Freeman and Company, New York, 1988.

27. Masmoudi M., Outils pour la conception optimale de formes, thèse d'Etat, Université de Nice, 1987.

28. S. Mehrez, Analyse et optimisation de forme de coque mince de révolution, thèse de l'Université de Nice, 1990.

29. S. Mehrez, B. rousselet, Analysis and optimization of a shell of revolution, in Computer aided optimum design of structures: applications, C. Brebbia & S. Hernandez, eds., Springer, 1989.

30. S. Moriano(1988) Optimisation de forme de coques, thèse de l'Université de NIce.

31. Z. Mroz, Design sensitivity of critical loads and vibrational frequencies of nonlinear structures, in Optimization of large structural systems G. Rozvany ed., pp 455-476, Kluwer Ac. Publ., 1993.

32. Murat F. & Simon J., Etude de problèmes d'optimum design, Proceedings of the 7th IFIP Conf., Springer Verlag, Lecture Notes in Computer Sciences, n 41, 1976, pp 54-62.

33. A. Myslinski, Shape Optimization of a Nonlinear lliptic System, Kybernetika Vol. 29, No 3, 1993, pp. 270 - 283.

34. A. Myslinski, J. Piekarski, B. Rousselet, Poutre courbe en grands deplacemnts; sensitivité par rapport à la ligne moyenne, in Colloque national en calcul de structures (11-14 mai 1993), Hermes pp 788-801, 1993 .

35. A. Myslinsli, B. Rousselet, Shape sensitivity of a hyperelastic rod, in preparation.

36. Palma Molina (1988). Module BSPLIN : manuel d'utilisation et deréférence. Rapport INRIA No 96 .

37. J. Piekarski, B. Rousselet, Computation of shape sensitivity of a hyperelastic rod, in preparation.

38. Pironneau O., Optimal shape design for elliptic systems, Springer series in computational physics, Springer Verlag 1984.

39. Rochette M., Conception optimale de formes appliquée aux résistances ajustables, thèse, Université de Nice Sophia-Antipolis, 1990.

40. Rousselet B., Etude de la régularité desz valeurs propres par rapport à des déformations bilipschitziennes du domaine, CRAS 283, série A, 1976, p 507.

41. Rousselet B., Shape design sensitivity of a membrane, pp 595-623, JOTA, 1983;

42. Rousselet B., Quelques résultats en optimisation de domaines, thèse d'état, Université de Nice, 1982.

43. Rousselet B., Shape optimization of structures with state constraints, pp 255-264 in

412

"Control of partial differential equations", A. Bermudez ed; Lect. notes in control and information sciences, vol. 114, Springer Verlag, 1989.

44. B. Rousselet, Shape design sensitivity, from partial differential equation to implementation, pp 151-171, Eng. Opt., vol 11, 1987

45. B. Rousselet, A Finite Strain Rod Model and its Design Sensitivity, Mechanical Structures and Machinery, vol.20, No 4, 1992.

46. B. Rousselet, Introduction to shape sensitivity; three dimensional and surface systems; in Optimization of large structural systems G. Rozvany ed., pp 397-432, Kluwer Ac. Publ., 1993.

47. Rousselet-Mehrez-Marro-Giuliano-Gauthier (1990) Developpement logiciel en optimisation de structures; rapport de synthèse DRET .

48. J.C. Simo, A Finite Strain Beam Formulation. The Three Dimensional Elastic Problem. Part I., Computer Methods in Applied Mechanics and Engineering, vol.49, pp. 50 – 70, 1985.

49. J.C. Simo, L. Vu-Quoc, A Three Dimensional Finite Strain Rod Model. Part II : computational aspects, Computer Methods in Applied Mechanics and Engineering, vol. 58, pp. 79 - 116, 1986.

50. J.C. Simo, J.E. Marsden, P.E. Krishnaprasad, The Hamiltonian Structure of Nonlinear Elasticity : The Material and Convective Representations of Solids, Rods and Plates, Archive for Rational Mechanics and Analysis, vol.104, pp. 125 – 183, 1988.

51. J. Sokolowski, J.P. Zolesio, Introduction to Shape Optimization. Shape Sensitivity Analysis, Springer Series in Computational Mathematics, vol. 16, Springer, Berlin 1992.

52. Timoshenko S.P. and Woinowski-Krieger, K., theory of plates and shells, second edition, Macgraw-Hill, 1959.

53. R. Valid, La mécanique des milieux continus et le calcul des structures, Eyrolles, Paris, 1977; Mechanics of continuous media and analysis of structures, North-holland, 1981.

54. R. Valid, Fondements de la théorie des coques: une présentation surfacique simple. J. meca. th. appl., vol 7, 1988, pp 135-156.

55. M. Zyczkowski, Recent advances in optimal structural design of shells, Eur. J. Mech., A/Solids,11, Special isssue, 5-24, 1992.

APPLICATION OF AUTOMATIC DIFFERENTIATION TO OPTIMAL SHAPE DESIGN

M. MASMOUDI
CERFACS, 42 Avenue G. Coriolis
31057 Toulouse cedex, France

AND

PH. GUILLAUME AND C. BROUDISCOU
Mathématiques pour l'Industrie et la Physique
Equations aux Dérivées partielles et Modélisation
Unité Mixte de Recherches CNRS - Université Paul Sabatier
Toulouse 3
UFR MIG, 118 route de Narbonne, 31062 Toulouse cedex,
France

1. Introduction

Automatic differentiation (AD) allows the differentiation of a function defined by its program (in Fortran, C, ADA, ...). Let f be a map from $D(f) \subset \mathbb{R}^n$ to $\mathbb{R}^m$. The AD process can be represented by the following scheme (fig. 1):

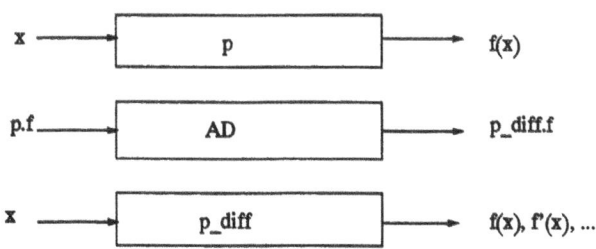

Figure 1. the AD scheme

413

J. Herskovits (ed.), Advances in Structural Optimization, 413–446.

If f is a composition of basic operations ($\times, /, +, -$, cos, sin, ...), the derivative of f can be computed using a few fundamental rules.

Unlike the derivative obtained by numerical methods (finite difference), the so obtained derivative is exact.

Symbolic differentiation used to be the only alternative to numerical differentiation if one wanted to avoid truncation errors. But symbolic differentiation leads to very complicated expressions with respect to the original function; a typical example is given in [Gri 89]. Moreover, symbolic differentiation needs a symbolic representation of the function to differentiate and can not process an existing fortran program. The application domain of AD seems to be far more general.

The AD was introduced for practical reasons. If we use a descent method or the Newton method to solve an optimization problem, AD can be used for computing the gradient and/or the Hessian exactly; for solving a system of nonlinear equations or a nonlinear least square problem we generally linearize the equations to use a variant of the Newton method.

The validation of numerical results is another important domain of application of AD. The rounding error estimate and analysis can be done by computing the derivative of the result with respect to any variable.

In shape optimization problems there are two well known methods: the direct method and the adjoint method. In both cases, AD can be used to compute the linearized direct problem or the adjoint problem. Then the cost of implementation of shape optimization methods in industrial softwares can be reduced.

In the field of shape optimization, the interest of AD was pointed out first by O. Pironneau in 1992, but the application of AD in this field remains rare. Automatic differentiation is used in many fields like atmospheric dynamics [Tal 91] [RoD 92], Astrodynamical Modelling [KaL 91], Satellite Simulation [Lay 91], ...

A researcher in shape optimization can understand easily that in AD we find also the two basic modes: the direct mode and the adjoint mode.

The natural mode called forward mode, bottom up mode, direct mode,... was the first to be introduced, apparently by R.E. Wengert [Wen 64].

The more sophisticated mode called reverse mode, top down mode, adjoint mode, backward mode, ...was published by G.E. Ostrovskii and al. [OWB 71]. B. Speelpenning gave in his dissertation [Spe 80] an implementation of the reverse mode.

The time complexity and the space complexity of AD algorithms have been studied by several authors [BaS 83], [Mor 85]. It seems that the paper of Strassen [BaS 83] was the first to point out the efficiency of the reverse mode when the number of variables is important. But similar results are well known in the field of shape optimization.

We shall give an overview on AD in section 2, and we shall show in section 3 how to apply AD to shape optimization problems. In section 4 we present the *high order derivative method* introduced by the authors, and some numerical results are given in sections 5 and 6.

2. Automatic Differentiation

This section is an adaptation of the work of J.C. Gilbert [Gil 91] to our context. We first introduce some models for computational programs and we apply the forward and the reverse modes to those models.

2.1. A FUNCTION DEFINED BY ITS PROGRAM

Let f be a map from $D(f) \subset \mathbb{R}^n$ to $\mathbb{R}^m$. We assume that $f(x)$ is the composition of a finite number of basic operations ($\times, /, +, -$, cos, sin, ...). We call $x = (x_1, \ldots, x_n)$ the *independent variables*. We may also use *intermediate variables*.

2.1.1. *A first model*
For instance, the computation of

$$f(x) = \frac{\cos(x_1 + x_2) + x_3}{1 + x_1 x_3 \exp(x_2)}$$

can be splitted in the following way:

$$x_4 := x_1 + x_2$$
$$x_5 := \cos(x_4)$$
$$x_6 := x_5 + x_3$$
$$x_7 := x_1 \times x_3$$
$$x_8 := \exp(x_2)$$
$$x_9 := x_7 \times x_8$$
$$x_{10} := x_9 + 1$$
$$f = x_{11} := x_6 / x_{10}$$

In this *straight-line algorithm*, we introduce one intermediate variable at each instruction. If the number of instructions is K, we use $N = K + n$ variables (n independent and K intermediate variables), the appropriate model is

$$\begin{cases} for\ i := n+1\ to\ N\ do\ x_i := \varphi_i(x_1, \ldots, x_{i-1}) \\ f := x_N \end{cases}$$

Usually the function φ_i depends only on a few number of the parameters x_j, $j < i$. We denote by $S_i \subset \{1, \ldots, i-1\}$ the set of indices corresponding to those parameters and we obtain an improved model

$$\text{(M1)} \qquad \begin{cases} for\ i := n+1\ to\ N\ do\ x_i := \varphi_i(x_{S_i}) \\ f := x_N. \end{cases}$$

2.1.2. *A more general model*

But the previous model is not general. For instance, in a scalar product program we redefine the intermediate variable p.

$$\begin{cases} p := 0 \\ for\ i := 1\ to\ n\ do\ p := p + x_i * a_i \\ f := p \end{cases}$$

If we allow the redefinition of independent and intermediate variables we obtain the following model (using the notations introduced in the previous section).

$$\text{(M2)} \qquad \begin{cases} for\ k := 1\ to\ K\ do\ x_{\mu_k} := \varphi_k(x_{S_k}) \\ f := x_N \end{cases}$$

For this model we need the consistency assumption: a variable x_j must be defined before it is used as a parameter of φ_k.

2.1.3. *The intermediate functions*

The functions φ_k must be regular in a neighbourhood of x_{S_k}. In most of the actual AD softwares, if the function φ_k depends on more than two parameters, it is splitted automatically. We first recall the direct method in shape optimization.

2.2. THE DIRECT METHODS

2.2.1. *The direct method in shape optimization*
Let

$$A(x)X(x) = B(x) \qquad (1)$$

be a linear system obtained by the finite elements discretization of a linear static structure. The matrix $A(x)$ is the stiffness matrix and $B(x)$ is the load. In this system $x = (x_1, \dots, x_n)$ is a parameter vector (shape, material properties, ...) and $X(x)$ is the displacement vector for a given x. We have to minimize a cost function

$$J(x, X(x)) \qquad (2)$$

and to compute the total derivative of $f(x) = J(x, X(x))$. If f is differentiable, its derivative is given by:

$$d_x f(x) = \partial_x J(x, X(x)) + \partial_X J(x, X(x)) d_x X(x) \qquad (3)$$

where d_x is the total derivative with respect to x, and ∂_u denotes the partial derivative with respect to u.

The derivative $d_x X(x)$ itself is obtained by solving the systems

$$A(x)\partial_{x_i} X(x) = \partial_{x_i} B(x) - \partial_{x_i} A(x) X(x) \quad 1 \leq i \leq n. \qquad (4)$$

We shall see in the following section that the forward mode of AD is similar to the direct mode of shape optimization: we just have to differentiate the equations of the model.

2.2.2. *The forward mode in AD*

We come back to AD and we denote by $d_x x_i$ for $1 \leq i \leq N$ the derivative of an independent or an intermediate variable x_i with respect to the independent variables $x_1, \ldots, x_n$. We have N gradient vectors and each vector has n components. It is obvious that

$$d_x x_i = \epsilon_i \text{ for } 1 \leq i \leq n$$

where $e_i = (0, \ldots, 0, 1.0, \ldots, 0)$ is the ith element of the canonical base of $\mathbb{R}^n$.

For $1 \leq k \leq K$, using the chain rule and the definition of x_{μ_k}, we obtain

$$\left\{ \begin{array}{l} for\ i := 1\ to\ n\ do\ d_x x_i = e_i \\ for\ k := 1\ to\ K\ do\ \{ \\ \quad d_{x_i} x_{\mu_k} := \sum_{j \in S_k} \partial_{x_j} \varphi_k \partial_{x_i} x_j \quad 1 \leq i \leq n \\ \quad x_{\mu_k} := \varphi_k(x_{S_k}) \\ \} \\ \partial_{x_i} f := \partial_{x_i} x_N \quad 1 \leq i \leq n \\ f := x_N \end{array} \right.$$

It appears that we just have to insert instructions related to AD in the initial program. In order to reduce the memory storage, we can compute the derivative with respect to only one variable at a time, but we increase the computation time by computing f n times with the same data:

$$\begin{cases} for\ i := 1\ to\ n\ do\ \{ \\ \quad for\ l := 1\ to\ n\ do\ u_l := 0 \\ \quad u_i := 1 \\ \quad for\ k := 1\ to\ K\ do\ \{ \\ \qquad u_{\mu_k} := \sum_{j \in S_k} \partial_{x_j} \varphi_k\, u_j \\ \qquad x_{\mu_k} := \varphi_k(x_{S_k}) \\ \quad \} \\ \quad \partial_{x_i} f := u_N \\ \quad f := x_N \\ \} \end{cases}$$

where $u_k = \partial x_k / \partial x_i$. This algorithm is quite similar to the numerical differentiation algorithm, but gives exact results. In both cases (in shape optimization and in AD), it is clear that the computation time is linear with respect to the number of parameters. For this reason, adjoint methods were introduced.

2.3. THE ADJOINT METHODS

2.3.1. *The adjoint method in shape optimization*
We first apply the well known adjoint method to the model problem (eq. (1) and (2) in section 2.2.1). We consider the Lagrange operator

$$L(x, X, P) = J(x, X) + (A(x)X - B(x), P)$$

where $P \in \mathbb{R}^K$ is the Lagrange multiplier; to each equation is associated a component of P.

If $X = X(x)$ is the solution of the linear system, we have

$$J(x, X(x)) = L(x, X(x), P) \quad \forall P \in \mathbb{R}^K.$$

Then (using $d_x P = 0$), the total derivative of f is given by

$$d_x J(x, X(x)) = \partial_x L(x, X(x), P) + \partial_X L(x, X(x), P) d_x X(x). \qquad (5)$$

In order to eliminate the unknown derivatives $d_x X$, the vector $P = P_X$ can be chosen in a such way that $\partial_X L(x, X(x), P) = 0$. This equation is equivalent to the *adjoint system*

$$A(x)^T P = -\partial_X J(x, X)^T, \qquad (6)$$

the solution of which is called the *adjoint state*.

Finally, the derivative of f is given by:

$$d_x J(x, X(x)) = \partial_x J(x, X(x)) + (\partial_x A(x)X - \partial_x B(x), P_X) \qquad (7)$$

We will show in the next section that the reverse mode used in AD is similar to this approach.

2.3.2. *The reverse mode in AD*

For the sake of simplicity, we introduce the reverse mode with the simple model (M1) (section 2.1.1). We associate a variable p_k to each equality of (M1). Then the Lagrange operator related to this model is

$$L(x, X, P) = f(x) + \sum_{k=n+1}^{K} (x_k - \varphi_k(x_{S_k}))p_k,$$

where X is the vector of intermediate variables: $X = (x_{n+1}, \ldots, x_N)$ and $P = (p_{n+1}, \ldots, p_N)$ is the Lagrange multiplier. Using the same method as in the previous section, we obtain the *adjoint problem*:

$$\partial_X L(x, X(x), P) = 0 \iff \partial_{x_i} L(x, X(x), P) = 0 \quad n + 1 \leq i \leq N,$$

leading to the following equations

$$\sum_{k=n+1}^{N} (\delta_{ij} - \partial_{x_i}\varphi_k)p_k = -\delta_{iN} \quad n + 1 \leq i \leq N.$$

Here δ is the Kronecker symbol, $\delta_{ij} = 1$ if $i = j$ and $\delta_{ij} = 0$ else. Since $\partial_{x_i}\varphi_k = 0$ if $i \geq k$, P is the solution of an upper triangular system. Moreover, this system has a unique solution, its ith diagonal term being equal to $\delta_{ii} = 1$. Finally the *adjoint state* is given by the following adjoint algorithm

$$\begin{cases} P_N = 1 \\ for \ i := N - 1 \ down \ to \ n + 1 \ do \\ \quad p_i := \sum_{k=n+1}^{N} \partial_{x_i}\varphi_k \, p_k \end{cases}$$

The adjoint state is computed from the last instruction to the first instruction of the program (*the reverse mode*). Using the same arguments as in the previous section, we obtain the derivative of f

$$d_{x_i}f = \partial_{x_i}L(x, X(x), P_x) \quad 1 \leq i \leq n.$$

So we have

$$d_{x_i}f = \sum_{k=n+1}^{N} \partial_{x_i}\varphi_k \, p_k \quad 1 \leq i \leq n.$$

As for the adjoint method in shape optimization, the computing time needed by the reverse mode in AD is independent of the number of independent variables. But in the reverse mode the size of the vector P is proportional to the number of instructions needed to compute the function f.

We will discuss the complexity problem in the following section.

2.4. THE COMPLEXITY OF AD ALGORITHMS

2.4.1. *The complexity of the gradient computation in shape optimization*

In a shape optimization problem, the derivatives of the cost function f involves the following operations:

the direct method	the adjoint method
computation of the right hand side of (4): $d_x B(x) - d_x A(x) X(x)$	computation of the right hand side of (6): $-\partial_X J(x, X(x))^T$
resolution of system (4)	resolution of the adjoint system (6)
computation of $d_x f$ (3)	computation of $d_x J$ (7)

Usually the resolution of the linear system (4) (resp. (6)) is far more expensive than the computation of its right hand side and of $d_x f$ by the expression (3) (resp. (7)). For this reason we only take into account the resolution of linear systems.

If we use a direct method to solve the state problem (1), the systems (4) or (6) are particularly easy to solve. But in the general case, the complexity is the same for the state and adjoint systems.

We give some well known complexity results for shape optimization problems in table 1 with the following notations:

$f : D(f) \subset \mathbb{R}^n \longrightarrow \mathbb{R}$: a function,
$F : D(F) \subset \mathbb{R}^n \longrightarrow \mathbb{R}^m$: an m function,
$\nabla f \in \mathbb{R}^n$: the gradient of f,
$DF \in \mathbb{R}^{m \times n}$: The Jacobian of F,
$y \in \mathbb{R}^n$: for directional derivative,
$u \in \mathbb{R}^m$: if $g := u^T F$, then $\nabla g = (DF)^T u$, we obtain the same complexity as for f
$L(\ldots)$: number of basic operations needed to compute $(\ldots)$,
$S(\ldots)$: work space needed to compute $(\ldots)$,

We shall compare the complexity of AD algorithms with the classical complexity results of shape optimization problems.

	direct method	adjoint method
$L(f, \nabla f)$ $L(f, \nabla f, H)$	$O(nL(f))$ $O(n^2 L(f))$	$O(L(f))$ $O(n L(f))$
$S(f, \nabla f)$ $S(f, \nabla f, H.y)$ $S(f, \nabla f, H)$	$O(S(f))$	$O(S(f))$
$L(F, DF)$ $L(F, u^T DF)$ $L(F, DF.y)$	$O(nL(F))$ $O(nL(F))$ $O(L(F))$	$O(m L(F))$ $O(L(F))$ $O(mL(F))$
$S(F, DF)$ $S(F, u^T DF)$ $S(F, DF.y)$	$O(S(F))$	$O(S(F))$

TABLE 1. The shape optimization complexities

2.4.2. *The complexity of AD algorithms*
The most important complexity results are given in table 2 [Kub 88], using the notations of the previous section:

	forward	reverse
$L(f, \nabla f)$ $L(f, \nabla f, H.y)$ $L(f, \nabla f, H)$	$\leq 4nL(f)$ $O(n L(f))$ $O(n^2 L(f))$	$\leq 4L(f)$ $\leq 14 L(f)$ $\leq (10n + 4) L(f)$
$S(f, \nabla f)$ $S(f, \nabla f, H.y)$ $S(f, \nabla f, H)$	$O(S(f))$	$O(S(f) + L(f))$
$L(F, DF)$ $L(F, u^T DF)$ $L(F, DF.y)$	$O(nL(F))$ $O(nL(F))$ $O(L(F))$	$\leq (3m + 1) L(F)$ $\leq 4L(F)$ $\leq 4mL(F)$
$S(F, DF)$ $S(F, u^T DF)$ $S(F, DF.y)$	$O(S(F))$	$O(S(F) + L(F))$

TABLE 2. The AD complexities

The computing complexities given in table 2 are similar to the classical results known in shape optimization (table 1). In both cases, the adjoint method is appropriate to problems with a large number of independent variables. But our shape optimization example does not need an important work space.

2.5. HOW TO USE AD

To perform automatic differentiation, we just have to modify the declarative part of an existing program: we define the *independent variables* (a subset of x components), *the dependent variables* (a subset of y components) and the *active section* of the program (the part to differentiate). The automatic differentiation can be achieved:

- by processing the source Fortran program (ex. Adifor, Jakef [8], Padre2, [RoD 92], ...) and generating a new one,

- by using the operator overloading (redefintion of $\times$, $+$, ...) which is available in recent languages like C++, ADA, ...(e.g. ADOL-C [GJS 90], ADOGEN [Roc 94]).

In the first case we use a type of precompiler. In the latter case, operator overloading works only when running the program. At each operation we store the operation and its derivative instructions in a file: in the actual implementation of ADOL-C, this file is a binary file, but ADOGEN generates a Fortran file program.

How does a Fortran programmer use ADOL-C or ADOGEN ?

It is quite easy to translate Fortran to C (a subset of C++), we have at least two software to do that:

- f2c, a public domain software available through the Netlib library,

- the actual Fortran 90 "compiler" of NAG is just a Fortran to C translator.

We have now a C program and we just have to follow the instructions of the User Guide of ADOL-C or ADOGEN.

The overloading technique is very flexible, but has a major restriction: the file generated by ADOL-C or ADOGEN is the trace of the execution for a given data. Let x_0 denote this data and let V_{x_0} be a neighbourhood of this data. If x is a new data the result of the new program is $\mathbf{f}(x)$ where $\mathbf{f}$ is the analytic extension of $f_{|V_{x_0}}$. But we shall see in the next section that we need only to differentiate analytic functions: elementary stiffness matrices and loads.

3. How to use AD in shape optimization problems

3.1. THE LINEAR CASE

It appears from the previous section that AD algorithms are quite expensive. We can say roughly that the computing time and the work space needed by AD algorithms are proportional to the number of basic operations of f.

As shown in section 2.4.1, the computing time of f or of its gradient is mainly the computing time needed to solve the involved linear system(s). But we can easily obtain the derivative of a linear equation (eq. (4)). Then for the most important part of the program we do not need AD tools to compute ∇f, and we will use AD only to differentiate a reduced number of instructions.

Moreover if we use the pivot Gauss method to solve the linear system, the map $x \longmapsto X(x)$ is differentiable, but its numeric implementation is not and AD does not work correctly. This gives another reason not to use AD to differentiate $x \longmapsto X(x)$.

Thus we only use AD to perform the following operations:

The direct method	The adjoint method
computation of the right hand side of (4): $d_x B(x) - d_x A(x) X(x)$	computation of the right hand side of (6): $-\partial_X J(x, X(x))^T$
computation of $d_x f$ (3)	computation of $d_x f$ (7)

We can use AD to compute $d_x A(x)$ and $d_x B(x)$ even if $A(x)$ and $B(x)$ are very complicated. For this aim we use the fact that $A = \sum_{K \in \mathcal{K}} A_K$ (resp. $B = \sum_{K \in \mathcal{K}} B_K$) and that $d_x A = \sum_{K \in \mathcal{K}} d_x A_K$ (resp. $d_x B = \sum_{K \in \mathcal{K}} d_x B_K$) where A_K is the elementary stiffness matrix of the element K and B_K is the load on the element K. The active section is then reduced to a small part of the software: the computation of A_K and B_K. The additional memory used does not exceed a few Kbytes.

3.2. THE NONLINEAR CASE

We obtain the same results if the state equation is nonlinear:

$$F(x, X) = 0. \tag{8}$$

But let us show how to apply AD to solve the state problem (8).

3.2.1. *Application of AD to solve the state problem*
Usually we use a variant of the Newton method to solve (8), thus we have to solve at each step k the linear system for a given x_k

$$\partial_X F(x_k, X_k) d_k = -F(x_k, X_k) \tag{9}$$

$$X_{k+1} = X_k + d_k.$$

We can use AD to compute $(x, X) \longmapsto \partial_X F(x, X)$ giving another field of application for AD.

3.2.2. *Application of AD to nonlinear shape optimization*

Here, in order to solve the state equation (8), we have to solve a sequence of linear systems (9). For complexity reasons, it is impossible to differentiate the process (9) (computing $x \longmapsto X(x)$) using actual AD implementations.

There is another restriction: $X(x)$ is obtained as the limit of a sequence (X_k). The application of AD to the computational process (9) will give a sequence $(\partial_x X_k)$, but we are not sure that $(\partial_x X_k)$ will converge to $\partial_x X(x)$ [Gil 92].

3.2.3. *The direct method*

Under classical assumptions, we use the implicit function theorem and the derivative of $x \longmapsto X(x)$ is given by solving the system:

$$\partial_X F(x, X(x)) d_x X = -\partial_x F(x, X(x)). \tag{10}$$

This system and system (9) have the same matrix. If we use a direct method to solve (9), then $d_x X$, the solution of system (10), can be computed by solving two triangular systems.

In order to compute

$$d_x f(x) = \partial_x J(x, X(x)) + \partial_X J(x, X(x)) d_x X,$$

automatic differentiation can be used to compute:
- $x \longmapsto \partial_X F(x, X(x))$,
- $x \longmapsto \partial_x F(x, X(x))$,
- $\partial_x J(x, X(x))$ and $\partial_X J(x, X(x)) d_x X$.

3.2.4. *The adjoint method*

If we use the adjoint method we obtain the following adjoint equation:

$$\partial_X F(x, X(x))^T P = -\partial_X J(x, X(x))^T. \tag{11}$$

The matrix of the system (11) is the transposite of the matrix of the system (9). If we use a direct method to solve (9), then the solution P of system (11) can be computed by solving two triangular systems.

In this case, in order to compute the derivative:

$$d_x J(x, X(x)) = \partial_x J(x, X(x)) + (\partial_x F(x, X(x)), P_x),$$

AD can be used to compute:
- $x \longmapsto \partial_X F(x, X(x))$,
- $x \longmapsto \partial_X J(x, X(x))$,
- $x \longmapsto \partial_x J(x, X(x))$ and $\partial_x F(x, X(x))$.

4. High order derivatives in shape optimization

Before the recent development of shape optimization methods [Céa 86], the designer had to make at each step an analysis including roughly the domain triangulation, the computation of a stiffness matrix, the resolution of a linear system and the computation of a cost function. This method is expensive in human and computer time. Shape optimization methods, allow the automatization of the previous process using the first derivative of the cost function.

The use of the second order derivative is a good way to reduce the number of analyses. Fujii [Fuj 86] [MaF 92] was the first to study this problem. In [NBC 89] the second order derivative is used to improve the linear search and the authors describe in [GuM 92] a method which gives an intrinsic expression of the first and second order derivatives on the boundary of the domain involved.

In this section we study higher order derivatives. But one can ask the following questions:
- are they expensive to calculate?
- are they complicated to use?
- are they imprecise?
- are they useless?

At first sight, the answers seem to be positive, but we shall see that higher order derivatives are not expensive to calculate, and can be computed automatically. In the references [GuM 93], [GuM 94] the autors gave an answer to the third question by proving that the higher order derivatives of a function can be computed with the same precision as the function itself. We prove also that the derivatives so computed are equal to the derivatives of the discrete problem. We call the discrete problem the finite dimensional problem processed by the computer. This result allows the use of automatic differentiation ([GuM 93], [GuM 94]) which works only on discrete problems. Furthermore, the computations of Taylor's expansions which are proposed in sections 5 and 6, give an answer to the last question.

The basic ideas of this work come from the following remarks:
- generally the results of an analysis are only used to update the data given to the computer. Thus **the true unknown of the structural problem is this data**,
- the results of the analysis (displacement) **depend analytically on the data even if the solution is not regular** (loaded crack). By computing the Taylor expansion of the solution, we give an explicit expression of the results with respect to the data.

We first see in 4.1 how to use AD to compute the Taylor expansion of the solution. In section 4.2 we give some properties of this method. In sections 4.3 and 4.4 we show how to use this Taylor expansion. Finally we give in sections 5 and 6 some numerical results.

4.1. TAYLOR EXPANSION IN SHAPE OPTIMIZATION

This powerfull tool can be used with efficiency in shape optimization. The method suggested in this paper is quite general but for a seek of clarity we only present its application to the static linear case.

Let x be a parameter (geometry, material property, ...), and let $Y = Y(x)$ be the solution of the linear static problem:

$$A(x)Y(x) = B(x)$$

We know that the first variation $\partial_x Y$ of Y with respect to x is the solution of the linear system:

$$A(x)\partial_x Y = d_x B(x) - d_x A(x)Y.$$

Automatic differentiation can be performed also to compute higher order derivatives of $A(x)$ and $B(x)$. Then $Y^{(d)}$ is the solution of the system

$$A(x)Y^{(d)} = B^{(d)} - \sum_{p=1}^{d} C_d^p A^{(p)} Y^{(d-p)},$$

and the solution $Y(x)$ is available from its Taylor expansion

$$Y(x+v) = Y(x) + Y'(x).v + \cdots + \frac{1}{d!} Y^{(d)}(x).v^d + \cdots$$

With our method, a single analysis is sufficient to know the solution in a neighbourhood of the parameter x. We refer to [GuM 93] and [Des 76] for the study of the continuous problem; but let us prove the analycity of the discrete problem $x \longmapsto Y(x)$:
- the map $x \longmapsto A_K(x)$ (resp. $x \longmapsto B_K(x)$) is analytic
- the map $(A_K)_{K \in \mathcal{K}} \longmapsto A = \sum_{K \in \mathcal{K}} A_K$ (resp. $(B_K)_{K \in \mathcal{K}} \longmapsto B = \sum_{K \in \mathcal{K}} B_K$) is analytic,
- the map $(A, B) \longmapsto Y = A^{-1}B$ is analytic if the matrix A is invertible.
The composition of those three maps is exactly the map $x \longmapsto Y(x)$.

4.2. SOME PROPERTIES OF THE METHOD

For reasonable perturbations of x, if we compare the solutions given by computing the Taylor expansion and by a direct computation of $Y(x)$ we obtain exactly the same results (the same binary representation [GuM 94]). We can explain this fact as follows:

- the automatic differentiation gives the exact derivatives and the Taylor expansion of the discrete problem,
- the discrete problem is analytic (equal to its Taylor series). For the same reasons, the previous property holds even if we use an approximate integration formula for computing A_K and B_K. Thus we can use our method with the boundary element method.

If we have at least two parameters, the suggested method is parallelizable: the computations of the derivatives in different directions are independent. The degree of parallelism increases quickly with the number of parameters and the degree of the Taylor expansion.

4.3. SHAPE OPTIMIZATION AND HIGH ORDER DERIVATIVES METHOD

Shape optimization allows the automatization of the design process by the use of the first (resp. the second) order Taylor expansion, leading to the computation of the first (resp. the first and the second) order derivative. In this case, the designer must define a mathematical programming problem including all the design constraints. Then many problems arise:
- some constraints are quite difficult to describe (esthetic, manufacturing constraints, ...)
- non realistic domains may be obtained,
- mathematical programming is often a strange activity to the designer.

The suggested method takes advantage of the matrix decomposition obtained in the first analysis. The computation of the Taylor expansion needs only to solve some already factorized systems. The designer can obtain in real time the solution of the modified domain (polynomial evaluation), and uses at each moment his own know-how for updating parameters.

This method is particularly powerful for large scale problems: the ratio
complexity of the factorization / complexity of the factorized system
increases with the dimension of the system.

Naturally, it is possible to use mathematical programming methods. Moreover, it is possible to use methods (not usually used in shape optimization) which need a great number of Y evaluations: bundle method, simulated annealing. ...

4.4. OPTIMIZATION

Generally we have to solve

$$\begin{cases} \text{minimize} & J(x, Y(x)) \\ \text{subject to} & G_i(x, Y(x)) \leq 0, \quad 1 \leq i \leq m \end{cases}$$

where J is the cost function and (G_i) are m constraint functions. We assume that $Y(x)$ is given by its Taylor expansion and that the constraints (G_i)

include the domain of validity of this approximation. If the solution of this approximated problem is not on the boundary of the validity domain, it is the final solution. Otherwise, we compute the Taylor expansion at this point and we solve a new mathematical programming problem.

4.4.1. *Non-differentiable problems*

Usually, the map $x \longmapsto Y(x)$ is analytic, but the maps $x \longmapsto J(x, Y(x))$ and $x \longmapsto G_i(x, Y(x)) 1 \leq i \leq m$ are not. Let us recall two classical examples:
- the cost function involves the Tresca criterion,
- minimize $J(x; Y(x)) = \max_{z \in \Omega_x} |\sigma(Y(x))(z)|$, where Ω_x is the domain and $\sigma(Y(x))$ is the Von Meases equivalent stress.
Since the map $x \longmapsto Y(x)$ is explicit, we can compute easily $J(x, Y(x))$ and $G_i(x, Y(x))$. The non-differentiable optimization algorithms, which need a great number of functions evaluations, can be used with efficiency in this case.

4.4.2. *Global optimization*

For the same reasons as in the previous section, we can use global optimization methods such as:
- simulated annealing,
- the tunneling method,
- the random search method,
- "biological" methods.
Moreover we can construct new global optimization methods based on the simultaneous resolution of all the solutions of a polynomial system. In fact, if the functions $x \longmapsto J(x, Y(x))$ and $x \longmapsto G_i(x, Y(x)) 1 \leq i \leq m$ are analytic, the use of their Taylor expansion leads to a polynomial condition of optimality. We can compute all the solutions, using classical methods. Then we choose the global solution.

4.5. HOW TO USE HIGH ORDER DERIVATIVES METHOD

This method can be used with different state equation types: high order elliptic problems, nonlinear problems, integral equations, ... In an industrial environment, we can obtain a satisfying solution in one, two or three steps:
- The first step: the designer explores the structure by "asking questions": what is the behavior of my structure if I modify this parameter? This step gives the significant parameters of the structure. If we find a satisfying solution we stop here.
- The second step: we compute a Taylor expansion involving all the parameters selected in the first step. If we find a satisfying solution we stop here.

- The third step: we define a mathematical programming problem and we solve it using an initial design obtained in the first or second steps.

5. Numerical results for the Laplacian equation

With intent to illustrate numerically the use of higher order derivatives, let us take the popular example of the torsional rigidity of an elastic bar, whose cross section is an open set $(I + V)(\Omega)$. Ω is the initial domain (fig. 2), and $(I + V)(\Omega) := \{x + V(x); x \in \Omega\}$ is a perturbation of this domain.

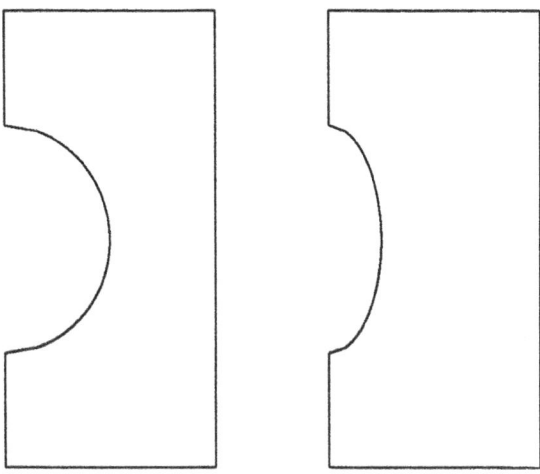

Figure 2. initial design   modified design $(I + V)(\Omega)$

The cost function (torsional rigidity) for a cross section $(I + V)(\Omega)$ is

$$j(I + V) = 2 \int_{(I+V)(\Omega)} y_{(I+V)(\Omega)} \, dx \, ,$$

where $y_{(I+V)(\Omega)} \in H_0^1\big((I+V)(\Omega)\big)$ is the solution to the Laplacian equation

$$-\Delta\, y_{(I+V)(\Omega)} = 2 \, .$$

This cost function j is of class C^∞.

5.1. FIRST DESIGN PERTURBATION

Here the modified design is given in fig. 2. We are using finite elements of degree 1, using smaller and smaller elements: at each step, the size of the elements becomes half of the previous ones, as shown on fig. 3.

We give in Table 3 the results of the computing of $j(I + V)$ obtained when using the Taylor's expansion of j at the point I, which have to be

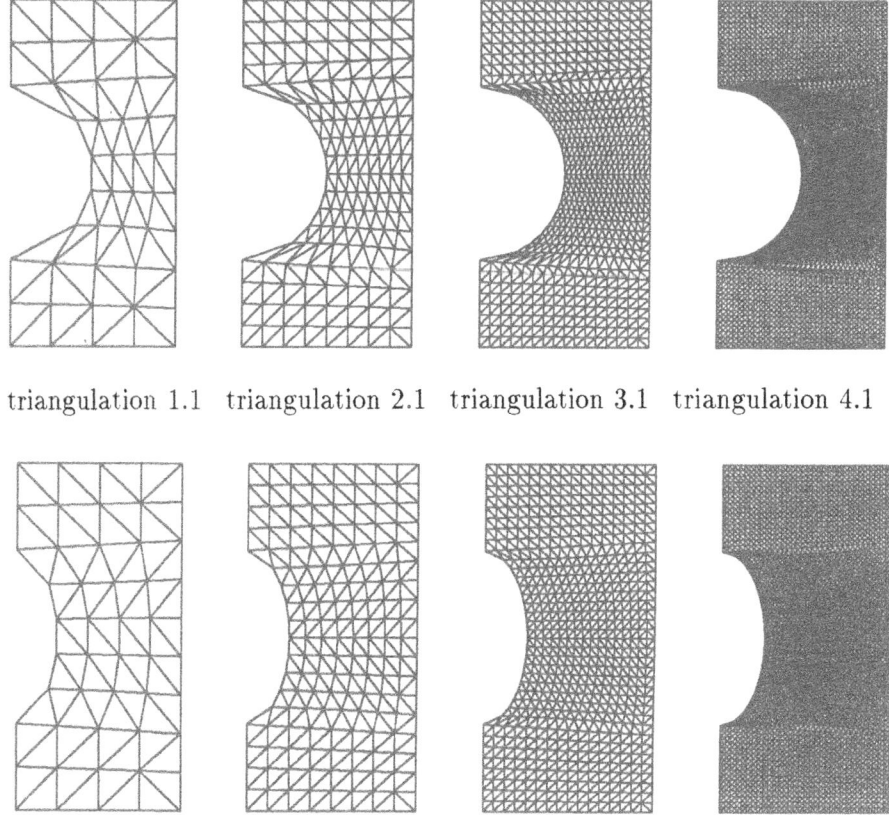

triangulation 1.1 triangulation 2.1 triangulation 3.1 triangulation 4.1

triangulation 1.2 triangulation 2.2 triangulation 3.2 triangulation 4.2

Figure 3. triangulations of initial design and modified designs

compared with the ones obtained when computing directly $j(I + V)$ on the modified domain.

Observe that the nodal table is made of the components of the map I in an appropriate basis. When V is a perturbation of the identity, the modified design nodal table is made of the components of the map $I + V$. Thus the derivatives of the map $V \longmapsto j(I + V)$ are exactly the derivatives of j with respect to the nodal table.

We give in Table 4 the relative error for the norm L^∞ between the solution y_{I+V} computed on the modified domain and the approximation of the latter by Taylor's expansion of $y_{(}I + V)$ at the point I.

The convergence of the series seems to depend only slightly on the size of the elements. On the other hand, when the number of nodes increases, it becomes more and more advantageous to solve some linear systems where

Order of expansion	triangulation 1	triangulation 2	triangulation 3	triangulation 4
0	.165754759809873	.17144321842656	.17416741916434	.1751508015
1	.169	.178	.181	.182
3	.1754	.1840	.1870	.1879
5	.174584	.183198	.18630	.18719
10	.1745975	.183210	.1862835	.1871720
20	.1745992758	.18320866	.186283765	.1871722234
30	.174599276026168	.18320864866	.18628376224	
40	.174599276026178	.1832086486832	.1862837622883	
50	.174599276026178	.18320864868347	.18628376228882	
j(I+V)	.174599276026178	.18320864868347	.18628376228882	.1871722233

TABLE 3. behaviour of Taylor's expansion of the cost function

Order of expansion	triangulation 1	triangulation 2	triangulation 3	triangulation 4
1	.073309714253548	.086073617452509	.093782161947154	.0968730219
3	.006718268316071	.013093233335998	.014979462707718	.0149311039
5	.001722788508621	.002539750030623	.004808110366787	.0053616606
10	.000028892586130	.000129843146119	.000248661883022	.0002795846
20	.000000002534305	.000000923092796	.000001335313315	.0000013535
30	.000000000000226	.000000002219957	.000000006721870	.0000000087
40	.000000000000001	.000000000010874	.000000000134758	
50	.000000000000001	.000000000000040	.000000000000581	

TABLE 4. behaviour of the solution

the LU decomposition has already been done, than to compute the LU decomposition of the new stiffness matrix at the point $I + V$; this means that the use of higher order derivatives is particularly valuable when solving large scale problems.

This is shown by Table 5; the CPU time spent to compute $j(I)$ (which is also the one spent to compute $j(I+V)$ directly on the new domain) appears in the column $j(I)$, and the additional CPU time spent to compute $j(I+V)$ when using Taylor's expansion of j at the order k appears in the columns $Tj(k)$. We have done those computations on a processor MISP 6000.

Triangulation	j(X)	Tj(1)	Tj(3)	Tj(5)	Tj(10)	Tj(20)	Tj(30)
1	.5	.5	2.5	3.5	6.5	11.5	21.5
2	5	5	9	12	22	60	109
3	46	9	33	50	127	311	527
4	543	68	159	248	756	1908	

TABLE 5. CPU time (s)

One can see that a Taylor's expansion at the order 3 or 4, which gives a sufficient precision for the engineer, leads to a shorter computation of the result on and after the third triangulation. In the numerical algorithm, we have taken into account the fact that the stiffness matrix is a band matrix, which decreases the ratio

cost of the LU decomposition / cost of solving of the linear system.

When solving a three dimensional problem, this ratio is larger (the band of the matrix is larger), as well as the size of the problem himself. It follows that the use of the higher order derivatives will be more efficient in dimension 3 than in dimension 2.

5.2. OTHER DESIGN PERTURBATIONS

We are now interested in the circle of convergence of Taylor's series, as well as in the domain of validity of the method. Recall that analyticity results have been proved in [Des 76] for a similar problem. The chosen perturbations can damage the triangulation, as shown on fig. 4.

A correct triangulation of the new domains would be given by fig. 5.

We give in Table 6 the results of the computing of $j(I + V)$ obtained when using Taylor's expansion of j at the point I for these different perturbations, which have to be compared on the one hand with the direct computation of $j(I+V)$ on those perturbations with bad triangulation (triangulations 2.3, 2.4, 2.5), on the other hand with the direct computation of $j(I+V)$ on the correct triangulation of those perturbations (triangulations 2.3.1, 2.4.1, 2.5.1), denoted by $jct(I + V)$. One can see that the Taylor's series converges for the perturbation 2.3, and is undoubtly divergent for the large perturbation 2.5; however, one can see that even in this extreme

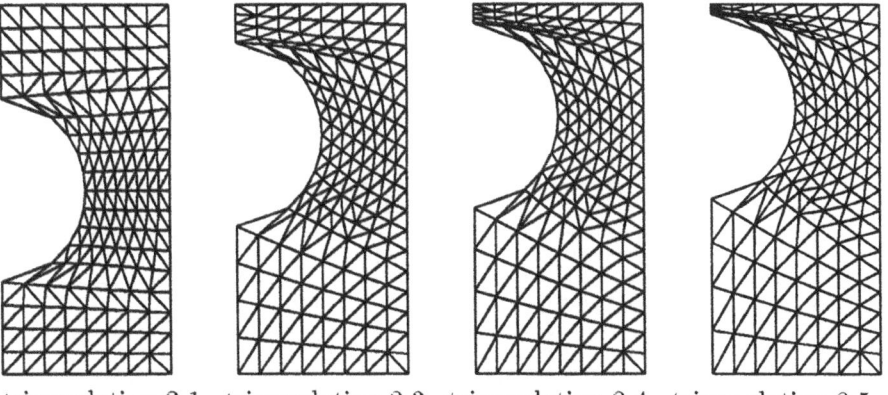

triangulation 2.1   triangulation 2.3   triangulation 2.4   triangulation 2.5

Figure 4.   damaged triangulations

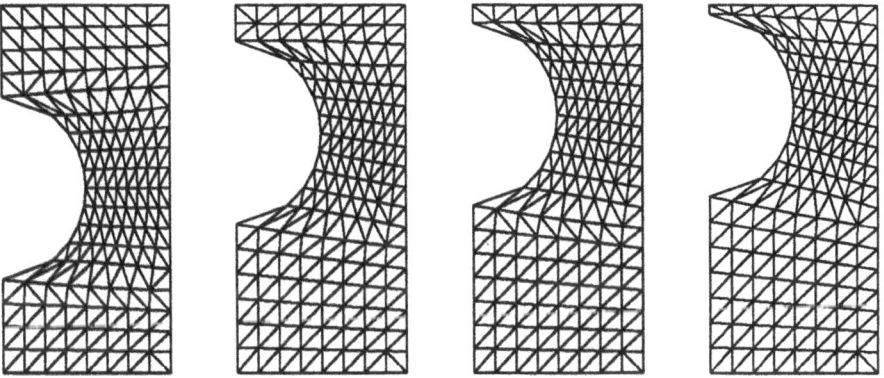

triangulation 2.1 triangulation 2.3.1 triangulation 2.4.1 triangulation 2.5.1

Figure 5.   correct triangulations

case, one can get a good approximation of $j(I + V)$ by choosing a correct order of Taylor's expansion (here between 5 and 10), and better, a good approximation of $jct(I + V)$.

We give in Table 7 the relative errors e1 and e2 for the norm L^∞:
- e1 is the error between Taylor expansion of $y_(I + V)$ at the point I and the solution y_{I+V} computed on the triangulation 2.5;
- e2 is the error between Taylor expansion of $y_(I + V)$ at the point I and the solution computed on the correct triangulation 2.5.1, the relative error between the solutions computed on the two triangulations beeing 0.060. One can see here that even in the case of an important perturbation of the triangulation, the use of the higher order derivatives leads to quite a good approximation (relative error of six per cent), which is often sufficient in

Order of expansion	triangulation 2.3	triangulation 2.4	triangulation 2.5
1	0.17	0.17	0.17
3	0.1951	0.2134	0.225
5	0.19471	0.2122	0.2228
10	0.1945934	0.21151	0.22119
20	0.1945949508	0.21153	0.22082
30	0.1945949526	0.21154	0.2196
50	0.19459494967	0.21153	0.187
100	0.194594949635326	0.21132	-178.7
j(I+V)	0.194594949635325	0.21158	0.22146
jct(I+V)	0.194553646999355	0.21287	0.22313

TABLE 6. behaviour of Taylor's series of the cost function with a bad triangulation

order of expansion	0	1	3	5	10	20	30	50	70
e1	0.408	0.102	0.054	0.016	0.016	0.034	0.127	3.990	126.4
e2	0.425	0.105	0.069	0.057	0.061	0.061	0.146	3.955	124.7

TABLE 7. behaviour of Taylor's series of the solution with a bad triangulation

practice.

6. Numerical tests for Maxwell's equations

6.1. INTRODUCTION

The sources of some Spatial Antennas are a network of waveguides. The topology of such a network is obtained using a recent patent [MPD 93]. The aim of this section is to study an element of this network, i.e. the junction of two rectangular waveguides. The shape of this junction is of most importance in a telecommunication satellite, where room, weight and performance are crucial.

We show that the use of higher order derivatives of the discrete problem (with respect here to the frequency and the shape of the junction)[GuM 94] leads to a very performant method for the numerical simulation of the waveguide.

6.2. THE 3D PROBLEM

Let us roughly describe the physical problem. Two rectangular waveguides G and G' meet together as shown on fig. 6.

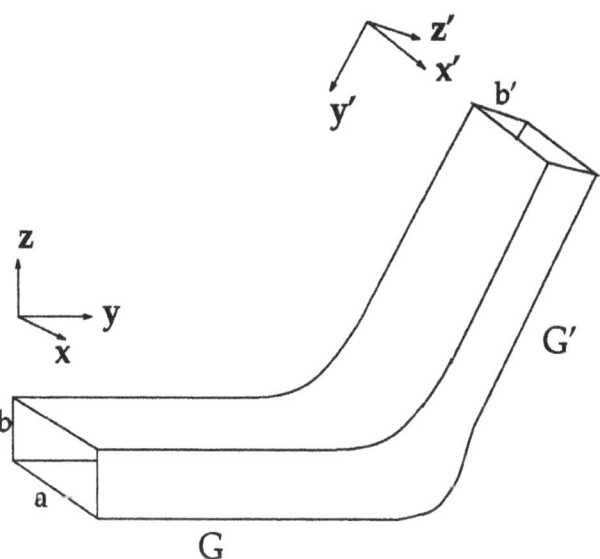

Figure 6. the 3D waveguide

We denote by $\mathcal{G}$ the inside of the complete waveguide. The inner boundary Γ is supposed to be a perfect conductor.

We state the problem as follows: an incident wave is given, which propagates in the waveguide G toward the junction; a part of it is reflected, the other part beeing transmitted in the waveguides G'. Our goal is to obtain a Taylor expansion of the reflected wave with respect to the frequency and the shape of the junction. This expansion will be used to perform shape optimization, in order to minimize the modulus of the reflected wave on a large frequency scale.

Following usual assumptions are made:

• The electromagnetic field is time-harmonic, i.e. the time dependence occurs through a factor $\exp(i\omega t)$ with $\omega = 2\pi f$ (f is the frequency); more precisely, the physical electromagnetic field is the real part of a complex field $E \exp(i\omega t)$, where E is time-independent.

• The incident wave is not perturbated by the scattered electromagnetic field. This defines in each waveguide G and G' an electric incident field E^i, which vanishes in G', and is expressed in G by

$$E^i = \sin \frac{\pi x}{a}(0, 0, e^{-iky})$$

where k is the wave number, defined as (c is the light speed in vacuum):

$$k = \sqrt{\beta^2 - \frac{\pi^2}{a^2}}, \quad \beta = \frac{2\pi f}{c} \tag{12}$$

It is the fundamental mode of the rectangular waveguide ([DaL 88], [Vas 85], [Ces]), and the only one which does not vanish at infinity if we assume that

$$\frac{\pi}{a} < \beta < \min(\frac{\pi}{b}, \frac{\pi}{b'}). \tag{13}$$

This hypothesis is not really necessary, and is just made to avoid heavy formulation.

The global electric field is a solution to the equations (which derive directly from Maxwell's equations):

$$(P) \quad \begin{cases} \text{curl curl } E - \beta^2 E = 0 & \text{in } \mathcal{G} \\ E \wedge n = 0 & \text{on } \Gamma \\ (\text{RC}) \text{ conditions} \end{cases}$$

We denote by n the outward unit normal to Γ, and the condition $E \wedge n = 0$ reflects the fact that the boundary is a perfect conductor. For $e_y = (0, 1, 0)$ in G and $e_{y'} = (0, 1, 0)$ in G', the radiation conditions (RC) (Sommerfeld conditions) are

$$(\text{RC}) \quad \lim_{y \to -\infty} \text{curl } (E - E^i) \wedge e_y - ik\,(E - E^i) = 0 \quad \text{in G}$$

$$\lim_{y' \to -\infty} \text{curl } E \wedge e_{y'} - ik\,E = 0 \quad \text{in G'}$$

These equations express that the wave $E - E^i$ is outgoing, and behaves at infinity like the fundamental mode.

THE 2D PROBLEM

The 2D waveguide (fig. 7) is the median plane of the 3D waveguide (i.e. the intersection of the 3D waveguide with the plane $x = a/2$). We still denote by $\mathcal{G}$ the inside of the 2D waveguide, and by Γ it's boundary. We also denote

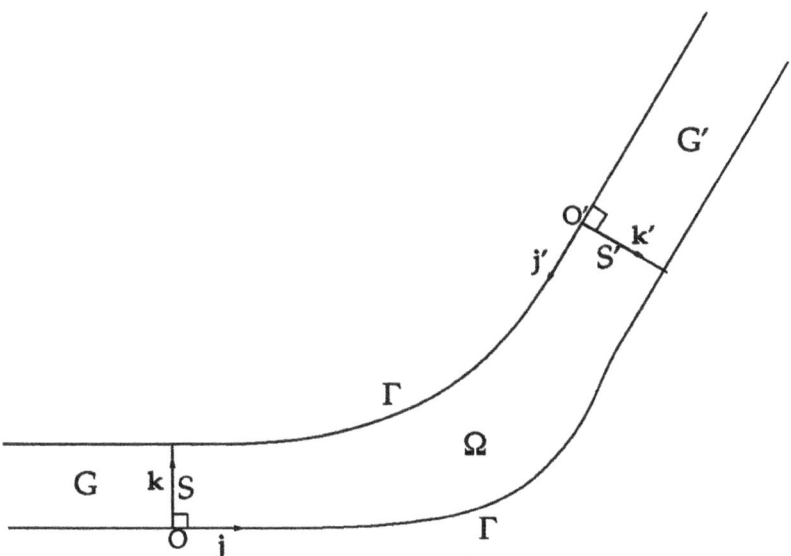

Figure 7. the domain Ω

by Ω a bounded part of the junction delimited by two cross-sections S and S'.

For all scalar field g and all vector-valued field $u = (u_y, u_z)$, let

$$\text{Curl}\, g = (\partial_z g, -\partial_y g)$$
$$\text{curl}\, u = \partial_y u_z - \partial_z u_y.$$

and denote by u_t the tangential component of u (i.e. $u = u_n\, n + u_t\, t$ where (n, t) is a direct oriented orthonormal basis).

It can be proved [Gui 94] that when the width a of the waveguide is constant, then the 3D problem reduces to a 2D problem. Moreover, the solution of the 2D problem is well approximated by the solution of the following problem:

$$(\text{P}) \qquad \begin{cases} \text{Curl curl}\, u - k^2 u = 0 & \text{in } \Omega \\ u_t = 0 & \text{on } \Gamma \\ \text{curl}\, u + ik\, u_t = \text{curl}\, u^i + ik\, u^i_t & \text{on } S \cup S \end{cases}$$

The incident wave u^i (which is the restriction of E^i to the median plane) vanishes in G', and is expressed in G by

$$u^i = (0,\ e^{-iky}).$$

We attempt to calculate the reflexion coefficient S_{11} and the transmission coefficient S_{12}. These are given by

$$S_{11} = \frac{1}{b} \int_S (u_t - u_t^i) \, ds$$

$$S_{12} = \frac{1}{b'} \int_{S'} u_t \, ds$$

where u is the solution to problem (P).

6.3. MULTI-FREQUENCY ANALYSIS

We use an H(curl) conforming finite element method [Ned 80]. Hence we have to solve a linear system of equations

$$A(k)X(k) = B(k) \tag{14}$$

Thanks to the form of the approximate problem, The matrix $A(k)$ and the vector $B(k)$ are polynomial in k, that is

$$A(k) = C - k^2 S + ikF$$

(C like curl, S like scalar and F like frontier) and

$$B(k) = kL$$

where C, S, F, L do not depend on k. Observe that higher than second order derivatives of A and B vanish.

Let k_g be a given frequency. The solution to eq. (14) with $k = k_g$ is computed with a direct method, using a LDL^T Cholesky-Crout factorization of the matrix $A(k)$.

In order to compute $X(k)$ for other values of k *without computing a new factorization of the matrix $A(k)$*, we use the Taylor expansion

$$X(k) = X(k_g) + \sum_{j \geq 1} \frac{X^{(j)}(k_g)}{j!}(k - k_g)^j. \tag{15}$$

For this purpose, we need to compute the derivatives of $X(k)$ with respect to k. This is done by solving the following systems, which are obtained by taking the successive derivatives of the system (14):

$$A(k)X'(k) = -A'(k)X(k) + L \tag{16}$$

$$A(k)X^{(n)}(k) = -A'(k)X^{(n-1)}(k) + SX^{(n-2)}(k) \quad \forall n \geq 2 \tag{17}$$

Note that second member of (17) is calculated from the non zero elements of the matrices $A'(k) = -2kS + iF$ and $A''(k) = -2S$ (these are at the most five on each line).

The approximation of the coefficients $S_{11}(k)$ and $S_{12}(k)$ becomes

$$S_{11}(k) = \frac{1}{2ikb}B.\overline{X}(k) - 1 = \frac{1}{2ikb}AX(k).\overline{X}(k) - 1$$

$$S_{12}(k) = \frac{1}{2ikb'}B'.\overline{X}(k)$$

where B' is defined in a similar way as B, and $X(k)$ is approximated through (15).

6.4. FIRST EXAMPLE

Our first example is a circular waveguide (see fig. 8), for which physical measures were supplied.

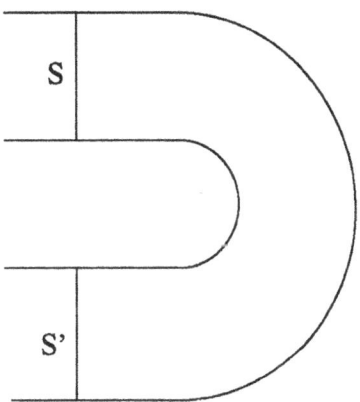

Figure 8. circular guide

Computation has been performed on a HP 9000/867S. Time spent is given in table 8. It can be seen that the multi-frequency analysis (i.e. the computation of the Taylor expansion, here up to the order 50) is about so expensive as a single ordinary analysis.

By classical methods, the knowledge of the reflexion coefficient needs about hundred analyses.

6.4.1. *Convergence of the Taylor expansion*
Results of convergence of the Taylor expansion are given in tables 9 and 10, with k_g corresponding to the frequency 12.3 GHz. Comparison is made

Number of degrees of freedom	CPU time (s) for Cholesky	CPU time (s) for Taylor(50)
799	1	18
3264	20	166
10853	571	678

TABLE 8. computation time

with the solution computed by solving directly the system (14) for different values of k (10, 11, 14 and 15 Ghz). Of course, the rate of convergence is better on 11-14 Ghz as on 10-15 Ghz.

	Frequency (GHz)	$|S_{11}|$ (dB)	$|S_{12}|$ (dB)	phase of S_{11} (deg)	phase of S_{12} (deg)
Taylor(30)	10	-30.918855	-0.248347	-136.899087	18.637913
Taylor(50)	10	-37.476134	-0.001559	-71.579726	18.761669
direct computation	10	-37.496066	0.000773	-71.204879	18.763322
Taylor(30)	15	-24.902031	0.029737	-100.536434	-11.838846
Taylor(50)	15	-33.595172	-0.001504	-99.458462	-9.562543
direct computation	15	-33.657307	-0.001960	-99.606644	-9.553712

TABLE 9. convergence of the Taylor expansion on the interval 10-15 GHz

6.4.2. *Comparison with experimental measures*

Comparison with the experimental measures (non smooth curves) is given on figures 9 and 10. Take into account that the reflected wave is very weak, thus computation as well as experimental measure are quite inaccurate (fig 9(a) and 10(a)). In contrast, the computation of the transmitted wave is so precise that the result of the computation of the phase (fig. 10(b)) is exactly the same as the one of the measures (the slight difference in the modulus (fig. 9(b)) comes from non perfect conduction of the real waveguide).

	Frequency (GHz)	$|S_{11}|$ (dB)	$|S_{12}|$ (dB)	phase of S_{11} (deg)	phase of S_{12} (deg)
Taylor(30)	11	-37.960780	-0.000695	22.602659	-67.361454
Taylor(50)	11	-37.960780	-0.000695	22.602660	-67.361454
direct computation	11	-37.960780	-0.000695	22.602660	-67.361454
Taylor(30)	14	-32.765963	-0.002320	-26.661061	63.463939
Taylor(50)	14	-32.765955	-0.002320	-26.660971	63.463938
direct computation	14	-32.765955	-0.002320	-26.660971	63.463938

TABLE 10. convergence of the Taylor expansion on the interval 11-14 GHz

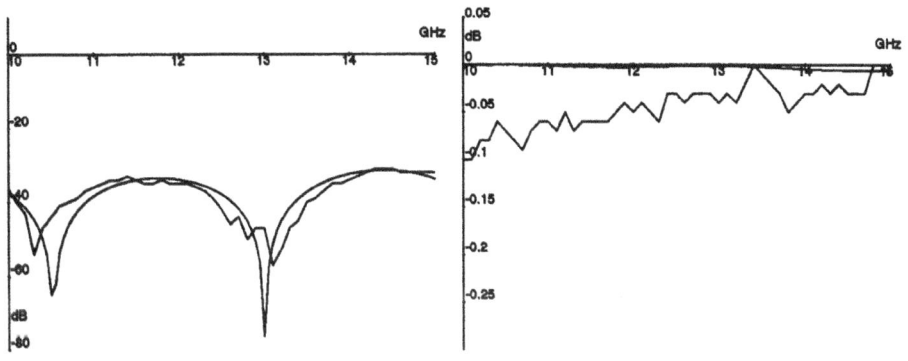

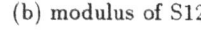

Figure 9. (a) modulus of S11   (b) modulus of S12

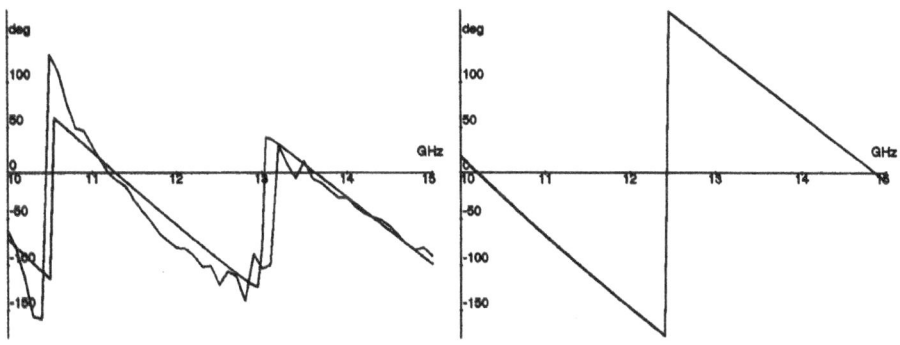

Figure 10. (a) phase of S11   (b) phase of S12

6.5. SECOND EXAMPLE

Our second example is shown on fig. 11. As in previous section, the coefficients S_{11} and S_{12} are computed by using a Taylor expansion, but now with respect to the frequency *and* the shape: the position of the middle facet depends upon a parameter t. The shapes corresponding to the values $t = -2$, $t = 0$ and $t = 2$ are shown respectively on fig. 11(a), fig. 11(b) and fig. 11(c).

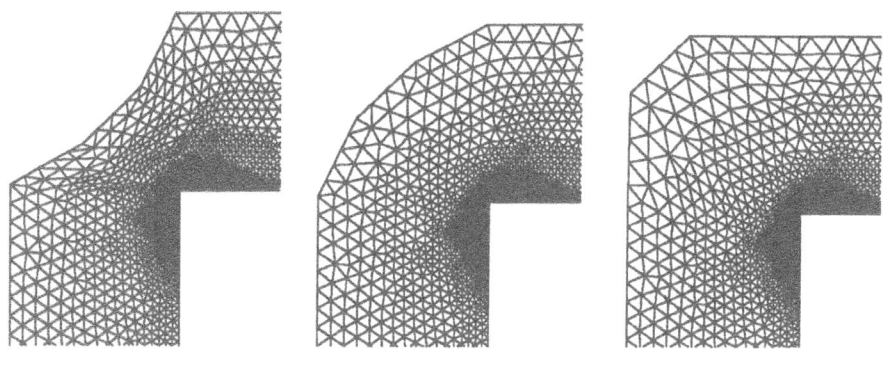

Figure 11.   (a) shape at $t = -2$ (b) shape at $t = 0$ (c) shape at $t = 2$

The difference here is that the dependence of the matrix $A(k, t)$ is no more polynomial in t. Thus the successive derivatives of $t \longmapsto A(k, t)$ are computed by using automatic differentiation on the elementary matrices [Mor 85], [Gri 89]. A polynomial in two variables $P(k, t)$ is obtained. The convergence of the Taylor expansion is given in table 11. Taylor(p, q) indicates a derivation at order p with respect to the frequency, and at order q with respect to the shape.

The graph of the map $(f, t) \longmapsto |P(k(f), t)|$ is shown on fig. 12 and fig. 13. It is worth to note that a simple view on fig. 13 gives the value of t solution to the non differentiable problem:

$$\text{minimize } j(t) \text{ with}$$
$$j(t) = \sup_{10\,GHz \leq f \leq 15\,GHz} |S_{11}(f, t)|$$

The curve $t \longmapsto j(t)$ is the superior envelope of all the curves, and the optimal t is near zero (which corresponds to the initial design!).

	Frequency (GHz)	$|S_{11}|$ (dB)	$|S_{12}|$ (dB)	phase of S_{11} (deg)	phase of S_{12} (deg)
Taylor(20,0)	10	-28.36168	0.06291	-5.23238	-48.04583
Taylor(20,3)	10	-11.11887	-0.29565	-135.08447	-44.57097
Taylor(20,5)	10	-11.07522	-0.34964	-135.87184	-44.63012
Taylor(20,10)	10	-11.09239	-0.35196	-135.92671	-44.59784
Taylor(20,20)	10	-11.09178	-0.35187	-135.92933	-44.59680
direct computation	10	-11.09184	-0.35185	-135.92960	-44.59692
Taylor(20,0)	15	-36.93388	-0.00209	13.10028	-117.14689
Taylor(20,3)	15	-6.01872	-1.31288	164.36942	-104.22770
Taylor(20,5)	15	-5.08636	-2.07613	172.37182	-99.06069
Taylor(20,10)	15	-4.83339	-1.75674	169.51543	-98.01031
Taylor(20,20)	15	-4.84270	-1.72714	169.23757	-98.05877
direct computation	15	-4.84289	-1.72714	169.23763	-98.05965

TABLE 11. convergence of the Taylor expansion at $t = -2$ and $f = 15GHz$

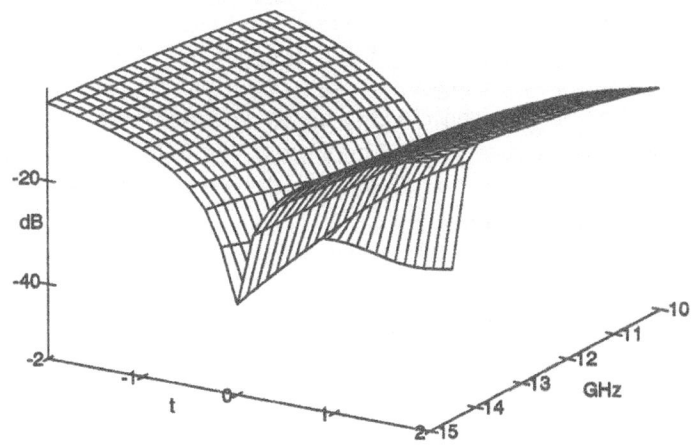

Figure 12. graph of $(f,t) \longmapsto |S_{11}(f,t)|$

444

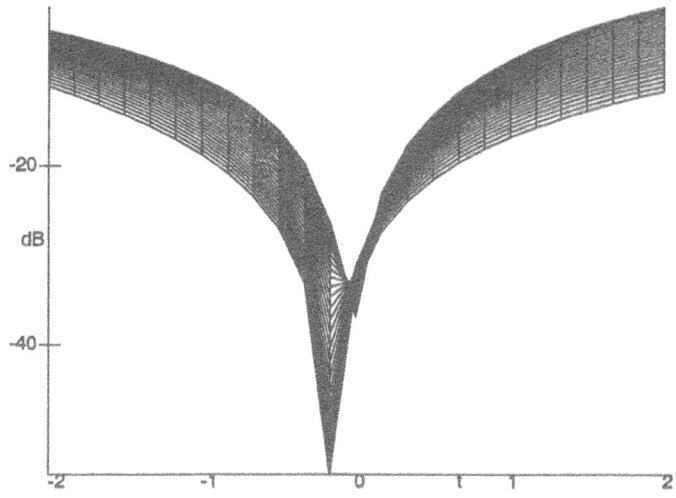

Figure 13. projection of the graph on the plane $f = 10\,GHz$

7. Conclusion

Those numerical results show clearly the efficiency of the high order derivatives method and automatic differentiation.

It is necessary, in order to introduce shape optimization methods into industry, to lower their cost of implementations. There is, at least, one mean to achieve this objective: the use of *Automatic Differentiation* to compute first and higher order derivatives.

The communication between CAD (Computer Aided Design) and computing environments needs a lot of human time; the *high order derivatives* method reduces the number of analysis and the number of conversions of CAD models to finite element models.

The designer can obtain in real time the solution of the modified domain (polynomial evaluation), and uses at each moment his own know-how for updating parameters.

In conclusion, using the suggested methods, the designer can obtain a satisfactory design in a short time.

REFERENCES

[BaS 83] W. BAUR AND V. STRASSEN *The complexity of partial derivatives*, Theoretical Computer Science, 22 (1983), pp. 317-330.

[Céa 86] J. CÉA, *Conception optimale ou identification de forme, calcul rapide de la dérivée directionnelle de la fonction coût*, M.A.A.N, 20 (1986), pp. 371-402.

[DaL 88] R. DAUTREY AND J.L. LIONS, *Analyse Mathématique et calcul numérique pour les sciences et les techniques*, Masson, Paris, 1988.

[Des 76] DESTUYNDER *Thèse de troisième cycle, Université PARIS VI, 1976.*

[Fuj 86] N. FUJII, *Necessary conditions for a domain optimization problem in elliptic boundary value problems*, SIAM.J. Control and Optimization, 24 (1986), pp. 346-360.

[Gil 92] J.C. GILBERT, *Automatic differentiation and iterative processes*, Optimization Methods and Softwares, 1 (1992), pp. 13-21.

[GVM 91] J.C. GILBERT, G. LE VEY, J. MASSE: *La differentiation automatique de fonctions représentées par des programmes*, INRIA, Rapports de recherhe, No 1557, 1991.

[Gri 89] A. GRIEWANK, *On automatic differentiation*, In M. Iri and K. Tanabe (editors), Mathematical programming: Recent developments and Applications, pp. 83-108, Kluwer Academic Publishers, Dordrecht, 1989.

[GJS 90] A. GRIEWANK, D. JUEDES, J. SRINIVASAN, ANS C. TYNER, *ADOL-C, a package for the automatic differentiation of algorithms written in C/C++*, ACM Trans. Math. Software.

[Gui 94] PH.GUILLAUME, *Dérivées d'ordre supérieur en conception optimale de forme* (1994), thèse de doctorat. Université Paul Sabatier, Toulouse, France.

[GuM 92] PH.GUILLAUME AND M. MASMOUDI, *Dérivées d'ordre supérieur en optimisation de domaines*, C.R. Acad. Sci. Paris, t.315, Série I (1992), pp. 859-862.

[GuM 93] PH.GUILLAUME AND M. MASMOUDI, *Calcul numérique des dérivées d'ordre supérieur en conception optimale de forme*, C.R. Acad. Sci. Paris, t.316, série 1 (1993), pp. 1091-1096.

[GuM 94] PH.GUILLAUME AND M. MASMOUDI, *Compulation of high order derivatives in optimal shape design*, Numerische Mathematik, 67 (1994), pp. 231-250.

[Hil 85] HILLSTROM: *User guide for Jakef. Technical Memorandum*, ANL/MCS TM-16, Argone National Laboratory, Argone, II 60439, 1985.

[KaL 91] D. KALMAN AND R. LINDELL, *Automatic differentiation in astrodynamical modeling*, Automatic Differentiation of Algorithms, A. Griewank, G.F. Corliss (editors), 1991, SIAM.

[Kub 99] K. KUBOTA *A preprocessor for fast automatic differentiation - Applications and difficulties on practical problems*, in RIMS Kokyuroku 648 "fundamental Numerical Algorithms and their Software".

[Lay 91] J.D. LAYNE, *Applying automatic differentiation and self-validation numerical methods in satellite simulations*, Automatic Differentiation of Algorithms, A. Griewank, G.F. Corliss (editors), 1991, SIAM.

[MaF 92] T. MASANAO AND N. FUJII, *Second order Necessary conditions for Domain Optimization Problems in Elastic Structures*, Journal of Optimization Theory and Applications, 72, n°2, (1992).

[MaG 93] M. MASMOUDI AND PH. GUILLAUME: *Conception optimale de formes et applications*, Colloque en l'honneur de Jean Céa, Ed. Désideri, Fézoui, Larrouturou, Rousselet, Cepadues, 1993.

[Mas 87] M. MASMOUDI, *Numerical Solution for Exterior Problems*, Numerische Mathematik, 51 (1987), pp. 87-101.

[MPD 93] M. MASMOUDI, PH. BRUNET, TH. DUSSEUX AND M. SAURY, *Décomposition d'un répartiteur orthogonal en coupleurs élémentaires*, Patent with Alcatel Espace No 19711, 1993.

[Mor 85] J. MORGENSTERN, *How to compute fast a function and all its derivatives, a variation on the theorem of Baur-Strassen*, Sigact News, 1985.

[NBC 89] F. NAVARRINA, E. BENDITO AND M. CASTELEIRO: *High Order Sensitivity in Shape Optimization Problems*, Computer methods in applied mechanics and engineering 75 267-281, North-Holland, 1989.

[Ned 80] J.C. NÉDÉLEC. *Mixed finite elements in* $\mathbb{R}^3$, Numerische Mathematik, 35 (1980), pp. 315-341.

[NeS 91] J.C. NÉDÉLEC AND F. STARLING, *Integral equation methods in quasi-periodic diffraction problem for the time-harmonic Maxwell's equations*, SIAM J. Math. Anal., 22 No 6 (1991), pp. 1679-1701.

[OWB 71] G.M. OSTROVSKII, J.M. VOLIN AND W.W. BORISOV, *Uber die Berechnung von Ableitungen*, Wissenschaftliche Zeitschcrift der tecnischen Hochschule fur Chemie, Leuna Merseburg, 13 (1971), pp.382-384.

[RoD 92] N. ROSTAING AND S. DALMAS, *Automatic differentiation analysis and transformation of fortran program using a typed functional language* International Conference on computing methods in Applied Sciences and Engineering, Feb. 11-14, 1992, Paris.

[Roc 94] M. ROCHETTE, *Le manuel d'utilisation d'Adogen*, (1994).

[Sim 89] J. SIMON, *Second variation for domain optimization problems*, International Series of Numerical Mathematics, 91, Birkhauser, (1989).

[Spe 80] B. SPEELPENNING *Computing fast partial derivatives of functions given by algorithms*, PhD thesis, Department of Computer Science, University of Illinois at Urbana-Champain, Urbana-Champain,IL 61801, Jan. 1980.

[Str 90] V. STRASSEN, *Algebraic complexity theory*, in J. van Leeuwen (Editeur), Handbook of Theoritical Computer Science, volume A: Algorithms and complexity (1990), Elsevier, Amsterdam.

[Tal 91] O. TALAGRAND *The use of adjoint equations in numerical modeling of the atmospheric circulation*, Automatic Differentiation of Algorithms, A. Griewank, G.F. Corliss (editors), 1991, SIAM.

[Vas 85] CH. VASSALLO, *Théorie des guides d'ondes électromagnétiques*, Collection Technique et Scientifique des Telécommunications, Eyrolles, France, 1985.

[Wen 64] R.E. WENGERT, it A simple automatic derivation evaluation program Comm. ACM, 7 (1964), pp. 463-464.

LARGE SCALE TRACKED VEHICLE CONCURRENT ENGINEERING ENVIRONMENT

Kyung K. Choi, J. Kirk Wu, Kuang-Hua Chang, Jun Tang, Jia-Yi Wang, and
Edward J. Haug
Center for Computer Aided Design
College of Engineering
The University of Iowa
Iowa City, Iowa 52242, U. S. A.

ABSTRACT. In this paper, a fully integrated Tracked Vehicle Concurrent Engineering environment that exploits CAD and CAE technologies in support of simulation-based design of large scale tracked vehicles is presented. The Tracked Vehicle Concurrent Engineering environment comprises a series of engineering workspaces that include CAD/CAE Services, Tracked Vehicle Workspace, Dynamic Stress and Life Prediction Workspace, and Design Sensitivity Analysis and Optimization Workspace. These engineering workspaces are the principal functional components of the integrated simulation-based design environment that utilizes mechanical system data modeling techniques and ROSE object-oriented database to facilitate and manage data sharing. Wrappers have been developed to integrate these engineering workspaces by providing bi-directional data access and translation capabilities between the engineering workspaces and the database. In addition to these engineering workspaces, the Iowa Driving Simulator is integrated into the environment to provide a customer-driven simulation and design environment.

1. Introduction

The fact that simultaneous consideration of disciplinary engineering activities must occur early in the product development process to achieve competitive product quality, and reduced life-cycle cost and acquisition time, has motivated active research in concurrent engineering [1-4]. On the other hand, computer-based simulation technologies are maturing to the point that simulation results are comparable to test results [5]. With these recent developments in simulation technologies and computer systems, realization of a computer simulation-based concurrent engineering environment is now feasible. However, achievement of this goal has numerous technical challenges in developing a product data model and its management and communication methodologies for integration and support of broad range of disciplinary simulation tools in a CAE environment. Moreover, the environment must be easily extendible to the full scope of disciplines that must be supported for concurrent engineering of large scale tracked vehicles.

The simulation and design tools such as tracked vehicle dynamic performance analysis, structural analysis and optimization, component dynamic stress and life prediction, and driving simulator that are being developed at the Center for Computer Aided Design have been integrated into the concurrent engineering environment supporting concurrent system simulation and design of large scale tracked vehicles.

447

J. Herskovits (ed.), Advances in Structural Optimization, 447–482.
© 1995 *Kluwer Academic Publishers.*

Software integration of these simulation and design tools focuses on data management, information flow, and utilization of application wrapper concepts that are being developed at the Center. The integrated environment utilizes object-oriented mechanical system data modeling techniques and ROSE database to manage engineering information.

The scope of the simulation-based design environment and the range of engineering analysis and design tools that have been integrated are presented in Section 2. The global data management and tool integration methodologies are also presented in this section. The environment is applied to a design process of a tracked vehicle component and presented in Section 3. Finally, conclusions and future challenges in developing a comprehensive concurrent engineering environment are presented in Section 4.

2. Simulation-Based Concurrent Engineering Environment

The Tracked Vehicle Concurrent Engineering (TVCE) environment is illustrated in Fig. 2.1. The functional engineering workspaces that comprise this environment consist of CAD/CAE Services (CCS), Tracked Vehicle Workspace (TVWS), Dynamic Stress and Life Prediction Workspace (DSLP), Design Sensitivity Analysis and Optimization Workspace (DSO), and Iowa Driving Simulator (IDS). A variety of commercially available CAD and CAE software, including Unigraphics [6], Pro/ENGINEER [7], PATRAN [8] geometric and finite element modeler, DADS [9] dynamic simulation code, and ANSYS [10], NASTRAN [11], and ABAQUS [12] finite element analysis (FEA) codes are supported and utilized by these workspaces. The concurrent engineering system integration is supported by workspace wrappers; i.e., bi-directional data translation and data transfer capabilities, and the ROSE [13] object-oriented database. These engineering workspaces do not exchange mechanical system data directly. Rather, they communicate with the global database through wrappers and a network communication channel. Data management is provided by unifying the global data model, data access manager, version control manager, and graphical user interface; all of which comprise a data management layer-- Design Data Server (DDS).

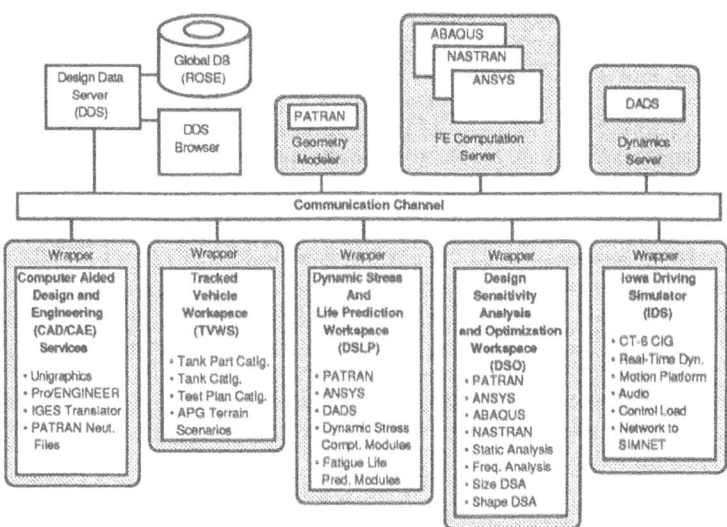

Figure 2.1 Tracked Vehicle Concurrent Engineering Environment

2.1 CAD/CAE SERVICES

2.1.1 CAE Model Creation The CAD/CAE Services (CCS) Workspace provides an environment for a multidisciplinary team to process CAD data for CAE applications. Within this environment, engineers can import CAD data, translate, and/or supplement with additional information to create CAE models for downstream CAE activities. Therefore, CCS is designed to serve as a bridge between design engineers, personnel who specify and draw various graphic/geometric objects, and simulation engineers, whose members test and evaluate models derived from CAD data. A CAE model created within this environment can be as large as a whole mechanical system or as small as a pin. A structural engineer can create finite element models of parts, and a dynamic simulation engineer can define mechanical systems, which contain bodies, joint/force connections, and assembly information for dynamic analysis.

In the CAD system, a mechanical system is described as a collection of discrete parts and part assemblies. A part has specific geometry and material properties. Several parts can be rigidly attached or assembled into a part assembly, which has composite mass property data. In this case, the parts of a part assembly are member parts of the assembly. However, from the dynamic analysis point of view, entities of a mechanical system (such as a tank) are, in general, classified into three types of functional entities: bodies (such as the hull and turret), joints (such as the bearing connecting the road arm and road wheel), and force elements (such as spring and hydraulic actuators), based on the functions of the entities in the multibody dynamic analysis. Each functional entity can be a part, a part assembly, or a collection of parts and part assemblies, from the CAD point of view.

To create CAE models from CAD models, engineers must be able to process mechanical system data created in the CAD system. For example, certain information such as mass properties, assembly, and connection have to be specified for each individual entity for dynamic analysis. Also, the geometric representation from CAD systems has to be translated into certain formats that can be interpolated by CAE tools for structural analysis or graphical animation.

It is important to define terminology in CCS that can be used by members of a multidisciplinary team. Definitions for some of the important terms used in CCS are given below:

A model is either an entity that already exists (which an engineer wants to analyze or modify) or an entity that does not exist (but an engineer wishes to create).

An elementary model is a fundamental object that cannot be further decomposed or broken down.

A composite model is a collection of several (at least two) member models which, in turn, can be either elementary or composite models. For this kind of model, assembly information, such as relative position and orientation between the member models, must be specified. The composite model can be defined only if all the related member models are already created. Therefore, models are constructed from the bottom-up in much the same way engineers build hardware.

Two assumptions are made based on the above definitions:

If a composite model is treated as a body in the dynamic analysis, then there should not be any joint/force connection between member models of the composite model.

If a composite model is treated as a mechanical system in the dynamic analysis, then joint/force connections between member models should be specified.

Geometric representations of models created by design engineers are accessible within CCS from the CAD system. Through CCS, engineers can launch the CAD system locally or remotely to retrieve CAD data and transfer the geometric representation to an IGES or PATRAN neutral file. Also, the geometric and finite element modeler, PATRAN, is accessible through CCS. In addition, an IGES translator has been developed and integrated into CCS. Functionalities provided by the IGES translator include verification of the geometry, reduction of information kept in IGES files, generation of different formats of geometric representation, and change of geometric representation.

The CAD system used in CCS is required to have a capability to calculate the mass, moments of inertia, and center of gravity from the geometry created and the density specified. The CAD system also has to be used to assemble member models of a composite model. The assembly information of a member model is defined by specifying the position and orientation of the geometric construction reference frame of the member model as a transformation matrix. The location and orientation of this reference frame are specified relative to the inertial reference frame of the composite model. Once all member parts of a composite model are defined in the CAD system, then correct model characteristics (such as the composite mass and the center of gravity) can be derived from individual member model information by assembling them.

The joint and force connections of a model specify features that are involved in assembling member models with a kinematic connector or a force element, respectively. For instance, a hole on a member model can be a kinematic interface, which specifies the geometric feature to assemble the model with another model by a pin. Joint and force reference frames of each mechanical system can be converted from a global coordinate system to the geometric construction reference frame of the member model of the mechanical system.

Using the CAD system, the CAD geometry can be translated into IGES or PATRAN neutral files. The IGES translator in CCS can then be used to examine the geometry translated from a CAD file and verify that it is equivalent to the original CAD model. The IGES translator can also be used to clean out unnecessary information created by the CAD system, change representation of the part, and translate geometry files to another format. Once a clean IGES file is generated using the IGES translator in CCS, or a PATRAN neutral file generated from the CAD system, PATRAN can be launched to create a finite element model which contains a solid model, finite element mesh, boundary condition, and material property.

2.1.2 Operational Scenario   The operational scenario of CCS is shown in Fig. 2.2. The runstream depends on the type of CAE model to be created. Different CAE models require different runstreams since different analysis codes require different input data. For example, structural analysis requires a finite element model, and dynamic analysis would require the entire mechanical system.

Dynamic analyst can use CCS to create both body and mechanical system definitions for multibody dynamic analysis. Geometrical representations for animation and mass properties can be generated to completely define the body. The mechanical system definition requires assembly information of each body in terms of a set of PQR coordinates [14], which describes the position and orientation of each individual body relative to a global reference frame. Connectivity information, describing the connection type between a pair of bodies and one or two sets of PQR coordinates, are also required for mechanical system model creation.

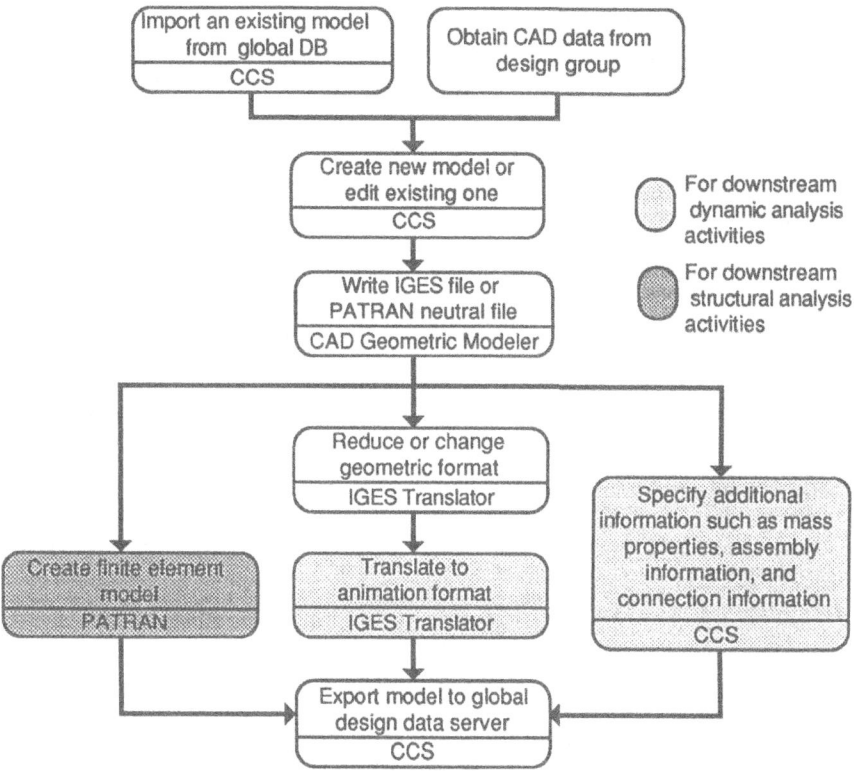

Figure 2.2 Flow Chart for CAD/CAE Services

To summarize, using CCS, the engineer can:

import existing design models or start new CAD data to create new models
translate CAD geometry output files so that PATRAN can generate finite
element models for structural analysis
translate CAD geometry output files to animation format for supporting
visualization of vehicle system in dynamic analysis
specify information required for dynamic analysis such as mass properties,
connection, and assembly information
export CCS generated data files to the global database via the Design Data
Server (DDS)

The CCS stores the CAD and CAE data in an organized manner by providing a
hierarchical structure for all data. The files and the location of the data file will be
controlled by CCS. The engineer can edit any information necessary for use in the CAE
application. The CCS is designed to be used by CAE engineers working in engineering
workspaces such as DSLP, DSO, or TVWS. It helps CAE engineers by informing data that
are required for a particular application, acting as an interface with commercial codes,
controlling the working directories whenever a commercial code is invoked, and
providing a communication channel to the global database. Therefore, common data
definition can be shared among team members.

2.1.3 Data Wrapper Through the CCS wrapper, engineering data are translated. For example, definition of the moments of inertia in the CAD system may be different from those defined in dynamic analysis code DADS. In CCS, the moments of inertia are specified relative to a geometric construction reference frame, which may not be the same as the centroidal reference frame used in DADS. When mass properties data is exported to the global database, CCS will automatically calculate the moments of inertia, following definitions specified in DADS, with respect to the centroidal reference frame. Using the same rule, when mass properties are imported from the global database to CCS, the moments of inertia, following the definition of the CAD system, with respect to geometric construction reference frame will be automatically calculated.

For another example, in DADS, the joint/force reference frames of regular bodies of a tracked vehicle, such as hull, gun, and turret, can be defined relative to the user defined non-centroidal body-fixed reference frame (NCBF). For this type of vehicle system, the default NCBF of the roadarm is defined as the geometric construction reference frame. In order to get correct loading histories, the geometric construction reference frame must be consistent with the definition of roadarm body-fixed reference frame inside DADS.

2.2 TRACKED VEHICLE WORKSPACE

The Tracked Vehicle Workspace (TVWS) [15,16] is a dynamic simulation software system designed to assist journeyman engineers to evaluate dynamic performance of vehicle systems and to estimate the dynamic loading histories and motions of vehicle components using a dynamic simulation code DADS.

The functional requirements of TVWS are to (1) provide a database system to manage dynamic characteristic data of vehicle components, (2) allow the engineer to change design by replacing vehicle parts with standard parts, (3) generate dynamic analysis model for several standard simulation scenarios without requiring too much user interaction, and (4) provide basic graphical visualization utilities for the engineer to review analysis results and extract desired data.

Initially, TVWS has been developed to directly integrate with a CAD system I/EMS [17]. When TVWS is integrated into the TVCE environment, its connection to the CAD system is modified to connect directly with the global database.

The TVWS has two basic functional modules: vehicle performance evaluation system and interface to the global database. The vehicle performance evaluation system is the main body of TVWS and is composed of a database object manager, vehicle performance evaluation controller, and dynamic simulation code DADS, as shown in Fig. 2.3. The dynamic simulation can be carried out on a computation intensive machine, while the remaining modules may run on an engineering workstation.

2.2.1 Database Object Manager The TVWS database object manager is composed of a database and manipulation functions that manipulate database objects. Vehicle and simulation characteristics, which are called database objects, are stored in the TVWS database through the manipulating functions. The TVWS database is an object data management system developed on top of a relational database system, Informix [18], using National Institutes of Health (NIH) object classes [19] and C++ programs. The TVWS database is a repository for all part characteristics and simulation parameters to generate simulation models, run through simulations, and examine the simulation results. Because most of the parameters for dynamic simulation are stored in the database, it is possible to make quick parametric changes to a vehicle model or the definition of a simulation scenario and rerun the simulation.

The database object manipulating functions are used to retrieve and present tracked vehicle and simulation characteristics. These functions include Main Menu Bar, Database Hierarchical Viewer (Browser), Table Editor, 2D Plotter, 3D Animation Utility, and Text Editor. The Main Menu Bar is the controlling device of the TVWS. Its primary functions are to command the TVWS to (1) import vehicle part information from the global database, (2) launch or stop dynamic simulation runs, and (3) execute database object functions. The Browser is a graphical user interface that reflects the layout of database objects as stored in the TVWS database. It allows engineers to view tracked vehicle data, which are organized hierarchically in the database. The Table Editor is used to view and edit ASCII information in a table format. The 2D Plotter represents sets of coordinate pairs in a two-dimensional curve format. The 3D animation utility displays three-dimensional representations of a tracked vehicle and a terrain condition, either animated or in a static frame, to assist the engineer to visualize vehicle initial conditions and simulation results in a meaningful form. The Text Editor is for special users, such as senior simulation engineers, to define standard test simulation or modify data files for advanced dynamic simulation.

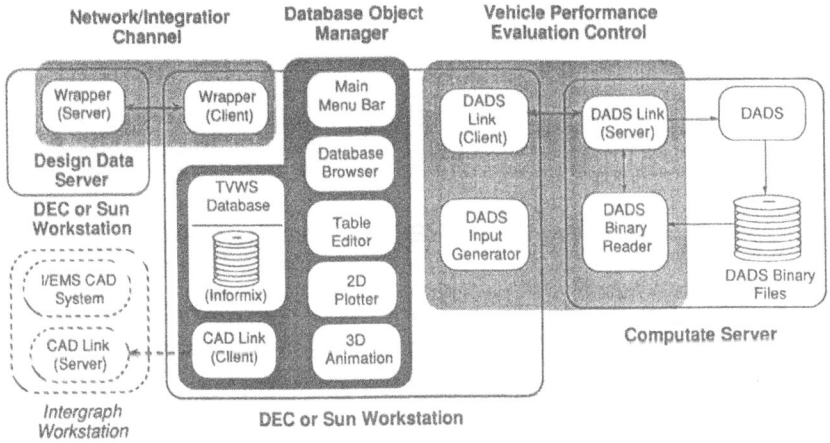

Figure 2.3 System Diagram of Tracked Vehicle Workspace

The information organization principle in the database is its object hierarchy. The database consists of a hierarchy of objects with data stored in the leaf nodes of the hierarchy [20]. Nodes in the database object hierarchy are called objects. The TVWS database provides six objects; VECTOR, CATALOG, RACK, POINTER, TABLE, and FILE. The CATALOG objects are containers for other TVWS objects and are used to form information hierarchy. The database root catalog contains four sub-catalogs: library, templates, user_catalogs, and TVWS information. The RACK objects are also containers with a fixed number of slots and is for certain kinds of database objects. Each slot has a name, and only a certain kind of object may fill a slot. The RACK objects are developed for engineers to group related pieces of engineering information to define a functional or logic unit of a vehicle system or simulation conditions, such as a suspension system and a fire-on-move test scenario.

To make the vehicle information stored in the database transparent to engineers and assist them to quickly get, review, and even modify the information, a graphical viewer of the database object hierarchy, called Browser, has been developed, as shown in Fig. 2.4. In Browser, engineers may browse through different levels of the database hierarchy, back up to a previous point, scroll through the hierarchy, edit the data using the Table Editor and the Text Editor, and view the data using the Table Editor, 2D Plotter, or 3D Animation.

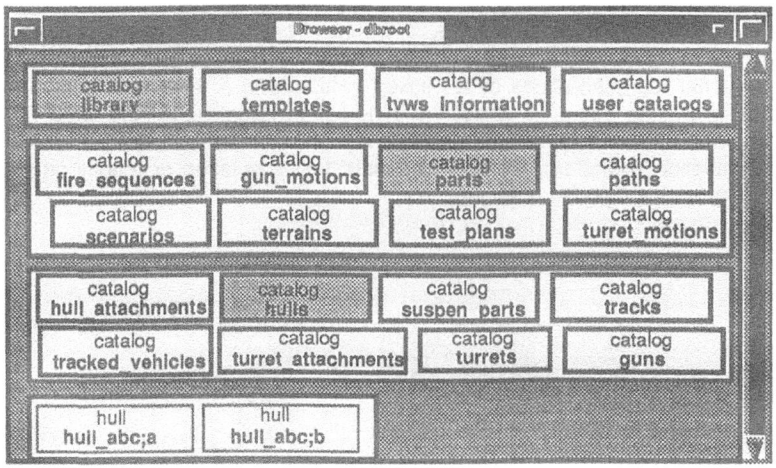

Figure 2.4 TVWS Database Browser From Root Catalog

To provide an easy way for simulation engineers to incorporate new design of tracked vehicle components in dynamic simulation models, database object Cut, Copy, and Paste operations have been developed. The engineer can select an object, invoke Cut or Copy to place the object in memory space, called clipboard, and paste the contents of the clipboard to a compatible container. Data checking capabilities are provided when pasting an object into racks, but not for catalogs. The hierarchical data structure and Cut, Copy, and Paste operations simplify the process to generate new vehicle models to evaluate alternative design. For instance, an engineer can combine a standard suspension system with a new design of a hull to generate a new tracked vehicle.

2.2.2 Vehicle Performance Evaluation Although the contents of TVWS objects in the database are sufficient for defining vehicle models for dynamic simulation, the model information needs to be retrieved from the TVWS database. To evaluate dynamic performance of a vehicle, the journeyman engineer needs not only the dynamic simulation models being generated automatically but also an efficient way to launch and control the simulation. The performance evaluation controller is developed for these purposes, which contains mainly DADS Link and DADS Input Generator.

The DADS Input Generator is a set of modules used to retrieve and assemble dynamic simulation model information from (1) the TVWS database and (2) DADS template file library and generate dynamic simulation input files. A default DADS input file of a vehicle is broken down into several DADS template data files. These template files are stored in a library in the template catalog of the TVWS database, called DADS template file library. The DADS input file generation flow, as depicted in Fig. 2.5, is divided into three steps: preliminary checking, loading necessary modifications (relative to the default DADS input file), and generating command files for DADS pre- and post-processors.

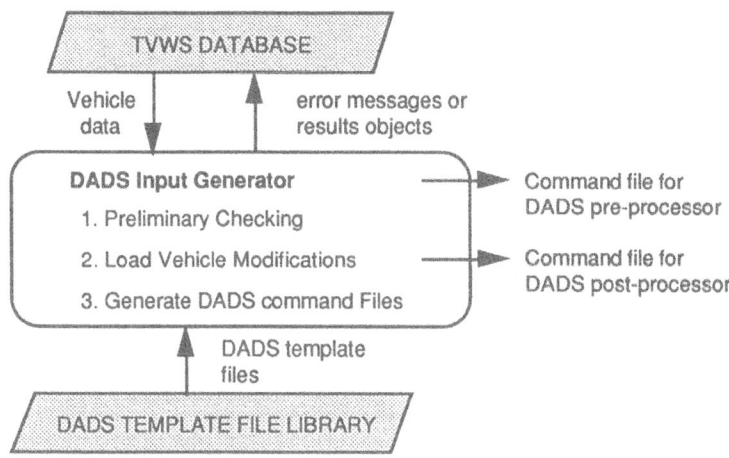

Figure 2.5 DADS Input Data Generation Flow

In the preliminary checking step, the DADS Input Generator checks (1) if a proper simulation test plan, test scenario, and tracked vehicle are selected; (2) if the engineer is authorized to perform simulation tests; (3) if all test required data are in the database; and (4) if the simulation test has already been executed. The functions of the second step are to create statements to (1) load template files from the DADS template file library, (2) load test plan and test scenario attributes that the engineer chose, and (3) load vehicle definitions that are stored in the database and overwrite the data in the template files. The last step in DADS Input file generation process is to create command files for DADS pre- and post-processors. In the DADS post-process command file, parameters for a simulation run that the engineer wants are defined. These parameters may be selected from standard vehicle performance parameters that the engineer prefers to compare with. The files generated in this step can then be loaded into DADS pre-processor and post-processor to analyze the dynamic behavior of the tracked vehicle.

The DADS Link provides communications between the workstation and machine where DADS runs. Hence the journeyman engineer can sit in front of the engineering workstation and conveniently monitor the ongoing simulations. If the model characteristics are not reasonably set up, or if simulation parameters (such as integration step size) are not appropriate for the vehicle configuration, the simulation results will not be useful. These flaws can be visually detected using the 2D Plotter or 3D animation Utility. If a simulation is not acceptable, the engineer can stop the simulation to save CPU time, and initiate new simulation activities.

2.2.3 Data Wrapper Figure 2.6 shows the data flow diagram of an entire design information importing and vehicle design evaluation activity from the TVWS user point of view. Parts of a tracked vehicle that are created in the CAD system and stored in the global database are imported into the TVWS database via the TVWS wrapper. The parts are then used to generate dynamic simulation models. Once a journeyman simulation engineer selects a test plan and scenario, the test plan and scenario information will be

456

added to the simulation model definition. The engineer can instruct TVWS to generate a dynamic simulation model using the vehicle component information stored in the TVWS database. Then the model is fed to a dynamic simulation code, and, finally, on-line or off-line simulation results are returned to the TVWS database for the engineer to review. Acceptable dynamic simulation results can be sent to the global database for vehicle component structural analysis and design through the TVWS wrapper.

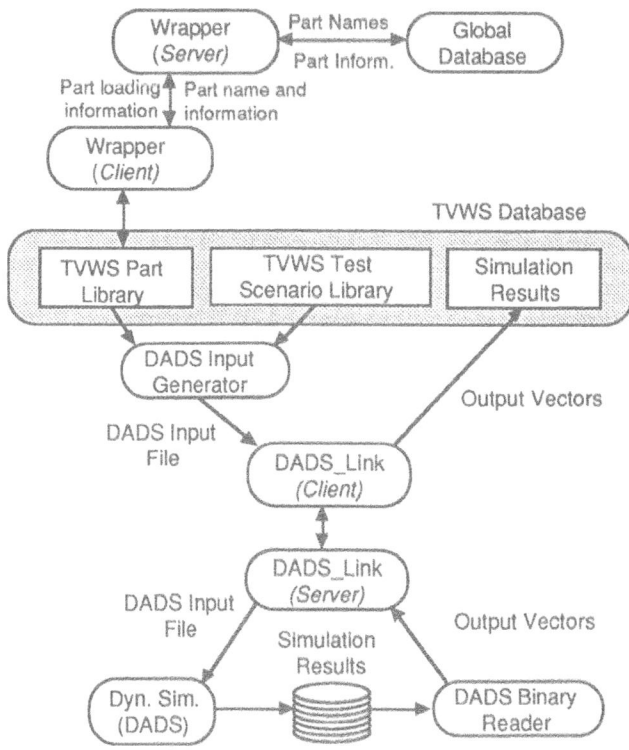

Figure 2.6 Data Flow of Tracked Vehicle Dynamic Simulation

2.3 DYNAMIC STRESS AND LIFE PREDICTION WORKSPACE

The Dynamic Stress and Life Prediction (DSLP) Workspace predicts mechanical component fatigue life. The DSLP can evaluate flexible body dynamic stress and fatigue life of a mechanical component, based on either crack initiation and propagation criteria. Two engineering computational phases, dynamic stress computation and fatigue life prediction, are implemented in DSLP to allow the engineer to carry out simulation-based design procedure. The DSLP is a framework that includes related CAE tools, user interfaces, local database, remote execution, and working flows that show engineers the procedure to operate and interact with the workspace.

The functionality of DSLP is decomposed into processes for dynamic stress computation and fatigue life prediction. The engineering analysis phases are depicted in

Fig. 2.7, where the ellipses indicate the input and output data, while the rectangles indicate the engineering computational phases.

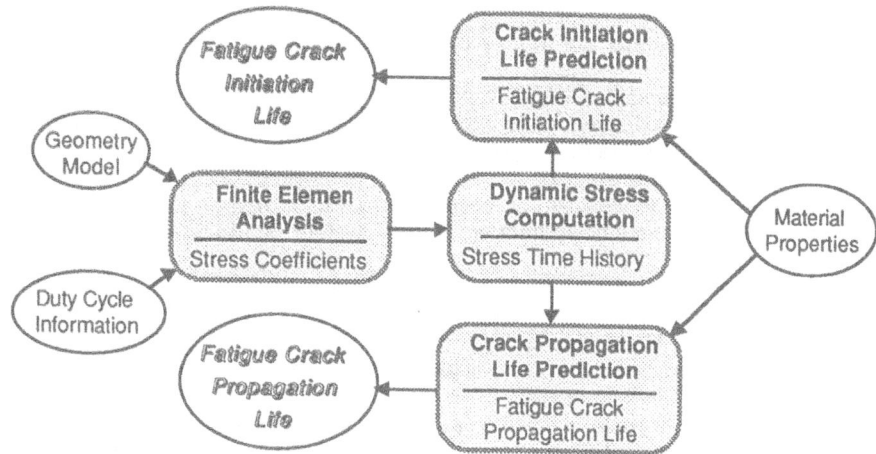

Figure 2.7 Engineering Analysis Phases

The dynamic stress computational phase calculates dynamic stress histories of vehicle components at finite element nodes based on the duty-cycle loads estimated from multibody dynamic simulation. A hybrid method [21-24] to couple nonlinear gross motion and linear elastic deformation for dynamic stress computation has been developed and integrated into DSLP. The algorithm allows dynamic stress computations to be carried out using modal stress and stress influence coefficients. Also, this algorithm uses the stress field due to inertia forces instead of calculating the stress field with unit load at each nodal point. Since the number of variables involved in the nodal acceleration expression is much less than the number of nodes in the finite element model, the number of stress fields needed to be computed can be significantly reduced.

The fatigue life prediction phase calculates values related to the fatigue life of the mechanical component. A local strain approach is used for crack initiation fatigue life prediction in DSLP, and failure is said to have occurred when a crack has grown to approximately 2 mm. A stress intensity approach based on linear elastic fracture mechanics (LEFM) is used for crack propagation fatigue life prediction. The crack growth rate and the crack length after certain cycles are the results describing the crack propagation. The rain flow counting method is also used for counting the number of cycles of dynamic stress time history, and by plotting the true stress-true strain diagram to find hysteresis loops. Finally, the effect of the individual cycles on the life of the component is calculated using the Palmgren-Miner cumulative damage rule in DSLP.

2.3.1 Dynamic Stress Computation For dynamic stress computation, the quasi-static method is used to calculate finite element nodal stresses due to joint reaction forces and inertia forces. The space-dependent part of inertia force and unit loads are applied to the mechanical component to obtain stress influence matrices using FEA. Then, the stress influence matrices, the time-dependent part of the inertia force, and dynamic loads (estimated from dynamic simulation) are superposed to calculate dynamic stress time histories. The computational procedure has five parts: (1) get geometry model and duty cycle information from global database through the Design Data Server; (2) generate unit joint reaction and quasi-inertial forces applied to the component; (3) extract stress influence coefficients of the component from FEA results; (4) extract the dynamic loading histories and dynamic parameter histories from dynamic simulation results; and (5) compute dynamic stress time history by superposing stress coefficients and dynamic parameters.

The inertia load vectors are formulated by multiplying the system lumped mass matrix and the space-dependent parameter matrix. Therefore, load vector generation deals with the space-dependent parameter using finite element information and lumped mass data. The unit loads are used to calculate nodal stress coefficients of joint reaction and external forces. The identifiers of nodes at which joint reaction forces or torques should be applied are determined from mechanical system simulation and finite element models.

The finite element model used to compute the stress influence coefficient matrix can be generated using a PATRAN neutral file or using ANSYS or NASTRAN directly. To obtain the stress coefficients for all load cases requires multiple FEA with different load conditions. In order to perform multiple analyses, the load vector in each case is specified in the finite element model.

A data extraction module is used for extracting joint reaction force, nodal coordinates, modal coordinate, body reference frame velocity and acceleration, as well as modal coordinate displacement, velocity, and acceleration from dynamic analysis. All these data are required by dynamic stress computation and should be relative to a local body-fixed coordinate.

The superposition principle is applied to formulate dynamic stress history. The stress influence matrix and duty cycle information are used to obtain dynamic stress histories at every finite element node of the component. The flow chart of dynamic stress computation in DSLP is shown in Fig. 2.8.

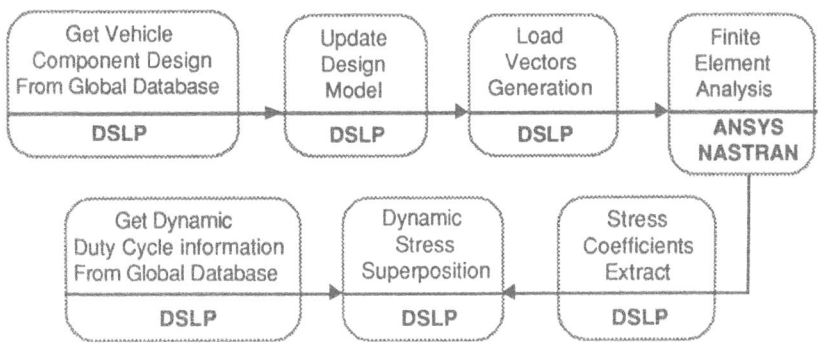

Figure 2.8 The Flow Chart of Dynamic Stress Computation in DSLP

2.3.2 Fatigue Life Prediction   There are two parts in the fatigue life prediction phase: crack initiation and crack propagation life predictions. The Morrow and Smith-Watson-Topper methods based on local strain approach have been implemented for crack initiation life prediction in DSLP, whereas a public domain software package FLAGRO [25] developed by NASA has been integrated in DSLP for crack propagation life prediction. The FLAGRO uses a modified Forman's equation based on a stress intensity approach. To use these methods, local stress/strain and stress spectra have to be calculated, and then the equivalent number of cycles has to be found. Therefore, the procedure for predicting fatigue life can be broken down into five parts: (1) calculate principal stress and von Mises stress histories using dynamic stress time history tensor; (2) compute local stress and strain by screening principal stress peak and valley; (3) count the equivalent cycle number using the rain flow counting method; (4) predict each cycle crack initiation life and add them up to get the total crack initiation life; and (5) launch FLAGRO to predict fatigue crack propagation life for a specified crack length.

The main input data is the nodal stress tensors that are obtained from dynamic stress computation. The first and second elastic principal stress histories can be calculated from these stress tensors. Screening both stress histories is carried out to capture the peak and valley values. A peak value is defined as a local maximum, whereas a valley is defined as a local minimum.

Using the magnitudes of both principal stresses, biaxial factors are computed for each pair. The local stress and strain magnitudes are computed by simultaneously solving Neuber's rule and the nonlinear stress-strain relationship using Newton-Raphson numerical method. During the computation, necessary material properties are automatically retrieved from the material property data library.

The rain flow counting algorithm used in this part assumes that all reversals in a block of true strain data are part of a cycle, so that there are no hanging reversals at the end of computation. This number can be converted into time by considering the simulation time of the mechanical system.

The constant amplitude fatigue damage curve represents a set of tests together with associated lives. Operation at a certain amplitude will result in failure in, say, N cycles. Operation over a spectrum of different levels results in a partial damage contribution from each cycle. Failure is then predicted when the sum of these partial damage fractions reaches unity. For the crack propagation, the system generates a script file to launch FLAGRO. This script file uses parameters provided by engineers through user interfaces. The flow chart of life-prediction computation in DSLP is depicted in Fig. 2.9.

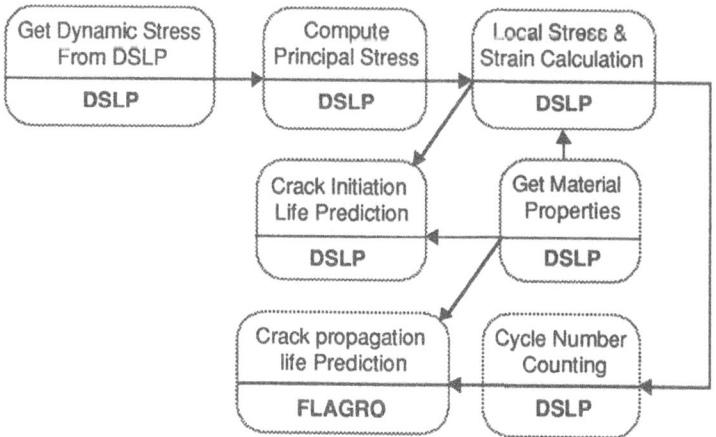

Figure 2.9 The Flow Chart of Life Prediction Computation in DSLP

2.3.3 Frameworks The DSLP is an integrated CAE environment that is designed to be user-friendly and does not require special knowledge of system operations or computer programming. DSLP integrates FEA, dynamic stress computation, and fatigue life prediction. This is done by integrating a number of specialized commercial software packages including PATRAN [8], ANSYS [10], NASTRAN [11], and FLAGRO [25].

Since the amount of information manipulated by DSLP is not very large or complex the DSLP database needs not be a full-fledged database system. It uses the host UNIX file system and data extractor software for local transient data. A heterogeneous computer system can be used for DSLP in executing different commercial packages. The C shell script files are used in DSLP for remote executions. Since CAE applications are highly interactive and model-oriented, the graphics user interface windows are developed in DSLP to visualize objects and events.

2.3.4 Data Wrapper The load characteristics, PATRAN geometric and finite element models, and material properties of mechanical components are imported from the global database through the wrapper. The loads applied to a component for fatigue life prediction are joint reaction, force element, and inertial forces and torques. After getting input files from the global database, the engineer can run DSLP following the runstream provided in a user interface window. The DSLP users can exported design modifications suggested from the fatigue life perspective to the global database for other concurrent engineering team members to consider.

2.4 DESIGN SENSITIVITY ANALYSIS AND OPTIMIZATION WORKSPACE

The Design Sensitivity Analysis and Optimization workspace (DSO) provides the design engineer with a visually driven design environment for efficient design optimization of mechanical components [26-28]. Three design stages, pre-processing, design sensitivity analysis (DSA), and post-processing, are implemented in DSO to allow the design engineer to carry out the design process systematically. A framework, that includes a local database, user interface, foundation class, and remote module, has been designed and implemented to facilitate software development for DSO. Also, a number of dedicated commercial software/packages have been integrated in DSO to support design activities, e.g., PATRAN [8] is used as a geometric modeler to create geometric and finite element models; ANSYS [10], NASTRAN [11], and ABAQUS [12] are used to perform FEA; and the HARWELL QP Solver [29] and VMA's Design Optimization Tool (DOT) [30] are used for design trade-off and optimization, respectively. Moreover, a wrapper incorporates DSO into the Concurrent Engineering environment.

2.4.1 Engineering Design Stages Organized into three design stages: pre-processing, DSA, and post-processing, DSO allows the design engineer to carry out the design process systematically. In DSO, instead of parameterizing a finite element model, design parameters are defined on a geometric model associated with physical quantities. The DSA capability in DSO uses the continuum DSA theory to compute design sensitivity coefficients using FEA results. Finally, a four-step interactive design process is developed to support effective design improvement.

2.4.2 Pre-Processing Design Stage The major goal in the pre-processing design stage is to formulate a design problem by creating a design model (including geometry and finite element models); parameterizing the model; carrying out FEA, finite element error analysis, and mesh adaptation [31]; and defining performance measures, cost, and constraints. Figure 2.10 illustrates the processes and activities in this stage.

Design Parameterization Design parameterization is a key step in the structural design process. The purpose of design parameterization is to define parameters to characterize the properties of the geometric entities for sizing design applications or to characterize the movements of geometric control points that govern the shape of the structural boundary. The engineer selects a subset of these parameters as design parameters to vary in the design process in order to improve structural performance. Design parameters must be defined based on both design and manufacturing considerations.

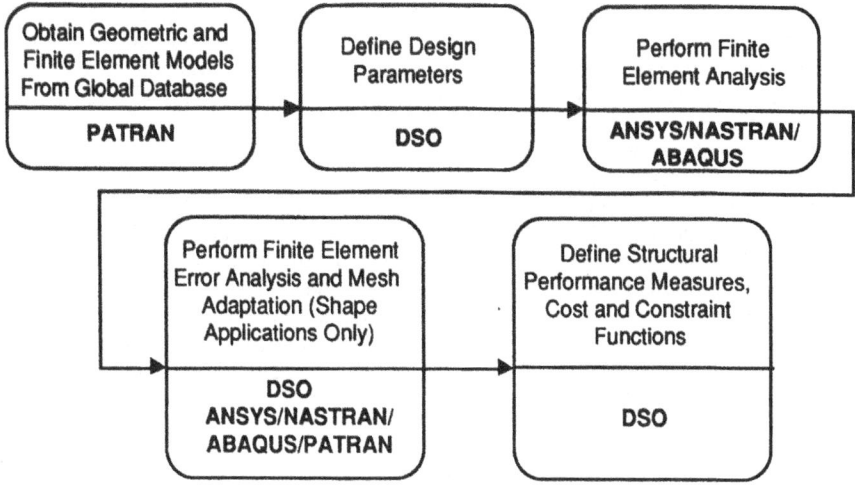

Figure 2.10 Pre-Processing Design Stage

Sizing Design Applications The DSO supports constant and linear sizing design parameterizations, as shown in Fig. 2.11. Geometric parameters are defined at end grid points of a line or at corner points of a patch (a spatial surface in PATRAN). Bilinear thickness distribution can be used to characterize a surface design entity, as shown in Fig. 2.11. Note that each dimension that defines the cross-sectional shape in Fig. 2.11, such as width or height, could be treated as a design parameter and be allowed to vary in the same amount as the corresponding parameter at the other end (constant parameterization), or in different amounts (linear parameterization). Moreover, through design parameter linking, design parameters can vary independently of or proportionally to certain parameters across design entities, to maintain design continuity for symmetric design or to reduce the number of design parameters.

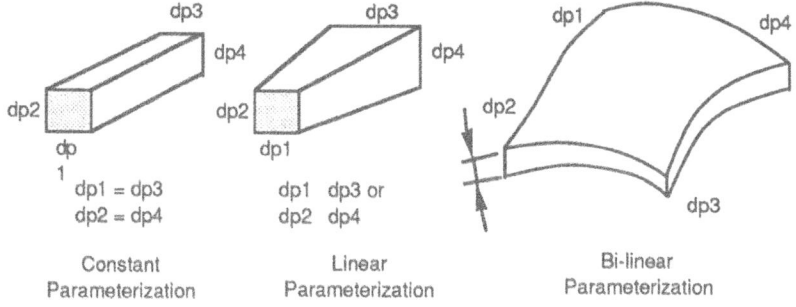

Figure 2.11 Line and Surface Design Parameterization

Shape Design Applications The shape design parameterization method developed in DSO parameterizes geometric features. A geometric feature is a subset of the geometric boundaries of a structural component. For example, a fillet or a circular hole is a geometric feature that has certain characteristics associated with it and may be chosen as the design. A geometric feature with design parameters defined is a parameterized geometric feature and is treated as a single entity in the shape design process. For example, a circular hole, with the radius and location of its center defined as design parameters, is a parameterized geometric feature. In accordance with design changes, the parameterized circular hole can be moved around in the structure, and its size can be varied. However, the shape of the circular hole is retained.

A three-step shape design parameterization procedure has been developed in DSO. The first step is to create a geometric feature by grouping a number of inter-connected geometric entities and defining the type of the geometric feature. The second step is to define design parameters within each geometric feature. To generate a parameterized geometric feature, the engineer can use the design parameter definition within the geometric entities and link design parameters across the entities. The third step is to link design parameters across parameterized geometric features, if necessary.

The DSO allows the engineer to place geometric features that are frequently used in construction of structural entities in a library of predefined geometric features. For example, the circular hole and tapered slot shown in Fig. 2.12 could be placed in the library.

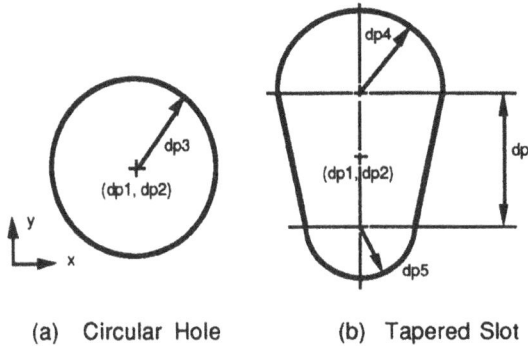

(a) Circular Hole   (b) Tapered Slot

Figure 2.12 Predefined Geometric Features

The shape design parameterization method developed for DSO uses the geometry representation defined in PATRAN. In PATRAN, all geometric entities are represented using parametric cubic (PC) lines, patches (surfaces), and hyperpatches (solids). A planar PC line is represented as [8]

$$p(u) = a_3 u^3 + a_2 u^2 + a_1 u + a_0$$

$$= [u^3 \ u^2 \ u \ 1] \begin{bmatrix} a_3 \\ a_2 \\ a_1 \\ a_0 \end{bmatrix}_{4 \times 2}$$

$$= \mathbf{U} \, \mathbf{A}, \qquad u \in [0,1] \tag{2.1}$$

where $\mathbf{p}(u) = [p_x, p_y]$, u is the parametric direction of the line with domain [0,1], and $\mathbf{a}_i$ = $[a_{ix}, a_{iy}]$, i = 0 to 3, are the algebraic coefficients of the curve.

Equation 2.1 indicates that any PC line can change the sign of its slope at most twice, and it can have only one inflection point. Consequently, PC entities such as PC lines and PC patches minimize the possibility of yielding oscillatory boundaries in the design process [32]. However, certain geometric features with predefined shapes or sophisticated geometry, such as a circular hole, cannot be represented by a single cubic geometric entity. To minimize modeling errors, it is necessary to model these boundaries by breaking them into small pieces. In the design process, these pieces must be "glued" together as one geometric feature, by linking design parameters appropriately. For shape design, planar PC lines and spatial parametric bicubic patches represent the design boundaries of 2-D and 3-D structural entities, respectively.

For 2-D structural shape design, the design boundaries are planar curves. In general, there are eight degrees of freedom for a planar parametric cubic curve. For 2-D shape design problems, DSO supports three curves: geometric, four-point, and Bezier.

For 3-D structural shape design, the design boundaries are surfaces in space. In general, there are forty-eight degrees of freedom for a parametric bicubic surface. The mathematical expression for a bicubic parametric surface is

$$
\begin{aligned}
\mathbf{p}(u,w) &= \sum_{i,j=0}^{3} a_{ij} \, u^i \, w^j \\
&= [u^3 \ u^2 \ u \ 1] \begin{bmatrix} a_{33} & a_{32} & a_{31} & a_{30} \\ a_{23} & a_{22} & a_{21} & a_{20} \\ a_{13} & a_{12} & a_{11} & a_{10} \\ a_{03} & a_{02} & a_{01} & a_{00} \end{bmatrix}_{4\times4\times3} \begin{bmatrix} w^3 \\ w^2 \\ w \\ 1 \end{bmatrix}_{4\times1} \\
&= \mathbf{U} \, \mathbf{A} \, \mathbf{W}^T , \quad (u,w) \in [0,1] \times [0,1]
\end{aligned}
\tag{2.2}
$$

where $\mathbf{p}(u,w) = [p_x, p_y, p_z]$, $a_{ij} = [a_{ijx}, a_{ijy}, a_{ijz}]$ are the algebraic coefficients of the surface, and u and w are the parametric directions of the geometric entity.

In DSO, geometric, 16-point, and Bezier surfaces are supported for 3-D shape design problems.

Definition of Performance Measures, Cost and Constraints The DSO supports seven types of performance measures: mass, volume, displacement, stress, compliance, frequency, and buckling. Among these, mass, volume, compliance, frequency, and buckling are global performance measures of the structural system. Displacement and stress, however, are defined at specific points or elements in the structure and are considered local performance measures. Displacement performance measures can be defined by selecting nodes, degrees of freedom, and load cases. Stress performance measures can be defined at Gauss points or averaged in an element. Also, for each loading case, stress measures are defined using material failure criteria, such as von Mises, maximum shear, or maximum or minimum principal stresses.

These structural performance measures can be combined to define the cost and constraint functions of the design optimization problem. That is,

$$
\text{Cost or Constraint} = \sum_{i=1}^{n} a_i \, \psi_i^{c_i}
\tag{2.3}
$$

where ψ_i is a performance measure; a_i and c_i are real and integer coefficients, respectively; and n is the number of performance measures employed to define the cost or constraint function. The cost function, constraint functions with bounds, and design parameters with bounds form a design problem that can be used for trade-off determination and/or design optimization.

2.4.3 Design Sensitivity Computation Stage In the design sensitivity computation stage shown in Fig. 2.13, the design sensitivity coefficient matrix is computed for performance measures with respect to the design parameters defined in the pre-processing stage. The sensitivity computation in DSO uses the continuum DSA method [33], which is very efficient, accurate, and general. Moreover, the sensitivity coefficients are computed outside the FEA codes using only post-processing data from the FEA codes. The sensitivity computation in DSO has been integrated and automated so that the program is executed and necessary data are transferred and accessed without the engineer's interaction. To compute design sensitivity coefficients, the engineer simply clicks a menu button.

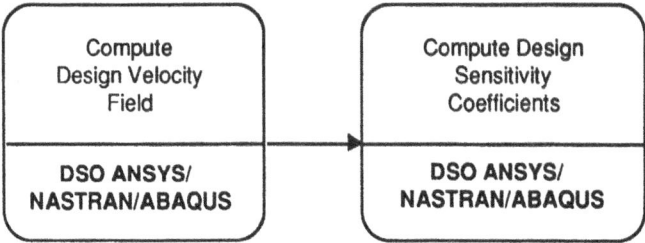

Figure 2.13 Design Sensitivity Computation Stage

2.4.4 Post-Processing Design Stage The post-processing design stage in DSO is a four-step interactive design process, which includes design sensitivity display, what-if study, trade-off determination, and design optimization. This interactive design environment, shown in Fig. 2.14, allows the engineer to improve designs, using the design sensitivity coefficients. The first three design steps help the engineer understand the structural behavior of the current design and guide the engineer toward obtaining better designs. The last design step is the design optimization, which allows the engineer to launch commercial optimization codes. The post-processing stage in DSO does not dictate to the engineer; instead, it provides sufficient design information and design suggestions for the engineer to make appropriate design decisions.

The engineer can use the four-step design process in various ways, depending on the objectives. For instance, if the engineer is looking for an improved design, he or she can go through the first three design steps to understand structural design trends of the current design and then change the design based on his or her understanding of structural behavior or suggestions made by DSO. Once changes are made, DSO updates the design model, analyzes it, computes its design sensitivity coefficients, and gets ready for the next design iteration. Instead of carrying out design interactively at each iteration, the engineer can also ask the design optimization module to perform a certain number of design iterations in batch mode. After several design iterations, the engineer can return to the first three design steps to display the design sensitivity coefficients and perform several what-if or trade-off studies, before deciding the next design direction. When an optimum design is obtained using the optimization algorithm, the engineer might again go through design sensitivity display, what-if, or trade-off design steps to get a thorough understanding of the structural behavior of the optimum design or to provide tolerance information to the manufacturer.

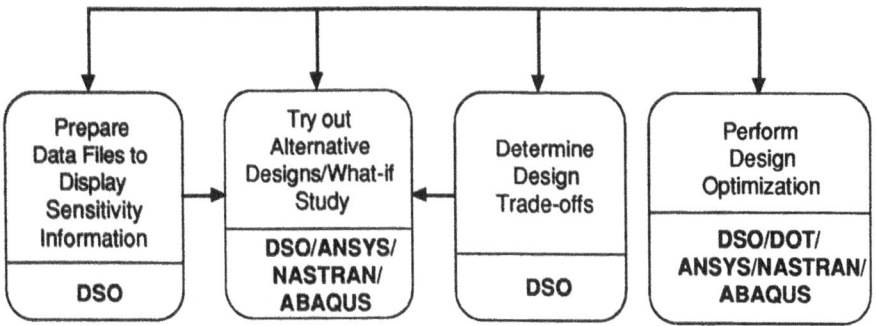

Figure 2.14 Four-Step Design Process In the DSO

2.4.5 Design Optimization in Post-Processing For design optimization, a commercial optimization code, DOT is integrated into DSO to serve as a nonlinear programming solver. Three algorithms in DOT, the Modified Feasible Direction Method, the Sequential Linear Programming Method, and the Sequential Quadratic Programming Method, are available to solve constrained optimization problems.

The DSO uses the procedure shown in Fig. 2.15 to update the design model. The DOT provides design vector values that defines the new design. This vector is then sent to DSO to update design parameterization information defined in the design model. The DSO computes finite element section properties of the new design for sizing design applications or creates new finite element mesh for shape design applications. The DSO then updates the finite element input data file for the FEA code used. The updated file is then sent to the FEA code to perform analysis. The finite element interface is executed again to retrieve structural responses. Once the structural responses are retrieved, performance measure values and cost and constraint function values can be updated. And finally, design sensitivity coefficients for the new design are computed. The same model update process is employed for perturbed designs obtained from the what-if studies for interactive-mode design optimization.

2.4.6 Frameworks A framework that includes the database, user interface, foundation class, and remote facility, has been designed and implemented to facilitate software development of DSO.

<u>User Interface</u> The backbone of the user interface developed in DSO is the engineering spreadsheet. The spreadsheet is utilized in DSO to allow the engineer to browse important design data, define and modify existing data, and create formulas to define design parameters [34]. The spreadsheets are developed using an object-oriented approach [35]. A general spreadsheet is developed as a framework, and specialized spreadsheets are constructed by inheriting from and adding properties to the general spreadsheet. This spreadsheet-based user interface is developed using OSF-Motif [36] based on the X Window System [37].

<u>Remote Facility</u> The DSO remote facility provides flexibility in configuring the design environment by permitting computationally intensive tasks, such as FEA, to be executed at the mainframe or supercomputer, instead of the workstation. In addition, the remote

facility permits data files to be transferred to the other graphical workstations so that the engineer can visualize model data. The remote facility thus enables DSO to better utilize the graphical capability of the workstation and the computational power of the mainframe. The remote configuration is illustrated in Fig. 2.16.

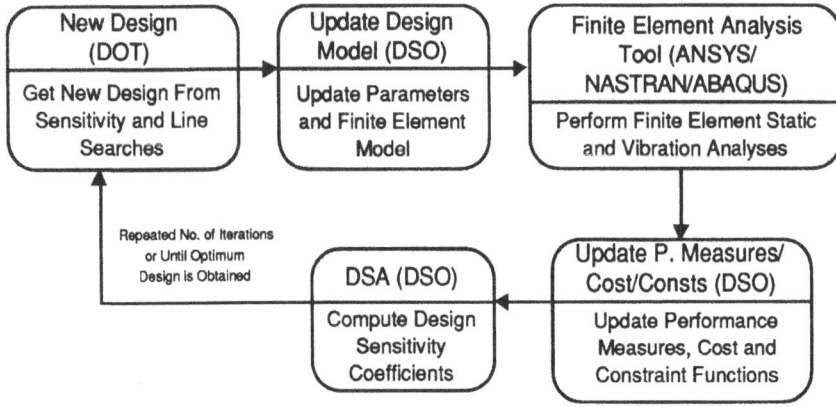

Figure 2.15 Design Model Update

With the remote facility, the DSO user interface runs on an engineering workstation that has an X server (defined as the local machine), while PATRAN runs at another workstation (called the geometric modeler machine) that provides excellent graphics, and FEA jobs are sent to a supercomputer or a mainframe (called the finite element machine). Once the remote configuration is defined, the engineer can request DSO to check the connection to make sure that the remote machines are available in the network.

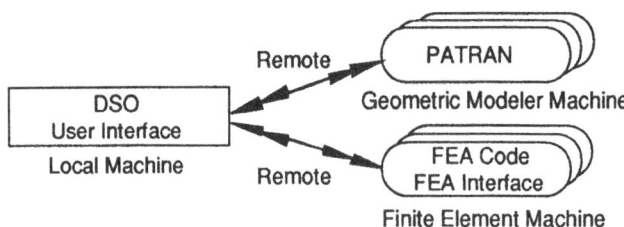

Figure 2.16 Remote Configuration

Foundation Classes In order to promote software reusability and to encourage a building block approach for software construction, common data structures that are used in developing DSO are grouped and implemented as foundation classes. This group consists of data structures for numerical computation, such as Vector, Matrix, and Complex Object; for persistent objects, such as cached hash record (CHR), cached hash file (CHF), and Table; and for object management, such as Sequence and List. The class hierarchy of foundation classes is shown in Fig. 2.17.

Database The DSO uses a dedicated data management system to provide consistent and efficient access to the large set of data manipulated by the computation modules. A table-oriented data management idea is employed to develop the database. A total of 54 tables support software development and data storage during engineering design.

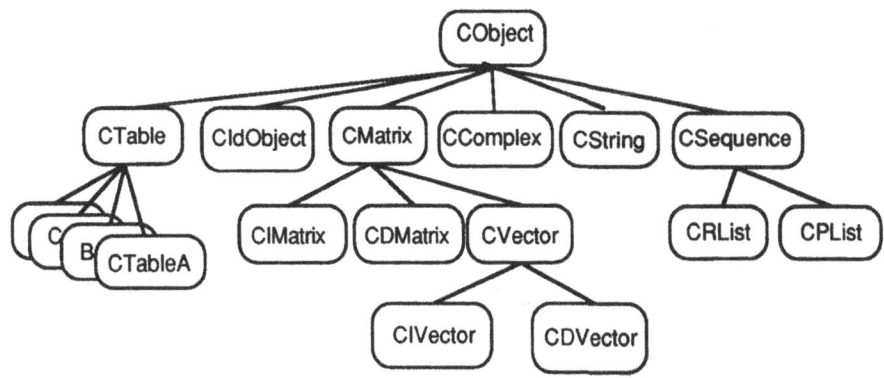

Figure 2.17 The Foundation Class Inheritance Hierarchy

Software Implementation and Tools Integration   Both object-oriented and modular approaches are employed in developing DSO. The object-oriented approach, which supports software reusability, extendibility, and maintainability, is used to design and implement the basic classes and user interface application codes. These codes are written in the C++ language [35]. The modular approach, on the other hand, is used to design and implement the design sensitivity computation codes. Fortran 77 is used for these codes because of its efficiency in number crunching.

2.4.6 Data Wrapper   The DSO allows the engineer to import, via the DSO wrapper, definitions of mechanical components that include geometry and material properties and loading histories on the component from the global database. Loads applied to structural systems for DSA and optimization correspond to those used for life prediction analysis; i.e., joint reactions, force elements, and inertial forces and torques. Structural sizing and shape design parameterizations are defined using the PATRAN geometric model of the structural system for design evaluation, sensitivity analysis, and design optimization. The engineer can report to the global database design change suggestions applied to the parameterized PATRAN geometric models. The engineer working on the DSLP can then quickly evaluate component life and carry out trade-off analysis, since the same PATRAN geometric model can be used in DSLP and DSO.

2.5 GLOBAL DATA MANAGEMENT AND WORKSPACE INTEGRATION

Integration of software systems include two aspects: data transferring and information management. Information management of the CAD and CAE applications in TVCE is the function of the global database management layer--Design Data Server (DDS). The data transferring for these applications is the main role of wrappers developed for each application in this concurrent engineering environment.

2.5.1 Information Management   The DDS is a database management layer developed on top of a persistent, object-oriented data server, called ROSE [13]. For TVCE users, the DDS is the logic global database. The components of the DDS include a global database (i.e., ROSE), a version manager, and an access controller.

The global data model defines the information to be managed by the DDS. The global database in the TVCE is not intended to store all vehicle information that every application would need or generate. Instead, the global database is designed to store shared vehicle definition and information to be transferred between applications. The basic shared vehicle information includes the geometries and material properties of the vehicle components. Other shared information includes the finite element models and loadings of the vehicle components. There is no direct data exchange among engineering workspaces. All data exchange are through the global database. Hence the global database also store information to be exchanged, such as loading histories of vehicle components that are computed by dynamic simulation and used in structural analysis and design.

Figure 2.18 shows the global data model. In the global database a mechanical system is composed of several interconnected bodies and connectors. The simplest mechanical system has a body. A body is a part assembly whose mass or motion in three-dimensional space and whose structural behavior are important from structural and dynamics points of view. A part assembly is composed of parts that are rigidly assembled, for instance through welding, rivets, and bolts and nuts. A part is an entity that has homogeneous material property and specific geometry or is an entity whose member entity information is not interesting to the above CAE applications. A connector is a part assembly whose function in a mechanical system is a kinematic joint or a force element. A joint can be a revolute, translational, cylindrical, universal, or spherical joint. A force element can be a spring, a damper, or an actuator that provides translational and/or rotational forces or toques. The connectivity between bodies and connectors are defined in a data entity called Assembly.

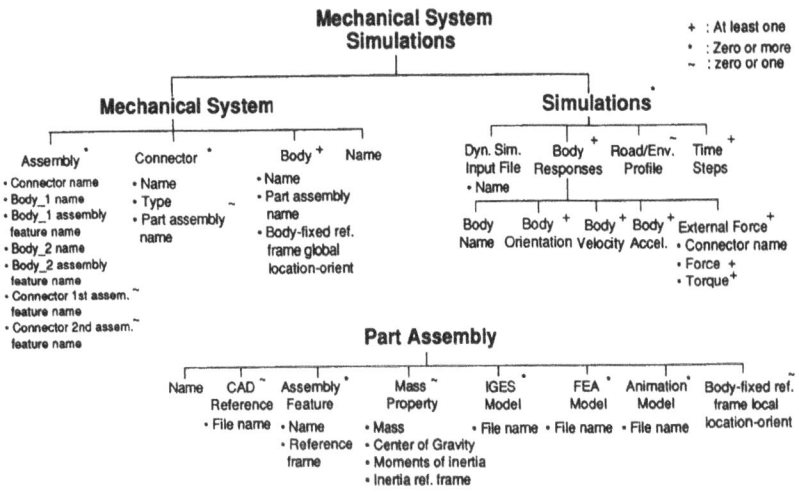

Figure 2.18 Global Data Model of Mechanical System Simulation

In the global database, the part assembly is the smallest (i.e., atomic) mechanical entity to be managed, although in the physical world a part assembly may contain several parts. For each assembly, the TVCE global database stores (1) the name of the part assembly; (2) the CAD reference of the part assembly (which is a file name and its path in a CAD system); (3) names of assembling geometry feature and reference frames of the assembly features that are used to connect the connectors and other bodies; (4) mass property data of the part assembly including total mass, moments of inertia, center of gravity, and the reference frame used to define the moments of inertia; (5) IGES representations of the part assembly; and (6) FEA representations of the part assembly which contain material property data and geometry that structural analysis applications would need.

Since a mechanical system will be used in several dynamic simulations, the global database stores the simulation results for a mechanical system. For each simulation, the global database stores a general message regarding the simulation, the dynamic analysis (DADS) input file, a road profile, total number of output time steps, the time step history, and body responses of each body of interest. For a body response, the global database stores the body name, body position and orientation histories, body velocity histories, body acceleration histories, and force and torque histories at each assembly feature (i.e. joint or force element reference frame).

To manage the information stored in the global database, two management units have been developed: mechanical system simulation catalogue and part assembly catalogue. A mechanical system simulation contains the definition of a mechanical system and dynamic simulation results of the mechanical system. The part assembly catalog stores the part assembly data.

One of the key features of the global database is the version management for entities stored in these two catalogs. A new version of a part assembly is created when a different design for the assembly is sent to the database. Changing geometry or material property (including density) is treated as creating a new version of the part assembly. For some CAE models, such as a finite element model and IGES file, there may exist multiple instances for a part assembly. These instances may be slightly different from each other, but basically, they are representations of the same part assembly. Sending additional representations of the part assembly to the global database will not produce a new version of the assembly. When sending data to the global database, engineers have to judge and indicate if they are sending an additional representation or creating a new version of the part assembly.

A new version of the mechanical system is created when a new version of any of its member entities (such as bodies and connectors) is created. In the new version of the mechanical system, only the part assembly that has been changed is new, and the remaining part assemblies are the same as those of the previous version of the mechanical system. When the engineer creates a new part assembly or a new mechanical system, he or she can submit a message, in a text form, to describe the changes in the new design.

Hence, with the version management capabilities, the design history of the vehicle can be captured. Before sending mechanical system data to the global database or obtaining CAE data of a part assembly, the engineer has to select a specific version of the mechanical system. Referring to the message specified for each versionable entity, the engineer can determine a correct version of the part assembly or mechanical system to use.

2.5.2 Application Integration Integrating CAE applications means creating a software environment to assist the applications to get the required data and send useful data to other applications in the environment, such that the environment can provide desired engineering capabilities in an integrated form. To make the CAE applications work in an integrated form, several issues must be considered: contents and format of the data exchanged or shared. The integration approach used in the TVCE environment is to (1) provide a global database to store and extract shared data for applications (i.e. the data content issue) and (2) provide a wrapper for an application to (a) allow engineers to specify the data to be sent and obtained and (b) translate data from a receiving format to a destination format, (i.e. the data content and format issues).

The basic function of application wrappers is to provide bi-directional data access capabilities between any engineering application and the global database of the integrated environment. Two types of application wrappers are developed for TVCE: open-application wrappers and closed-application wrappers. If the integration group of the TVCE does not have an access to source code, the application is treated as a closed-application. The closed-application wrapper provides only the data extracted from the global database that the application needs in specific forms or retrieves desired information from the output of the application for the global database. In this approach, the application is loosely integrated into the environment. On the other hand, if the integration group has the application source code, the application is treated as an open application. In this case, function calls to the global database are linked with the application code. Figure 2.19 shows the general concept of these two wrappers. In TVCE, the wrappers for CCS and TVWS are open-application wrappers, whereas those for DSO and DSLP are closed-application wrappers.

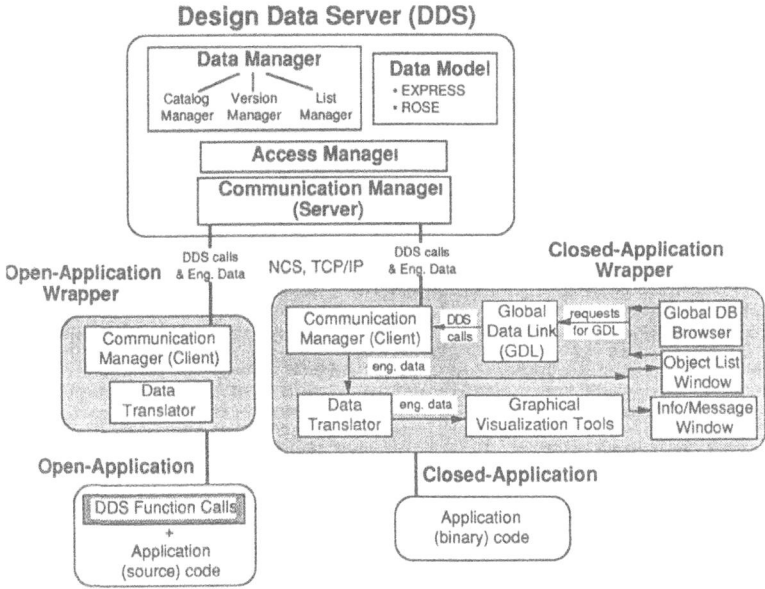

Figure 2.19 Open and Closed Application Wrappers

A typical closed-application wrapper has functional modules including the client side of Communication Manager (CM), Global Data Link, User Interface, and data translators. The CM is the communication channel shown in Fig. 2.19. The CM is the agent through which the other components of the wrapper communicate with the DDS. The CM is composed of two separate modules. One of these, the client side, is linked to the wrapper. The other, server side, is linked to the DDS. These two sides use the Remote Procedure Call (RPC) network communication mechanism of Network Computing System (NCS) [38] as the means by which various data and requests are passed from wrapper to the DDS and back. The CM provides a set of functions that allow the wrapper to treat the DDS as if it were just another local component. In reality, the DDS is a separate process that may be executed on a remote machine.

The Global Data Link provides a view of the global database for the application and allows the engineer on the machine where the application is running to set context for data to be sent into or obtained from the global database. The Global Data Link is a tree representation of the data model in the global database. It shows the portion of data stored in the global database that is of interest to the application and presents the data in a way that is meaningful to the application. The application view of the global database objects can be tailored by modifying data configuration files used by the Global Data Link.

The User Interface part of the wrapper consists of a menu, Database Browser, Database Object List Window, and Information/Message Window, as shown in Fig. 2.20. The menu provides users with options such as Run, Export, Exit, and Plot.

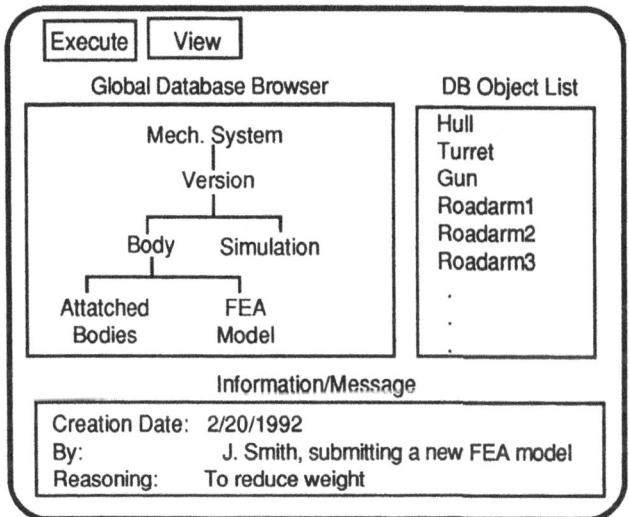

Figure 2.20   Wrapper Windows

The Database Browser serves as the interface between the user and the Global Data Link. The engineer can use the Browser to select the mechanical system, version, and specific part assembly to be sent or obtained. All the nodes of the Browser form a local (application) view of the global data model that is of interest to an application. Once the engineer selects a node in the Browser, the selected object or version will be obtained from the global database through the CM, and displayed in the Object List Window. For example, when the body node of a version of a tracked vehicle is selected, the Global Data Link uses the CM to send appropriate DDS commands to the DDS, which sends back body names of the tracked vehicle to the Global Data Link, including Hull, Turret, Gun, and Roadarms 1 through 14. The wrapper then displays the body names in the Object List Window, as shown in Fig. 2.20. The Information/Message Window displays messages associated with the object selected from the Object List. The message helps the engineer recognize if he or she has selected the desired object or if additional steps must be completed before he or she can use the application.

The data translators [39] are developed to (1) translate data produced by an application to specific forms for the global database or (2) translate data stored in the global database to specific forms that an application can use. A typical data translation converts data such as forces and geometry. For instance, the loading histories of a body are reported by DADS [9] relative to the body fixed reference frame of the body. These forces are used in the finite element model of the body in structural analysis. But, the finite element model may be defined relative to a reference frame that is different from the body fixed reference frame. In this case, data translation is needed.

An open-application wrapper has functional components that include the client side of the Communication Manager, data translators, and DDS function calls. Calls to the CM and the DDS that are needed to manipulate or create objects are coded directly into the application. The CM in this case serves as a networking mechanism to send and get function calls and engineering data from a remote machine on which DDS is running. Open application wrappers vary according to the needs of the applications and DDS. Particularly, the Global Data Link and User Interface of a wrapper may not be needed or can be implemented slightly differently for two reasons. First, the implementation of these two functionalities should be consistent with the software concepts of the open application. Second, the engineering application may already have similar functionalities, and it would be easier if these functionalities can be reused for wrapper development.

The advantages of closed-application wrappers over open-application wrappers include: (1) developments of the wrapper and applications can be carried out more independently, (2) the wrapper development does not need application source codes, and (3) data are exchanged in the secondary memory space instead of the main memory.

3. An Example of Tracked Vehicle Component Design Process

The Tracked Vehicle Concurrent Engineering (TVCE) environment has been tested using a military experimental tracked vehicle system, which is a tank model. The vehicle model used to test the TVCE has a hull, a turret, a gun, and two suspension systems, as shown in Fig. 3.1. Each suspension system has a sprocket, an idler, seven roadarms, and seven roadwheels. Vehicle information that are given to the TVCE team includes the dynamic characteristics of the vehicle system and the design drawings of the seventh roadarm of the right suspension system. The test scenario starts from creating the roadarm CAD model and vehicle CAE model and evaluating the vehicle loading under various vehicle operational scenarios. Design focus is on the seventh roadarm of the right suspension system, the one near the sprocket.

The CAD model of the roadarm, roadarm.prt, has been generated using Unigraphics [6] and translated using UGII-PATRAN-interface to a PATRAN neutral file. Then, in CCS, the DSO and DSLP engineers use PATRAN to read the neutral file and create a PATRAN geometric and finite element model of the roadarm. The roadarm CAD model is generated using the same reference frame as that of the dynamic model. Note that in DADS formulation, the roadarm is assumed to be a straight bar. However, a profile of the arm shown in the design drawings is composed of two straight lines connected by a circular arc. The roadarm CAD model is shown in Fig. 3.2.

In CCS, the tracked vehicle system is created using DADS input file as a reference. The procedure of using CCS is summarized as follows:

(1) Run CCS at the home directory of a testbed account on SUN SPARC station. Create a new roadarm model, named ArmR7, with roadarm.prt as the CAD geometry file and save the model under the Roadarm catalog.

 (a) Invoke Unigraphics remotely from CCS to read the geometry data file, calculate the mass properties, and transfer the CAD geometry file into IGES format and PATRAN neutral file format. Use the menu option of the CCS user interface to add the generated IGES file to the roadarm model.

 (b) Invoke the IGES translator of CCS to clean up the IGES file generated using Unigraphics.

 (c) Create a finite element model of the roadarm using PATRAN. Next, copy the PATRAN neutral file with complete finite element model information of the roadarm to the proper directory. Use the menu option of the CCS user interface to add the PATRAN neutral file to the roadarm model.

(2) The previous step is repeated to create other tank components. Specify mass properties of each component using the CCS user interface windows.

(3) Create a tank mechanical system by defining assembly information between the hull, turret, gun, roadarms, and roadwheels.

(4) Export member models of the tank to the DDS part assembly catalog through the CCS wrapper. Also, export the tank mechanical system to the DDS mechanical system catalog.

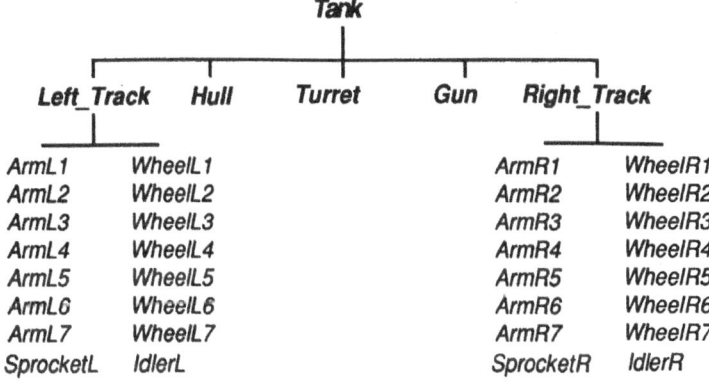

Figure 3.1 Tank Components Created Using CAD/CAE Services

 In CCS, the assembly information of each member model is defined according to the geometric construction reference frame relative to the global reference frame of the composite model. The connection (joint or force) reference frame between bodies is defined relative to the global inertial reference frame of the mechanical system. When the assembly and connection information is exported to the global database, CCS will calculate the joint/force reference frame relative to the geometric construction reference frame, or the so-called non-centroidal body-fixed reference frame (NCBF) used in DADS, based on both assembly and connection information. The global database contains only the joint/force reference frame relative to the NCBF for each body. In DADS, the joint/force reference frames of regular bodies of the track vehicle, such as hull, gun, and turret, are defined relative to the NCBF. However, the body-fixed reference frame of the roadarm is assumed to be at a certain position and orientation in DADS code. In order to obtain consistent loading information and finite element models, the geometric construction reference frame must be consistent with the definition of roadarm body-fixed reference frame in DADS.

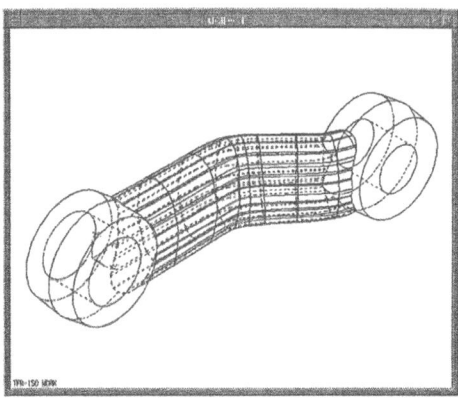

Figure 3.2 Roadarm CAD Model

Using CCS the following information and data are generated: (1) a mechanical system in the global database, (2) Model directory that contains CCS data files, (3) IGES file of the roadarm in the global database, (4) PATRAN neutral file of the roadarm in the global database, and (5) assembly information of each body in the global database. All vehicle component and vehicle system data are sent to DDS through the CCS wrapper.

The TVWS user generates a generic tank DADS input file based on the dynamics characteristic data of the tank. The DADS input file is then divided into several template files. Once the vehicle definition is available in the global database, the TVWS user imports the the mass property data of the hull, turret, and gun and stores it under a tracked vehicle hierarchy in the TVWS database. After the user selects several test scenarios from test scenario catalogs, road profiles from road profile catalog, and bodies of interest that include the right side seventh roadarm, hull, and others, the corresponding dynamic simulation models for DADS are then generated by TVWS.

Following that, the TVWS user launches DADS simulation runs. When dynamic analyses are completed, the TVWS user checks the analysis results. If the analysis results are acceptable, the loading history and motion characteristic data of the roadarm and other selected bodies are exported to the DDS through the TVWS wrapper.

Once the finite element models, loading histories, and velocity and acceleration histories of the roadarm for different dynamic simulations are available in the global database, the DSLP user imports these information from the DDS through the DSLP wrapper. Before importing the roadarm information, the DSLP user needs to specify the mechanical system, a version, and the body of interest (i.e. roadarm) using the DSLP wrapper. If there exist more than one finite element model for the body, the finite element models can be displayed in the wrapper window for the engineer to select one to be imported. In this example, the finite element model has 1913 nodes and 310 20-node isoparametric elements, as shown in Fig. 3.3. Figure 3.4 shows the loading histories that are applied at wheel end joint.

After the engineer specifies the desired data, and if the data are available, the wrapper will retrieve selected information, translate the retrieved data for DSLP, and transfer it to the specified places. The DSLP can then be used to compute dynamic stress time history and predict fatigue life of the roadarm under the loading histories provided by TVWS.

The procedure of using DSLP is summarized as the following:

Dynamic stress computation

- FEA to get lump mass matrix
- Load vector (24 load cases for the roadarm example) generation using lump mass and joint location
- Stress coefficients computation using FEA
- Dynamic stress superposition using stress coefficients and dynamic parameters
Fatigue life prediction
- Principal stress history computation using dynamic stress history tensor
- Local stress and strain calculation using principal stress
- Cycle counting using rain flow counting method
- Damage assessment and crack initiation and propagation life prediction

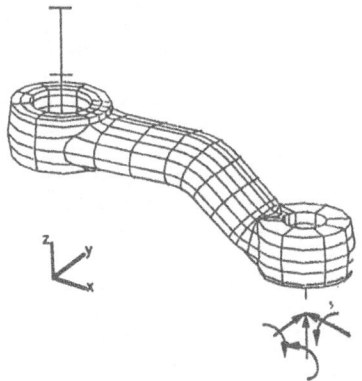

Figure 3.3 Finite Element Model of Roadarm

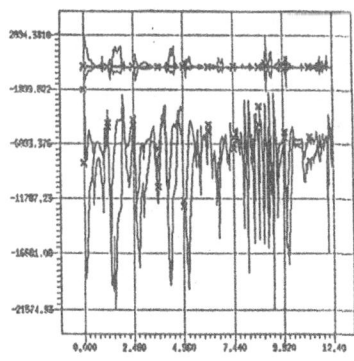

Figure 3.4 Roadarm Loading History at the Wheel End

The results of DSLP are crack initiation life and crack growth rate (or crack propagation life). Roadarm life due to crack initiation at the corner nodes of critical element No. 239 is shown in Fig. 3.5. The unit life is a load block. Since total dynamic simulation time is 12 sec, the load block is 12 sec. Thus, the crack initiation (at node 1) life of the roadarm is 318 days (12 sec × 2.29E6) if the tracked vehicle runs 24 hours a day.

Node No.	Original Design
34	6.79E9
35	8.04E6
1	2.29E6
3	2.77E9
38	2.19E12
40	1.98E8
11	3.19E6
39	5.31E8

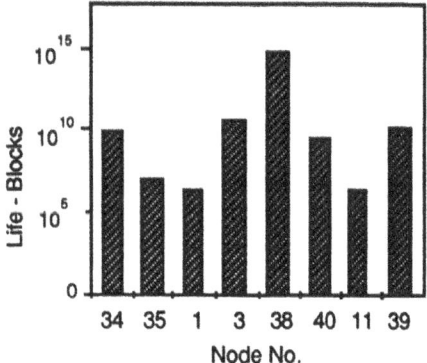

Figure 3.5 Roadarm Crack Initiation Life

In crack propagation calculation, it is assumed that the initial crack length is 0.01 inch. Figure 3.6 depicts the crack length along with the life.

No. of Cycles	Crack Length
300	1.13320E-2
600	1.28682E-2
900	1.46444E-2
1200	1.67037E-2
1500	1.90979E-2
1800	2.18897E-2
2100	2.51481E-2
2400	2.88091E-2
2700	3.31379E-2
3000	3.82776E-2

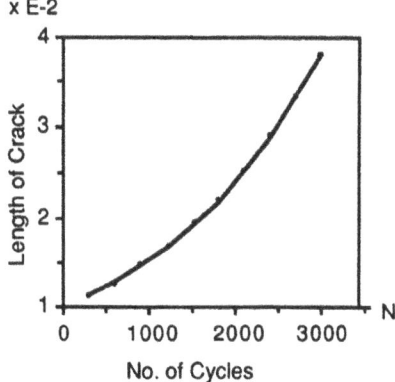

Figure 3.6 Crack Propagation Life of Roadarm

Similar to the use of the DSLP wrapper, once the finite element models, loading histories, and velocity and acceleration histories of the roadarm for different dynamic simulations are available in the global database, the DSO user can import this information from the DSO wrapper, after he or she specifies the mechanical system, a version, and the body of interest. If there exist more than one finite element model for the body, the existing finite element models can be displayed in the wrapper window for the engineer to select one to be imported.

The DSO uses peak loads in analysis instead of dynamic loading history. In this case the engineer can use a 2-D plotter in the wrapper to display the loading histories. The engineer can search for the desired loading information from the loading history plots. In many cases the body has more than one joint or force element, hence the engineer needs to select a specific joint or force element before the loading history of the selected joint (or force element) can be displayed in the 2-D plotter. For the roadarm, the 2-D plot has been used to identify the peak load at time step 9.4 sec. These forces are added to the wheel end of the roadarm in the PATRAN model as shown in Fig. 3.7.

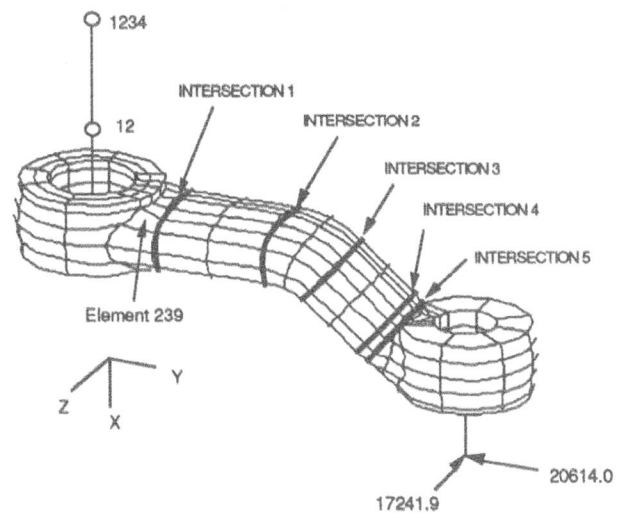

Figure 3.7 Five Intersections and Element 239 of the Roadarm

While DSLP computes dynamic stress and estimates roadarm fatigue life, DSO is used to perform DSA for the roadarm to obtain an improved design from a stress distribution perspective. The roadarm was parameterized by defining 10 shape design parameters characterizing five cross-section movements in the X and Z-directions, as shown in Fig. 3.7. The volume and maximum von Mises stresses at integration points of each finite element are defined as the performance measures. There are 310 stress and 1 volume performance measures defined. The maximum stress is found at element 239. The finite element model and eight corner node numbers of the element 239 are sent to the global database for DSLP to calculate the fatigue life. The DSLP user then predicts the roadarm fatigue life at these nodes.

In DSO, the design velocity field is computed using the isoparametric mapping method [40]. For shape DSA, the direct differentiation method is used. The engineer can understand the structural behavior by visualizing the sensitivity results; e.g., the bar chart of sensitivity coefficients of the stress at element 239 shown in Fig. 3.8 suggests that stress at element 239 will be reduced by moving the first cross-section in the positive X- and Z-directions.

A what-if study has been performed in the steepest descent direction of the stress performance measure at element 239 with a step size of 0.8 in. as the design change. The current and predicted stress values of 310 finite elements are shown in Fig. 3.9.

It can be seen that stresses are more evenly distributed in the new design. A new PATRAN neutral file was then generated by DSO which contains the new geometric and finite element models of the roadarm as shown in Fig. 3.10.

The new PATRAN neutral file has been exported to the DDS using the DSO wrapper and new versions of the roadarm and the vehicle system are generated in the global database. After design change proposed by DSO, fatigue life of the modified roadarm is analyzed by DSLP. Note that the loadings are not influenced significantly by the design change of the roadarm. Thus, for dynamic load computation, it is not necessary to carry out a new dynamic analysis using TVWS. Figure 3.11 shows the improved lives of the modified designs: 2nd and 3rd. For the 3rd design, the crack initiation life at node 40 is 9180 days (12 sec × 6.61E7) if the tracked vehicle runs 24 hours a day. This is an improvement of 28.9 times in fatigue life.

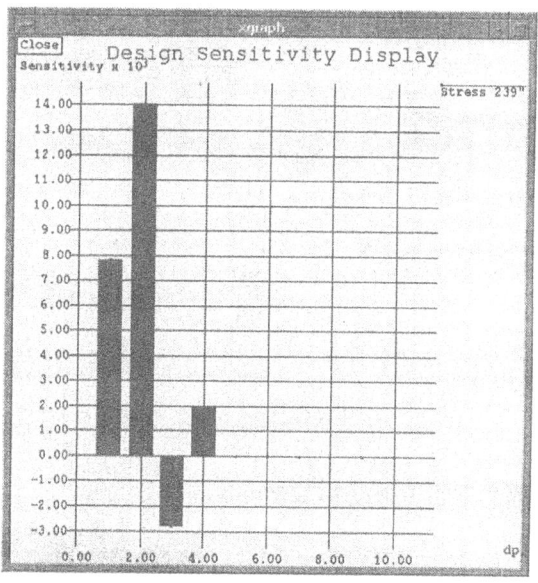

Figure 3.8 Bar Chart of Sensitivity Coefficients of Stress Performance at Element 239

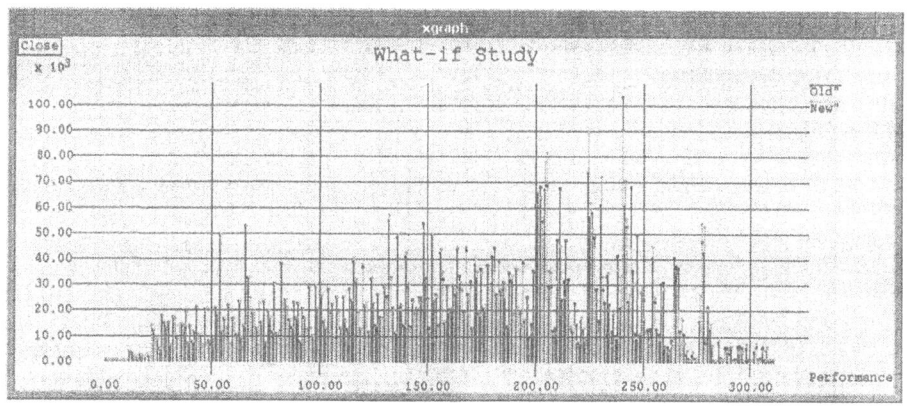

Figure 3.9 Bar Chart of Stresses at Current and New Designs

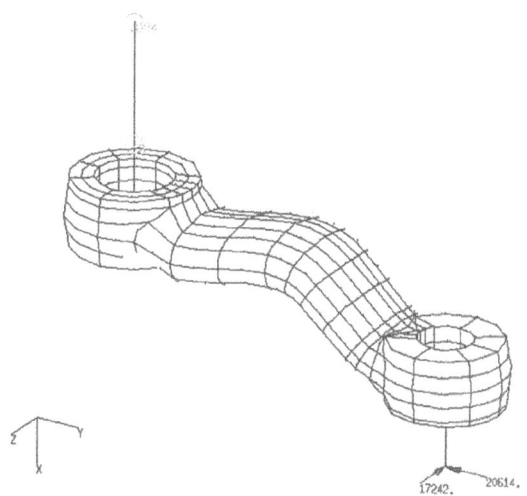

Figure 3.10 PATRAN Finite Element Model of a New Roadarm Design

Node No.	2nd Design	3rd Design
34	8.51E9	7.61E9
35	1.55E10	1.61E9
1	3.72E8	2.60E11
3	1.72E13	9.37E16
38	2.04E12	1.19E12
40	1.35E8	6.61E7
11	3.57E7	5.49E8
39	7.77E9	1.17E12

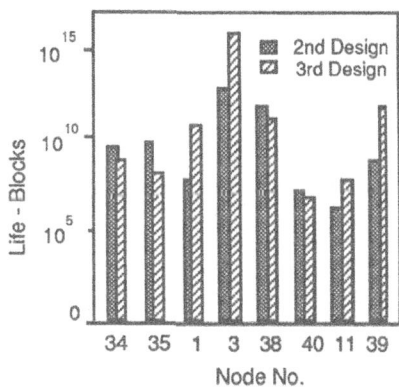

Figure 3.11 Crack Initiation Life of Modified Roadarm

4. Conclusions and Future Challenges

In this paper, a fully integrated Tracked Vehicle Concurrent Engineering environment has been implemented by developing a data model that uniquely defines a mechanical system and supports multidisciplinary computer-aided simulation and design activities. To facilitate and manage data sharing among multidisciplinary engineers, ROSE object-oriented database has been utilized. By enforcing multidisciplinary engineers exchange mechanical system data through the global database and wrappers, a version control in the design process can be achieved. Moreover, the unified data model allows engineers to carry out quick multidisciplinary trade-off analysis and design optimization to produce products that satisfy multi-functionality in reduced time.

However, there are several challenges to overcome to develop a simulation-based design environment that allows engineers to carry out multidisciplinary design optimization to obtain manufacturable designs: weak CAD-CAE connection; lack of simulation modeling techniques that can quickly provide simulation results of the modified design in the data model; lack of DSA of simulation results with respect to the unified design parameters that are defined in the data model; and data management of the global database.

Due to weak CAD-FEA connection, in this paper, PATRAN is used to support the data model. However, a CAD system must be used to create geometric models that can support the data model with unified design parameters so that CAE and manufacturing process simulations can be carried out for the same physical model. Moreover, simulation models must be created so that when design engineers modify design parameters in the data model, quick simulation results of the modified designs can be obtained. Only when the simulation models are tied with design parameters defined in the data model, can DSA capabilities be developed to link product performance measures with the design parameters.

The data management scheme implemented in this paper may have several difficulties. First, the version of the mechanical system simulation may grow rapidly. Second, automatically carrying over the unchanged part assembly to the new version of the mechanical system may create an incompatible mechanical system definition. This is because changes in the geometry of the part assembly may make the geometries of the other part assemblies no longer compatible. Similar issues may happen to the part assemblies.

Acknowledgement   Research supported by the Defense Advanced Research Projects Agency's Initiative in Concurrent Engineering (DICE) Phase 4 project.

REFERENCES

1. Port, O., Schiller, and King, R.W., "A Smart Way to Manufacture," Business Week, April 30, 1990.
2. Derotuzos, M.L., Lester, R.K., and Solow, R.M., Made in America, MIT Press, Cambridge, MA, 1990.
3. Haug, E.J., "Concurrent Engineering of Mechanical Systems," Proc. of First Annual Symposium on Mechanical System Design in a Concurrent Engineering Environment, Vol. I, University of Iowa, Iowa City, IA, October, 1989.
4. Cleetus, J., Functional Specifications for the DICE Architecture, Concurrent Engineering Research Center, West Virginia University, Morgantown, WV, 1988
5. Baek, W.K. and Stephens, R.I., "Fatigue Life Prediction and Experimental Verification for an Automotive Suspension Component Using Dynamic Simulation and Finite Element Analysis," Advances in Fatigue Lifetime Prediction Techniques, ASTM STP 1122, Mitchell, M.R. and Landgraf, R.W. Eds., pp. 354-368, Philadelphia, PA, 1992.

6. McDonnel Douglas Corporation, UNIGRAPHICS II Concepts and Common Functions, 5701 Katella Avenue, Cypress, CA, 1991.

7. Parametric Technology Corporation, Pro/ENGINEER User's Guide, Release 9.0, Parametric Technology Corporation, Waltham, MA, 1993.

8. PDA Engineering, PATRAN Plus User's Manuals, Vols I and II, Software Products Division, 1560 Brookhollow Drive, Santa Ana, CA, 1987.

9. CADSI Inc., DADS User's Manual, Rev. 6.0, Oakdale, IA, 1988.

10. DeSalvo, G. J. and Swanson, J. A., ANSYS Engineering Analysis System, User's Manual Vols I and II, Swanson Analysis Systems, Inc., P.O. Box 65, Houston, PA, 1989.

11. MSC/NASTRAN, MSC/NASTRAN User's Manual, Vol. I and II, The MacNeal-Schwendler Co., 815 Colorado Boulevard, Los Angeles, CA, 1988.

12. ABAQUS, ABAQUS User's Manual, The Hibbit, Karlsson & Sorensen, Inc., 100 Medway Street, Providence, RI, 1990.

13. The STEP Programmer's Tool Kit, STEP Tool Inc., Troy, NY, 1992.

14. Haug, E.J., Computer-Aided Kinematics and Dynamics of Mechanical Systems, Vol. I: Basic Methods, Allyn and Bacon. 1989

15. Wu, J. K., Koppes, E. A., Ciarelli, K. J., Calbeck, D., and Smuda, W. J., An Integrated CAD/CAE System for Vehicle Design and Evaluation, Technical Report R-128, Center for Simulation and Design Optimization, The University of Iowa, Iowa City, IA, June 1992.

16. Ciarelli, K. J., "Integrated CAE System for Military Vehicle Applications," 16th ASME Design Automation Conference, pp. 15-24, Chicago, IL, Sep., 1990.

17. Intergraph Co., Intergraph Engineering Modeling System Reference Manual, Huntsville, AL, 1988.

18. Informix Software, Inc., Informix-ESQL/C: Embedded SQL and Tools for C, Menlo Park, CA, 1987.

19. Gorlen, K. E., Orlow, S. M., and Plexico, P. S., Data Abstraction and Object-Oriented Programming in C++, John Wiley and Sons, Inc., New York, NY, 1990.

20. Chou, M. P. and Pudloski, S., Tracked Vehicle Workstation Data Model, Technical Report, R-127, Center for Simulation and Design Optimization, The University of Iowa, Iowa City, IA, 1992.

21. Yim, H. J., Haug, E. J., and Dopker, B., Computational Methods for Stress Analysis of Mechanical Components in Dynamic Systems, Technical Report No. R-79, Center for Simulation and Design Optimization, The University of Iowa, Iowa City, IA, October, 1989.

22. Tang, J., "An Integrated System for Dynamic Stress and Fatigue Life Prediction of Mechanical Systems", ASME Pressure Vessel & Piping Conference, Denver, CO, July 25-29, 1993.

23. Shin, S. H., Yoo, W. S., Tang, J., Theoretical Development and Computer Implementation of the DADS Intermediate Processor, Technical Report No. R-135, Center for Simulation and Design Optimization, The University of Iowa, Iowa City, IA, May, 1992.

24. Shin, S. H., Yoo, W. S., Tang, J., "Effects of Mode Selection, Scaling, and Orthogonalization in the Dynamic Analysis of Flexible Multibody Systems," Mechanics of Structures and Machines, Vol. 21, No. 4, 1993.

25. Fatigue Crack Growth Computer Program, NASA/FLAGRO, Rev. Mar., 1989, Structures and Mechanics Division, NASA, L. J. Space Center, Houston, TX, 1989.

26. Choi, K. K. and Chang, K-H., "Shape Design Sensitivity Analysis and Optimization of Elastic Solids," Structural Optimization: Status & Promise, AIAA Progress in Astronautics and Aeronautics Series, Chapter 21, pp. 569-609, 1993.

27. Chang, K-H., Choi, K. K., and Perng J-H., "Design Sensitivity Analysis and Optimization Tool (DSO) for Sizing Design Applications," 4th AIAA/USAF/NASA/OAI Symposium on Multidisciplinary Analysis and Optimization, Cleveland, OH, September 21-23, 1992.

28. Choi, K. K. and Chang, K-H., "Design Sensitivity Analysis and Optimization Tool for Concurrent Engineering," Concurrent Engineering Tools and Technologies for Mechanical System Design, (Ed. E.J. Haug), pp. 587-626, Springer-Verlag, 1993.

29. AEA Industrial Technology, Harwell Laboratory, Oxfordshire, OX11 ORA, England.

30. Vanderplaats, G. N. and Hansen, S. R., DOT User's Manual, VMA Engineering, 5960 Mandarin Ave., Suite F, Goleta, CA, 1990.

31. Chang, K. H. and Choi, K. K., "An Error Analysis and Mesh Adaptation Method For Shape Design of Elastic Solids," Computers and Structures, Vol. 44, No. 6, pp. 1275-1289, 1992.

32. Ding, Y., "Shape Optimization of Structures - A Literature Survey," Computers & Structures, Vol. 24, No. 6, pp. 985-1004, 1986.

33. Haug, E. J., Choi, K. K., and Komkov, V., Design Sensitivity Analysis of Structural Systems, Academic Press, New York, NY, 1986.

34. Mohan, G. M., Koch, P., and Santos, J. L. T., "Development of a User Interface for a Design Sensitivity Analysis and Optimization Workstation," 16th ASME Design Automation Conference, pp. 117-126, ASME Conference, 1990.

35. Meyer, B., Object-Oriented Software Construction, Prentice Hall, Englewood Cliffs, NJ, 1988.

36. Open Software Foundation, User Environment Specification, Prentice-Hall, Englewood Cliffs, NJ, 1990.

37. X-Window, X-Window System User's Guide, O'Reilly & Associates, CA, 1989.

38. Apollo Computer Inc., Network Computing System Reference Manual, 330 Billerica Road, Chelmsford, MA, 1987.

39. Center for Computer Aided Design, Tracked Vehicle Workstation Users Manual, The University of Iowa, Iowa City, IA, April, 1993.

40. Chang, K. H. and Choi, K. K., "A Geometry Based Shape Design Parameterization Method For Elastic Solids," Mechanics of Structures and Machines, Vol. 20, No. 2, pp. 215-252, 1992.

MULTIDISCIPLINARY DESIGN OPTIMIZATION: AN EMERGING NEW ENGINEERING DISCIPLINE

by

Jaroslaw Sobieszczanski-Sobieski
NASA Langley Research Center
Hampton, Virginia

ABSTRACT

This paper attempts to define the Multidisciplinary Design Optimization (MDO) as a new field of research endeavor and as an aid in the design of engineering systems. It examines the MDO conceptual components in relation to each other and defines their functions.

1. INTRODUCTION

During the last decade, a set of previously disjointed ideas and tools has crystallized into a new technology imparted with cohesiveness and distinct character that give it characteristics of a new engineering discipline. This discipline is unique in its role in engineering design where it acts as an agent binding together the other engineering disciplines. The new discipline is often called the Multidisciplinary Design Optimization (MDO), e.g., Ref. 1, and will be referred to by this name herein. Alternative names such as Multidisciplinary Analysis and Optimization (MAO) and Multidisciplinary Design Methodology or Technology, (MDM or MDT) have also been proposed.

The purpose of this paper is to introduce the definition of MDO and of its principal conceptual components which are Design-Oriented Analysis, Approximation Concepts, System Mathematical Modeling, Decomposition, Design Space Search, Optimization Procedures, and Human Interface as shown in Fig. 1. The paper describes the functions and mutual relationships of these components and examines the subtopics within each component as depicted in Fig. 1. Without attempting a comprehensive survey, selected references are quoted to support the above discussion.

Having its roots in structural synthesis (ref. 2), MDO came into being because analysis and optimization of an engineering system differ significantly from those commonly encountered in a single discipline or in application to a single part of the system. The most obvious is the qualitative difference of the problem dimensionality: the number of the behavior unknowns, the number of the design variables, and the number of the constraint functions, all tend to be very large, typically in the range of tens of thousands. In addition there are several important qualitative differences. Typically, there is no single mathematical model available for the entire engineering system at hand. Instead, there is an assemblage of mathematical models, each corresponding to a physical phenomenon, an engineering discipline, or a physical part of the system. These models implemented as computer codes collectively form a modular software system that becomes a mathematical model of the engineering system. The modules depend on each other for input, hence the engineering system analysis is usually iterative. Often, the above modules are being operated by separate groups of specialists in different organizations that may be dispersed

J. Herskovits (ed.), Advances in Structural Optimization, 483–496.
© 1995 *Kluwer Academic Publishers.*

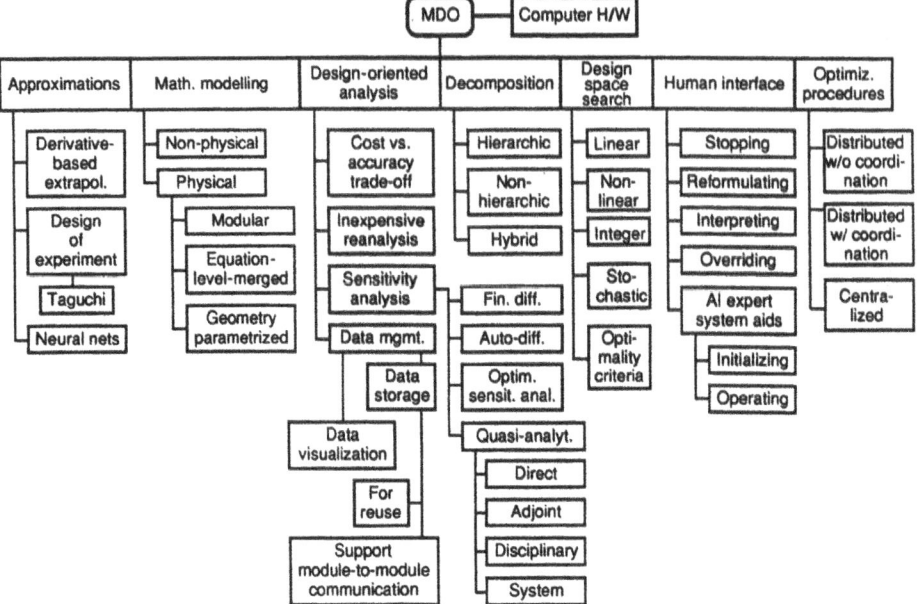

Figure 1. MDO principal conceptual components and their breakdown.

geographically. Because of this, the engineering system analysis is very expensive, time consuming, and poses a non-trivial managerial task.

Furthermore, many functions and variables are discrete and burdened with uncertainties, there may be many conflicting objectives, and the very formulations of the analysis and optimization problems to be solved change as the design process moves on. Consequently, human interaction with both the analysis and optimization is essential for successful design of an optimal engineering system.

2. DEFINITION OF MDO

The term "methodology" is defined by Webster's Dictionary as "a body of methods, procedures, working concepts, and postulates, etc." Consistent with that generic definition, MDO can be described as *a methodology for design of complex engineering systems that are governed by mutually interacting physical phenomena and made up of distinct interacting subsystems.* The MDO methodology exploits the state of the art in each contributing engineering discipline and emphasizes the synergism of the disciplines and subsystems.

An aircraft, a car, a ship, or an electric power generation plant are all good examples of engineering systems. If one were to name their single common characteristic, it is that in their design **everything influences everything else.**

3. MDO COMPONENTS

Several conceptual components coalesced to form MDO. They are listed and briefly characterized in this section of the paper. They form the top layer in the diagram in Fig. 1 that shows their internal breakdown which is also examined in this section. Referring to Fig. 1 often as the discussion unfolds will help in developing a broad perspective on the subject.

3.1. Design-Oriented Analysis

Designers often need to get rough solution estimates quickly and inexpensively, so typically they apply analysis repetitively to answer the "what if" questions. They also need to know sensitivity of the solution to design changes. Furthermore, in a multidisciplinary system, disciplinary analyses must interact, and the computing process produces an enormous volume of data that must be interpreted and saved to build an easily accessible data base.

A new brand of analysis has been developing in the last two decades, called the Design-Oriented Analysis to meet the needs mentioned above. Its basic features are summarized below.

3.1.1. Sensitivity analysis (SA). An important concern in design is about the "what if" questions. Derivatives of the output variables (state or behavior variables) with respect to (w.r.t.) the input variables (design or decision variables) constitute a precise measure of sensitivity and quantify answers to the above questions. They are referred to as sensitivity derivatives (SD's) or sensitivity coefficients (SC's).

Finite differencing (FD) on an analysis code is the simplest way to calculate SD. Even though adequate in many applications, the FD approach has shortcomings. Its accuracy deteriorates with the step length in nonlinear problems, but making the step too short may incur excessive round-off errors (refs. 3, 4, and 5 offer a technique for optimal setting of the step length, and an adaptive step length technique is described in ref. 6). If analysis is iterative, then an FD may produce meaningless results (a remedy is given in ref. 7). Finally, the FD cost grows linearly with the number of the design variables; hence it may become excessive in large problems, even if the inexpensive reanalysis techniques discussed later in the paper are used.

Motivated by the above shortcomings, several alternative techniques have evolved. Because in the field of engineering analytical closed-form solutions amenable to symbolic differentiation are usually not available, an analytical approach based on the Implicit Function Theorem is one popular technique. When applied to a disciplinary analysis, the technique differentiates the governing equations of the analysis to obtain the companion sensitivity equations that are always linear, simultaneous algebraic equations in which the SD's appear as unknowns. The above approach results in two versions of SA. The direct version generates SD's of all the unknowns in the analysis. An indirect version referred to as the adjoint variable method computes the derivatives of only a selected subset of the unknowns.

The SA methodology has been most fully developed in computational chemistry, control theory, and in structural analysis. In the latter the capability to obtain derivatives of displacement, stress, and vibration frequency w.r.t. cross-sectional dimensions and shape variables is now routinely available. Reviews of the subject may be found in Refs. 8, 9, and 10. Similar developments are currently under way in aerodynamic analysis; e.g., Refs. 11 and 12, and the beginning is being made in other engineering disciplines as well.

To avoid the costs of new coding required when analytical techniques are retrofitted in the existing codes, an "automatic differentiation" method has recently become available. The method applies a line-by-line symbolic differentiation to an existing code and stores numerical values of the dependent variables for each code line. Moving from one line to the next, the algorithm links the derivatives in a chain-differentiation manner as required by the variable dependencies from the beginning to the end of the code. The result is a set of the derivatives of the output w.r.t. input. The method is implemented in the form of an automatic differentiator code that reads the user's existing source code and produces a new source code that retains exactly the same capability as the original but is enhanced with an option of computing the SD's. A survey of the differentiator codes currently available is given in Ref. 13, and examples of engineering applications may be found in Refs. 14 and 15.

Typically, the above SA techniques apply to a code representative of an engineering discipline. In a complete vehicle analysis, these codes form a system coupled by the output-to-input data cross-flow that simulates behavior of the entire vehicle system. Since neither of the above SA techniques is practical in a direct application to a modular system of codes, a special technique for such systems has been developed (refs. 16 and 17). That technique, called the System SA (SSA), uses the SD's obtained for the system modules by any of the above SA methods as coefficients to build a set of the linear, algebraic equations also derived from the Implicit Function Theorem. These equations yield the SD's of the system behavior with respect to the design variables, with the interactions among the disciplines accounted for. Examples of applications of the above technique are discussed in Ref. 18.

The SD's may be used directly to support engineering judgment and intuition, or they may be entered into a design space search algorithm to guide a formal optimization. Implications of these SD uses in the design process are discussed in Ref. 19.

Often, a "what if" question arises at the end of an optimization process in which certain quantities, e.g., structural material stress allowable, were held constant. The question is about the influence of these quantities (optimization parameter, or OP) on the objective and on the design variables at the constrained optimum. Instead of repeating the optimization with an OP perturbed, one may obtain an answer in the form of the derivatives of the objective and of the optimal design variables w.r.t. OP using an algorithm described in Refs. 20, 21, 22, 23, and 24. The above derivatives are valid at the constrained minimum and depend on the critical constraint subset. When they are used in a linear extrapolation away from that minimum, the membership of the critical constraint subset may change with the attendant abrupt perturbations in the extrapolation error. This may bound the constrained minimum neighborhood in which such extrapolation is useful (ref. 25).

3.1.2. Inexpensive reanalysis. The objective here is to make the analysis code organization flexible enough so that a new solution reflecting a change to the input variables may be obtained by repeating as small a subset of the analysis as possible. To achieve this in a rigidly structured code that offers a menu of different execution options set up a priori in anticipation of the "what if" questions that user might be asking (where each question is defined by a subset of the input variables that might be changed) is very difficult. It is practically impossible to anticipate all such questions that might arise in the operation of a code with input that is diverse and voluminous. Even if perfect anticipation was possible, a combinatorial explosion of the number of the execution options would set in.

One remedy is to organize a program as a system of modules and to establish the data dependence information that for each module input identifies a source either in the output of another module or in the input from outside of the system. That information, tabulated and recorded as a part of the system, enables the system executive module to find out the following: 1.) what modules and output data will be affected and, therefore, what modules will have to be re-executed when there is a change to particular input data (a forward chaining mode), or 2.) to determine what modules must be executed to compute particular output data (a backward chaining mode).

To satisfy a user's request for particular output data, the stored output (output from each execution is routinely saved) is searched first to see if the requested data are readily available there. If not, the executive module scans the data dependence data and the stored output to determine the smallest subset of modules and its execution sequence which will produce the required data. Then, it commands the data calculation. If the user changes the particular input data, the executive module activates a computational sequence which will produce the data that are affected by the change. In either case, the user request is satisfied with a minimum of a computational effort. This type of a system is called nonprocedural because the user does not have to choose among the execution options or to code the execution procedure appropriate to the requests. A pioneering example of such a non-procedural system may be found in Ref. 26.

3.1.3. Computational cost-accuracy trade. An example of a technique that enables one to trade the solution accuracy for its computational cost in repetitive applications is given in Ref. 27. This technique uses two mathematical models for the same physical phenomenon: 1.) a refined model, (R), costly to analyze, and, therefore, to be invoked sparingly; and 2.) a simplified, inexpensive model, (S), to be used often. The ratio of an element of the R output to the corresponding element of the S output is termed a correlation factor and those factors form a correlation vector. Assuming that the same design variables govern both models, derivatives of the above vector w.r.t. those variables may be computed in terms of the SD's of the R and S outputs w.r.t. the same variables. Then, the state of R for modified design variables may be approximated by analyzing S and multiplying its output vector by the correlation vector updated by the derivative-based linear extrapolation. Periodically, the R model has to be reanalyzed to recover from the extrapolation errors. As demonstrated in Ref. 27, the accuracy of such extrapolation is better than the one obtained with constant correlation factors.

3.1.4. Data management and visualization. Because of its repetitive nature, the Design-Oriented Analysis produces a very large volume of data. Some of these data has to be stored for future use, some of it must be immediately transferred among the modules, and some of it should be made available to the user for interpretation. The data storage requires a data base facility, preferably based on the relational storage scheme supporting the data recall by attributes. However, the overhead required by such a storage facility may be unacceptable for data transfers that have to be re-executed often, e.g., in an iterative loop; hence a separate, direct data transmission mechanism may be preferred for that purpose as an option. An example of such a two-tier data handling system is described in Ref. 28. A recent trend toward the object-oriented software systems (ref. 29) appears to have a potential for making the management of data and programs much simpler for engineers.

To support the user judgment, intuition, and the continuity of the train of thought in design, it is imperative to present the data visually. The use of color as the additional dimension has begun to be regarded as indispensable.

3.2. Approximation Concepts

Direct coupling of the design space search code (DSS) to the analysis is impractical when the analysis is expensive because the number of calls from DSS to analysis for new information may be excessive. To gain control over that number, it became customary in large applications to refer the DSS calls to an approximate analysis (AA), while invoking the full analysis infrequently at the rate required by the approximation error control. The approximation concepts underlying AA range from the Taylor series approximations based on the SD's (ref. 30) to the Design of Experiments with a fitted response surface (ref. 31). This class of methods has recently been enhanced by the use of orthogonal arrays (ref. 32) and the application of the neural nets in lieu of the fitted response surface (ref. 33).

3.3. Mathematical Modeling of a System

It is axiomatic that a mathematical model of an engineering system is a modular system of codes, rather than a monolithic code. Because codes in a modular software system send data to each other, the overall computational efficiency will benefit from having the mathematical models embodied in those codes set up so as to minimize the additional data processing required for the data transfers among the codes. An example of this is an aircraft wing design where the aerodynamic analysis that uses a 3D grid interacts with a finite element model that uses its own nodal mesh. Ideally, the aerodynamic analysis module should output forces at the structural mesh nodes, and the structural analysis module should produce the displacements at the surface aerodynamic grid points. Furthermore, to support the shape optimization both grids should be parametrized in terms of the shape variables of interest so that when the wing shape changes both grids adjust without having to be regenerated from scratch.

When mathematical models of two or more disciplines interact very often in a multidisciplinary system analysis and send large volumes of data to each other, it may be necessary to reduce the volume of the data exchange by a condensation technique. An elastic wing system provides an example. Instead of sending all the thousands of displacements from the structural finite element analysis module to the aerodynamic module, one may transmit a relatively small set (typically, 20 to 40) displacement functions (natural vibration modes are often used) in exchange for a comparably curtailed set of functions representing the distribution of aerodynamic forces over the wing finite element model (ref. 18).

Alternatively, such closely coupled disciplines may be candidates for merging at the equation level. An example of this is recently developed heat transfer analysis blended with structural analysis using a shared finite element model (ref. 34).

Driven by the increasing concern of cost as the dominant consideration in design, the notion of mathematical modeling has recently begun to extend from the design phase (concerned primarily with physical phenomena) to the phases of the product specification development, manufacture, and operation (dominated by man-made, nonphysical processes). Once these four principal phases of the product life are mathematically modeled at the same level of fidelity now common in the design phase and their couplings are accounted for, the entire life cycle will become amenable to optimization for minimal total cost as forecast in Ref. 35.

3.4. Decomposition

In the time-honored "divide-and-conquer" approach to large engineering tasks, a number of techniques have been developed to partition large engineering design optimization tasks into smaller tasks that remain coordinated so that their coupling and therefore their synergism are not lost. The systems amenable to decomposition are usually categorized as hierarchic, non hierarchic, and hybrid, as depicted in Fig. 2. The distinguishing feature is the data flow among the modules that collectively simulate the system. In the hierarchic system, the modules form a pyramid with the data flow starting from the top, so that several "children" modules may receive input data from the same "parent" module; however, the children modules do not communicate directly with each other. A classical analysis of an airframe by substructuring illustrates this case in Fig. 2 (left-hand side). The data transmitted from the parent (the assembled structure analysis) to its children (the individual substructure analyses) consist of the substructure boundary forces, and no data are being exchanged among the individual substructure analyses. In contrast to that, the modules in a non-hierarchic system may communicate with each other without any restrictions so that one cannot identify a parent to a set of children. This is illustrated in Fig. 2 (middle) by a flexible wing with active control under aerodynamic loads. In this case, the mathematical models of aerodynamics, structures, and control send data to each other, and none of these models may be singled out as a parent. A mixture of the two categories constitutes a hybrid system. For instance, one may break down the airframe structure in the above non- hierarchic case into a hierarchic set of substructures, as shown in Fig. 2 (right-hand side).

It is often obvious at a glance how to organize a given collection of modules into a system of one of the above categories on the basis of the examination of the data flow among the modules. However, for large collections of modules, the system organization may not be so obvious and an aid such as the N-square method embodied in a code described in Ref. 36 may be needed. Examples of engineering system optimization based on decomposition may be found in Refs. 37 and 38 for a hierarchic case, and in Refs. 39, 40, and 41 for non-hierarchic cases.

In a fully coupled system of n modules, the number of data transmission links is n^2-n. However, in most of the systems occurring in practice that number is far less because many of the potentially possible data links represent negligible influence or do not exist at all. Consequently, the matrices in analysis and optimization are sparse and that sparseness may be exploited to lower the computational cost. This approach has been recently attracting a considerable attention, e.g., Refs. 42, 43, 44, and 45, and in some applications it may be an alternative to decomposition.

3.5. Design Space Search

The purpose of the search is to produce a design improvement and, ultimately, to find a constrained minimum. Vigorous research and development in Operations Research has

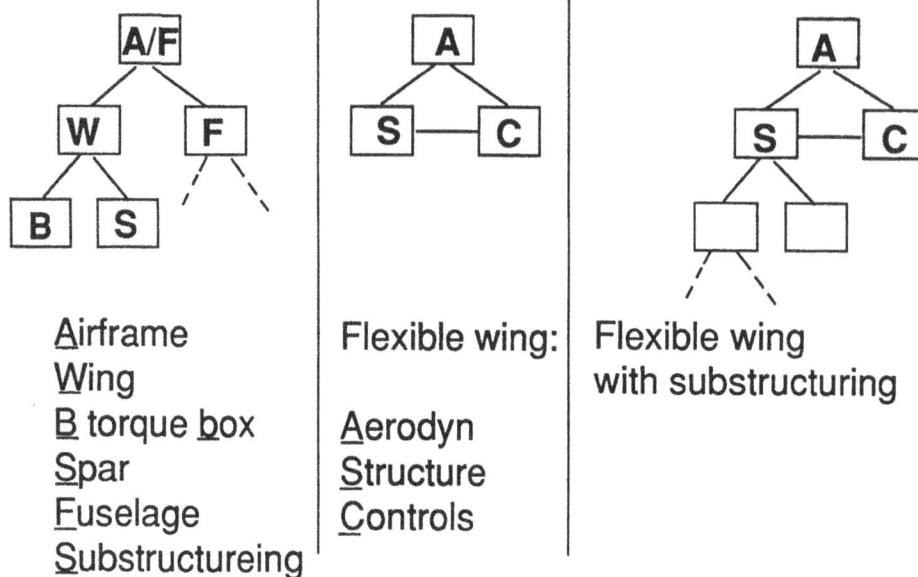

Figure 2. Three categories of system decomposition, left to right: hierarchic, non-hierarchic, hybrid.

produced a large variety of DSS algorithms and codes (search algorithms, optimizers) that may be broadly classified in the categories of Linear Programing, Non-linear Programing, Integer Programing, and Stochastic Programing. Many of these algorithms have been collected in software packages tailored for practical, diverse applications, e.g., Ref. 46. Surveys of DSS algorithms and codes may be found in Refs. 47, 48, and 49.

An alternative approach to locating a constrained or unconstrained minimum in a design space is provided by the optimality criteria methods. An optimality criterion and a resizing algorithm are two principal elements in each of these methods. Several different criteria and the algorithms associated with them have been formulated (a review may be found in refs. 50 and 51). The criterion may be based on the problem-dependent physics, e.g., the fully stressed design or the uniform strain energy density design in structures. Criteria that are physics-independent have been derived from the analytical conditions necessary for a constrained optimum to occur. A resizing algorithm, e.g., the stress ratio formula in the fully stressed design method modifies the design variables to converge to a state that satisfies a particular criterion. The optimality criteria methods have proved to be effective primarily in applications to structures. Under certain conditions, the optimality criteria methods become mathematically equivalent to nonlinear programming (ref. 52).

3.6. Optimization Procedures

Contrary to a common misconception, optimization in a large-scale application cannot be made practical by simply coupling an optimizer directly to the analysis. Instead, one needs an optimization procedure to tie together the MDO elements discussed in this paper into an execution sequence. An optimizer is only one of many components in such procedure. Many different procedure organizations have evolved for various engineering application-computer hardware combinations. Examples for structural optimization on a serial computer are provided in Ref. 53. Optimization procedures for engineering systems may be classified in three categories: 1. Procedures that distribute optimization among the system parts and the disciplines, without solving a coordination problem at the system level; 2. Procedures that distribute optimization as above but do solve a coordination problem at the system level; 3. Procedures that distribute sensitivity analysis but centralize optimization as a single problem to be solved at the system level. An early example of category 1 for a hierarchic system is Ref. 54. In category 2, one finds examples for a hierarchic system in Ref. 38, and for a non-hierarchic system in Refs. 55 and 56. Finally, Refs. 18, 40, and 57 provide examples of category 3 for a non-hierarchic system.

An optimization procedure of category 3 for a non-hierarchic system is illustrated by a flowchart in Fig. 3. The procedure provides an opportunity for a macro-scale concurrent processing (modules executing concurrently on separate processors or separate computers) and it engages human judgment before each major operation. A procedure of this type was used in Ref. 18. One may expect that further progress in the heterogeneous, massively parallel computing technology together with new developments in the system optimization theory will spawn more development of innovative optimization procedures.

3.7. Human Interface

The MDO is definitely not a "pushbutton" approach to design; hence it relies heavily on human participation. To facilitate that participation an MDO software system typically incorporates the means for the user to review and judge the intermediate results, to intervene and override the algorithmic decisions, to reformulate the problem, and to decide when to stop (the "Human decision" boxes in Fig. 3). Artificial intelligence tools such as the expert systems often are found useful here, either as guides to the software usage (ref. 58) or as aids in making design decisions (ref. 59).

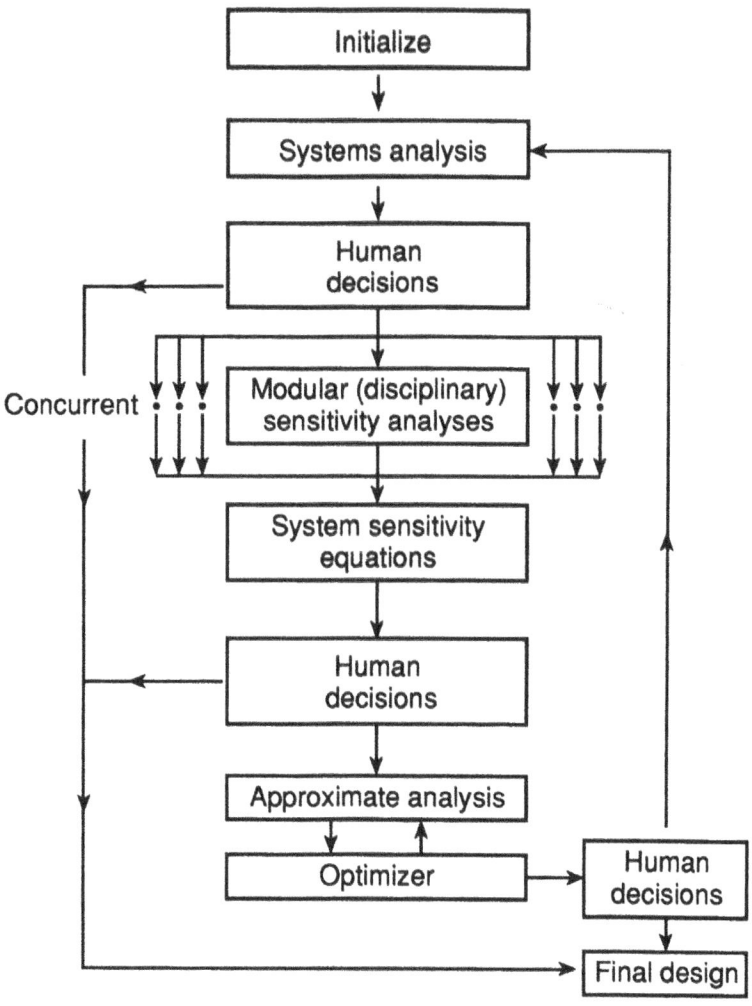

Figure 3. An optimization procedure for a non-hierarchic system (note an opportunity for concurrent processing at the level of disciplinary sensitivity analyses).

4. CONCLUDING REMARKS

Multidisciplinary Design Optimization (MDO) has developed into a methodology comprising a number of components. The principal ones have been identified as Design-Oriented Analysis, Approximation Concepts, System Mathematical Modeling, Decomposition, Design Space Search, Optimization Procedure, and Human Interface. These components were examined, including their breakdown into subcomponents and their relationships in the MDO methodology framework. As indicated in a review in Ref. 60, the application experience with MDO is accumulating and showing promise of significant pay-offs. In its rapidly growing maturity and diversity, and in its offering of numerous opportunities for future research, MDO exhibits the attributes of a new discipline whose definition is now evolving.

5. REFERENCES

1. Current State of the Art in Multidisciplinary Design Optimization. An AIAA White Paper, September 1991, Washington, D.C.

2. Schmit, L. A.: Structural Synthesis—Its Genesis and Development. AIAA J., Vol. 19, No. 10, 1981, pp. 1249–1263.

3. Hull, D. G.: Numerical Derivatives for Parameter Optimization. J. Guidance and Control, Vol. 2, No. 2, March-April 1979, pp.158–160.

4. Hallman, W. P.: "Numerical Derivative Techniques for Trajectory Optimation," in 3rd Air Force/NASA Symposium on recent Advances in Multidisciplinary Analysis and Optimization. San Francisco, CA, Sept. 1990, pp. 418–424.

5. Gill, P. E.; Murray, W.; Saunders, M. A.; and Wright, M. H: "Computing Forward-Difference Intervals for Numerical Optimization," SIAM Journal on Scientific and Statistical Computing, Vol. 4, pp. 310–321, June 1993.

6. Beltracchi, T. J.: "An Overview of the Current State of the Art Optimization Used for Trajectory Design in the GTS System," in proceedings of third Air Force/NASA Symposium on recent advances in Multidisciplinary Analysis and Optimization, Sept. 1990, San Francisco, CA, pp. 545–552.

7. Haftka, R. T.: "Sensitivity Calculations for Iteratively Solved Problems," International Journal for Numerical Methods in Engineering, Vol. 21, 1985, pp. 1535–1546.

8. Adelman, H. A.; and Haftka, R. T.: Sensitivity Analysis of Discrete Structural Systems. AIAA J., Vol. 24, No. 5, May 1986, pp. 823–832.

9. Haftka, R. T.; and Adelman, H. M.: Recent Developments in Structural Sensitivity Analysis. Structural Optimization, 1, 1989, pp. 137–151.

10. Proceedings of the Symposium on Sensitivity Analysis in Engineering. NASA Langley Research Center, Hampton, VA, Sept. 1986; Adelman, H. M.; and Haftka, R. T.—editors. NASA CP-2457, 1987.

11. Baysal, O.; and Eleshaky, M. E.: Aerodynamic Design Optimization Using Sensitivity Anslysis and Computational Fluid Dynamics. AIAA Paper 91-0471, AIAA 29th Aerospace Sciences Meeting and Exhibit, Reno, NV, Jan. 1991.

12. Newman, P. A.; Hou, G. J.-W; Jones, H. E.; Taylor, A. C. III; and Korivi, V. M.: Observations on Computational Methodologies for Use in Large Scale Gradient Based Multidisciplinary Design Incorporating Advanced CFD Codes. NASA TM-104206, NASA Langley Research Center, February, 1992.

494

13. Griewank, A.: The Chain Rule Revisited in Scientific Computing. SIAM News, Part 1, May 1991, pp. 20–21, part 2, July 1990, pp. 9–24.

14. Barthelemy, J.-F. M.; and Hall, L. E.: Automatic Differentiation as a Tool for Engineering Design. 4th AIAA/Air Force/NASA/OAI Symposium on Multidisciplinary Analysis and Optimization, Cleveland, Ohio, September 1992. Proceedings of.

15. Green, L. L.; Newman P. E.; Haigler, C. Y.: Sensitivity Derivatives for Advance CFD Algorithm and Viscous Modelling Parameters Via Automatic Differentiation. AIAA Paper 93-3321, 11th AIAA Computational Fluid Dynamic Conference, July 1993, Orlando, FL.

16. Sobieszczanski-Sobieski, J.: On the Sensitivity of Complex, Internally Coupled Systems. AIAA/ASME/ASCE/AHS 29th Structures, Structural Dynamics and Materials Conference, Williamsburg, VA, April 1988. AIAA Paper No. CP-88-2378, and AIAA J., Vol. 28, No. 1, Jan. 1990, also published as NASA TM-100537, January 1988.

17. Sobieszczanski-Sobieski, J.: Higher Order Sensitivity Analysis of Complex, Coupled Systems. AIAA Journal, Vol. 28, No. 4, Apr. 1990.

18. Barthelemy, J-F.; Wrenn, G.; Dovi, A.; and Coen, P.: Integrating Aerodynamics and Structures in the Minimum Weight Design of a Supersonic Transport Wing. AIAA 92-2372, Presented at AIAA/ASME/ASCE/AHS/ASC 33rd Structures, Structural Dynamics, and Materials Conference, Dallas, TX, April 1992, Proceedings of.

19. Sobieszczanski-Sobieski, J.: Sensitivity Analysis and Multidisciplinary Optimization for Aircraft Design: Recent Advances and Results. Int'l Council for Aeronautical Sc., 16th Congress, Jerusalem, Aug.–Sept., 1988; Proceedings of, Vol. 2, pp. 953–964.

20. Sobieszczanski-Sobieski, J.; Barthelemy, J. F.; and Riley, K. M.: Sensitivity of Optimum Solutions to Problem Parameters. AIAA J., Vol. 21, Sept. 1992, pp. 1291–1299.

21. Barthelemy, J.-F. M.; and Sobieszczanski-Sobieski, J.: Optimum Sensitivity Derivatives of Objective Function in Nonlinear Programing. AIAA J. Vol. 21, No. 6, June 1983, pp. 913–915.

22. Hallman, W. P.: "Sensitivity Analysis for Trajectory Optimization Applications," AIAA Paper 90-0471.

23. Fiacco, A. V.: "Introduction to Sensitivity and Stability Analysis in Nonlinear Programming," Academic Press 1983.

24. Beltracchi, T. J.; and Gabriele, G. A.: "An Investigation of Using an RQP-Based Method to Calculate Parameter Sensitivity Derivatives," in NASA CP-3031 Recent Advances in Multidisciplinary Analysis and Optimization, Sept. 28–30, 1988, Hampton, VA, pp. 673–698.

25. Barthelemy, J.-F. M.; and Sobieszczanski-Sobieski, J.: Extrapolation of Optimum Design Based on Sensitivity Derivatives. AIAA J., Vol. 21, No. 5, May 1983, pp. 797-799.

26. Gage, P.; and Kroo, I.: Development of the Quasi-Procedural Method for Use in Aircraft Configuration Optimization. AIAA 92-4693, 4th AIAA/USAF/NASA/OAI Symposium on Multidisciplinary Analysis and Optimization, September 1992, Cleveland, Ohio.

27. Chang, K. J.; Haftka, R. T.; Giles, G. L.; and Kao, P.-J.: Sensitivity Based Scaling for Correlating Structural Response from Different Analytical Models. AIAA Paper No. 91-925. Presented at AIAA/ASME/ASCE/AHS/ASC 32nd Structures, Structural Dynamics, and Materials Conference, Baltimore, MD, 1991; Proceedings of.

28. Dovi, A. R.: ISSYS—An Integrated Synergistic Synthesis System, NASA CR 159221, February 1980.

29. Stefik, M.; and Bobrow, D. G.: Object-Oriented Programming: Themes and Variations. The AI Magazine, Vol. 6, No. 4, Winter 1986, pp. 40-62.

30. Barthelemy, J.-F. M.; and Haftka, R. T.: Approximation Concepts for Optimum Structureal Design—A Review Structural Optimization, 5, 1993, pp. 129–144.

31. Montgomery, D. C.: Design and Analysis of Experiments. Published by John Wiley & Sons, 1991.

32. Stanley, D.; Unal, R.; and Joyner, R.: Application of Taguchi Methods to Dual Mixture Ratio Propulsion System Optimization for SSTO Vehicles. AIAA Paper 92-0213, AIAA 30th Aerospace Sciences Meeting and Exhibit, Reno, NV, January 1992.

33. Hajela, P.; and Berke, L.: Neural Networks in Structural Analysis and Design: An Overview. Int'l J. for Computing Systems in Engineering, Vol. 3, No. 1–4, 1992, pp. 525–539.

34. Thornton, E. A.: Thermal Structures: Four Decades of Progress. Journal of Aircraft, No. 29, 1992, pp. 485–498.

35. Tulinius, J.: Multidisciplinary Optimization is Key to Integrated Product Development Process. 4th AIAA/USAF/NASA/OAI Symposium on Multidisciplinary Analysis and Optimization, September 1992, Cleveland, Ohio.

36. Rogers, J. L.: A. Knowledge-Based Tool for Multilevel Decomposition of a Complex Design Problem. NASA TP-2903, 1989.

37. Weisshaar, T. A.; Newsom, J. R.; Gilbert, M. G.; and Zeiler, T. A.: Integrated Structure/Control Design-Present Methodology and Future Opportunities. ICAS-86-4.8.1, Sept. 1986.

38. Wrenn, G. A.; and Dovi, A. R.: Multilevel Decomposition Approach to the Preliminary Sizing of a Transport Aircraft Wing. AIAA Journal of Aircraft, Vol. 25, No. 7, July 1988, pp. 632–638.

39. Unger, E. R.; Hutchinson, M.G.; Rais-Rohani, M.; Haftka, R. T.; and Grossman, B.: Variable-Complexlty Multidisciplinary Design of a Transport Wing. Int'l J. of Systems Automation: Research and Applications (SARA) 2, 1992, pp. 87–113.

40. Coen, P. G.; Sobieszczanski-Sobieski, J; and Dollyhigh, S. M.: Preliminary Results from the High-Speed Airframe Integration Research Project. AIAA Paper No. 92-1004, 1992 Aerospace Design Conference, February 1992, Irvine, CA.

41. Kornbold, J.; Gabriele, G.; Renaud, J.; and Kott, G.: "Application of Multidisciplinary Design Optimization to Electronics Package Design," AIAA Paper 92-4704 4th AIAA/USAF/NASA/OAI Symp. on Multi. Analysis and Optimization.

42. Cramer, E. J.; Huffman, W. P.; Mastro, R. A.: "Sparse Optimization for Aircraft Design," AIAA Paper No. 93-3935, AIAA Aircraft Design, Systems and Operations Meeting, Aug. 11–13, Monterey, CA, 1993.

43. Betts, J. T.; and Huffman, W. P.: "The Application of Sparse Nonlinear Programming to Trajectory Optimization," AIAA Paper 90-3448 presented at the AIAA Guidance Navigation and Control Conference.

44. Betts, J. T.; Elderscele, S.; and Huffman, W. P.: "A Performance Comparison of Nonlinear Programming Algorithms for Large Sparse Problems," AIAA Paper (3-3751 presented at the AIAA Guidance Navigation and Control Conference, Monterey CA, Aug. 1993.

45. Brenan, K. E.; Hallman, W. P.; Yeung, W. K.: "The Design of a Large Scale NLP Code for Trajectory Optimization Problems," The Aerospace Corporation Report ART 92(8189)-1.

46. Vanderplaats, G. N.: ADS—A Fortran Program for Automated Design Synthesis. NASA CR-177985, Sept. 1985.

47. Lasdon, L. S.: "Nonlinear Programming Algorithms—Applications, Software, and Comparisons," in Numerical Optimization 1984, SIAM.

48. Arora, J. S.: Introduction to Optimum Design. McGraw-Hill, New York, 1989.

49. Haftka, R. T.; and Gurdal, Z.: Elements of Structural Optimization. Kluwer Academic Publishers, 1992, pp. 242–243.

50. Venkayya, V. B.: "Optimality Criteria: A Basis for Multidisciplinary Optimization," Computational Mechanics, Vol. 5, pp. 1–21, 1989.

51. Berke, L.; and Khot, N. S.: "Structural Optimization Using Optimality Criteria," Computer Aided Structural Design: Structural and Mechanical Systems. (C.A. Mota Soares, Editor), Springer Verlag, 1987.

52. Morris, A. J.: "Fundamentals of Structural Optimization," in NATO AGARD Lecture Series 186, Integrated Design Analysis and Optimization of Aircraft Structures. (pp. 1–7 and 1–8), AGARD-LS-186, May 1992, pp. 1–8.

53. Sobieszczanski-Sobieski, J.: From a Black-Box to a Programing System, Ch. 11 in Foundations for Structural Optimization—A Unified Approach. Morris, A. J., ed. John Wiley & Sons, 1982.

54. Sobieszczanski-Sobieski, J.; and Loendorf, D.: A Mixed Optimization Method for Automated Design of Fuselage Structures. Journal of Aircraft, Vol. 9, Dec. 1972, pp. 805–811.

55. Sobieszczanski-Sobieski, J.: Optimization by Decomposition: A Step from Hierarchic to Non-Hierarchic Systems, Recent Advances in Multidisciplinary Analysis and Optimization, NASA CP-3031, 1988, Part 1.

56. Bloebaum, C. L.: Formal and Heuristic System Decomposition Methods in Multidisciplinary Synthesis. NASA CR 4413, Dec. 1991.

57. Consoli, R. D.; and Sobieszczanski-Sobieski, J.: Application of Advanced Multidisciplinary Analysis and Optimization Methods to Vehicle Design Systems. Journal of Aircraft, Vol. 29, No. 5, Sept–Oct. 1992, pp. 811–818.

58. Rogers, J. L.; and Barthelemy, J.-F.M.: An Expert System for Choosing the Best Combination of Options in a General Purpose Program for Automated Design Synthesis. Microcomputer-Based Expert Systems; ed.: Gupta, A.; and Prosad, B. E. IEEE Press, N.Y., 1988, pp. 225–235.

59. Gilmore, J. F.; Pulaski, K.; and Howard, C.: A Comprehensive Evaluation of Expert System Tools. Microcomputer-Based Expert Systems ed.: Gupta, A.; and Prasad, B. E. IEEE Press, N.Y., 1988, pp. 208–222.

60. Sobieszczanski-Sobieski, J.: Aircraft Optimization by a System Approach: Achievements and Trends, International Council of Aeronautical Sciences, 18th Congress. Beijing, China, September, 1992, Proceedings of.

Dr. Jaroslaw Sobieski
NASA Langley Research Center
Hampton, Virginia 23681

Mechanics

SOLID MECHANICS AND ITS APPLICATIONS

Series Editor: G.M.L. Gladwell

Aims and Scope of the Series

The fundamental questions arising in mechanics are: *Why?, How?*, and *How much?* The aim of this series is to provide lucid accounts written by authoritative researchers giving vision and insight in answering these questions on the subject of mechanics as it relates to solids. The scope of the series covers the entire spectrum of solid mechanics. Thus it includes the foundation of mechanics; variational formulations; computational mechanics; statics, kinematics and dynamics of rigid and elastic bodies; vibrations of solids and structures; dynamical systems and chaos; the theories of elasticity, plasticity and viscoelasticity; composite materials; rods, beams, shells and membranes; structural control and stability; soils, rocks and geomechanics; fracture; tribology; experimental mechanics; biomechanics and machine design.

1. R.T. Haftka, Z. Gürdal and M.P. Kamat: *Elements of Structural Optimization*. 2nd rev.ed., 1990   ISBN 0-7923-0608-2
2. J.J. Kalker: *Three-Dimensional Elastic Bodies in Rolling Contact*. 1990
                                                                           ISBN 0-7923-0712-7
3. P. Karasudhi: *Foundations of Solid Mechanics*. 1991   ISBN 0-7923-0772-0
4. *Not published*
5. *Not published.*
6. J.F. Doyle: *Static and Dynamic Analysis of Structures*. With an Emphasis on Mechanics and Computer Matrix Methods. 1991   ISBN 0-7923-1124-8; Pb 0-7923-1208-2
7. O.O. Ochoa and J.N. Reddy: *Finite Element Analysis of Composite Laminates*.
                                                                           ISBN 0-7923-1125-6
8. M.H. Aliabadi and D.P. Rooke: *Numerical Fracture Mechanics*.   ISBN 0-7923-1175-2
9. J. Angeles and C.S. López-Cajún: *Optimization of Cam Mechanisms*. 1991
                                                                           ISBN 0-7923-1355-0
10. D.E. Grierson, A. Franchi and P. Riva (eds.): *Progress in Structural Engineering*. 1991
                                                                           ISBN 0-7923-1396-8
11. R.T. Haftka and Z. Gürdal: *Elements of Structural Optimization*. 3rd rev. and exp. ed. 1992
                                                            ISBN 0-7923-1504-9; Pb 0-7923-1505-7
12. J.R. Barber: *Elasticity*. 1992   ISBN 0-7923-1609-6; Pb 0-7923-1610-X
13. H.S. Tzou and G.L. Anderson (eds.): *Intelligent Structural Systems*. 1992
                                                                           ISBN 0-7923-1920-6
14. E.E. Gdoutos: *Fracture Mechanics*. An Introduction. 1993   ISBN 0-7923-1932-X
15. J.P. Ward: *Solid Mechanics*. An Introduction. 1992   ISBN 0-7923-1949-4
16. M. Farshad: *Design and Analysis of Shell Structures*. 1992   ISBN 0-7923-1950-8
17. H.S. Tzou and T. Fukuda (eds.): *Precision Sensors, Actuators and Systems*. 1992
                                                                           ISBN 0-7923-2015-8
18. J.R. Vinson: *The Behavior of Shells Composed of Isotropic and Composite Materials*. 1993
                                                                           ISBN 0-7923-2113-8

Kluwer Academic Publishers – Dordrecht / Boston / London

Mechanics

SOLID MECHANICS AND ITS APPLICATIONS
Series Editor: G.M.L. Gladwell

19. H.S. Tzou: *Piezoelectric Shells*. Distributed Sensing and Control of Continua. 1993
ISBN 0-7923-2186-3
20. W. Schiehlen (ed.): *Advanced Multibody System Dynamics*. Simulation and Software Tools.
1993 ISBN 0-7923-2192-8
21. C.-W. Lee: *Vibration Analysis of Rotors*. 1993 ISBN 0-7923-2300-9
22. D.R. Smith: *An Introduction to Continuum Mechanics*. 1993 ISBN 0-7923-2454-4
23. G.M.L. Gladwell: *Inverse Problems in Scattering*. An Introduction. 1993 ISBN 0-7923-2478-1
24. G. Prathap: *The Finite Element Method in Structural Mechanics*. 1993 ISBN 0-7923-2492-7
25. J. Herskovits (ed.): *Advances in Structural Optimization*. 1995 ISBN 0-7923-2510-9
26. M.A. González-Palacios and J. Angeles: *Cam Synthesis*. 1993 ISBN 0-7923-2536-2
27. W.S. Hall: *The Boundary Element Method*. 1993 ISBN 0-7923-2580-X
28. J. Angeles, G. Hommel and P. Kovács (eds.): *Computational Kinematics*. 1993
ISBN 0-7923-2585-0
29. A. Curnier: *Computational Methods in Solid Mechanics*. 1994 ISBN 0-7923-2761-6
30. D.A. Hills and D. Nowell: *Mechanics of Fretting Fatigue*. 1994 ISBN 0-7923-2866-3
31. B. Tabarrok and F.P.J. Rimrott: *Variational Methods and Complementary Formulations in
Dynamics*. 1994 ISBN 0-7923-2923-6
32. E.H. Dowell, E.F. Crawley, H.C. Curtiss Jr., D.A. Peters, R. H. Scanlan and F. Sisto (eds.):
A Modern Course in Aeroelasticity. Third Revised and Enlarged Edition. (forthcoming)
ISBN 0-7923-2788-8; Pb: 0-7923-2789-6
33. A. Preumont: *Random Vibration and Spectral Analysis*. 1994 ISBN 0-7923-3036-6
34. J.N. Reddy (ed.): *Mechanics of Composite Materials*. Selected works of Nicholas J. Pagano.
1994 ISBN 0-7923-3041-2
35. A.P.S. Selvadurai (ed.): *Mechanics of Poroelastic Media*. 1995 (forthcoming)
ISBN 0-7923-3329-2
36. Z. Mróz, D. Weichert, S. Dorosz (eds.): *Inelastic Behaviour of Structures under Variable
Loads*. 1995 ISBN 0-7923-3397-7
37. R. Pyrz (ed.): *IUTAM Symposium on Microstructure-Property Interactions in Composite
Materials*. Proceedings of an IUTAM Symposium held in Aalborg, Denmark. 1995
(forthcoming) ISBN 0-7923-3427-2
38. M.I. Friswell and J.E. Mottershead: *Finite Element Model Updating in Structural Dynamics*.
1995 ISBN 0-7923-3431-0

Kluwer Academic Publishers – Dordrecht / Boston / London

Mechanics

FLUID MECHANICS AND ITS APPLICATIONS

Series Editor: R. Moreau

Aims and Scope of the Series

The purpose of this series is to focus on subjects in which fluid mechanics plays a fundamental role. As well as the more traditional applications of aeronautics, hydraulics, heat and mass transfer etc., books will be published dealing with topics which are currently in a state of rapid development, such as turbulence, suspensions and multiphase fluids, super and hypersonic flows and numerical modelling techniques. It is a widely held view that it is the interdisciplinary subjects that will receive intense scientific attention, bringing them to the forefront of technological advancement. Fluids have the ability to transport matter and its properties as well as transmit force, therefore fluid mechanics is a subject that is particularly open to cross fertilisation with other sciences and disciplines of engineering. The subject of fluid mechanics will be highly relevant in domains such as chemical, metallurgical, biological and ecological engineering. This series is particularly open to such new multidisciplinary domains.

Kluwer Academic Publishers – Dordrecht / Boston / London

Mechanics

FLUID MECHANICS AND ITS APPLICATIONS
Series Editor: R. Moreau

21. J.P. Bonnet and M.N. Glauser (eds.): *Eddy Structure Identification in Free Turbulent Shear Flows.* 1993   ISBN 0-7923-2449-8
22. R.S. Srivastava: *Interaction of Shock Waves.* 1994   ISBN 0-7923-2920-1
23. J.R. Blake, J.M. Boulton-Stone and N.H. Thomas (eds.): *Bubble Dynamics and Interface Phenomena.* 1994   ISBN 0-7923-3008-0
24. R. Benzi (ed.): *Advances in Turbulence V.* 1995 (forthcoming)   ISBN 0-7923-3032-3
25. B.I. Rabinovich, V.G. Lebedev and A.I. Mytarev: *Vortex Processes and Solid Body Dynamics.* The Dynamic Problems of Spacecrafts and Magnetic Levitation Systems. 1994
    ISBN 0-7923-3092-7
26. P.R. Voke, L. Kleiser and J.-P. Chollet (eds.): *Direct and Large-Eddy Simulation I.* Selected papers from the First ERCOFTAC Workshop on Direct and Large-Eddy Simulation. 1994
    ISBN 0-7923-3106-0
27. J.A. Sparenberg: *Hydrodynamic Propulsion and its Optimization.* Analytic Theory. 1995
    ISBN 0-7923-3201-6
28. J.F. Dijksman and G.D.C. Kuiken (eds.): *IUTAM Symposium on Numerical Simulation of Non-Isothermal Flow of Viscoelastic Liquids.* Proceedings of an IUTAM Symposium held in Kerkrade, The Netherlands. 1995   ISBN 0-7923-3262-8
29. B.M. Boubnov and G.S. Golitsyn: *Convection in Rotating Fluids.* 1995   ISBN 0-7923-3371-3
30. S.I. Green (ed.): *Fluid Vortices.* 1995   ISBN 0-7923-3376-4
31. S. Morioka and L. van Wijngaarden (eds.): *IUTAM Symposium on Waves in Liquid/Gas and Liquid/Vapour Two-Phase Systems.* 1995   ISBN 0-7923-3424-8

Kluwer Academic Publishers – Dordrecht / Boston / London

Mechanics

From 1990, books on the subject of *mechanics* will be published under two series:
FLUID MECHANICS AND ITS APPLICATIONS
   Series Editor: R.J. Moreau
SOLID MECHANICS AND ITS APPLICATIONS
   Series Editor: G.M.L. Gladwell

Prior to 1990, the books listed below were published in the respective series indicated below.

MECHANICS: DYNAMICAL SYSTEMS
Editors: L. Meirovitch and G.Æ. Oravas

1. E.H. Dowell: *Aeroelasticity of Plates and Shells.* 1975   ISBN 90-286-0404-9
2. D.G.B. Edelen: *Lagrangian Mechanics of Nonconservative Nonholonomic Systems.*
   1977   ISBN 90-286-0077-9
3. J.L. Junkins: *An Introduction to Optimal Estimation of Dynamical Systems.* 1978
                                            ISBN 90-286-0067-1
4. E.H. Dowell (ed.), H.C. Curtiss Jr., R.H. Scanlan and F. Sisto: *A Modern Course in*
   Aeroelasticity.   *Revised and enlarged edition see under Volume 11*
5. L. Meirovitch: *Computational Methods in Structural Dynamics.* 1980
                                            ISBN 90-286-0580-0
6. B. Skalmierski and A. Tylikowski: *Stochastic Processes in Dynamics.* Revised and
   enlarged translation. 1982   ISBN 90-247-2686-7
7. P.C. Müller and W.O. Schiehlen: *Linear Vibrations.* A Theoretical Treatment of Multi-
   degree-of-freedom Vibrating Systems. 1985   ISBN 90-247-2983-1
8. Gh. Buzdugan, E. Mihăilescu and M. Radeş: *Vibration Measurement.* 1986
                                            ISBN 90-247-3111-9
9. G.M.L. Gladwell: *Inverse Problems in Vibration.* 1987   ISBN 90-247-3408-8
10. G.I. Schuëller and M. Shinozuka: *Stochastic Methods in Structural Dynamics.* 1987
                                            ISBN 90-247-3611-0
11. E.H. Dowell (ed.), H.C. Curtiss Jr., R.H. Scanlan and F. Sisto: *A Modern Course in*
   Aeroelasticity. Second revised and enlarged edition (of Volume 4). 1989
                           ISBN Hb 0-7923-0062-9; Pb 0-7923-0185-4
12. W. Szemplińska-Stupnicka: *The Behavior of Nonlinear Vibrating Systems.* Volume I:
   Fundamental Concepts and Methods: Applications to Single-Degree-of-Freedom
   Systems. 1990   ISBN 0-7923-0368-7
13. W. Szemplińska-Stupnicka: *The Behavior of Nonlinear Vibrating Systems.* Volume II:
   Advanced Concepts and Applications to Multi-Degree-of-Freedom Systems. 1990
                                            ISBN 0-7923-0369-5
                           Set ISBN (Vols. 12–13) 0-7923-0370-9

MECHANICS OF STRUCTURAL SYSTEMS
Editors: J.S. Przemieniecki and G.Æ. Oravas

1. L. Frýba: *Vibration of Solids and Structures under Moving Loads.* 1970
                                            ISBN 90-01-32420-2
2. K. Marguerre and K. Wölfel: *Mechanics of Vibration.* 1979   ISBN 90-286-0086-8

Mechanics

3. E.B. Magrab: *Vibrations of Elastic Structural Members.* 1979   ISBN 90-286-0207-0
4. R.T. Haftka and M.P. Kamat: *Elements of Structural Optimization.* 1985
   Revised and enlarged edition see under Solid Mechanics and Its Applications, Volume 1
5. J.R. Vinson and R.L. Sierakowski: *The Behavior of Structures Composed of Composite Materials.* 1986   ISBN Hb 90-247-3125-9; Pb 90-247-3578-5
6. B.E. Gatewood: *Virtual Principles in Aircraft Structures.* Volume 1: Analysis. 1989
   ISBN 90-247-3754-0
7. B.E. Gatewood: *Virtual Principles in Aircraft Structures.* Volume 2: Design, Plates, Finite Elements. 1989   ISBN 90-247-3755-9
   Set (Gatewood 1 + 2) ISBN 90-247-3753-2

MECHANICS OF ELASTIC AND INELASTIC SOLIDS
Editors: S. Nemat-Nasser and G.Æ. Oravas

---

1. G.M.L. Gladwell: *Contact Problems in the Classical Theory of Elasticity.* 1980
   ISBN Hb 90-286-0440-5; Pb 90-286-0760-9
2. G. Wempner: *Mechanics of Solids with Applications to Thin Bodies.* 1981
   ISBN 90-286-0880-X
3. T. Mura: *Micromechanics of Defects in Solids.* 2nd revised edition, 1987
   ISBN 90-247-3343-X
4. R.G. Payton: *Elastic Wave Propagation in Transversely Isotropic Media.* 1983
   ISBN 90-247-2843-6
5. S. Nemat-Nasser, H. Abé and S. Hirakawa (eds.): *Hydraulic Fracturing and Geothermal Energy.* 1983   ISBN 90-247-2855-X
6. S. Nemat-Nasser, R.J. Asaro and G.A. Hegemier (eds.): *Theoretical Foundation for Large-scale Computations of Nonlinear Material Behavior.* 1984   ISBN 90-247-3092-9
7. N. Cristescu: *Rock Rheology.* 1988   ISBN 90-247-3660-9
8. G.I.N. Rozvany: *Structural Design via Optimality Criteria.* The Prager Approach to Structural Optimization. 1989   ISBN 90-247-3613-7

MECHANICS OF SURFACE STRUCTURES
Editors: W.A. Nash and G.Æ. Oravas

---

1. P. Seide: *Small Elastic Deformations of Thin Shells.* 1975   ISBN 90-286-0064-7
2. V. Panc: *Theories of Elastic Plates.* 1975   ISBN 90-286-0104-X
3. J.L. Nowinski: *Theory of Thermoelasticity with Applications.* 1978
   ISBN 90-286-0457-X
4. S. Łukasiewicz: *Local Loads in Plates and Shells.* 1979   ISBN 90-286-0047-7
5. C. Firt: *Statics, Formfinding and Dynamics of Air-supported Membrane Structures.* 1983   ISBN 90-247-2672-7
6. Y. Kai-yuan (ed.): *Progress in Applied Mechanics.* The Chien Wei-zang Anniversary Volume. 1987   ISBN 90-247-3249-2
7. R. Negruţiu: *Elastic Analysis of Slab Structures.* 1987   ISBN 90-247-3367-7
8. J.R. Vinson: *The Behavior of Thin Walled Structures.* Beams, Plates, and Shells. 1988
   ISBN Hb 90-247-3663-3; Pb 90-247-3664-1

Mechanics

MECHANICS OF FLUIDS AND TRANSPORT PROCESSES
Editors: R.J. Moreau and G.Æ. Oravas

1. J. Happel and H. Brenner: *Low Reynolds Number Hydrodynamics*. With Special Applications to Particular Media. 1983   ISBN Hb 90-01-37115-9; Pb 90-247-2877-0
2. S. Zahorski: *Mechanics of Viscoelastic Fluids*. 1982   ISBN 90-247-2687-5
3. J.A. Sparenberg: *Elements of Hydrodynamics Propulsion*. 1984   ISBN 90-247-2871-1
4. B.K. Shivamoggi: *Theoretical Fluid Dynamics*. 1984   ISBN 90-247-2999-8
5. R. Timman, A.J. Hermans and G.C. Hsiao: *Water Waves and Ship Hydrodynamics*. An Introduction. 1985   ISBN 90-247-3218-2
6. M. Lesieur: *Turbulence in Fluids*. Stochastic and Numerical Modelling. 1987   ISBN 90-247-3470-3
7. L.A. Lliboutry: *Very Slow Flows of Solids*. Basics of Modeling in Geodynamics and Glaciology. 1987   ISBN 90-247-3482-7
8. B.K. Shivamoggi: *Introduction to Nonlinear Fluid-Plasma Waves*. 1988   ISBN 90-247-3662-5
9. V. Bojarevičs, Ya. Freibergs, E.I. Shilova and E.V. Shcherbinin: *Electrically Induced Vortical Flows*. 1989   ISBN 90-247-3712-5
10. J. Lielpeteris and R. Moreau (eds.): *Liquid Metal Magnetohydrodynamics*. 1989   ISBN 0-7923-0344-X

MECHANICS OF ELASTIC STABILITY
Editors: H. Leipholz and G.Æ. Oravas

1. H. Leipholz: *Theory of Elasticity*. 1974   ISBN 90-286-0193-7
2. L. Librescu: *Elastostatics and Kinetics of Aniosotropic and Heterogeneous Shell-type Structures*. 1975   ISBN 90-286-0035-3
3. C.L. Dym: *Stability Theory and Its Applications to Structural Mechanics*. 1974   ISBN 90-286-0094-9
4. K. Huseyin: *Nonlinear Theory of Elastic Stability*. 1975   ISBN 90-286-0344-1
5. H. Leipholz: *Direct Variational Methods and Eigenvalue Problems in Engineering*. 1977   ISBN 90-286-0106-6
6. K. Huseyin: *Vibrations and Stability of Multiple Parameter Systems*. 1978   ISBN 90-286-0136-8
7. H. Leipholz: *Stability of Elastic Systems*. 1980   ISBN 90-286-0050-7
8. V.V. Bolotin: *Random Vibrations of Elastic Systems*. 1984   ISBN 90-247-2981-5
9. D. Bushnell: *Computerized Buckling Analysis of Shells*. 1985   ISBN 90-247-3099-6
10. L.M. Kachanov: *Introduction to Continuum Damage Mechanics*. 1986   ISBN 90-247-3319-7
11. H.H.E. Leipholz and M. Abdel-Rohman: *Control of Structures*. 1986   ISBN 90-247-3321-9
12. H.E. Lindberg and A.L. Florence: *Dynamic Pulse Buckling*. Theory and Experiment. 1987   ISBN 90-247-3566-1
13. A. Gajewski and M. Zyczkowski: *Optimal Structural Design under Stability Constraints*. 1988   ISBN 90-247-3612-9

Mechanics

MECHANICS OF ELASTIC AND INELASTIC SOLIDS
Editors: S. Nemat-Nasser and G.Æ. Oravas

MECHANICS OF SURFACE STRUCTURES
Editors: W.A. Nash and G.Æ. Oravas

The manufacturer's authorised representative in the EU is Springer
Nature Customer Service Centre GmbH, Europaplatz 3, 69115 Heidelberg,
Germany. If you have any concerns regarding our products, please
contact ProductSafety@springernature.com

Printed and bound by CPI Group (UK) Ltd, Croydon, CR0 4YY
23/04/2026
02095624-0010